Genetic Programming IV

GENETIC PROGRAMMING SERIES

Series Editor

John Koza

Stanford University

Also in the series:

GENETIC PROGRAMMING AND DATA STRUCTURES: Genetic Programming + Data Structures = Automatic Programming! *William B. Langdon*; ISBN: 0-7923-8135-1

AUTOMATIC RE-ENGINEERING OF SOFTWARE USING GENETIC PROGRAMMING, *Conor Ryan*; ISBN: 0-7923-8653-1

DATA MINING USING GRAMMAR BASED GENETIC PROGRAMMING AND APPLICATIONS, *Man Leung Wong and Kwong Sak Leung*; ISBN: 0-7923-7746-X

GRAMMATICAL EVOLUTION: Evolutionary Automatic Programming in an Arbitrary Language, *Michael O'Neill and Conor Ryan*; ISBN: 1-4020-7444-1

The cover image was created by Jerry Jager

Genetic Programming IV

Routine Human-Competitive Machine Intelligence

John R. Koza
Martin A. Keane
Matthew J. Streeter
William Mydlowec
Jessen Yu
Guido Lanza

Genetic Programming IV

Authors:
John R. Koza
Martin A. Keane
Matthew J. Streeter
William Mydlowec
Jessen Yu
Guido Lanza

ISBN: 978-0-387-25067-0 (Paperback edition, 2005)

e-book ISBN: 978-0-387-26417-2

Library of Congress Control Number: 2005275613

9 8 7 6 5 4 3

springer.com

Advance Praise for
Genetic Programming IV: Routine Human-Competitive Machine Intelligence

"In 1992, John Koza published his first book on genetic programming and forever changed the world of computation. At the time, many researchers, myself included, were skeptical about whether the idea of using genetic algorithms directly to evolve programs would ever amount to much. But scores of conquered problems and three additional books makes the case utterly persuasive. The latest contribution, *Genetic Programming IV: Routine Human-Competitive Machine Intelligence*, demonstrates the everyday solution of such 'holy grail' problems as the automatic synthesis of analog circuits, the design of automatic controllers, and the automated programming of computers. This would be impressive enough, but the book also shows how to evolve whole families of solutions to entire classes of problems in a single run. Such *parametric GP* is a significant achievement, and I believe it foreshadows generalized evolution of complex contingencies as an everyday matter. To artificial evolutionaries of all stripes, I recommend that you read this book and breath in its thoughtful mechanism and careful empirical method. To specialists in any of the fields covered by this book's sample problem areas, I say read this book and discover the computer-augmented inventions that are your destiny. To remaining skeptics who doubt the inventive competence of genetics and evolution, I say read this book and change your mind or risk the strong possibility that your doubts will soon cause you significant intellectual embarrassment."
—David E. Goldberg, University of Illinois

"The research reported in this book is a tour de force. For the first time since the idea was bandied about in the 1940s and the early 1950s, we have a set of examples of human-competitive automatic programming."
—John H. Holland, University of Michigan

"The adaptive filters and neural networks that I have worked with over many years are self-optimizing systems where the relationship between performance (usually mean-square-error) and parameter settings (weights) is continuous. Optimization by gradient methods works well for these systems. Now, this book describes a wider class of optimization problems where the relationship between performance (fitness) and parameters is highly disjoint, and self-optimization is achieved by nature-inspired genetic algorithms involving random search (mutation) and crossover (sexual reproduction). John Koza and his colleagues have done remarkable work in advancing the development of genetic programming and applying this to practical problems such as electric circuit design and control system design. What is ingenious about their work is that they have found ways to approach design problems by parameterizing both physical and topological variables into a common code that can be subjected to genetic programming for optimization. It is amazing how this approach finds optimized solutions that are not obvious to the best human experts. This fine book gives an accounting of the latest work in genetic programming, and it is 'must reading' for

those interested in adaptive and learning systems, neural networks, fuzzy systems, artificial intelligence, and neurobiology. I strongly recommend it.
—Bernard Widrow, Electrical Engineering Department, Stanford University

"John Koza's genetic programming approach to machine discovery can invent solutions to more complex specifications than any other I have seen."
—John McCarthy, Computer Science Department, Stanford University

To our parents—all of whom were best-of-generation individuals

Acknowledgments

Forrest H Bennett III, William Comisky, and Oscar Stiffelman participated in some of the work described in this book (as noted in the appropriate places throughout this book).

Frank Dunlap of Enabling Technology Inc. of Sunnyvale, California, made numerous helpful comments on the chapter on the post-2000 inventions.

Edgar Sheh of Western Digital Corporation made numerous helpful comments on the chapters on control.

David E. Goldberg of the University of Illinois at Urbana-Champaign, William Langdon of University College London, and Lee Spector of Hampshire College made extensive helpful suggestions concerning this book.

Kalyanmoy Deb of the Indian Institute of Technology at Kanpur; Adrian Stoica of JPL and NASA in Pasadena, California; and Stuart T. Smith of the University of North Carolina at Charlotte provided useful information for this book.

Jeffrey Schwarz of San Francisco, California, made numerous comments for improving earlier drafts of this book.

Douglas B. Kell of the University of Wales made comments on the description of our work on metabolic pathways.

We especially thank Ray Prill and Jon O. Satre of Gordon-Prill Inc. of Mountain View, California, for designing, constructing, and maintaining the site for our 1,000-Pentium computer.

Biography of the Authors

John R. Koza received his Ph.D. in Computer Science from the University of Michigan in 1972 under the supervision of John Holland. He was co-founder, Chairman, and CEO of Scientific Games Inc. from 1973 through 1987 where he co-invented the rub-off instant lottery ticket used by state lotteries. He has taught a course on genetic algorithms and genetic programming at Stanford University since 1988. He is currently a consulting professor in the Biomedical Informatics Program in the Department of Medicine at Stanford University and a consulting professor in the Department of Electrical Engineering at Stanford University.
John R. Koza [koza@Stanford.edu]

Martin A. Keane received a Ph.D. in Mathematics from Northwestern University in 1969. He worked for Applied Devices Corporation until 1972, in the Mathematics Department at General Motors Laboratory until 1976, and was Vice-President for Engineering of Bally Manufacturing Corporation until 1986. He is currently chief scientist of Econometrics Inc. of Chicago and a consultant to various computer-related and gaming-related companies.
Martin A. Keane [martinkeane@ameritech.net]

Matthew J. Streeter received a Masters degree in Computer Science from Worcester Polytechnic Institute in 2001. His Masters thesis applied genetic programming to the automated discovery of numerical approximation formulae for functions and surfaces. His primary research interest is applying genetic programming to problems of real-world scientific or practical importance. He is currently working at Genetic Programming Inc. as a systems programmer and researcher.
Matthew J. Streeter [matt@genetic-programming.com]

William Mydlowec is Chief Executive Officer and co-founder of Pharmix Corporation, a venture-funded computational drug discovery company in Silicon Valley. He received his B.S. degree in Computer Science from Stanford University in 1998. He formerly did research at Genetic Programming Inc. with John Koza between 1997 and 2000.
William Mydlowec [bill@pharmix.com]

Jessen Yu is Director of Engineering of Pharmix Corporation. He received a B.S. degree in Computer Science and Chemistry from Stanford University. He formerly did research at Genetic Programming Inc. with John Koza between 1998 and 2000.
Jessen Yu [jyu@pharmix.com]

Guido Lanza is Vice President of Biology and co-founder of Pharmix Corporation. He received his B.A. degree in 1998 from the University of California at Berkeley from the Department of Molecular and Cell Biology and Department of Integrative Biology. He received an M.Sc. in 1999 in Bioinformatics from the University of Manchester, UK. He formerly did research at Genetic Programming Inc. with John Koza in 2000.
Guido Lanza [guido@Pharmix.com]

High-Level Table of Contents

Full Table of Contents

1

Introduction

The goal of getting computers to automatically solve problems is central to artificial intelligence, machine learning, and the broad area encompassed by what Turing called "machine intelligence" (Turing 1948, 1950).

Genetic programming is a systematic method for getting computers to automatically solve a problem. Genetic programming starts from a high-level statement of what needs to be done and automatically creates a computer program to solve the problem.

The most important point of this book is: Genetic programming now routinely delivers high-return human-competitive machine intelligence.

There are now 36 instances where genetic programming has produced a human-competitive result. In section 1.1, we define "routine," "high-return," "human-competitive," and "machine intelligence" and outline the evidence supporting each claimed human-competitive result.

The second of this book's four main points is: Genetic programming is an automated invention machine.

There are now 23 instances where genetic programming has duplicated the functionality of a previously patented invention, infringed a previously issued patent, or created a patentable new invention. Specifically, there are 15 instances where genetic programming has created an entity that either infringes or duplicates the functionality of

Table 1.1 Four main points of this book

	Main point
1	• Genetic programming now routinely delivers high-return human-competitive machine intelligence.
2	• Genetic programming is an automated invention machine.
3	• Genetic programming can automatically create a general solution to a problem in the form of a parameterized topology.
4	• Genetic programming has delivered a progression of qualitatively more substantial results in synchrony with five approximately order-of-magnitude increases in the expenditure of computer time.

a previously patented 20th-century invention, six instances where genetic programming has done the same with respect to an invention patented after January 1, 2000, and two instances where genetic programming has created a patentable new invention. The two new inventions are general-purpose controllers that outperform controllers employing tuning rules that have been in widespread use in industry for most of the 20th century.

Novelty and creativity are prerequisites for patentability. A new idea that can be logically deduced from facts that are known in a field, using transformations that are known in a field, is not considered to be patentable by the Patent Office. A new idea is patentable only if there is an "illogical step" (that is, a logically unjustified step) that distinguishes the proposed invention from that which is readily deducible from what is already known. As we discuss in section 1.2, genetic programming often unearths novel solutions to problems because it does not travel along the well-trod paths of previous human thinking. The inventions generated by genetic programming exhibit the kind of illogical discontinuity from previous human work that is required to obtain a patent.

The third main point of this book is: Genetic programming can automatically create a general solution to a problem in the form of a parameterized topology.

Eleven problems in this book demonstrate that genetic programming can automatically create, in a single run, a general (parameterized) solution to a problem in the form of a graphical structure whose nodes or edges represent components and where the parameter values of the components are specified by mathematical expressions containing free variables. Section 1.3 previews the automatic creation of such parameterized topologies.

This book's fourth main point is: Genetic programming has delivered a progression of qualitatively more substantial results in synchrony with five approximately order-of-magnitude increases in the expenditure of computer time.

Section 1.4 discusses the progression of results produced by genetic programming over the 15-year period from 1987 to 2002, including

- solving toy problems,
- producing human-competitive results not involving previously patented inventions,
- duplicating 20th-century patented inventions,
- duplicating 21st-century patented inventions, and
- creating patentable new inventions.

Table 1.1 shows the four main points of this book.

In addition to solving numerous problems involving analog electrical circuits (chapters 4, 5, 10, 11, 14, and 15) and controllers (chapters 3, 9, 12, and 13), the book presents results involving the automatic synthesis of networks of chemical reactions (chapter 8), antennas (chapter 6), and genetic networks (chapter 7).

Chapter 2 provides general background on genetic programming. Chapters 9, 10, 11, and 13 discuss parameterized topologies. Chapter 16 discusses the characteristics that may make certain problems better suited for genetic algorithms or genetic programming. Chapter 17 discusses issues of parallelization and computer time. Chapter 18 provides a historical perspective on computer speed and the succession of qualitatively more substantial results produced by genetic programming.

As far as we know, genetic programming is, at the present time, unique among methods of artificial intelligence and machine learning in terms of its duplication of numerous previously patented results, unique in its generation of patentable new results, unique in the breadth and depth of problems solved, unique in its demonstrated ability to produce parameterized topologies, and unique in its delivery of routine high-return, human-competitive machine intelligence.

1.1 Genetic Programming Now Routinely Delivers High-Return Human-Competitive Machine Intelligence

Focusing on this book's first main point (i.e., that genetic programming now routinely delivers high-return human-competitive machine intelligence), the next four sub-sections explain what we mean by the terms

- human-competitive (section 1.1.1),
- high-return (section 1.1.2),
- routine (section 1.1.3), and
- machine intelligence (section 1.1.4).

Then, four additional sub-sections outline the evidence that supports the claim that genetic programming now delivers results with these four characteristics.

1.1.1 What We Mean by "Human-Competitive"

In attempting to evaluate an automated problem-solving method, the question arises as to whether there is any real substance to the demonstrative problems that are published in connection with the method. Demonstrative problems in the fields of artificial intelligence and machine learning are often contrived toy problems that circulate exclusively inside academic groups that study a particular methodology. These problems typically have little relevance to any issues pursued by any scientist or engineer outside the fields of artificial intelligence and machine learning.

In his 1983 talk entitled "AI: Where It Has Been and Where It Is Going," machine learning pioneer Arthur Samuel said:

> "[T]he aim [is]... to get machines to exhibit behavior, which if done by humans, would be assumed to involve the use of intelligence."

Samuel's statement reflects the common goal articulated by the pioneers of the 1950s in the fields of artificial intelligence and machine learning. Indeed, getting machines to produce human-like results is *the* reason for the existence of the fields of artificial intelligence and machine learning.

To make this goal more concrete, we say that a result is "human-competitive" if it satisfies one or more of the eight criteria in table 1.2.

The eight criteria in table 1.2 have the desirable attribute of being at arms-length from the fields of artificial intelligence, machine learning, and genetic programming. That is, a result cannot acquire the rating of "human-competitive" merely because it is endorsed by researchers *inside* the specialized fields that are attempting to create

Table 1.2 Eight criteria for saying that an automatically created result is human-competitive

	Criterion
A	The result was patented as an invention in the past, is an improvement over a patented invention, or would qualify today as a patentable new invention.
B	The result is equal to or better than a result that was accepted as a new scientific result at the time when it was published in a peer-reviewed scientific journal.
C	The result is equal to or better than a result that was placed into a database or archive of results maintained by an internationally recognized panel of scientific experts.
D	The result is publishable in its own right as a new scientific result—independent of the fact that the result was mechanically created.
E	The result is equal to or better than the most recent human-created solution to a long-standing problem for which there has been a succession of increasingly better human-created solutions.
F	The result is equal to or better than a result that was considered an achievement in its field at the time it was first discovered.
G	The result solves a problem of indisputable difficulty in its field.
H	The result holds its own or wins a regulated competition involving human contestants (in the form of either live human players or human-written computer programs).

machine intelligence. Instead, a result produced by an automated method must earn the rating of "human-competitive" *independent* of the fact that it was generated by an automated method.

These eight criteria are the same as those presented in *Genetic Programming III: Darwinian Invention and Problem Solving* (Koza, Bennett, Andre, and Keane 1999a).

1.1.2 What We Mean by "High-Return"

What is delivered by the actual automated operation of an artificial method in comparison to the amount of knowledge, information, analysis, and intelligence that is pre-supplied by the human employing the method?

We define the *AI ratio* (the "artificial-to-intelligence" ratio) of a problem-solving method as the ratio of that which is delivered by the automated operation of the *artificial* method to the amount of *intelligence* that is supplied by the human applying the method to a particular problem.

The AI ratio is especially pertinent to methods for getting computers to automatically solve problems because it measures the value added by the artificial problem-solving method. Manifestly, the aim of the fields of artificial intelligence and machine learning is to generate human-competitive results with a high AI ratio.

Deep Blue: An Artificial Intelligence Milestone (Newborn 2002) describes the 1997 defeat of the human world chess champion Garry Kasparov by the Deep Blue computer system. This outstanding example of machine intelligence is clearly a human-competitive result (by virtue of satisfying criterion H of table 1.2). Feng-Hsiung Hsu (the system architect and chip designer for the Deep Blue project) recounts the intensive work on the Deep Blue project at IBM's T. J. Watson Research Center between 1989 and 1997 (Hsu 2002). The team of scientists and engineers spent years developing the software and the specialized computer chips to efficiently evaluate large numbers of alternative moves as part of a massive parallel state-space search. In short, the human developers invested an enormous amount of "I" in the project. In spite of the fact that Deep Blue delivered a high (human-competitive) amount of "A," the project has a low return when measured in terms of the A-to-I ratio. The builders of Deep Blue convincingly demonstrated the high level of intelligence of the humans involved in the project, but very little in the way of machine intelligence.

The Chinook checker-playing computer program is another impressive human-competitive result. Jonathan Schaeffer recounts the development of Chinook by his eight-member team at the University of Alberta between 1989 and 1996 in his book *One Jump Ahead: Challenging Human Supremacy in Checkers* (Schaeffer 1997). Schaeffer's team began with analysis. They recognized that the problem could be profitably decomposed into three distinct subproblems. First, an opening book controls the play at the beginning of each game. Second, an evaluation function controls the play during the middle of the game. Finally, when only a small number of pieces are left on the board, an endgame database takes over and dictates the best line of play. Perfecting the opening book entailed an iterative process of identifying "positions where Chinook had problems finding the right move" and looking for "the elusive cooks" (Schaeffer 1997, page 237). By the time the project ended, the opening book had over 40,000 entries. In a chapter entitled "A Wake-Up Call," Schaeffer refers to the repeated difficulties surrounding the evaluation function by saying "the thought of rewriting the evaluation routine... and tuning it seemed like my worst nightmare come true." Meanwhile, the endgame database was painstakingly extended from five, to six, to seven, and eventually eight pieces using a variety of clever techniques. As Schaeffer (page 453) observes,

> "The significant improvements to Chinook came from the knowledge added to the program: endgame databases (computer generated), opening book (human generated but computer refined), and the evaluation function (human generated and tuned). We, too, painfully suffered from the knowledge-acquisition bottleneck of artificial intelligence. Regrettably, our project offered no new insights into this difficult problem, other than to reemphasize how serious a problem it really is."

Chinook defeated world champion Marion Tinsley. However, because of the enormous amount of human "I" invested in the project, Chinook (like Deep Blue) has a low return when measured in terms of the A-to-I ratio.

The aim of the fields of artificial intelligence and machine learning is to get computers to automatically generate human-competitive results with a high AI ratio—not to have humans generate human-competitive results themselves.

1.1.3 What We Mean by "Routine"

Generality is a precondition to what we mean when we say that an automated problem-solving method is "routine." Once the generality of a method is established, "routineness" means that relatively little human effort is required to get the method to successfully handle new problems within a particular domain and to successfully handle new problems from a different domain. The ease of making the transition to new problems lies at the heart of what we mean by "routine."

What fraction of Deep Blue's and Chinook's highly specialized software, hardware, databases, and evaluation techniques can be brought to bear on different games? For example, can Deep Blue's massive parallel state-space search or Chinook's three-way decomposition be gainfully applied to a game, such as Go, with a significantly larger number of possible alternative moves at each point in the game? What fraction of these systems can be applied to a game of incomplete information, such as bridge? What more broadly applicable principles are embodied in these two systems? For example, what fraction of these methodologies can be applied to the problem of getting a robot to mop the floor of an obstacle-laden room? Correctly recognizing images or patterns? Devising an algorithm to solve a mathematical problem? Automatically synthesizing a complex structure?

A problem-solving method cannot be considered routine if its executional steps must be substantially augmented, deleted, rearranged, reworked, or customized by the human user for each new problem.

1.1.4 What We Mean by "Machine Intelligence"

We use the term "machine intelligence" to refer to the broad vision articulated in Alan Turing's 1948 paper entitled "Intelligent Machinery" and his 1950 paper entitled "Computing Machinery and Intelligence."

In the 1950s, the terms "machine intelligence," "artificial intelligence," and "machine learning" all referred to the *goal* of getting "machines to exhibit behavior, which if done by humans, would be assumed to involve the use of intelligence" (to again quote Arthur Samuel).

However, in the intervening five decades, the terms "artificial intelligence" and "machine learning" progressively diverged from their original goal-oriented meaning. These terms are now primarily associated with particular *methodologies* for attempting to achieve the goal of getting computers to automatically solve problems. Thus, the term "artificial intelligence" is today primarily associated with attempts to get computers to solve problems using methods that rely on knowledge, logic, and various analytical and mathematical methods. The term "machine learning" is today primarily associated with attempts to get computers to solve problems that use a particular small and somewhat arbitrarily chosen set of methodologies (many of which are statistical in nature). The narrowing of these terms is in marked contrast to the broad field envisioned by Samuel at the time when he coined the term "machine learning" in the 1950s, the charter of the original founders of the field of artificial intelligence, and the broad vision encompassed by Turing's term "machine intelligence."

Of course, the shift in focus from broad goals to narrow methodologies is an all-too-common sociological phenomenon in academic research.

Turing's term "machine intelligence" did not undergo this arteriosclerosis because, by accident of history, it was never appropriated or monopolized by any group of academic researchers whose primary dedication is to a particular methodological approach. Thus, Turing's term remains catholic today. We prefer to use Turing's term because it still communicates the broad *goal* of getting computers to automatically solve problems in a human-like way.

In his 1948 paper, Turing identified three broad approaches by which human-competitive machine intelligence might be achieved.

The first approach was a logic-driven search. Turing's interest in this approach is not surprising in light of Turing's own pioneering work in the 1930s on the logical foundations of computing.

The second approach for achieving machine intelligence was what he called a "cultural search" in which previously acquired knowledge is accumulated, stored in libraries, and brought to bear in solving a problem—the approach taken by modern knowledge-based expert systems.

Turing's first two approaches have been pursued over the past 50 years by the vast majority of researchers using the methodologies that are today primarily associated with the term "artificial intelligence."

However, most pertinently for this book, Turing also identified a third approach to machine intelligence in his 1948 paper entitled "Intelligent Machinery" (Turing 1948, page 12; Ince 1992, page 127; Meltzer and Michie 1969, page 23), saying:

> "There is the genetical or evolutionary search by which a combination of genes is looked for, the criterion being the survival value."

Turing did not specify in 1948 how to conduct the "genetical or evolutionary search" for solutions to problems and, in particular, did not mention the concept of a population or recombination. However, he did point out in his 1950 paper "Computing Machinery and Intelligence" (Turing 1950, page 456; Ince 1992, page 156):

> "We cannot expect to find a good child-machine at the first attempt. One must experiment with teaching one such machine and see how well it learns. One can then try another and see if it is better or worse. There is an obvious connection between this process and evolution, by the identifications
> "Structure of the child machine = Hereditary material
> "Changes of the child machine = Mutations
> "Natural selection = Judgment of the experimenter"

Thus, Turing correctly perceived in 1948 and 1950 that machine intelligence might be achieved by an evolutionary process in which a description of a computer program (the hereditary material) undergoes progressive modification (mutation) under the guidance of natural selection (i.e., selective pressure in the form of what is now usually called "fitness" by practitioners of genetic and evolutionary computation).

Of course, the measurement of fitness in modern genetic and evolutionary computation is usually performed by automated means (as opposed to a human passing judgment on each candidate individual, as suggested by Turing). In addition, modern work generally employs a population (i.e., not just a point-to-point evolutionary progression) and sexual recombination—two key aspects of John Holland's seminal work on genetic algorithms, *Adaptation in Natural and Artificial Systems* (Holland 1975).

1.1.5 Human-Competitiveness of Results Produced by Genetic Programming

The previous four sub-sections defined the terms "human-competitive," "high-return," "routine," and "machine intelligence." In this sub-section (and the next three sub-sections), we evaluate the results produced by genetic programming in light of these four definitions.

Starting with human-competitiveness, table 1.3 lists the 36 human-competitive instances (of which we are aware) where genetic programming has produced

Table 1.3 Thirty-six human-competitive results produced by genetic programming

	Claimed instance	Basis for claim of human-competitiveness	Reference
1	Creation of a better-than-classical quantum algorithm for the Deutsch-Jozsa "early promise" problem	B, F	Spector, Barnum, and Bernstein 1998
2	Creation of a better-than-classical quantum algorithm for Grover's database search problem	B, F	Spector, Barnum, and Bernstein 1999
3	Creation of a quantum algorithm for the depth-two AND/OR query problem that is better than any previously published result	D	Spector, Barnum, Bernstein, and Swamy 1999; Barnum, Bernstein, and Spector 2000
4	Creation of a quantum algorithm for the depth-one OR query problem that is better than any previously published result	D	Barnum, Bernstein, and Spector 2000
5	Creation of a protocol for communicating information through a quantum gate that was previously thought not to permit such communication	D	Spector and Bernstein 2003
6	Creation of a novel variant of quantum dense coding	D	Spector and Bernstein 2003
7	Creation of a soccer-playing program that won its first two games in the Robo Cup 1997 competition	H	Luke 1998
8	Creation of a soccer-playing program that ranked in the middle of the field of 34 human-written programs in the Robo Cup 1998 competition	H	Andre and Teller 1999
9	Creation of four different algorithms for the transmembrane segment identification problem for proteins	B, E	Sections 18.8 and 18.10 of *Genetic Programming II* and sections 16.5 and 17.2 of *Genetic Programming III*
10	Creation of a sorting network for seven items using only 16 steps	A, D	Sections 21.4.4, 23.6, and 57.8.1 of *Genetic Programming III*
11	Rediscovery of the Campbell ladder topology for lowpass and highpass filters	A, F	Section 25.15.1 of *Genetic Programming III* and section 5.2 of this book
12	Rediscovery of the Zobel "*M*-derived half section" and "constant *K*" filter sections	A, F	Section 25.15.2 of *Genetic Programming III*
13	Rediscovery of the Cauer (elliptic) topology for filters	A, F	Section 27.3.7 of *Genetic Programming III*
14	Automatic decomposition of the problem of synthesizing a crossover filter	A, F	Section 32.3 of *Genetic Programming III*
15	Rediscovery of a recognizable voltage gain stage and a Darlington emitter-follower section of an amplifier and other circuits	A, F	Section 42.3 of *Genetic Programming III*
16	Synthesis of 60 and 96 decibel amplifiers	A, F	Section 45.3 of *Genetic Programming III*

(Continued)

Table 1.3 *(Continued)*

	Claimed instance	Basis for claim of human-competitiveness	Reference
17	Synthesis of analog computational circuits for squaring, cubing, square root, cube root, logarithm, and Gaussian functions	A, D, G	Section 47.5.3 of *Genetic Programming III*
18	Synthesis of a real-time analog circuit for time-optimal control of a robot	G	Section 48.3 of *Genetic Programming III*
19	Synthesis of an electronic thermometer	A, G	Section 49.3 of *Genetic Programming III*
20	Synthesis of a voltage reference circuit	A, G	Section 50.3 of *Genetic Programming III*
21	Creation of a cellular automata rule for the majority classification problem that is better than the Gacs-Kurdyumov-Levin (GKL) rule and all other known rules written by humans	D, E	Andre, Bennett, and Koza 1996 and section 58.4 of *Genetic Programming III*
22	Creation of motifs that detect the D–E–A–D box family of proteins and the manganese superoxide dismutase family	C	Section 59.8 of *Genetic Programming III*
23	Synthesis of topology for a PID-D2 (proportional, integrative, derivative, and second derivative) controller	A, F	Section 3.7 of this book
24	Synthesis of an analog circuit equivalent to Philbrick circuit	A, F	Section 4.3 of this book
25	Synthesis of a NAND circuit	A, F	Section 4.4 of this book
26	Simultaneous synthesis of topology, sizing, placement, and routing of analog electrical circuits	A, F, G	Chapter 5 of this book
27	Synthesis of topology for a PID (proportional, integrative, and derivative) controller	A, F	Section 9.2 of this book
28	Rediscovery of negative feedback	A, E, F, G	Chapter 14 of this book
29	Synthesis of a low-voltage balun circuit	A	Section 15.4.1 of this book
30	Synthesis of a mixed analog-digital variable capacitor circuit	A	Section 15.4.2 of this book
31	Synthesis of a high-current load circuit	A	Section 15.4.3 of this book
32	Synthesis of a voltage-current conversion circuit	A	Section 15.4.4 of this book
33	Synthesis of a Cubic function generator	A	Section 15.4.5 of this book
34	Synthesis of a tunable integrated active filter	A	Section 15.4.6 of this book
35	Creation of PID tuning rules that outperform the Ziegler-Nichols and Åström-Hägglund tuning rules	A, B, D, E, F, G	Chapter 12 of this book
36	Creation of three non-PID controllers that outperform a PID controller that uses the Ziegler-Nichols or Åström-Hägglund tuning rules	A, B, D, E, F, G	Chapter 13 of this book

human-competitive results. Each entry in the table is accompanied by the criteria (from table 1.2) that establish the basis for the claim of human-competitiveness.

This book reports in detail on the last 14 of the human-competitive results in table 1.3. The rating of "human-competitive" is justified for each such result (sections 3.7.3, 4.3.3, 4.4.3, 5.2.3, 9.2.3, 12.4, 13.3, 14.3, and 15.6).

Twenty-three of the instances in table 1.3 involve patents (as indicated by an "A" in column 3). Eleven of the automatically created results infringe previously issued patents and 10 duplicate the functionality of previously patented inventions in a non-infringing way. The 29th through 34th entries in table 1.3 relate to patents for analog circuits that were issued after January 1, 2000. The last two entries are patentable new inventions. Tables C.1 and C.2 in appendix C provides additional details on the 23 patent-related results produced by genetic programming.

1.1.6 High-Return of the Results Produced by Genetic Programming

Ascertaining the return of a problem-solving method requires measuring the amount of "A" that is delivered by the method in relation to the amount of "I" that is supplied by the human user.

Because each of the 36 results in table 1.3 is a human-competitive result, it is reasonable to say that genetic programming delivered a high amount of "A" for each of them.

The question thus arises as to how much "I" was supplied by the human user in order to produce these 36 results. Answering this question requires the discipline of carefully identifying the amount of analysis, intelligence, information, and knowledge that was supplied by the intelligent human user prior to launching a run of genetic programming.

In this book (and our previous books and papers on genetic programming), we make a clear distinction between the problem-specific preparatory steps and the problem-independent executional steps of a run of genetic programming.

The *preparatory steps* are the problem-specific and domain-specific steps that are performed by the human user prior to launching a run of the problem-solving method. The preparatory steps establish the "I" component of the AI ratio (i.e., the denominator).

The *executional steps* are the problem-independent and domain-independent steps that are automatically executed during a run of the problem-solving method. The *executional steps* of genetic programming are defined by the flowchart in figure 2.1 of this book. The results produced by the executional steps provide the "A" component of the AI ratio (i.e., the numerator).

The five major preparatory steps for the basic version of genetic programming require the human user to specify

(1) the set of terminals (e.g., the independent variables of the problem, zero-argument functions, and random constants) for each branch of the to-be-evolved computer program,
(2) the set of primitive functions for each branch of the to-be-evolved computer program,
(3) the fitness measure (for explicitly or implicitly measuring the fitness of candidate individuals in the population),
(4) certain parameters for controlling the run, and
(5) a termination criterion and method for designating the result of the run.

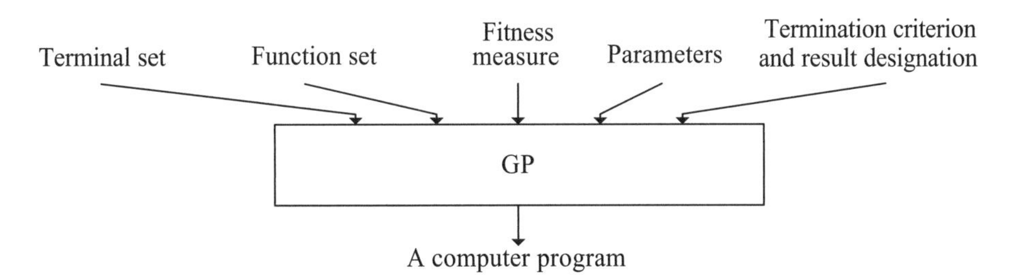

Figure 1.1 Five major preparatory steps for the basic version of genetic programming.

Figure 1.1 shows the five major preparatory steps for the basic version of genetic programming. The preparatory steps (shown at the top of the figure) are the input to the genetic programming system. A computer program (shown at the bottom) is the output of the genetic programming system. The program that is automatically created by genetic programming may solve, or approximately solve, the user's problem.

Genetic programming requires a set of primitive ingredients to get started. The first two preparatory steps specify the primitive ingredients that are to be used to create the to-be-evolved programs. The universe of allowable compositions of these ingredients defines the search space for a run of genetic programming.

The identification of the function set and terminal set for a particular problem (or category of problems) is often a mundane and straightforward process that requires only *de minimus* knowledge and platitudinous information about the problem domain.

For example, if the goal is to get genetic programming to automatically program a robot to mop the entire floor of an obstacle-laden room, the human user must tell genetic programming that the robot is capable of executing functions such as moving, turning, and swishing the mop. The human user must supply this information prior to a run because the genetic programming system does not have any built-in knowledge telling it that the robot can perform these particular functions. Of course, the necessity of specifying a problem's primitive ingredients is not a unique requirement of genetic programming. It would be necessary to impart this same basic information to a neural network learning algorithm, a reinforcement learning algorithm, a decision tree, a classifier system, an automated logic algorithm, or virtually any other automated technique that is likely to be used to solve this problem.

Similarly, if genetic programming is to automatically synthesize an analog electrical circuit, the human user must supply basic information about the ingredients that are appropriate for solving a problem in the domain of analog circuit synthesis. In particular, the human user must inform genetic programming that the components of the to-be-created circuit may include transistors, capacitors, and resistors (as opposed to, say, neon bulbs, relays, and doorbells). Although this information may be second nature to anyone working with electrical circuits, genetic programming does not have any built-in knowledge concerning the fact that transistors, capacitors, and resistors are the workhorse components for nearly all present-day electrical circuits. Once the human user has identified the primitive ingredients, the same function set can be used to automatically synthesize amplifiers, computational circuits, active filters, voltage reference circuits, and any other circuit composed of these basic ingredients.

Likewise, genetic programming does not know that the inputs to a controller include the reference signal and plant output and that controllers are composed of integrators, differentiators, leads, lags, gains, adders, subtractors, and the like. Thus, if genetic programming is to automatically synthesize a controller, the human user must give genetic programming this basic information about the field of control.

The third preparatory step concerns the fitness measure for the problem. The fitness measure specifies what needs to be done. The result that is produced by genetic programming specifies "how to do it." The fitness measure is the primary mechanism for communicating the high-level statement of the problem's requirements to the genetic programming system. If one views the first two preparatory steps as defining the search space for the problem, one can then view the third preparatory step (the fitness measure) as specifying the search's desired direction.

The fitness measure is the means of ascertaining that one candidate individual is better than another. That is, the fitness measure is used to establish a partial order among candidate individuals. The partial order is used during the executional steps of genetic programming to select individuals to participate in the various genetic operations (i.e., crossover, reproduction, mutation, and the architecture-altering operations).

The fitness measure is derived from the high-level statement of the problem. Indeed, for many problems, the fitness measure may be almost identical to the high-level statement of the problem. The fitness measure typically assigns a single numeric value reflecting the extent to which a candidate individual satisfies the problem's high-level requirements. For example:

- If an electrical engineer needs a circuit that amplifies an incoming signal by a factor of 1,000, the fitness measure might assign fitness to a candidate circuit based on how closely the circuit's output comes to a target signal whose amplitude is 1,000 times that of the incoming signal. In comparing two candidate circuits, amplification of 990-to-1 would be considered better than 980-to-1.
- If a control engineer wants to design a controller for the cruise control device in a car, the fitness measure might be based on the time required to bring the car's speed up from 55 to 65 miles per hour. When candidate controllers are compared, a rise time of 10.1 seconds would be considered better than 10.2 seconds.
- If a robot is expected to mop a room, the fitness measure might be based on the percentage of the area of the floor that is cleaned within a reasonable amount of time.
- If a classifier is needed for protein sequences (or any other objects), the fitness measure might be based on the correlation between the category to which the classifier assigns each protein sequence and the correct category.
- If a biochemist wants to find a network of chemical reactions or a metabolic pathway that matches observed data, the fitness measure might assign fitness to a candidate network based on how closely the network's output matches the data.

The fitness measure for a real-world problem is typically multiobjective. That is, there may be more than one element that is considered in ascertaining fitness. For example, the engineer may want an amplifier with 1,000-to-1 gain, but may also want low distortion, low bias, and a low parts count. In practice, the elements of a multiobjective fitness measure usually conflict with one another. Thus, a multiobjective fitness measure must prioritize the different elements so as to reflect the tradeoffs that the engineer is willing to accept. For example, the engineer may be willing to tolerate an additional 1% of distortion in exchange for the elimination of one part from the circuit. One approach is to blend the distinct elements of a fitness measure into a single numerical value (often merely by weighting them and adding them together).

The fourth and fifth preparatory steps are administrative.

The fourth preparatory step entails specifying the control parameters for the run. The major control parameters are the population size and the number of generations to be run. Some analytic methods are available for suggesting optimal population sizes for runs of the genetic algorithm on particular problems. However, the practical reality is that we generally do not use any such analytic method to choose the population size. Instead, we determine the population size such that genetic programming can execute a reasonably large number of generations within the amount of computer time we are willing to devote to the problem. As for other control parameters, we have, broadly speaking, used the same (undoubtedly non-optimal) set of minor control parameters from problem to problem over a period of years. Although particular problems in this book could possibly be solved more efficiently by means of a different choice of control parameters, we believe that our policy of substantial consistency in the choice of control parameters helps the reader eliminate superficial concerns that the demonstrated success of genetic programming depends on shrewd or fortuitous choices of the control parameters. As can be seen in this book (and our previous books), we frequently make only one run (or, at most, only a few runs) of each major new problem.

The fifth preparatory step consists of specifying the termination criterion and the method of designating the result of the run.

We have now identified that which is supplied by the human user of genetic programming. For the problems in this book, we believe that it is generally fair to say that only a *de minimus* amount of "I" is contained in the problem's primitive ingredients (the first and second preparatory steps), the problem's fitness measure (the third preparatory step containing the high-level statement of what needs to be done), and the run's control parameters and termination procedures (the administrative fourth and fifth preparatory steps).

In any event, the amount of "I" required by genetic programming is certainly not greater than that required by any other method of artificial intelligence and machine learning of which we are aware. Indeed, we know of no other problem-solving method (automated or human) that does not start with primitive elements of some kind, does not incorporate some method for specifying what needs to be done to guide the method's operation, does not employ administrative parameters of some kind, and does not contain a termination criterion of some kind.

In view of the numerous human-competitive results produced by genetic programming (table 1.3), it can be seen that genetic programming is capable of delivering a large amount of "A." That is, its AI ratio is high.

Throughout this book, there are numerous sections where the AI ratio is qualitatively evaluated for particular problems (sections 3.7.4, 3.8.4, 3.9.4, 3.10.4, 4.2.4, 4.3.5, 4.4.5, 4.5.4, 4.6.4, 4.7.4, 5.2.5, 5.3.4, 6.7, 7.4.2, 8.6.2, 8.7.2, 9.1.4, 9.2.5, 10.2.4, 10.3.4, 10.4.4, 10.5.4, 11.1.4, 11.2.4, 12.6, 13.5, 14.5, and 15.8).

1.1.7 Routineness of the Results Produced by Genetic Programming

This book demonstrates the generality of genetic programming by solving illustrative problems from several fields, including problems involving

- control,
- analog electrical circuits (including six post-2000 patented circuits),
- placement and routing of circuits,
- antennas,
- genetic networks, and
- metabolic pathways.

Our previous publications (and previous publications by others) additionally demonstrate that genetic programming is capable of solving problems in numerous other areas.

The bright line distinction between that which is delivered by genetic programming and that which is supplied by the intelligent human user (in section 1.1.6) additionally helps make it clear that genetic programming is a systematic general problem-solving method.

As will be seen in this book, relatively little effort is required to make the transition to new problems within a particular domain or to new problems from an entirely different domain.

For example, after discussing the first problem of automatically synthesizing both the topology and tuning of a controller in chapter 3, the transition to each subsequent problem of controller synthesis in that chapter mainly involves providing genetic programming with a different specification of what needs to be done—that is, a different fitness measure. Because virtually all controllers are built from the same primitive ingredients (e.g., integrators, differentiators, gains, adders, subtractors, and signals representing the plant output and the reference signal), additional problems of controller synthesis can be handled merely by changing the statement of what needs to be done.

Similarly, after discussing the first problem of automatically synthesizing both the topology and sizing of an analog electrical circuit in chapter 4, the transition to each subsequent problem of circuit design in that chapter mainly involves providing genetic programming with a different specification of what needs to be done.

The routineness of the transition from problem to problem is especially clear in chapter 15 involving six circuits that were patented after January 1, 2000. All six problems were run consecutively over a period of about two months intentionally using the very same computer, the very same software, and the very same settings of the minor control parameters. All six circuits were composed of the workhorse ingredients of present-day electronics (i.e., resistors, capacitors, and transistors). As we move from one problem to the next in chapter 15, the only substantial change is the specification of what needs to be done. This specification is based on each inventor's statement of performance of each patented circuit. As stated in *Genetic Programming: On the Programming of Computers by Means of Natural Selection* (Koza 1992a), "Structure arises from fitness."

The transition from one problem domain to another becomes especially clear by comparing the work concerning the automatic synthesis of controllers, analog electrical circuits, antennas, genetic networks, and networks of chemical reactions.

In making the transition from problems of automatic synthesis of controllers to problems of automatic synthesis of circuits, the primitive ingredients change from integrators, differentiators, gains, adders, subtractors, and the like to transistors, resistors, capacitors, and the like. The fitness measure changes from one that minimizes a controller's integral of time-weighted absolute error, minimizes overshoot, and maximizes disturbance rejection to one that is based on the circuit's amplification, suppression or passage of a signal, elimination of distortion, and the like.

In making the transition from problems of automatic synthesis of circuits to problems of automatic synthesis of networks of chemical reactions (metabolic pathways), the primitive ingredients change to functions that represent chemical reactions that consume chemical substrates (inputs to chemical reactions) and produce reaction products (outputs), at certain rates, in the presence of certain catalysts (enzymes). The fitness measure compares the quantity of product that is produced by a candidate network to the observed data.

Of course, although the preparatory steps change from one problem to another and from one domain to another, the main executional steps (i.e., the flowchart) of genetic programming remain unchanged.

In numerous places throughout this book, we demonstrate

- the routineness of the transition from one problem to another problem in the same domain (sections 3.8.3, 3.9.3, 3.10.3, 4.3.4. 4.4.4, 4.5.3, 4.6.3, 4.7.3, 5.3.3, 8.7.1, 9.2.4, 10.3.3. 10.4.3. 10.5.3, 11.2.3, 13.4. 14.4. and 15.7),
- the routineness of the transition from one domain to the next (sections 4.2.3, 5.2.4, 6.6, 7.4.1, 8.6.1, and 12.5),
- the routineness of the transition from a non-parameterized version of a problem to a parameterized version (sections 9.1.3 and 10.2.3), and
- the routineness of the transition from a problem involving parameterized topologies without conditional developmental operators to a problem involving parameterized topologies with them (section 11.1.3).

1.1.8 Machine Intelligence

As will be seen throughout this book, genetic programming does indeed succeed in getting computers to automatically solve problems from a high-level statement of what needs to be done.

1.2 Genetic Programming Is an Automated Invention Machine

In a commencement address at Worcester Polytechnic Institute in 2000, C. Michael Armstrong, CEO of American Telephone & Telegraph, recounted:

> "On a sweltering summer morning in August 1927, a young man was seated on a passenger ferry as it churned across Upper New York Bay toward Manhattan.

He was gazing idly at the Statue of Liberty when suddenly he jumped from his seat and began frantically searching his pockets for a scrap of paper.

"Coming up empty, he raced to the newsboy on deck and bought a copy of *The New York Times*. The man tore through the pages until he found one that was nearly free of type. He uncapped his fountain pen, sketched a couple of crude diagrams, and surrounded them with mathematical equations."

Holding up the now-famous page from *The New York Times* (figure 1.2), C. Michael Armstrong continued:

"When the ferryboat docked at Manhattan, he raced to his office at Bell Laboratories. He showed his diagrams and equations to one of his coworkers who

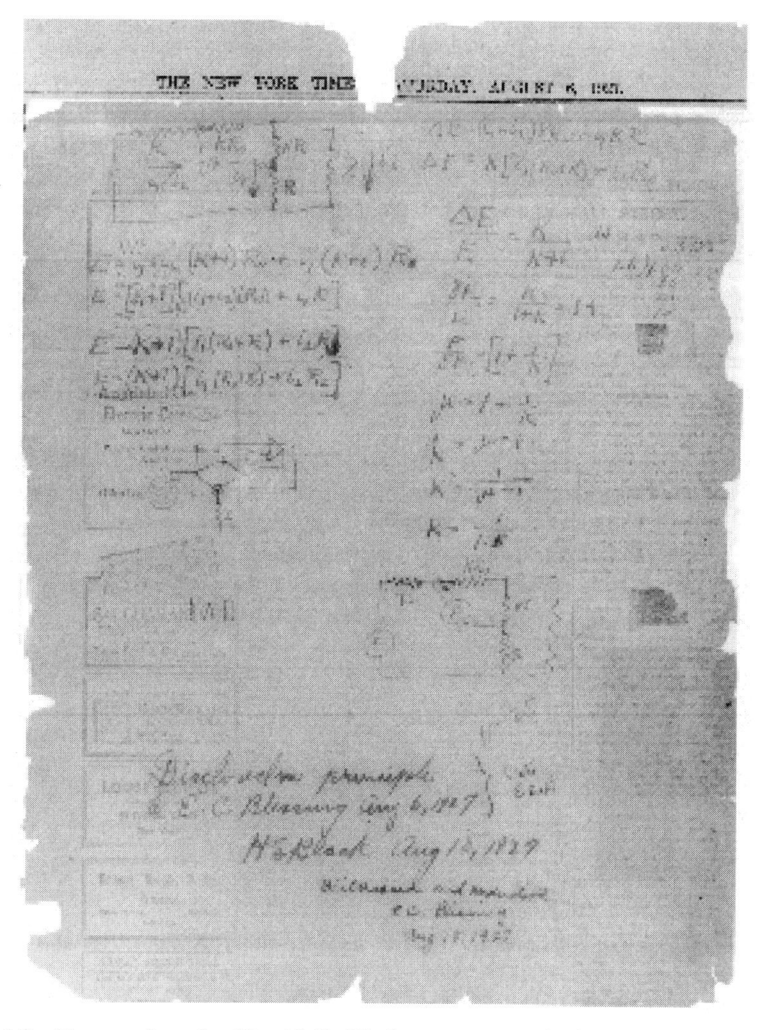

Figure 1.2 Notes written by Harold S. Black on a page of *The New York Times* while commuting on the Lackawanna Ferry. Reproduced here by kind permission of Lucent Technologies.

read them carefully. Then his friend let out a big whoop and they both scrawled their initials on the newspaper page.

"The young man on the ferryboat was Harold Black, Worcester Polytechnic Institute Class of 1921. And the scribblings on his newspaper were the blueprint for the negative-feedback amplifier, a device that played a vital role in 20th century electronics."

Referring to the scribblings on this newspaper page, Mervin Kelly, then president of Bell Labs, said in 1957 (Black 1977):

"Although many of Harold's inventions have made great impact, that of the negative feedback amplifier is indeed the most outstanding. It easily ranks coordinate with De Forest's invention of the audion as one of the two inventions of broadest scope and significance in electronics and communications of the past 50 years... It is no exaggeration to say that without Black's invention, the present long-distance telephone and television networks which cover our entire country and the transoceanic telephone cables would not exist. The application of Black's principle of negative feedback has not been limited to telecommunications.... [T]he entire explosive extension of the area of control, both electrical and mechanical, grew out of an understanding of the feedback principle."

Lee (1998) recounts the history that predated Black's 1927 invention of negative feedback. Earlier work on feedback included rocket pioneer Robert Goddard's 1915 patent for a vacuum tube oscillator using *positive* feedback (Goddard 1915) and Edwin Howard Armstrong's 1914 patent on amplifiers again using positive feedback (Armstrong 1914). As Lee observes,

"[P]rogress in electronics in those early years was largely made possible by Armstrong's regenerative [positive feedback] amplifier, since there was no other economical way to obtain large amounts of gain from the primitive (and expensive) vacuum tubes of the day."...

"Armstrong was able to get gain from a single stage that others could obtain only by cascading several. This achievement allowed the construction of relatively inexpensive, high-gain receivers and therefore also enabled dramatic reductions in transmitter power because of the enhanced sensitivity provided by this increased gain. In short order, *the positive feedback (regenerative) amplifier became a nearly universal idiom*, and Westinghouse (to whom Armstrong had assigned patent rights) kept its legal staff quite busy trying to make sure that only licensees were using this revolutionary technology." (Emphasis added.)

However, Westinghouse's "nearly universal idiom" did not solve a major problem facing American Telephone & Telegraph at the time, namely distortion in amplifiers. As Lee (1998, page 387) further points out:

"Although Armstrong's regenerative amplifier pretty much solved the problem of obtaining large amounts of gain from vacuum tube amplifiers, a different problem preoccupied the telephone industry. In trying to extend communications distances, amplifiers were needed to compensate for transmission-line attenuation. Using amplifiers available in those early days, distances of a few hundred miles were routinely achievable and, with great care, perhaps 1,000–2,000 miles was possible, but the quality was poor. After a tremendous amount of work, a crude transcontinental telephone service was inaugurated in 1915, with a 68-year-old Alexander Graham Bell making the first call to his former assistant, Thomas Watson, but this feat was more of a stunt than a practical achievement.

"The problem wasn't one of insufficient amplification; it was trivial to make the signal at the end of the line quite loud. Rather the problem was distortion. Each amplifier contributed some small (say, 1%) distortion. Cascading a hundred of these things guaranteed that what came out didn't very much resemble what went in.

"The main 'solution' at the time was to (try to) guarantee 'small signal' operation of the amplifiers. That is, by restricting the dynamic range of the signals to a tiny fraction of the amplifier's overall capability, more linear operation could be achieved. Unfortunately, this strategy is quite inefficient since it requires the construction of, say, 100-W amplifiers to process milliwatt signals. Because of the arbitrary distance between a signal source and an amplifier (or possibly between amplifiers), though, it was difficult to guarantee that the input signals were always sufficiently small to satisfy linearity."

Such was the state of affairs when Harold S. Black started working in 1921 at AT&T on the problem of reducing amplifier distortion. After considerable work, Black reached the conclusion in 1923 (Black 1977):

"There was just no way to meet our ambitious goal."

As Black recounts:

"This might have been the end of it, except that, on March 16, 1923, I was fortunate enough to attend a lecture by the famous scientist and engineer, Charles Proteus Steinmetz."...

"I no longer remember the subject, but I do remember the clarity and logic of his presentation."...

"I was so impressed by how Steinmetz got down to the fundamentals that when I returned home at 2 A.M., I restated my own problems as follows: Remove all distortion products from the amplifier output. In doing this, I was accepting an imperfect amplifier and regarding its output as composed of what was wanted plus what was not wanted. I considered what was not wanted to be distortion (regardless of whether it was due to nonlinearity, variation in the tube gain, or whatever), and I asked myself how to isolate and then eliminate this distortion. I immediately observed that by reducing the output to the same amplitude as the input, and subtracting one from the other, only the distortion would remain. This distortion could then be amplified in a separate amplifier and used to cancel out the distortion in the original amplifier....

"The next day, March 17, I sketched two such embodiments and thereby invented the feed-forward amplifier....

"Later that day, I set up each embodiment in the laboratory. Both worked as expected."

Black applied for a patent on his 1923 invention of the feed-forward amplifier and the patent was issued in 1928 (Black 1928). Unfortunately, his 1923 invention did not turn out to be practical. As Black (1977) laments,

> "[T]he invention required precise balances and subtractions that were hard to achieve and maintain with the amplifiers available at that time....
>
> "Over the next four years, I struggled with the problem of turning my intention into an amplifier that was practical....
>
> "[F]or my purpose the gain had to be absolutely perfect."
>
> "For example, every hour on the hour—24 hours a day—somebody had to adjust the filament current to its correct value....
>
> "In addition, every six hours it became necessary to adjust the B battery voltage, because the amplifier gain would get out of hand.
>
> "There were other complications too, but these were enough!"

The bottom line concerning the feed-forward amplifier that Black invented in 1923 was:

> "Nothing came of my efforts, however, because every circuit I devised turned out to be far too complex to be practical."
>
> In spite of this false start, Black continued to work on the problem.
>
> After working on the problem for a total of six years:
>
> "Then came the morning of Tuesday, August 2, 1927, when the concept of the negative feedback amplifier came to me in a flash while I was crossing the Hudson River on the Lackawanna Ferry, on my way to work. For more than 50 years, I have pondered how and why the idea came, and I can't say any more today than I could that morning. All I know is that after several years of hard work on the problem, I suddenly realized that if I fed the amplifier output back to the input, in reverse phase, and kept the device from oscillating (singing, as we called it then), I would have exactly what I wanted: a means of canceling out the distortion of the output. I opened my morning newspaper and on a page of *The New York Times* I sketched a simple canonical diagram of a negative feedback amplifier plus the equations for the amplification with feedback. I signed the sketch, and 20 minutes later, when I reached the laboratory at 463 West Street, it was witnessed, understood, and signed by the late Earl C. Blessing."

Numerous other inventors have reported similar singular moments when their previous thinking about a vexatious problem crystallized into an invention.

1.2.1 The Illogical Nature of Invention and Evolution

Most computer scientists unquestioningly assume that any effective problem-solving process must be logically sound and deterministic.

The consequence of this unproven assumption is that virtually all conventional approaches to artificial intelligence and machine learning possess these characteristics. Yet the reality is that logic does not govern two of the most important processes for solving complex problems—namely the invention process (performed by creative humans) and the evolutionary process (occurring in nature).

Moreover, neither the invention process nor the evolutionary process is deterministic.

A new idea that can be logically deduced from facts that are known in a field, using transformations that are known in a field, is not considered to be inventive by the Patent Office. A new idea is patentable only if there is what the courts have called an "illogical step" (i.e., a logically unjustified step). The required illogic distinguishes the proposed invention from that which is readily deducible from what is already known. The required illogical step is also sometimes referred to as a "flash of genius." In other words, logical thinking is not the key ingredient for one of the most significant human problem-solving activities, namely the invention process. Interestingly, everyday usage parallels the law concerning the point that a lack of logic is a precondition for inventiveness: People who mechanically apply existing facts in well-known ways are summarily dismissed as being uncreative.

Of course, when we say that the invention process is inherently illogical, we do not mean that logical thinking is not helpful to inventors or that inventors are oblivious to logic. Logical thinking often plays the important role of setting the stage for an invention. "[S]everal years of hard work on the problem" brought Black's thinking into the proximity of a solution (Black 1977). Then, at the critical moment, Black made the illogical leap during his now-famous ferryboat ride. Although logical thinking may play a role in invention and creativity, at the end of the day, the critical element is a logical discontinuity from established ideas.

The design of complex entities by the evolutionary process in nature is another important type of problem-solving that is not governed by logic. In nature, solutions to design problems are discovered by means of evolution and natural selection. The evolutionary process is probabilistic, rather than deterministic. Moreover, it is certainly not guided by mathematical logic. Indeed, one of the most important characteristics of the evolutionary process is that it intentionally creates and actively maintains inconsistent and contradictory alternatives. Logically sound systems do not do that. The active maintenance of inconsistent and contradictory alternatives (called *genetic diversity*) is a precondition for the success of the evolutionary process.

1.2.2 Overcoming Established Beliefs

As previously mentioned, Edwin Howard Armstrong's approach to amplification using positive feedback was "a nearly universal idiom" during the early part of the 20th century.

In spite of the elegance and manifest effectiveness of negative feedback, Armstrong's approach was so entrenched in the thinking of electrical engineers that there was widespread resistance to Black's concept of negative feedback for many years after its invention. As Black (1977) recalls:

> "Although the invention had been submitted to the U.S. Patent Office on August 8, 1928, more than nine years would elapse before the patent was issued on December 21, 1937.... One reason for the delay was that *the concept was so contrary to established beliefs*." (Emphasis added.)

The British Patent Office was even more resistant. As Black (1977) recounted:

> "... our patent application was treated in the same manner as one for a perpetual motion machine."

The British Patent Office continued to maintain that negative feedback would not work in spite of the fact that AT&T had "70 amplifiers working successfully in the telephone building at Morristown" for a number of years.

We believe that one reason why it took an inordinate amount of time for negative feedback to gain acceptance was that human thinking often becomes channeled along the well-traveled paths of "established beliefs."

One of the virtues of genetic programming is that it is not aware, much less concerned, about whether a solution is "contrary to established beliefs." Genetic programming approaches a problem in an open-ended way that is not encumbered by previous human thinking. For this reason, genetic programming often unearths solutions that might have never occurred to human scientists and engineers who are steeped in the thinking of the day.

In the section entitled "Genetic Programming Takes a Ride on the Lackawanna Ferry" (section 14.1), genetic programming is used to reinvent negative feedback. As will be seen, if one begins with a high-level statement of the problem that Black was trying to solve, Black's solution flows almost effortlessly from a run of genetic programming. It does so because Black's solution is a correct solution to the problem and, as they say, necessity is the mother of invention.

For the 20 other instances in table 1.3 where genetic programming created an entity that infringes a previously issued patent or duplicates the functionality of a previously patented invention in a non-infringing or novel way, the solution similarly flowed directly from a high-level statement of the problem.

The 23 instances where genetic programming has duplicated the functionality of a previously patented invention, infringed a previously issued patent, or created a patentable new invention are shown in tables C.1 and C.2 in appendix C.

1.2.3 Automating the Invention Process

For over 200 years, the U.S. Patent Office has been in the business of receiving written descriptions of human-designed inventions and judging whether the purported inventions are

- "new,"
- "improved,"
- "useful," and
- "[un]obvious ... to a person having ordinary skill in the art to which said subject matter pertains." (35 *United States Code* 103a)

When the Patent Office passes judgment on a patent application, it generally works from written documents and operates at arms length from the inventor. When an automated method duplicates the detailed structure of a previously patented human-created invention, the fact that the human-designed version originally satisfied the Patent Office's criteria for patent-worthiness means that an automatically created duplicate would also have satisfied the Patent Office's criteria for patent-worthiness had it arrived at the Patent Office prior to the human inventor's submission.

When genetic programming is applied to a problem whose solution is a previously patented invention, there are three possible outcomes:

- failure of the run to solve the problem,
- creation of a solution that infringes a previously issued patent, or
- creation of a non-infringing solution that duplicates the functionality of a previously patented invention.

There are two sub-cases associated with the third case.

First, a non-infringing solution may be a previously known solution (i.e., prior art). The previously known solution may or may not have been patented in the past.

Second, a non-infringing solution may be a new solution to the problem.

In this second sub-case, a new, genetically evolved, non-infringing solution may be patentable if it satisfies the additional requirements of being "useful," "improved," and "unobvious."

A genetically evolved solution would generally be deemed to be "useful" for the same reasons that the originally patented invention was deemed to be "useful."

Almost every alternative solution to a particular problem usually has some attribute that can be reasonably viewed (from some standpoint) as being "improved" in some respect or to some degree.

Because genetically evolved solutions often contain features that would never occur to human scientists or engineers, a genetically evolved alternative solution will often be "unobvious" to someone "having ordinary skill in the art."

U.S. law suggests that inventions created by automated means are patentable by saying:

> "Patentability shall not be negatived by the manner in which the invention was made." (35 *United States Code* 103a)

1.2.4 Patentable New Inventions Produced by Genetic Programming

Given that genetic programming has solved problems whose solutions were previously patented, it is a natural extension to try to use genetic programming to generate patentable new inventions.

Chapters 12 and 13 of this book describe a patent application filed on July 12, 2002, for improved PID (proportional, integrative, and derivative) tuning rules and non-PID controllers that were automatically created by means of genetic programming (Keane, Koza, and Streeter 2002a). The genetically evolved tuning rules and controllers outperform controllers tuned using the widely used Ziegler-Nichols tuning rules (1942) and the recently developed Åström-Hägglund tuning rules (1995). The applicants believe that the new tuning rules and controllers satisfy the statutory requirement of being "improved" and "useful." They are certainly "new." Because they contain features that would never occur to an experienced control engineer, they are certainly "unobvious" to someone "having ordinary skill in the art."

If (as expected) a patent is granted, it will (we believe) be the first patent granted for an invention created by genetic programming. For further discussion of the potential of genetic programming as an invention machine, see Koza, Keane, and Streeter 2003.

1.3 Genetic Programming Can Automatically Create Parameterized Topologies

Eleven problems in this book illustrate this book's third main point, namely that genetic programming can automatically create what we call *parameterized topologies*. That is, genetic programming can automatically create, in a single run, a general (parameterized) solution to a problem in the form of a graphical structure whose nodes or edges represent components and where the parameter values of the components are specified by mathematical expressions containing free variables.

In a parameterized topology, the genetically evolved graphical structure represents a complex structure (e.g., electrical circuit, controller, network of chemical reactions, antenna, genetic network). In the automated process, genetic programming determines the graph's size (its number of nodes) as well as the graph's connectivity (specifying which nodes are connected to each other). Genetic programming also assigns, in the automated process, component types to the graph's edges nodes or edges. In a circuit, the component types are usually transistors, resistors, and capacitors. In a controller, the components are integrators, differentiators, gain blocks, adders, subtractors, and the like. In the automated process, genetic programming also creates mathematical expressions that establish the parameter values of the components (e.g., the capacitance of a capacitor in a circuit, the amplification factor of a gain block in a controller). Some of these genetically created mathematical expressions contain free variables. The free variables confer generality on the genetically evolved solution by enabling a single genetically evolved graphical structure to represent a general (parameterized) solution to an entire category of problems. The important point about parameterized topologies is that genetic programming *can do all the above in an automated way in a single run.*

As an example, suppose the goal is to design a circuit to feed the woofer speaker of a hi-fi system. That is, the desired circuit is intended to pass signals below a certain frequency at full power into the woofer, but to suppress all higher frequencies. Moreover, suppose that you want a general solution to this design problem. That is, suppose you want a solution that works for any cutoff frequency *f*—not just a solution that works for, say, 1,000 Hertz. A genetically evolved general solution to this problem is shown later in this book (in figure 10.9). The general solution produced by genetic programming includes the circuit's topology. The genetically evolved parameterized circuit has nine components. The general solution includes the type of each of the nine components. There are five capacitors and four inductors in figure 10.9. The connections between the nine components are automatically created during the run of genetic programming. The problem's free variable, *f*, is an input to the genetically evolved solution. The genetically evolved solution is general because the capacitance of the five capacitors and the inductance of the four inductors are not constant, but instead, are functions of the free variable *f*. That is, the general solution produced by genetic programming includes nine different mathematical expressions—each containing the free variable f. For example, one of the nine mathematical expressions is

$$C2 = \frac{1.6786 \times 10^5}{f}$$

When all nine mathematical expressions are instantiated with a particular value of the free variable, f, the resulting circuit is a lowpass filter whose passband boundary is f. The numerical values of certain components could be constant (although none of the values happen to be constant in this particular case). The advantage of a parameterized topology is it is a general solution to the problem—not just a solution to a single instance of the problem.

If the genetically evolved program additionally contains conditional developmental operators, different graphical structures will, in general, be produced for different instantiations of the free variable. That is, the genetically evolved program will operate as a genetic switch. Depending on the values of the free variable, different graphical structures will result from the execution of the best-of-run program. The numerical values for all parameterized components in the graphical structure will also be established by the execution of the program.

The capability of genetic programming to automatically create parameterized topologies is demonstrated in this book by the automatic synthesis of

- a parameterized controller for controlling a three-lag plant whose time constant is specified by a free variable (section 9.1),
- a parameterized controller for controlling plants belonging to two different families (section 9.2),
- three parameterized controllers for controlling industrially representative plants (chapter 13),
- a parameterized circuit-constructing program tree containing two free variables that yields a Zobel network (section 10.2),
- a parameterized circuit-constructing program tree that yields a passive third-order elliptic lowpass filter whose modular angle is specified by a free variable (section 10.3),
- a parameterized circuit-constructing program tree that yields an active lowpass filter whose passband boundary is specified by a free variable (section 10.5),
- a parameterized circuit-constructing program tree that yields a passive lowpass filter whose passband boundary is specified by a free variable (section 10.4),
- a parameterized circuit-constructing program tree containing conditional developmental operators and free variables that yields either a lowpass or highpass passive filter (section 11.1),
- a parameterized circuit-constructing program tree containing conditional developmental operators and free variables that yields either a lowpass passive filter with a variable passband boundary or a highpass passive filter with a variable passband boundary (section 11.2),
- a parameterized circuit-constructing program tree containing conditional developmental operators and free variables that yields either a quadratic or cubic computational circuit (section 11.3), and

- a parameterized circuit-constructing program tree containing conditional developmental operators and free variables that yields either a 40 dB or 60 dB amplifier (section 11.4).

These 11 examples establish this book's third main point.

1.4 Historical Progression of Qualitatively More Substantial Results Produced by Genetic Programming in Synchrony with Increasing Computer Power

Numerous questions naturally arise in connection with any proposed approach to machine intelligence.

- Is the method formulated with sufficient precision to enable it to be implemented (or is it vagueware)?
- Has the method been successfully demonstrated on a specific single problem (or is it promiseware)?
  - Was the method applied to a difficult demonstrative problem (or is it toyware)?
  - Did the method top out after succeeding on a single demonstrative problem?
- Has the method solved multiple problems (or is it soloware)?
  - Are the multiple problems difficult?
  - Did the method top out at this stage?
- Has the method solved problems from multiple domains (or is it nicheware)?
  - Are the domains difficult?
  - Did the method top out at this stage?
- Were the results human-competitive?
- Can the method profitably take advantage of the increased computational power available by means of parallel processing (or is it serialware)?
- Or, is the method Mooreware—able to take advantage of the exponentially increasing computational power made available by the relentless iteration of Moore's law?

Genetic Programming: On the Programming of Computers by Means of Natural Selection (Koza 1992a) demonstrated that genetic programming is not vagueware, promiseware, soloware, or nicheware.

The numerous human-competitive results discussed in this book establish that it is not toyware.

This book's fourth main point is based on a historical perspective on the progression of results produced by genetic programming over the 15-year period between 1987 and 2002.

Table 1.4 (described in greater detail in section 18.1) lists the five computer systems used to produce our group's reported work on genetic programming in the 15-year period between 1987 and 2002. Column 7 shows the number of human-competitive results (as defined in table 1.2 and itemized in table 1.3) generated by each computer system.

The first entry in the table is a serial computer. The four subsequent entries are parallel computer systems. The presence of four increasingly powerful parallel computer systems in the table reflects the fact that genetic programming has successfully taken advantage of the increased computational power available by means of parallel processing. That is, genetic programming is not serialware.

Table 1.4 shows the following:

- There is an order-of-magnitude speed-up (column 4) between each successive computer system in the table. Note that, according to Moore's law (Moore 1996), exponential increases in computer power correspond approximately to constant periods of time.
- There is a 13,900-to-1 speed-up (column 5) between the fastest and most recent machine (the 1,000-node parallel computer system used for most of the work in this book) and the slowest and earliest computer system in the table (the serial LISP machine).
- The slower early machines generated few or no human-competitive results, whereas the faster more recent machines have generated numerous human-competitive results.

Four successive order-of-magnitude increases in computer power are explicitly shown in table 1.4. An additional order-of-magnitude increase was achieved by the expedient of making extraordinarily long runs on the largest machine in the table (the 1,000-node Pentium® II parallel machine). The length of the run that produced

Table 1.4 Number of human-competitive results produced by genetic programming with five computer systems between 1987 and 2002

System	Period of usage	Petacycles (10^{15} cycles) per day for entire system	Speed-up over previous system	Speed-up over first system in this table	Used for work in book	Human-competitive results
Serial Texas Instruments LISP machine	1987–1994	0.00216	1 (base)	1 (base)	*Genetic Programming I* and *Genetic Programming II*	0
64-node Transtech transputer parallel machine	1994–1997	0.02	9	9	A few problems in *Genetic Programming III*	2
64-node Parsytec parallel machine	1995–2000	0.44	22	204	Most problems in *Genetic Programming III*	12
70-node Alpha parallel machine	1999–2001	3.2	7.3	1,481	A minority (8) of problems in this book	2
1,000-node Pentium II parallel machine	2000–2002	30.0	9.4	13,900	A majority (28) of the problems in this book	12

the genetically evolved controller described in section 13.2.3 was 28.8 days—almost an order-of-magnitude increase (9.3 times) over the 3.4-day average for other problems described in this book (table 17.1). A patent application was filed for the controller produced by this four-week run (Keane, Koza, and Streeter 2002a). This genetically evolved controller outperforms controllers employing the widely used Ziegler-Nichols tuning rules and the recently developed Åström-Hägglund tuning rules. If the final 9.3-to-1 increase in table 1.5 is counted as an additional speed-up, the overall speed-up between the first and last entries in the table is 130,660-to-1.

Table 1.5 (described in greater detail in sections 18.2 and 18.3) is organized around the five just-explained order-of-magnitude increases in the expenditure of computing power. Column 4 of table 1.5 characterizes the qualitative nature of the

Table 1.5 Progression of qualitatively more substantial results produced by genetic programming in relation to five order-of-magnitude increases in computational power

System	Period of usage	Speed-up over previous row in this table	Qualitative nature of the results produced by genetic programming
Serial Texas Instruments LISP machine	1987–1994	1 (base)	• Toy problems of the 1980s and early 1990s from the fields of artificial intelligence and machine learning
64-node Transtech transputer parallel machine	1994–1997	9	• Two human-competitive results involving one-dimensional discrete data (not patent-related)
64-node Parsytec parallel machine	1995–2000	22	• One human-competitive result involving two-dimensional discrete data • Numerous human-competitive results involving continuous signals analyzed in the frequency domain • Numerous human-competitive results involving 20th-century patented inventions
70-node Alpha parallel machine	1999–2001	7.3	• One human-competitive result involving continuous signals analyzed in the time domain • Circuit synthesis extended from topology and sizing to include routing and placement (layout)
1,000-node Pentium II parallel machine	2000–2002	9.4	• Numerous human-competitive results involving continuous signals analyzed in the time domain • Numerous general solutions to problems in the form of parameterized topologies • Six human-competitive results duplicating the functionality of 21st-century patented inventions
Long (4-week) runs of 1,000-node Pentium II parallel machine	2002	9.3	• Generation of two patentable new inventions

results produced by genetic programming. The table shows the progression of qualitatively more substantial results produced by genetic programming in terms of five order-of-magnitude increases in the expenditure of computational resources.

The order-of-magnitude increases in computer power shown in table 1.5 correspond closely (albeit not perfectly) with the following progression of qualitatively more substantial results produced by genetic programming:

- toy problems,
- human-competitive results not related to patented inventions,
- 20th-century patented inventions,
- 21st-century patented inventions, and
- patentable new inventions.

In other words, genetic programming is able to take advantage of the exponentially increasing computational power made available by iterations of Moore's law—that is, it is Mooreware.

These results (explained in greater detail in chapter 18) establish this book's fourth main point: Genetic programming has delivered a progression of qualitatively more substantial results in synchrony with five approximately order-of-magnitude increases in the expenditure of computer time.

2

Background on Genetic Programming

This chapter provides basic background information on genetic programming.

Genetic programming is a domain-independent method that genetically breeds a population of computer programs to solve a problem. Specifically, genetic programming iteratively transforms a population of computer programs into a new generation of programs by applying analogs of naturally occurring genetic operations. The genetic operations include crossover (sexual recombination), mutation, reproduction, gene duplication, and gene deletion. Analogs of developmental processes that transform an embryo into a fully developed entity are also employed. Genetic programming is an extension of the genetic algorithm (Holland 1975) into the arena of computer programs.

This chapter describes

- the preparatory steps of genetic programming (section 2.1),
- the executional steps (flowchart) of genetic programming (section 2.2),
- advanced features of genetic programming (section 2.3),
- the main points of our previous books on genetic programming (section 2.4), and
- sources of additional information about genetic programming (section 2.5).

2.1 Preparatory Steps of Genetic Programming

Genetic programming starts from a high-level statement of the requirements of a problem and attempts to produce a computer program that solves the problem.

The human user communicates the high-level statement of the problem to the genetic programming system by performing certain well-defined preparatory steps.

The five major preparatory steps for the basic version of genetic programming are described in *Genetic Programming: On the Programming of Computers by Means of Natural Selection* (Koza 1992a). They require the human user to specify

(1) the set of terminals (e.g., the independent variables of the problem, zero-argument functions, and random constants) for each branch of the to-be-evolved program,
(2) the set of primitive functions for each branch of the to-be-evolved program,

(3) the fitness measure (for explicitly or implicitly measuring the fitness of individuals in the population),
(4) certain parameters for controlling the run, and
(5) the termination criterion and method for designating the result of the run.

Figure 1.1 shows the five major preparatory steps for the basic version of genetic programming. The preparatory steps (shown at the top of the figure) are the human-supplied input to the genetic programming system. The computer program (shown at the bottom) is the output of the genetic programming system.

The first two preparatory steps specify the ingredients that are available to create the computer programs. A run of genetic programming is a competitive search among a diverse population of programs composed of the available functions and terminals.

The identification of the function set and terminal set for a particular problem (or category of problems) is usually a straightforward process. For some problems, the function set may consist of merely the arithmetic functions of addition, subtraction, multiplication, and division as well as a conditional branching operator. The terminal set may consist of the program's external inputs (independent variables) and numerical constants.

For many other problems, the ingredients include specialized functions and terminals. For example, if the goal is to get genetic programming to automatically program a robot to mop the entire floor of an obstacle-laden room, the human user must tell genetic programming what the robot is capable of doing. For example, the robot may be capable of executing functions such as moving, turning, and swishing the mop.

If the goal is the automatic creation of a controller, the function set may consist of integrators, differentiators, leads, lags, gains, adders, subtractors, and the like. The terminal set may consist of signals such as the reference signal and plant output.

If the goal is the automatic synthesis of an analog electrical circuit, the function set may enable genetic programming to construct circuits from components such as transistors, capacitors, and resistors. Once the human user has identified the primitive ingredients for a problem of circuit synthesis, the same function set can be used to automatically synthesize an amplifier, computational circuit, active filter, voltage reference circuit, or any other circuit composed of these ingredients.

The third preparatory step concerns the fitness measure for the problem. The fitness measure specifies what needs to be done. The fitness measure is the primary mechanism for communicating the high-level statement of the problem's requirements to the genetic programming system. For example, if the goal is to get genetic programming to automatically synthesize an amplifier, the fitness function is the mechanism for telling genetic programming to synthesize a circuit that amplifies an incoming signal (as opposed to, say, a circuit that suppresses the low frequencies of an incoming signal or that computes the square root of the incoming signal). The first two preparatory steps define the search space whereas the fitness measure implicitly specifies the search's desired goal.

The fourth and fifth preparatory steps are administrative. The fourth preparatory step entails specifying the control parameters for the run. The most important control parameter is the population size. In this book, we normally choose a population size that will produce a reasonably large number of generations in the amount of computer time we are willing to devote to a problem (as opposed to analytically choosing the population size by somehow analyzing a problem's fitness landscape). Other control parameters include the probabilities of performing the genetic operations, the maximum size for programs, and other details of the run.

The fifth preparatory step consists of specifying the termination criterion and the method of designating the result of the run. The termination criterion may include a maximum number of generations to be run as well as a problem-specific success predicate. Our practice for all problems in this book is to manually monitor and manually terminate the run when the values of fitness for numerous successive best-of-generation individuals appear to have reached a plateau. The single best-so-far individual is then harvested and designated as the result of the run.

2.2 Executional Steps of Genetic Programming

After the user has performed the preparatory steps for a problem, the run of genetic programming can be launched. Once the run is launched, a series of well-defined, problem-independent steps is executed.

Genetic programming typically starts with a population of randomly generated computer programs composed of the available programmatic ingredients (as provided by the human user in the first and second preparatory steps).

Genetic programming iteratively transforms a population of computer programs into a new generation of the population by applying analogs of naturally occurring genetic operations. These operations are applied to individual(s) selected from the population. The individuals are probabilistically selected to participate in the genetic operations based on their fitness (as measured by the fitness measure provided by the human user in the third preparatory step). The iterative transformation of the population is executed inside the main generational loop of the run of genetic programming.

The executional steps of genetic programming are as follows:

(1) Randomly create an initial population (generation 0) of individual computer programs composed of the available functions and terminals.
(2) Iteratively perform the following sub-steps (called a *generation*) on the population until the termination criterion is satisfied:
    (a) Execute each program in the population and ascertain its fitness (explicitly or implicitly) using the problem's fitness measure.
    (b) Select one or two individual program(s) from the population with a probability based on fitness (with reselection allowed) to participate in the genetic operations in (c).
    (c) Create new individual program(s) for the population by applying the following genetic operations with specified probabilities:
        (i) *Reproduction:* Copy the selected individual program to the new population.
        (ii) *Crossover:* Create new offspring program(s) for the new population by recombining randomly chosen parts from two selected programs.
        (iii) *Mutation:* Create one new offspring program for the new population by randomly mutating a randomly chosen part of one selected program.

(iv) *Architecture-altering operations:* Choose an architecture-altering operation from the available repertoire of such operations and create one new offspring program for the new population by applying the chosen architecture-altering operation to one selected program.

(3) After the termination criterion is satisfied, the single best program in the population produced during the run (the best-so-far individual) is harvested and designated as the result of the run. If the run is successful, the result may be a solution (or approximate solution) to the problem.

Figure 2.1 is a flowchart of genetic programming showing the genetic operations of crossover, reproduction, and mutation as well as the architecture-altering operations. This flowchart shows a two-offspring version of the crossover operation (used in the example run in section 2.2.1). One-offspring crossover is used for all the other problems this book.

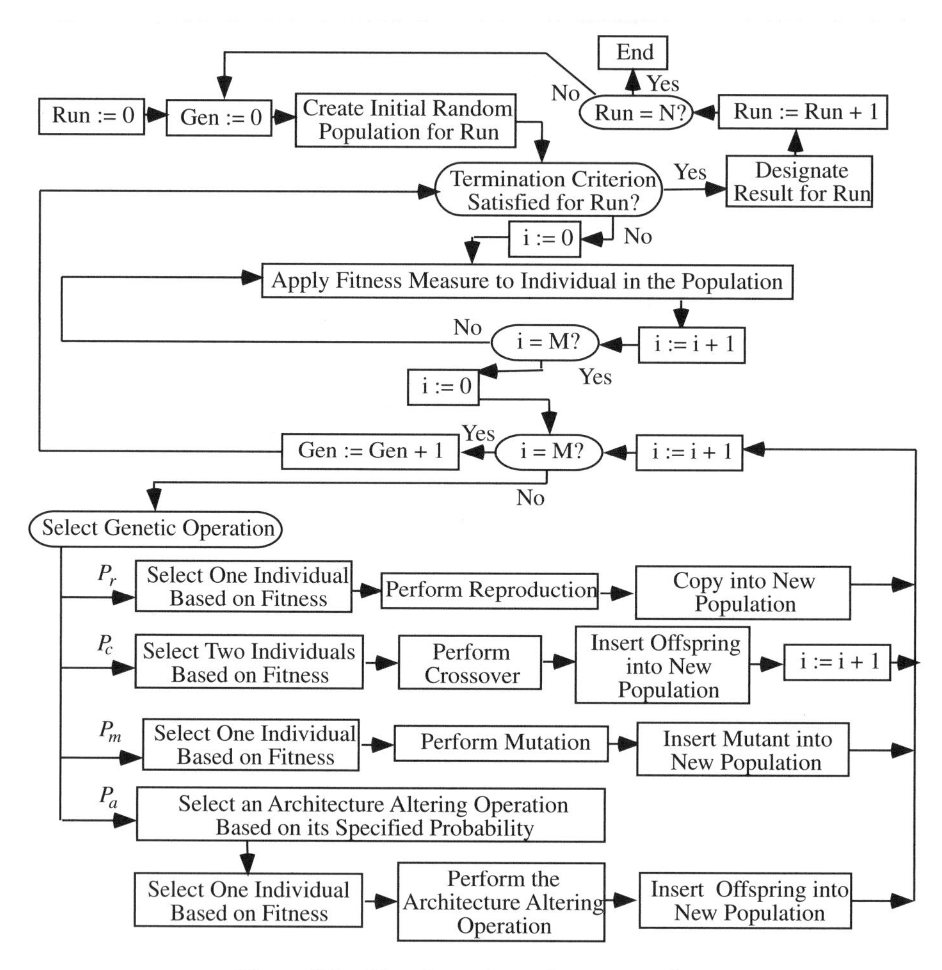

Figure 2.1 Flowchart of genetic programming.

The preparatory steps specify what the user must provide in advance to the genetic programming system. Once the run is launched, the executional steps as shown in the flowchart (figure 2.1) are executed. Genetic programming is problem-independent in the sense that the flowchart specifying the basic sequence of executional steps is not modified for each new run or each new problem.

There is usually no discretionary human intervention or interaction during a run of genetic programming (although a human user may exercise judgment as to whether to terminate a run).

Genetic programming starts with an initial population of computer programs composed of functions and terminals appropriate to the problem. The individual programs in the initial population are typically generated by recursively generating a rooted point-labeled program tree composed of random choices of the primitive functions and terminals (provided by the user as part of the first and second preparatory steps). The initial individuals are usually generated subject to a pre-established maximum size (specified by the user as a minor parameter as part of the fourth preparatory step). In general, the programs in the population are of different size (number of functions and terminals) and of different shape (the particular graphical arrangement of functions and terminals in the program tree).

Each individual program in the population is executed. Then, each individual program in the population is either measured or compared in terms of how well it performs the task at hand (using the fitness measure provided in the third preparatory step). For many problems (including all problems in this book), this measurement yields a single explicit numerical value, called *fitness*. The fitness of a program may be measured in many different ways, including, for example, in terms of the amount of error between its output and the desired output, the amount of time (fuel, money, etc.) required to bring a system to a desired target state, the accuracy of the program in recognizing patterns or classifying objects into classes, the payoff that a game-playing program produces, or the compliance of a complex structure (such as an antenna, circuit, or controller) with user-specified design criteria. The execution of the program sometimes returns one or more explicit values. Alternatively, the execution of a program may consist only of side effects on the state of a world (e.g., a robot's actions). Alternatively, the execution of a program may produce both return values and side effects.

The fitness measure is, for many practical problems, multiobjective in the sense that it combines two or more different elements. The different elements of the fitness measure are often in competition with one another to some degree.

For many problems, each program in the population is executed over a representative sample of different *fitness cases*. These fitness cases may represent different values of the program's input(s), different initial conditions of a system, or different environments. Sometimes the fitness cases are constructed probabilistically.

The creation of the initial random population is, in effect, a blind random search of the search space of the problem. It provides a baseline for judging future search efforts. Typically, the individual programs in generation 0 all have exceedingly poor fitness. Nonetheless, some individuals in the population are (usually) more fit than others. The differences in fitness are then exploited by genetic programming. Genetic programming applies Darwinian selection and the genetic operations to create a new population of offspring programs from the current population.

The genetic operations include crossover (sexual recombination), mutation, reproduction, and the architecture-altering operations. These genetic operations are applied to individual(s) that are probabilistically selected from the population based on fitness. In this probabilistic selection process, better individuals are favored over inferior individuals. However, the best individual in the population is not necessarily selected and the worst individual in the population is not necessarily passed over.

After the genetic operations are performed on the current population, the population of offspring (i.e., the new generation) replaces the current population (i.e., the now-old generation). This iterative process of measuring fitness and performing the genetic operations is repeated over many generations.

The run of genetic programming terminates when the termination criterion (as provided by the fifth preparatory step) is satisfied. The outcome of the run is specified by the method of result designation. The best individual ever encountered during the run (i.e., the best-so-far individual) is typically designated as the result of the run.

All programs in the initial random population (generation 0) of a run of genetic programming are syntactically valid, executable programs. The genetic operations that are performed during the run (i.e., crossover, mutation, reproduction, and the architecture-altering operations) are designed to produce offspring that are syntactically valid, executable programs. Thus, every individual created during a run of genetic programming (including, in particular, the best-of-run individual) is a syntactically valid, executable program.

There are numerous alternative implementations of genetic programming that vary from the preceding brief description.

2.2.1 Example of a Run of Genetic Programming

To provide concreteness, this section contains an illustrative run of genetic programming in which the goal is to automatically create a computer program whose output is equal to the values of the quadratic polynomial x^2+x+1 in the range from -1 to $+1$. That is, the goal is to automatically create a computer program that matches certain numerical data. This process is sometimes called *system identification* or *symbolic regression*.

We begin with the five preparatory steps.

The purpose of the first two preparatory steps is to specify the ingredients of the to-be-evolved program.

Because the problem is to find a mathematical function of one independent variable, the terminal set (inputs to the to-be-evolved program) includes the independent variable, x. The terminal set also includes numerical constants. That is, the terminal set, T, is

$$\mathsf{T} = \{\mathtt{X}, \Re\},$$

where $\Re$ denotes constant numerical terminals in some reasonable range (say from -5.0 to $+5.0$).

The preceding statement of the problem is somewhat flexible in that it does not specify what functions may be employed in the to-be-evolved program. One possible choice for the function set consists of the four ordinary arithmetic functions of addition, subtraction, multiplication, and protected division. This choice is reasonable because mathematical expressions typically include these functions. Thus, the function set, F, for this problem is

$$\mathsf{F} = \{+, -, *, \%\}.$$

The two-argument +, −, *, and % functions add, subtract, multiply, and divide, respectively. The protected division function % returns a value of 1 when division by 0 is attempted (including 0 divided by 0), but otherwise returns the quotient of its two arguments.

Each individual in the population is a composition of functions from the specified function set and terminals from the specified terminal set.

The third preparatory step involves constructing the fitness measure. The purpose of the fitness measure is to specify what the human wants. The high-level goal of this problem is to find a program whose output is equal to the values of the quadratic polynomial x^2+x+1. Therefore, the fitness assigned to a particular individual in the population for this problem must reflect how closely the output of an individual program comes to the target polynomial x^2+x+1. The fitness measure could be defined as the value of the integral (taken over values of the independent variable x between -1.0 and $+1.0$) of the absolute value of the differences (errors) between the value of the individual mathematical expression and the target quadratic polynomial x^2+x+1. A smaller value of fitness (error) is better. A fitness (error) of zero would indicate a perfect fit.

For most problems of symbolic regression or system identification, it is not practical or possible to analytically compute the value of the integral of the absolute error. Thus, in practice, the integral is numerically approximated using dozens or hundreds of different values of the independent variable x in the range between -1.0 and $+1.0$.

The population size in this small illustrative example will be just four. In actual practice, the population size for a run of genetic programming consists of thousands or millions of individuals. In actual practice, the crossover operation is commonly performed on about 90% of the individuals in the population; the reproduction operation is performed on about 8% of the population; the mutation operation is performed on about 1% of the population; and the architecture-altering operations are performed on perhaps 1% of the population. Because this illustrative example involves an abnormally small population of only four individuals, the crossover operation will be performed on two individuals and the mutation and reproduction operations will each be performed on one individual. For simplicity, the architecture-altering operations are not used for this problem.

A reasonable termination criterion for this problem is that the run will continue from generation to generation until the fitness of some individual gets below 0.01. In this contrived example, the run will (atypically) yield an algebraically perfect solution (for which the fitness measure attains the ideal value of zero) after merely one generation.

Now that we have performed the five preparatory steps, the run of genetic programming can be launched. That is, the executional steps shown in the flowchart of figure 2.1 are now performed.

Genetic programming starts by randomly creating a population of four individual computer programs. The four programs are shown in figure 2.2 in the form of trees.

The first randomly constructed program tree (figure 2.2a) is equivalent to the mathematical expression $x+1$. A program tree is executed in a depth-first way, from left to right, in the style of the LISP programming language. Specifically, the addition function (+) is executed with the variable x and the constant value 1 as its two arguments. Then, the two-argument subtraction function (−) is executed. Its first argument is the value returned by the just-executed addition function. Its second argument is the constant value 0. The overall result of executing the entire program tree is thus $x+1$.

The first program (figure 2.2a) was constructed by first choosing the subtraction function for the root (top point) of the program tree. The random construction process continued in a depth-first fashion (from left to right) and chose the addition function to be the first argument of the subtraction function. The random construction process then chose the terminal x to be the first argument of the addition function (thereby terminating the growth of this path in the program tree). The random construction process then chose the constant terminal 1 as the second argument of the addition function (thereby terminating the growth along this path). Finally, the random construction process chose the constant terminal 0 as the second argument of the subtraction function (thereby terminating the entire construction process).

The second program (figure 2.2b) adds the constant terminal 1 to the result of multiplying x by x and is equivalent to $x^2 + 1$. The third program (figure 2.2c) adds the constant terminal 2 to the constant terminal 0 and is equivalent to the constant value 2. The fourth program (figure 2.2d) is equivalent to x.

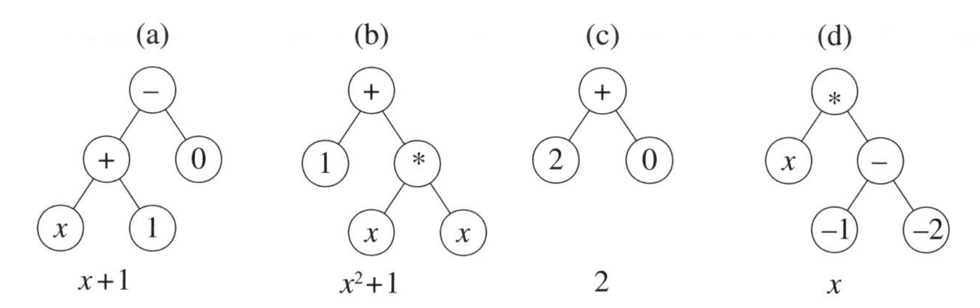

Figure 2.2 Initial population of four randomly created individuals of generation 0.

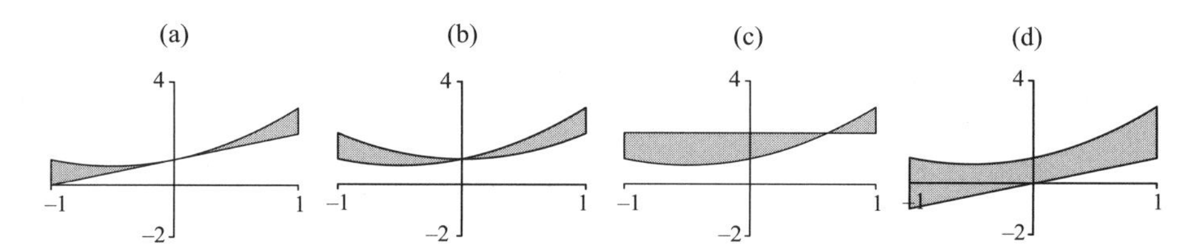

Figure 2.3 The fitness of each of the four randomly created individuals of generation 0 is equal to the area between two curves.

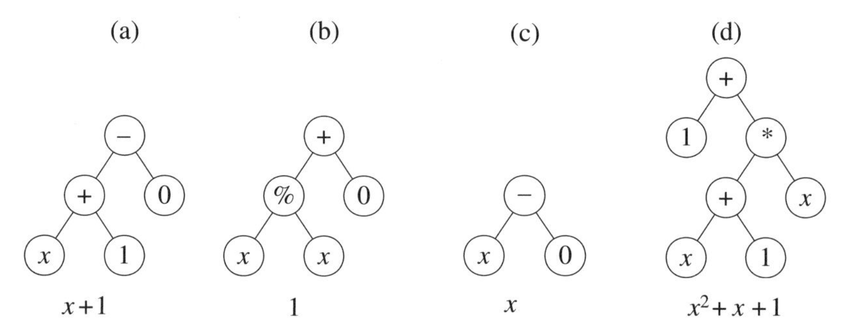

Figure 2.4 Population of generation 1 (after one reproduction, one mutation, and one two-offspring crossover operation).

Randomly created computer programs will, of course, typically be very poor at solving the problem at hand. However, even in a population of randomly created programs, some programs are better than others. The four random individuals from generation 0 in figure 2.2 produce outputs that deviate from the output produced by the target quadratic function x^2+x+1 by different amounts. In this particular problem, fitness can be graphically illustrated as the area between two curves. That is, fitness is equal to the area between the parabola x^2+x+1 and the curve representing the candidate individual. Figure 2.3 shows (as shaded areas) the integral of the absolute value of the errors between each of the four individuals in figure 2.2 and the target quadratic function x^2+x+1. The integral of absolute error for the straight line $x+1$ (the first individual) is 0.67 (figure 2.3a). The integral of absolute error for the parabola x^2+1 (the second individual) is 1.0 (figure 2.3b). The integral of the absolute errors for the remaining two individuals are 1.67 (figure 2.3c) and 2.67 (figure 2.3d), respectively.

As can be seen in figure 2.3, the straight line $x+1$ (figure 2.3a) is closer to the parabola x^2+x+1 in the range from -1 to $+1$ than any of its three cohorts in the population. This straight line is, of course, not equivalent to the parabola x^2+x+1. This best-of-generation individual from generation 0 is not even a quadratic function. It is merely the best candidate that happened to emerge from the blind random search of generation 0. In the valley of the blind, the one-eyed man is king.

After the fitness of each individual in the population is ascertained, genetic programming then probabilistically selects relatively more fit programs from the population. The genetic operations are applied to the selected individuals to create offspring programs. The most commonly employed methods for selecting individuals to participate in the genetic operations are tournament selection (used throughout this book) and fitness-proportionate selection. In both methods (described in Koza 1992a), the emphasis is on selecting relatively fit individuals. An important feature common to both methods is that the selection is not greedy. Individuals that are known to be inferior will be selected to a certain degree. The best individual in the population is not guaranteed to be selected. Moreover, the worst individual in the population will not necessarily be excluded. Anything can happen and nothing is guaranteed.

We first perform the reproduction operation. Because the first individual (figure 2.2a) is the most fit individual in the population, it is very likely to be selected to participate in a genetic operation. Let's suppose that this particular individual is, in fact, selected for reproduction. If so, it is copied, without alteration, into the next generation (generation 1). It is shown in figure 2.4a as part of the population of the new generation.

We next perform the mutation operation. Because selection is probabilistic, it is possible that the third best individual in the population (figure 2.2c) is selected. One of the three points of this individual is then randomly picked as the site for the mutation. In this example, the constant terminal 2 is picked as the mutation site. This program is then randomly mutated by deleting the entire subtree rooted at the picked point (in this case, just the constant terminal 2) and inserting a subtree that is randomly grown in the same way that the individuals of the initial random population were originally created. In this particular instance, the randomly grown subtree (shown in figure 2.4b) computes the quotient of x and x using the protected division operation %. This particular mutation changes the original individual from one having a constant value of 2 into one having a constant value of 1. This particular mutation improves fitness from 1.67 to 1.00.

Finally, we perform the crossover operation. Because the first and second individuals in generation 0 are both relatively fit, they are likely to be selected to participate in crossover. In fact, because of its high fitness, the first individual ends up being selected twice to participate in a genetic operation. In contrast, the unfit fourth individual is not be selected at all. The reselection of relatively more fit individuals and the exclusion and extinction of unfit individuals is a characteristic feature of Darwinian selection. The first and second programs are mated sexually to produce two offspring (using the two-offspring version of the crossover operation). One point of the first parent (figure 2.2a), namely the + function, is randomly picked as the crossover point for the first parent. One point of the second parent (figure 2.2b), namely its leftmost terminal x, is randomly picked as the crossover point for the second parent. The crossover operation is then performed on the two parents. The two offspring are shown in figures 2.4c and 2.4d. One of the offspring (figure 2.4c) is equivalent to x and is not noteworthy. However, the other offspring (figure 2.4d) is equivalent to x^2+x+1 and has a fitness (integral of absolute errors) of zero. Because the fitness of this individual is below 0.01, the termination criterion for the run is satisfied and the run is automatically terminated. This best-so-far individual (figure 2.4d) is designated as the result of the run. This individual is an algebraically correct solution to the problem.

Note that the best-of-run individual (figure 2.4d) incorporates a good trait (the quadratic term x^2) from the second parent (figure 2.2b) with two other good traits (the linear term x and constant term of 1) from the first parent (figure 2.2a). The crossover operation produced a solution to this problem by recombining good traits from these two relatively fit parents into a superior (indeed, perfect) offspring.

In summary, genetic programming has, in this example, automatically created a computer program whose output is equal to the values of the quadratic polynomial x^2+x+1 in the range from -1 to $+1$.

Additional details of the operation of basic genetic programming are found in *Genetic Programming: On the Programming of Computers by Means of Natural Selection* (Koza 1992a).

2.3 Advanced Features of Genetic Programming

Various advanced features of genetic programming are not covered by the foregoing illustrative problem and the foregoing discussion of the preparatory and executional steps of genetic programming.

2.3.1 Constrained Syntactic Structures

For certain simple problems (such as the illustrative problem in section 2.2.1), the search space for a run of genetic programming consists of the unrestricted set of possible compositions of the problem's functions and terminals.

However, for many problems (including every problem in this book), a constrained syntactic structure imposes restrictions on how the functions and terminals may be combined.

Consider, for example, a function that instructs a robot to turn by a certain angle. In a typical implementation of this hypothetical function, the function's first argument may be required to return a numerical value (representing the desired turning angle) and its second argument may be required to be a follow-up command (e.g., move, turn, stop). In other words, the functions and terminals permitted in the two argument subtrees for this particular function are restricted. These restrictions are implemented by means of syntactic rules of construction.

A *constrained syntactic structure* (sometimes called *strong typing*) is a grammar that specifies the functions or terminals that are permitted to appear as a specified argument of a specified function in the program tree.

When a constrained syntactic structure is used, there are typically multiple function sets and multiple terminal sets. The rules of construction specify where the different function sets or terminal sets may be used.

When a constrained syntactic structure is used, all the individuals in the initial random population (generation 0) are created so as to comply with the constrained syntactic structure. All genetic operations (i.e., crossover, mutation, reproduction, and the architecture-altering operations) that are performed during the run are designed to produce offspring that comply with the requirements of the constrained syntactic structure. Thus, all individuals (including, in particular, the best-of-run individual) that are produced during the run of genetic programming will necessarily comply with the requirements of the constrained syntactic structure.

2.3.2 Automatically Defined Functions

Human computer programmers organize sequences of reusable steps into subroutines. They then repeatedly invoke the subroutines—typically with different instantiations of the subroutine's dummy variables (formal parameters). Reuse eliminates the need to "reinvent the wheel" on each occasion when a particular sequence of steps may be needed. Reuse makes it possible to exploit a problem's modularities, symmetries, and regularities (and thereby potentially accelerate the problem-solving process).

Programmers commonly organize their subroutines into hierarchies.

The automatically defined function (ADF) is one of the mechanisms by which genetic programming implements the parameterized reuse and hierarchical invocation of evolved code. Each automatically defined function resides in a separate function-defining branch within the overall multi-part computer program. When automatically defined functions are being used, a program consists of one (or more) function-defining branches (i.e., automatically defined functions) as well as one or more main result-producing branches. An automatically defined function may possess zero, one, or more dummy variables (formal parameters). The body of an automatically defined function contains its work-performing steps. Each automatically defined function belongs to a particular program in the population. An automatically defined function may be called by the program's main result-producing branch, another automatically defined function, or another type of branch (such as those described in section 2.3.3).

Recursion may be allowed. Typically, the automatically defined functions are invoked with different instantiations of their dummy variables.

The work-performing steps of the program's main result-producing branch and the work-performing steps of each automatically defined function are automatically and simultaneously created during the run of genetic programming.

The program's main result-producing branch and its automatically defined functions typically have different function and terminal sets. A constrained syntactic structure (section 2.3.1) is used to implement automatically defined functions.

Automatically defined functions are the focus of *Genetic Programming II: Automatic Discovery of Reusable Programs* (Koza 1994a) and the videotape *Genetic Programming II Videotape: The Next Generation* (Koza 1994b). See also Koza 1990a, 1992a, 1992c; Koza and Rice 1991, 1992b, 1994a; and Koza, Bennett, Andre, and Keane 1999a.

2.3.3 Automatically Defined Iterations, Automatically Defined Loops, Automatically Defined Recursions, and Automatically Defined Stores

Automatically defined iterations (ADIs), automatically defined loops (ADLs), and automatically defined recursions (ADRs) provide means (in addition to automatically defined functions) to reuse code.

Automatically defined stores (ADSs) provide means to reuse the result of executing code.

Automatically defined iterations, automatically defined loops, automatically defined recursions, and automatically defined stores are described in *Genetic Programming III: Darwinian Invention and Problem Solving* (Koza, Bennett, Andre, and Keane 1999a).

2.3.4 Program Architecture and Architecture-Altering Operations

The architecture of a program consists of

- the total number of branches,
- the type of each branch (e.g., result-producing branch, automatically defined function, automatically defined iteration, automatically defined loop, automatically defined recursion, or automatically defined store),
- the number of arguments (if any) possessed by each branch, and
- if there is more than one branch, the nature of the hierarchical references (if any) allowed among the branches.

There are three ways by which genetic programming can arrive at the architecture of the to-be-evolved computer program:

- The human user may prespecify the architecture of the overall program (i.e., perform an additional architecture-defining preparatory step). That is, the number of preparatory steps is increased from the five itemized in section 2.1 to six.
- The run may employ evolutionary selection of the architecture (as described in *Genetic Programming II*), thereby enabling the architecture of the overall program

to emerge from a competitive process during the run of genetic programming. When this approach is used, the number of preparatory steps remains at the five itemized in section 2.1.

- The run may employ the architecture-altering operations (Koza 1994c, 1995a, 1995b, 1995c; Koza, Andre, and Tackett 1994, 1998; Koza, Bennett, Andre, and Keane 1999a), thereby enabling genetic programming to automatically create the architecture of the overall program dynamically during the run. When this approach is used, the number of preparatory steps remains at the five itemized in section 2.1.

2.3.5 Genetic Programming Problem Solver (GPPS)

The Genetic Programming Problem Solver (GPPS) is described in part 4 of the 1999 book *Genetic Programming III: Darwinian Invention and Problem Solving* (Koza, Bennett, Andre, and Keane 1999a, 2003).

If GPPS is being used, the user is relieved of performing the first and second preparatory steps (concerning the choice of the terminal set and the function set). The function set for GPPS consists of the four basic arithmetic functions (addition, subtraction, multiplication, and division) and a conditional operator (i.e., functions found in virtually every general-purpose digital computer that has ever been built). The terminal set for GPPS consists of numerical constants and a set of input terminals that are presented in the form of a vector.

By employing this generic function set and terminal set, GPPS reduces the number of preparatory steps from five to three.

GPPS relies on the architecture-altering operations to dynamically create, duplicate, and delete subroutines and loops during the run of genetic programming. Additionally, in version 2.0 of GPPS, the architecture-altering operations are used to dynamically create, duplicate, and delete recursions and internal storage. Because the architecture of the evolving program is automatically determined during the run, GPPS eliminates the need for the user to specify in advance whether to employ subroutines, loops, recursions, and internal storage in solving a given problem. It similarly eliminates the need for the user to specify the number of arguments possessed by each subroutine. And, GPPS eliminates the need for the user to specify the hierarchical arrangement of the invocations of the subroutines, loops, and recursions. That is, the use of GPPS relieves the user of performing the preparatory step of specifying the program's architecture. The results produced by GPPS include one human-competitive result (section 23.6 of *Genetic Programming III*). However, most of the problems solved using GPPS have, as of this writing, been toy problems comparable in difficulty to those found in the 1992 book *Genetic Programming* (Koza 1992a). See also Koza, Bennett, Andre, and Keane 1999c.

2.3.6 Developmental Genetic Programming

Developmental genetic programming is used for problems of synthesizing analog electrical circuits, as described in chapters 4, 5, 10, 11, 14, and 15 of this book and in part 5 of *Genetic Programming III*. When developmental genetic programming is used, a complex structure (such as an electrical circuit) is created from a simple initial structure (the embryo). If the user desires to specify the embryo, the number of preparatory steps is increased from five to six. On the other hand, the user can be relieved of specifying a particular embryo for a problem by using a generic floating embryo (section 4.7.1.1). Early work on development includes Kitano's (1990) use of genetic algorithms to evolve neural networks, Gruau's work on cellular encoding (developmental genetic programming) to evolve neural networks (1992a, 1992b), work by Spector and Stoffel on ontogenetic programming (1996a, 1996b), and work involving the evolution of Lindenmayer rules for creating structures (Koza 1993).

2.3.7 Computer Code for Implementing Genetic Programming

Genetic programming has been implemented in numerous programming languages. Most present-day versions of genetic programming are written in C, C++, or Java. LISP code for implementing genetic programming is available in *Genetic Programming* (Koza 1992a). Web sites such as `www.genetic-programming.org` contain links to computer code in various other programming languages.

2.4 Main Points of Four Books on Genetic Programming

Table 2.1 shows the main points of our four books on genetic programming. The two main points of the first book and the eight main points of the second book have been paraphrased for purposes of this table (but are quoted in full later in this section). As can be seen, the main points of this book fit into a progression of results starting with the 1992 book.

Table 2.1 Main points of our four books on genetic programming

Book	Main points
1992	• Virtually all problems in artificial intelligence, machine learning, adaptive systems, and automated learning can be recast as a search for a computer program. • Genetic programming provides a way to successfully conduct the search for a computer program in the space of computer programs.
1994	• Scalability is essential for solving non-trivial problems in artificial intelligence, machine learning, adaptive systems, and automated learning. • Scalability can be achieved by reuse. • Genetic programming provides a way to automatically discover and reuse subprograms in the course of automatically creating computer programs to solve problems.
1999	• Genetic programming possesses the attributes that can reasonably be expected of a system for automatically creating computer programs.
2003	• Genetic programming now routinely delivers high-return human-competitive machine intelligence. • Genetic programming is an automated invention machine. • Genetic programming can automatically create a general solution to a problem in the form of a parameterized topology. • Genetic programming has delivered a progression of qualitatively more substantial results in synchrony with five approximately order-of-magnitude increases in the expenditure of computer time.

The 1992 book *Genetic Programming: On the Programming of Computers by Means of Natural Selection* (Koza 1992a) and its accompanying videotape *Genetic Programming: The Movie* (Koza and Rice 1992a) established the principle that genetic programming is capable of automatically creating a computer program that solves a problem. The two main points of this book are:

- A wide variety of seemingly different problems from many different fields can be recast as requiring the discovery of a computer program that produces some desired output when presented with particular inputs. That is, many seemingly different problems can be reformulated as problems of program induction.
- The recently developed genetic programming paradigm described in this book provides a way to do program induction. That is, genetic programming can search the space of possible computer programs for an individual computer program that is highly fit in solving (or approximately solving) the problem at hand. The computer program (i.e., structure) that emerges from the genetic programming paradigm is a consequence of fitness. That is, fitness begets the needed program structure.

These points were demonstrated on a wide range of "toy" (proof of principle) problems. Most of these problems were benchmark problems taken from the literature of the 1980s and early 1990s from the fields of artificial intelligence, machine learning, neural networks, decision trees, and reinforcement learning. The 1992 book made the fundamental point that these seemingly different problems could all be recast as a search for a computer program. Of course, virtually all practitioners of artificial intelligence and machine learning subscribe to the Church-Turing thesis (and therefore believe that the solutions to such problems can be represented as computer programs). So, in one sense, our first book's first main point (about recasting problems as a search for a computer program) was platitudinous. Yet, in a practical sense, this point was radical. Indeed, as far as we know, genetic programming was (at the time of its invention) unique among techniques of artificial intelligence, machine learning, adaptive systems, and automated learning in that it searches a space of ordinary computer programs. Since then, work on automatic program synthesis has included Olsson's ADATE system (Olsson 1994a, 1994b).

In any event, once the toy problems from various fields of artificial intelligence and machine learning were recast as a search for a computer program, the book *Genetic Programming* demonstrated that they could all be solved with a *single uniform* method, namely genetic programming. The problems solved in *Genetic Programming* include problems of symbolic regression (system identification, empirical discovery, modeling, forecasting, data mining), classification, control, optimization, equation solving, game playing, induction, problems exhibiting emergent behavior, problems involving co-evolution, cellular automata programming, randomizer construction, image compression, symbolic integration and differentiation, inverse problems, decision tree induction, and many others.

The 1994 book *Genetic Programming II: Automatic Discovery of Reusable Programs* (Koza 1994a) and its accompanying videotape *Genetic Programming II Videotape: The Next Generation* (Koza 1994b) made the key point that the reuse of code is a critical ingredient to scalable automatic programming. The book discusses scalability in terms of the rate (e.g., linearly, exponentially) at which the computational effort required to yield a solution to differently sized instances of a particular problem (e.g., n^{th}-order parity problem, lawnmower problem for an $n \times m$ lawn) changes as a function of problem size. The book demonstrated that one way of achieving scalability is by reusing code by means of subroutines (automatically defined functions) and iterations (automatically defined iterations). The book's eight main points are:

- Automatically defined functions enable genetic programming to solve a variety of problems in a way that can be interpreted as a decomposition of a problem into subproblems, a solving of the subproblems, and an assembly of the solutions to the subproblems into a solution to the overall problem (or that can alternatively be interpreted as a search for regularities in the problem environment, a change in representation, and a solving of a higher-level problem).
- Automatically defined functions discover and exploit the regularities, symmetries, homogeneities, similarities, patterns, and modularities of the problem environment in ways that are very different from the style employed by human programmers.
- For a variety of problems, genetic programming requires less computational effort (fewer fitness evaluations to yield a solution with a satisfactorily high probability) with automatically defined functions than without them, provided the difficulty of the problem is above a certain relatively low break-even point.
- For a variety of problems, genetic programming usually yields solutions with smaller overall size (lower average structural complexity) with automatically defined functions than without them, provided the difficulty of the problem is above a certain break-even point.
- For the three problems in *Genetic Programming II* for which a progression of several scaled-up versions is studied, the average size of the solutions produced by genetic programming increases as a function of problem size at a lower rate with automatically defined functions than without them.
- For the three problems in *Genetic Programming II* for which a progression of several scaled-up versions is studied, the number of fitness evaluations required by genetic programming to yield a solution (with a specified high probability) increases as a function of problem size at a lower rate with automatically defined functions than without them.
- For the three problems in *Genetic Programming II* for which a progression of several scaled-up versions is studied, the improvement in computational effort and average structural complexity conferred by automatically defined functions increases as the problem size is scaled up.
- Genetic programming is capable of simultaneously solving a problem and selecting the architecture of the overall program (consisting of the number of automatically defined functions and the number of their arguments).

In addition, *Genetic Programming II* demonstrated that it is possible to

- automatically create multibranch programs containing an iteration-performing branch as well as a main program and subroutines (e.g., the transmembrane identification problem and the omega loop problem),

- automatically create multibranch programs containing multiple iteration-performing branches and iteration-terminating branches (e.g., the look-ahead version of the transmembrane problem), and
- automatically determine the architecture for a multibranch program in an architecturally diverse population by means of evolutionary selection.

Genetic Programming II contains solutions to problems from the fields of symbolic regression, control, pattern recognition, classification, computational molecular biology, and discovery of the impulse response function for an electrical circuit.

The 1999 book *Genetic Programming III: Darwinian Invention and Problem Solving* (Koza, Bennett, Andre, and Keane 1999a) and its accompanying videotape *Genetic Programming III Videotape: Human-Competitive Machine Intelligence* (Koza, Bennett, Andre, Keane, and Brave 1999) identified 16 attributes (table 2.2 of this book) that can reasonably be expected of a system for automatically creating computer programs. *Genetic Programming III* contains solutions to problems from the fields of system identification, time-optimal control, classification, synthesis of cellular automata rules, synthesis of minimal sorting networks, multi-agent programming, and synthesis of both the topology and sizing for analog electrical circuits. *Genetic Programming III* made the point that genetic programming unconditionally possesses the first 13 of these 16 attributes and that genetic programming also possesses the remaining three attributes (namely wide applicability, scalability, and human-competitiveness) to a substantial degree.

Genetic Programming III presented 14 specific instances where genetic programming automatically created a computer program that is competitive with a human-produced result. The 14 instances include two classification problems from the field of computational molecular biology, a long-standing problem involving cellular automata, a problem of synthesizing the design of a minimal sorting network, and 10 problems of synthesizing the design of both the topology and sizing of analog electrical circuits. Ten of the 14 human-competitive results in *Genetic Programming III* involve previously patented inventions.

2.5 Sources of Additional Information about Genetic Programming

Sources of information about genetic programming include

- *Genetic Programming: On the Programming of Computers by Means of Natural Selection* (Koza 1992a) and the accompanying videotape *Genetic Programming: The Movie* (Koza and Rice 1992a);
- *Genetic Programming II: Automatic Discovery of Reusable Programs* (Koza 1994a) and the accompanying videotape *Genetic Programming II Videotape: The Next Generation* (Koza 1994b);
- *Genetic Programming III: Darwinian Invention and Problem Solving* (Koza, Bennett, Andre, and Keane 1999a) and the accompanying videotape *Genetic Programming III Videotape: Human-Competitive Machine Intelligence* (Koza, Bennett, Andre, Keane, and Brave 1999);

Table 2.2 Sixteen attributes of a system for automatically creating computer programs

1	Starts with "What needs to be done"	It starts from a high-level statement specifying the requirements of the problem.
2	Tells us "How to do it"	It produces a result in the form of a sequence of steps that satisfactorily solves the problem.
3	Produces a computer program	It produces an entity that can run on a computer.
4	Automatic determination of program size	It has the ability to automatically determine the number of steps that must be performed and thus does not require the user to prespecify the exact size of the solution.
5	Code reuse	It has the ability to automatically organize useful groups of steps so that they can be reused.
6	Parameterized reuse	It has the ability to reuse groups of steps with different instantiations of values (formal parameters or dummy variables).
7	Internal storage	It has the ability to use internal storage in the form of single variables, vectors, matrices, arrays, stacks, queues, lists, relational memory, and other data structures.
8	Iterations, loops, and recursions	It has the ability to implement iterations, loops, and recursions.
9	Self-organization of hierarchies	It has the ability to automatically organize groups of steps into a hierarchy.
10	Automatic determination of program architecture	It has the ability to automatically determine whether to employ subroutines, iterations, loops, recursions, and internal storage, and to automatically determine the number of arguments possessed by each subroutine, iteration, loop, and recursion.
11	Wide range of programming constructs	It has the ability to implement analogs of the programming constructs that human computer programmers find useful, including macros, libraries, typing, pointers, conditional operations, logical functions, integer functions, floating-point functions, complex-valued functions, multiple inputs, multiple outputs, and machine code instructions.
12	Well-defined	It operates in a well-defined way. It unmistakably distinguishes between what the user must provide and what the system delivers
13	Problem-independent	It is problem-independent in the sense that the user does not have to modify the system's executable steps for each new problem.
14	Wide applicability	It produces a satisfactory solution to a wide variety of problems from many different fields.
15	Scalability	It scales well to larger versions of the same problem.
16	Competitive with human-produced results	It produces results that are competitive with those produced by human programmers, engineers, mathematicians, and designers.

- *Genetic Programming—An Introduction* (Banzhaf, Nordin, Keller, and Francone 1998);
- *Genetic Programming and Data Structures: Genetic Programming + Data Structures = Automatic Programming!* (Langdon 1998) in the series on genetic programming from Kluwer Academic Publishers;
- *Automatic Re-engineering of Software Using Genetic Programming* (Ryan 1999) in the series on genetic programming from Kluwer Academic Publishers;
- *Data Mining Using Grammar Based Genetic Programming and Applications* (Wong and Leung 2000) in the series on genetic programming from Kluwer Academic Publishers;
- *Grammatical Evolution: Evolutionary Automatic Programming in an Arbitrary Language* (O'Neill and Ryan 2003) in the series on genetic programming from Kluwer Academic Publishers;
- *Principia Evolvica: Simulierte Evolution mit Mathematica* (Jacob 1997, in German) and *Illustrating Evolutionary Computation with Mathematica* (Jacob 2001);
- *Genetic Programming* (Iba 1996, in Japanese);
- *Evolutionary Program Induction of Binary Machine Code and Its Application* (Nordin 1997);
- *Foundations of Genetic Programming* (Langdon and Poli 2002);
- *Emergence, Evolution, Intelligence: Hydroinformatics* (Babovic 1996);
- *Theory of Evolutionary Algorithms and Application to System Synthesis* (Blickle 1997);
- edited collections of papers such as the three *Advances in Genetic Programming* books from the MIT Press (Kinnear 1994; Angeline and Kinnear 1996; Spector, Langdon, O'Reilly, and Angeline 1999);
- the proceedings of the Genetic Programming Conferences held between 1996 and 1998 (Koza, Goldberg, Fogel, and Riolo 1996; Koza, Deb, Dorigo, Fogel, Garzon, Iba, and Riolo 1997; Koza, Banzhaf, Chellapilla, Deb, Dorigo, Fogel, Garzon, Goldberg, Iba, and Riolo 1998);
- the proceedings of the annual Genetic and Evolutionary Computation Conference (GECCO) (combining the formerly annual Genetic Programming Conference and the formerly biannual International Conference on Genetic Algorithms) operated by the International Society for Genetic and Evolutionary Computation (ISGEC) and held starting in 1999 (Banzhaf, Daida, Eiben, Garzon, Honavar, Jakiela, and Smith 1999; Whitley, Goldberg, Cantu-Paz, Spector, Parmee, and Beyer 2000; Spector Goodman, Wu, Langdon, Voigt, Gen, Sen, Dorigo, Pezeshk, Garzon, and Burke 2001; Langdon, Cantu-Paz, Mathias, Roy, Davis, Poli, Balakrishnan, Honavar, Rudolph, Wegener, Bull, Potter, Schultz, Miller, Burke, and Jonoska 2002);
- the proceedings of the annual Euro-GP conferences held starting in 1998 (Banzhaf, Poli, Schoenauer, and Fogarty 1998; Poli, Nordin, Langdon, and Fogarty 1999; Poli, Banzhaf, Langdon, Miller, Nordin, and Fogarty 2000; Miller, Tomassini, Lanzi, Ryan, Tettamanzi, and Langdon 2001; Foster, Lutton, Miller, Ryan, and Tettamanzi 2002);
- the proceedings of the Workshop of Genetic Programming Theory and Practice organized by the Center for Study of Complex Systems of the University of Michigan (Riolo and Worzel 2003),
- the *Genetic Programming and Evolvable Machines* journal (from Kluwer Academic Publishers) started in April 2000;

- web sites such as www.genetic-programming.org and www.genetic-programming.com;
- early papers on genetic programming, such as the Stanford University Computer Science Department technical report *Genetic Programming: A Paradigm for Genetically Breeding Populations of Computer Programs to Solve Problems* (Koza 1990a), the paper "Hierarchical Genetic Algorithms Operating on Populations of Computer Programs," presented at the 11th International Joint Conference on Artificial Intelligence in Detroit (Koza 1989), and Koza 1988;
- an annotated bibliography of the first 100 papers on genetic programming (other than those of which John Koza was the author or co-author) in appendix F of *Genetic Programming II: Automatic Discovery of Reusable Programs* (Koza 1994a); and
- William Langdon's bibliography on genetic programming at http://www.cs.bham.ac.uk/~wbl/biblio/ or http://liinwww.ira.uka.de/bibliography/Ai/genetic.programming.html. This bibliography is the most extensive in the field and contains over 3,034 papers (as of January 2003) by over 880 authors. It provides on-line access to many of the papers.

3

Automatic Synthesis of Controllers

Engineers are often called upon to design complex structures (e.g., controllers, circuits, antennas, networks of chemical reactions) that satisfy certain prespecified high-level design goals. In fact, design is a major activity of practicing engineers.

The design of a complex structure typically involves tradeoffs between competing considerations. The end product of the design process is usually a satisfactory, as opposed to a perfect, design. The design process is usually viewed as requiring human intelligence. Consequently, the field of design is a source of challenging problems for methods that purport to be able to automatically solve problems.

Design problems are an especially appropriate area for the application of automated problem-solving methods (such as genetic programming) because design problems are usually couched in terms of measurable behaviors or characteristics of the to-be-designed entity. Moreover, the tradeoffs that the engineer is willing to accept concerning a problem's (usually competing) considerations are typically explicitly addressed as part of the statement of a design problem.

Section 3.1 provides basic background information on controllers.

Section 3.2 discusses design considerations for controllers.

Section 3.3 describes the representation of a controller by means of a block diagram.

Section 3.4 discusses various possible techniques for designing controllers (with emphasis on search techniques).

Section 3.5 describes our approach to the automatic creation of both the topology and tuning of controllers by means of genetic programming.

Section 3.6 describes the representation of a controller by means of a transfer function, a LISP symbolic expression, a program tree, an expression in the Mathematica® programming language, a connection list, and a SPICE netlist.

The remaining sections of this chapter (sections 3.7 through 3.10) demonstrate the automatic synthesis of both the topology and tuning of controllers by means of genetic programming. Specifically, genetic programming is used to automatically synthesize the design of controllers for

- a two-lag plant (section 3.7),
- a three-lag plant (section 3.8),

- a three-lag plant with a five-second delay (section 3.9), and
- a non-minimal-phase plant (section 3.10).

Later, chapter 12 demonstrates the capability of genetic programming to automatically create tuning rules for a PID (proportional, integrative, and derivative) controller for controlling plants belonging to four families of industrially representative plants.

Chapter 9 demonstrates the capability of genetic programming to automatically create, in a single run, a general controller in the form of a graphical structure whose nodes represent signal-processing blocks and where the values of the parameters of the blocks are specified by mathematical expressions containing free variables. The capacity of genetic programming to create such parameterized topologies is demonstrated by the automatic synthesis of

- a parameterized controller for controlling a three-lag plant whose time constant is specified by a free variable (section 9.1),
- a parameterized controller for controlling plants belonging to two different families of plants (section 9.2), and
- parameterized controllers for controlling plants belonging to four families of industrially representative plants (chapter 13).

3.1 Background on Controllers

The purpose of a controller is to force, in a meritorious way, the response of a system (conventionally called the *plant*) to match a specified response (called the *reference signal, command signal*, or *setpoint*).

Controllers are ubiquitous. They are found in almost every electronic product and every mechanical, thermodynamic, hydraulic, pneumatic, hydrodynamic, electrical, or aeronautical system (Åström and Hägglund 1995; Dorf and Bishop 1998; Bryson and Ho 1975; Boyd and Barratt 1991; Ogata 1997).

The discovery of meritorious controllers is important because a small percentage improvement in the operation of plants may translate into large economic savings or other (e.g., environmental) benefits.

The cruise control device in an automobile is an example of a controller. The driver may want an automobile to travel at 65 miles per hour. The controller causes fuel to flow into the automobile's engine (the plant) so as to reduce the difference between the car's current speed (the *plant response*) and the desired speed (the reference signal). The car's speed does not, of course, jump instantaneously to 65 miles per hour. Instead, the controller monitors the continuously changing difference between the car's actual speed and the reference signal and continuously adjusts the flow of fuel to the plant based on this difference. If the automobile were traveling slower than 65 when the reference signal is set to 65, the controller would initially increase the flow of fuel to the engine. On the other hand, if the car's speed starts off (or ever becomes) greater than 65, the controller decreases the flow of fuel. Because the controller monitors the plant output (and continuously compares it to the reference signal), we say that there is *feedback* from the plant to the controller. When the plant output is subtracted from the reference signal (to obtain the *error*), we say that there is *negative feedback*.

Figure 3.1 shows elements of a system composed of a plant 140, closed-loop feedback 160, and controller 120. The plant 140 has one output 150 (the plant response). The controller 120 has output 130 (sometimes called the *control variable*). The controller's output 130 is the input to the plant 140. The controller 120 has two inputs, namely the externally supplied reference signal 110 and the plant response 150. The controller receives the plant response 150 by means of the feedback loop 160.

The overall system in the figure is called a *closed-loop system* because there is feedback 160 of the plant's output 150 into the controller 120. The feedback loop 160 is referred to as an external feedback because the feedback loop is external to the controller. Systems without such external feedback of the plant's output into the controller are called *open loop systems*. Open loop controllers are considerably simpler (and generally less useful in real-world applications) than closed-loop controllers. All the genetically evolved solutions to control problems in this book will turn out to be closed-loop controllers.

The plant response 150 is typically compared to the externally supplied reference signal 110. This comparison is typically implemented by subtracting these two time-domain signals. The purpose of a closed-loop controller is to produce, given the reference signal 110 and the feedback signal 160, a value for the controller output 130 that causes the plant response to approach (and ideally match) the reference signal 110 in a meritorious way.

Many controllers in the real world are manual. However, this book emphasizes automatic controllers—that is, controllers that automatically process information in order to generate a signal to control the plant. The information that is processed typically includes reference signal(s) and plant response(s) and may also include additional information such as the plant's internal state(s).

Controllers usually incorporate devices from several different engineering domains, making the field of control interdisciplinary. For example, the variable of interest for a home heating system is the room's current temperature—a thermodynamic variable. The setting of the thermostat is usually mechanical. In a home heating system, this mechanical variable is typically converted into an electrical signal.

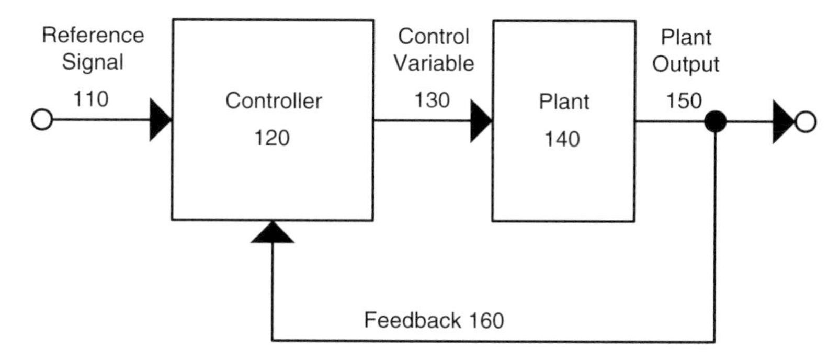

Figure 3.1 A system composed of a plant and a controller with closed-loop feedback.

The electrical signals are usually processed by an electrical controller (analog or digital). The controller's electrical output is typically then converted into yet another mechanical variable in order to adjust the valve that controls the flow of fuel to the furnace. The furnace itself is a thermodynamic system (involving combustion).

The underlying principles of controllers are broadly the same whether the system is mechanical, electrical, thermodynamic, hydraulic, pneumatic, hydrodynamic, aeronautical, biological, or economic and whether the variable of interest is temperature, pressure, velocity, voltage, interest rates, heart rate, or humidity (Dorf and Bishop 1998).

It is often convenient to analyze controllers in purely electrical terms. One reason is that electrical terminology provides a common language for modeling most plants and controllers (regardless of whether they are actually electrical). A second reason is that many present-day sensors and controllers are, in fact, entirely electrical. Another reason is that there are sophisticated electrical circuit simulators that can be used to simulate controllers and plants.

3.2 Design Considerations for Controllers

Numerous (usually conflicting) considerations bear on the design of a satisfactory controller.

One of the most common considerations in designing a controller is the minimization of the time required to produce the desired plant response. Sometimes, this goal is couched in terms of the rise time or the settling time. The *rise time* is the time required for the plant output to first reach a specified percentage (e.g., 90%) of the reference signal (assuming that the reference signal is greater than the starting value of the plant response). The *settling time* is the first time at which the plant response reaches and thereafter stays within a specified percentage (e.g., 2%) of the reference signal. Control engineers frequently use the integral of the time-weighted absolute error (ITAE) to measure a controller's performance. This measure favors the rapid reduction of the discrepancy (error) between the reference signal and the plant response because the weighting by time imposes ever-greater penalties on discrepancies that occur later.

In addition to wanting to minimize the time required to produce the desired plant response, it is common to simultaneously want to avoid significantly overshooting the desired plant response. In the case where the reference signal is initially greater than the plant response, the *overshoot* is the maximum percentage by which the plant response exceeds the reference signal. Returning to the example of the cruise control device in an automobile, it is desirable for the automobile to quickly reach the desired speed of 65 miles per hour. However, the driver certainly does not want to achieve this goal by providing so much fuel that the car first surges to 150 miles per hour. Such a strategy would undoubtedly reduce the time required to reach 65; however, this strategy would drastically overshoot the reference signal. Overshoot is generally undesirable. The driver may get arrested. In many practical situations, the plant may be destroyed by excessive overshoot (e.g., a boiler exceeds its safe operating temperature).

Aside from catastrophic outcomes, overshoot is typically costly in terms of energy. In fact, the cost of the energy required to bring about the desired plant response is frequently an independent additional consideration in measuring a controller's merit.

Plants in the real world can only handle inputs that lie within certain limits. For example, a furnace has a maximum rate of heat production and a motor has a maximum armature current. The limitation can be easily implemented by placing a limiter between the controller's output and the plant's input. Note that a limiter alone usually introduces significant nonlinearity into a system (thus complicating the task of controller design). A plant's internal state variables are often similarly constrained (thus introducing additional nonlinearity). Controller design in the real world must take such limitations into account.

In the real world, disturbances are felt at a variety of places throughout the overall system. A practical controller must operate robustly in the face of such disturbances. Such disturbances are typically modeled by adding a disturbance signal to the controller's output. Disturbance rejection is usually an important requirement of an effective real-world controller.

All sensors in the real world are imperfect. The controller may not receive an accurate reading of the plant's output because of sensor noise. The deviation in the plant's output from its correct value is often modeled by means of an additive noise signal. It is desirable that a controller operates robustly in the face of sensor noise.

In addition, the plant's characteristics are usually not known with certainty. They may be difficult to measure and they may not be measured accurately. The plant's characteristics may fluctuate at random over the short term and may drift systematically over the long term as the plant ages. Thus, a practical controller must operate robustly in the face of (often significant) variation in the plant's characteristics from their nominal values.

A good controller should operate in a consistent manner in the face of variations in the step size of the reference signal.

Stability is another very important consideration in measuring the merit of a controller. A system whose performance varies radically in response to small changes in the plant (e.g., an engine that doubles the car's speed in response to a miniscule change in temperature) would, at the minimum, be useless and, at worst, be dangerous.

3.3 Representation of a Controller by a Block Diagram

Automatic controllers are constructed from a variety of signal-processing functions (called *blocks*) that process time-domain signals. A signal-processing block may have one or more inputs; however, a block always has a single output. Examples of signal-processing functions that are commonly found in controllers include, but are not limited to,

- gain (i.e., multiplication of a signal by a scalar value),
- integrator,
- differentiator,
- adder,

- subtractor (i.e., an adder in which one input is negated),
- lead,
- lag,
- second-order lag,
- delay,
- inverter (i.e., negation of the input),
- absolute value,
- limiter,
- multiplier (i.e., multiplication of two time-domain signals),
- divider, and
- conditional operators (switches).

One or more parameter values are required to completely specify many of these signal-processing blocks. For example, the complete specification of a gain block requires specification of its *amplification factor*. The setting of the (typically numerical) parameter values for a signal-processing block is sometimes called the *tuning* of the block.

Figure 3.2 is a block diagram of a system composed of a plant 592, an external feedback loop 596, and a PID controller 500 composed of a proportional (P) block, an integrative (I) block, and a derivative (D) block. The output 590 from the controller is the input to the plant 592. The plant has one output (plant response) 594. The plant response is fed back (externally as signal 596) and becomes one of the controller's two inputs. The controller's other input is the externally supplied reference signal 508. The fedback plant response 596 and the reference signal 508 are compared (by subtractor 510).

The input to a controller typically consists of reference signal(s) and plant response(s). A controller typically computes the difference (error) between each

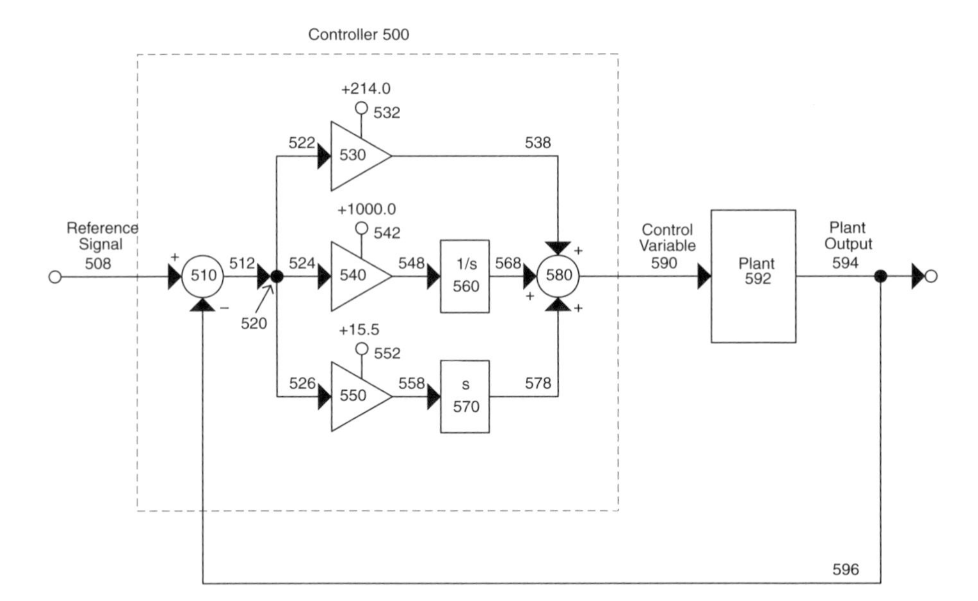

Figure 3.2 Block diagram of a plant and a PID controller.

reference signal and the corresponding plant response. The input to a controller sometimes also includes additional information such as the plant's internal state variable(s). The input to a controller sometimes also includes output(s) of the controller itself. The output of one of a controller's own blocks is sometimes fed back as an input to the controller. The providing of such outputs to the controller is called *internal feedback.*

The output(s) of a controller are passed from the controller to the plant.

The individual signal-processing blocks of a controller are coupled to one another in a particular topological arrangement. The *topology of a controller* entails the specification of

- the total number of processing blocks in the controller,
- the type of each block (e.g., gain, integrator, differentiator, lead, lag, adder, subtractor),
- the connections (directed lines) that may exist between the output of each block in the controller and the inputs of other blocks in the controller, and
- the connections (directed lines) that may exist between the blocks in the controller and
  - the controller's external input point(s) (e.g., reference signal and plant output), and
  - the controller's external output point(s).

Block diagrams are useful for representing the flow of information in controllers and systems containing controllers. Block diagrams contain signal-processing function blocks, external input and output points, and directed lines carrying signals.

Lines in a block diagram represent time-domain signals. Lines are directional in that they represent the flow of information. The line(s) pointing toward a block represent signal(s) coming into the block. The single line pointing away from a block represents the block's single output.

Note that a line in a block diagram differs from a wire in an electrical circuit in that a line in a block diagram is directional. Also, note that a function block in a block diagram differs from an electrical component in a circuit in that there is exactly one designated output from a function block.

In a block diagram, an external input is represented by an external point with a directed line pointing away from that point. Similarly, an external output point is represented by an external point with a line pointing toward that point.

Adders are conventionally represented by circles in block diagrams. Each input to an adder is labeled with a positive or negative sign.

In figure 3.2, adder block 510 performs the function of subtracting the plant output 596 (which has a minus sign) from the externally supplied reference signal 508 (which has a plus sign). This particular subtraction block implements negative feedback.

Takeoff points (conventionally represented in block diagrams by a large dot) provide a way to disseminate a signal to more than one other function block in a block diagram. Takeoff point 520 in figure 3.2 receives signal 512 and disseminates signal 512 to function blocks 530, 540, and 550.

In figure 3.2, the subtractor's output 512 is passed (through takeoff point 520) into a `GAIN` block 530. A `GAIN` block (shown in this figure as a triangle) multiplies (amplifies) its input by a specified constant amplification factor (i.e., the numerical constant 214.0). The amplified result 538 becomes the first of the three inputs to addition block 580. This portion of the figure implements proportional (P) control. Proportional control is the simplest type of control. P-type control is responsive to the difference (error) between the reference signal and the plant output and generates a time-domain signal that is proportional to the error. However, for all but the simplest situations, proportional control leads to overshoot, instability, and inefficiency. These known deficiencies of P-type control are, in turn, addressed by integrative (I-type) control and derivative (D-type) control.

The subtractor's output 512 is also passed (through takeoff point 520) into `GAIN` block 540 (with a gain of 1,000.0). The amplified result 548 is passed into `INTEGRATOR` block 560. The integrator is shown in figure 3.2 by a rectangle containing the expression $1/s$, where s is the Laplace transform variable. The result 568 of this integration (with respect to time) becomes the second input to three-argument addition block 580. This portion of the figure implements integral (I) control based on the integral, over time, of the difference between the plant output and reference signal.

The subtractor's output 512 is also passed (through takeoff point 520) into `GAIN` block 550 (with a gain of 15.5). The amplified result 558 is passed into `DIFFERENTIATOR` block 570 (shown in the figure by a rectangle containing the expression s) and becomes the third input to three-argument addition block 580. This portion of the figure implements derivative (D) control based on the derivative, over time, of the difference between the plant output and reference signal.

The output 590 of the controller in figure 3.2 is the sum of a proportional (P) term, an integrative (I) term, and a differentiating (D) term. This type of controller is called a *PID controller*.

Each of the three parts of a PID controller contributes to its overall performance.

The proportional part of the controller provides direct feedback of the error between the plant output and the externally supplied reference signal.

The integral part of the controller enables the controller to make the plant output agree with the reference signal in the steady state. Even a small positive (or negative) error will, when integrated over time, result in a noticeable non-zero contribution to the controller's output.

A proportional controller is almost always late in compensating for an error between the plant output and the reference signal. The derivative part of the controller enables the controller to make a prediction of future error (based on extrapolating the slope of the tangent to the error curve).

The PID controller was patented in 1939 by Albert Callender and Allan Stevenson of Imperial Chemical Limited of Northwich, England. The PID controller was a significant improvement over earlier and simpler control techniques (which often were merely proportional).

In discussing the problems of "hunting and instability," Callender and Stevenson (1939) state (in their patent),

> "If the compensating effect V is applied in direct proportion to the magnitude of the deviation Θ, over-compensation will result. To eliminate the consequent hunting and instability of the system, the compensating effect is additionally regulated in accordance with other characteristics of the deviation in order to bring the system

back to the desired balanced condition as rapidly as possible. These characteristics include in particular the rate of deviation (which may be indicated mathematically by the time-derivative of the deviation) and also the summation or quantitative total change of the deviation over a given time (which may be indicated mathematically by the time-integral of the deviation)."

Callender and Stevenson (1939) also say,

"A specific object of the invention is to provide a system which will produce a compensating effect governed by factors proportional to the total extent of the deviation, the rate of the deviation, and the summation of the deviation during a given period..."

Claim 1 of Callender and Stevenson (1939) covers what is now called a "PI" (proportional-integrative) controller:

"A system for the automatic control of a variable characteristic comprising means proportionally responsive to deviations of the characteristic from a desired value, compensating means for adjusting the value of the characteristic, and electrical means associated with and actuated by responsive variations in said responsive means, for operating the compensating means to correct such deviations in conformity with the sum of the extent of the deviation and the summation of the deviation."

Claim 3 of Callender and Stevenson (1939) covers the PID (proportional-integrative-derivative) controller:

"A system as set forth in claim 1 in which said operation is additionally controlled in conformity with the rate of such deviation."

The PID controller was an enormous improvement over previous methods for control. Consequently, PID controllers are in widespread use in industry today. As Åström and Hägglund (1995, pages 1 and 2) noted,

"PID controllers are found in large numbers in all industries. The controllers come in many different forms. There are stand-alone systems in boxes for one or a few loops, which are manufactured by the hundred thousands yearly..."

"In process control, more than 95% of the control loops are of PID types."

As Åström and Hägglund (1995, page 4) observe,

"Several studies...indicate the state of the art of industrial practice of control. The Japan Electric Measuring Instrument Manufacturing Association conducted a survey of the state of process control systems in 1989...According to the survey, more than 90% of the control loops were of the PID type."

Although PID controllers are in widespread use in industry today, the need for better controllers is widely recognized. As Åström and Hägglund (1995, page 4) further observe,

> "...audits of paper mills in Canada [show] that a typical mill has more than 2,000 control loops and that 97% use PI control. Only 20% of the control loops were found to work well."

As Åström and Hägglund (1995, page 2) also observe,

> "A large cadre of instrument and process engineers are familiar with PID control. There is a well-established practice of installing, tuning, and using the controllers. In spite of this there are substantial potentials for improving PID control."

Boyd and Barratt state in *Linear Controller Design: Limits of Performance* (Boyd and Barratt 1991),

> "The challenge for controller design is to productively use the enormous computing power available. Many current methods of computer-aided controller design simply automate procedures developed in the 1930's through the 1950's, for example, plotting root loci or Bode plots. Even the 'modern' state-space and frequency-domain methods (which require the solution of algebraic Riccati equations) greatly underutilize available computing power."

3.4 Possible Techniques for Designing Controllers

The design (synthesis) of a controller requires specification of both the topology (i.e., the block diagram) and the parameter values (tuning) such that the controller satisfies certain user-specified high-level design requirements.

The design process for controllers today is generally channeled along lines established by analytical techniques. These mathematical techniques almost always presume a choice for the controller's overall topology (often the conventional PID-type controller).

It would be desirable to have an automatic system for synthesizing the design of a controller that is open-ended in the sense that it does not require the human user to prespecify the controller's topology. Such a system would start only with a high-level statement of requirements and automatically produce both the appropriate controller topology and tuning.

In controller design, high-level statements of the requirements are typically couched in terms of multiple conflicting considerations such as

- optimization requirements (e.g., minimizing the integral of the time-weighted absolute error),
- time-domain constraints (e.g., minimizing overshoot, maximizing disturbance rejection),
- frequency domain constraints (e.g., attenuation of sensor noise, system bandwidth),
- robustness requirements,
- stability requirements, and
- limits on the values of various signals.

It is often difficult or impossible to use conventional analytical techniques to design a controller that satisfies arbitrary combinations of these conflicting considerations.

Indeed, there is no preexisting general-purpose analytic method for starting with a high-level statement of requirements and automatically creating both the topology and tuning of a controller for an arbitrary plant.

Search techniques offer a potential alternative to analytical methods for designing a controller. Search techniques do not find solutions by analysis, logic, or proof. Instead, they simply search a space for ever-better points.

There are numerous techniques for searching a space of candidate entities for an optimal or near-optimal entity, including, but not limited to,

- hill climbing (section 3.4.1),
- gradient methods (section 3.4.2),
- simulated annealing (section 3.4.3), and
- evolutionary methods, such as the genetic algorithm and genetic programming (section 3.4.4).

A search is an iterative process that generally involves

- starting with one or more entities (points) from the search space of the problem at hand,
- ascertaining the merit of each of the initial entity(ies),
- creating new candidate entity(ies) by modifying existing entity(ies),
- ascertaining the merit of the new candidate entity(ies),
- using the measure of merit to select among entity(ies), and
- repeating the previous three steps.

All search methods are driven by a measure of merit. The measure of merit is frequently called the "objective function" or "payoff" by practitioners of hill climbing or gradient methods, the "energy level" by practitioners of simulated annealing, and the "fitness measure" by practitioners of genetic algorithms or genetic programming. For consistency herein, we use the term "fitness" to refer to the measure of merit.

The individual steps of the iterative search process are called "cycles" or "time-steps" by practitioners of hill climbing, gradient methods, or simulated annealing. Typically, one new entity (point) in the search space is produced from one preexisting entity on each cycle.

Because both the genetic algorithm and genetic programming manipulate a population of entities, practitioners of genetic algorithms or genetic programming typically use the biologically oriented term "generations." In the most commonly used implementations of the genetic algorithm and genetic programming, a large number of new entities (usually equal to the full population size) are produced for each generation; however, in the "steady state" version of the genetic algorithm and genetic programming, only one new entity is produced in each generation.

Because search techniques do not rely on analysis, logic, or proof, they usually require large amounts of computational resources.

3.4.1 Search by Hill Climbing

Hill climbing is a search technique that starts with a single initial entity (point) in the search space, ascertains the fitness of the entity, creates a new candidate entity, ascertains the fitness of the new candidate entity, and uses the fitness measure to select between the preexisting entity and the new candidate entity. The new candidate entity is created by a problem-specific modification operation (often a probabilistic operation) that modifies the current entity (point) in the search space in order to obtain a new (often nearby) candidate entity in the search space. In some variations of hill climbing, more than one new candidate point may be created by the modification operation. Hill climbing is a point-to-point search technique in the sense that the search proceeds from a single point in the search space of the problem to another single point. Hill climbing is a greedy search in that, at each cycle, it unconditionally rejects any non-improving new candidate and unconditionally selects the best available new candidate. If the problem-specific modification operation is deterministic, the search necessarily ends on the first occasion when the best new candidate is no better than the preexisting point in the search. If the problem-specific modification operation is probabilistic, a subsequent new candidate may be an improvement (and therefore be accepted).

When hill climbing is applied to nontrivial problems, it often becomes trapped on a local optimum point (thus never finding the global optimum point of the search space). If the problem-specific modification operation is probabilistic or different starting points are used, different runs can produce different outcomes. In hill climbing (and many other probabilistic search techniques), it is often necessary to make multiple runs in order to solve a problem.

The starting point of a search by hill climbing (as well as gradient methods, simulated annealing, the genetic algorithm, or genetic programming) is often random. However, knowledge about the search space may be used to choose promising starting point(s) in any of these methods.

It should be noted that there are parallel versions of hill climbing (as well as parallel versions of gradient search and simulated annealing). However, there is usually no transfer of information among the different threads of such parallel searches. That is, each thread of the parallel search stands alone.

There are also versions of hill climbing (and gradient search and parallel simulated annealing) where the search is iteratively restarted after the search becomes trapped on a local optimum point. However, when the search is restarted, there is usually no forward transfer of information gained from the earlier searches. That is, each restarted run stands alone.

3.4.2 Search by Gradient Methods

If all dimensions of a search space are numerically valued, it may be possible to employ gradient methods. In gradient search, the new candidate entity is computed using the observed gradient of the fitness measure.

Like hill climbing, gradient search is usually implemented as a point-to-point search. Gradient search often becomes trapped on a local optimum point (thus never finding the global optimum for nontrivial problems). If different starting points are used or if the modification operation probabilistically chooses different step sizes for computing the gradient, different runs can produce different outcomes. When gradient search is applied to nontrivial problems, it (like hill climbing) often becomes trapped on a local optimum point.

3.4.3 Search by Simulated Annealing

Simulated annealing (Kirkpatrick, Gelatt, and Vecchi 1983; Aarts and Korst 1989; Salamon, Sibani, and Frost 2002) resembles hill climbing and gradient search in that it is usually implemented as a point-to-point search. Simulated annealing employs a problem-specific probabilistic modification operation (usually called a "mutation") for modifying the current entity in the search space in order to obtain a new candidate entity.

Like hill climbing and gradient search, simulated annealing starts with a single initial entity (point) in the search space, ascertains the entity's fitness, creates a new candidate entity, and ascertains the fitness of the new candidate entity.

Simulated annealing unconditionally selects the new candidate entity if it is better than the preexisting entity. That is, it operates in the same way as hill climbing in this particular case. However, simulated annealing is not entirely greedy. It differs from hill climbing in the way it handles the case when the new candidate entity is worse than the preexisting entity. In this case, the Metropolis algorithm and the Boltzmann equation are applied to determine whether to accept a non-improving candidate entity. Simulated annealing avoids the obvious weakness of hill climbing and gradient search (namely, the tendency to become trapped on a local optimum point) by making a mathematically principled choice between the preexisting entity and the new candidate entity.

A run of simulated annealing is governed by an annealing schedule in which a temperature T changes as the run proceeds. Typically, the temperature changes in an exponentially monotonically decreasing way. A run of simulated annealing behaves differently as the temperature changes. The effect of the Metropolis algorithm and the Boltzmann equation are that the probability of acceptance of a non-improving modification is greater if the fitness difference is small or if the temperature T is high. Thus, large non-improving modifications are likely to be accepted early in the run (when the temperature is high). As a result, simulated annealing resembles blind random search in early stages of the run (when the system is "hot") in that it usually accepts whatever the modification operation happens to produce. However, later in the run (when the system has "cooled"), only small non-improving modifications are likely to be accepted. Thus, simulated annealing resembles hill climbing in later stages of the run. However, even as the system cools, there is always some probability that a non-improving modification will be accepted. If a modification is not accepted at any step of the run of simulated annealing, the modification operation (assuming it is probabilistic) is re-invoked to generate another candidate for consideration.

It should be noted that the modification operation for simulated annealing and hill climbing is usually shrewdly chosen in light of the specific nature of the problem at hand. For example, if the problem is a traveling salesman problem (where the goal is to minimize total distance while visiting N cities exactly one time each), the modification operation may exchange sub-tours.

3.4.4 Search by the Genetic Algorithm and Genetic Programming

The genetic algorithm and genetic programming differ from hill climbing and gradient search in that they are not greedy. In addition, the recombination operation works in conjunction with the population in a way that makes genetic search (as opposed to hill climbing, gradient search, or simulated annealing) particularly advantageous in solving the problems of interest in this book. These two points are discussed in detail in section 16.3.

3.4.5 Previous Work on Controller Synthesis by Means of Genetic and Evolutionary Computation

There has been extensive previous work on the problem of automating the design of controllers using genetic and evolutionary computation.

The genetic algorithm has been used for synthesizing controllers composed of continuous-time signal-processing blocks. The books *Genetic Algorithms for Control and Signal-Processing* (Man, Tang, Kwong, and Halang 1997) and *Genetic Algorithms: Concepts and Designs* (Man, Tang, Kwong, and Halang 1999) are particularly noteworthy. See also Jamshidi, Coelho, Krohling, and Fleming 2003.

Genetic programming has been used for synthesizing controllers that have mutually interacting continuous-time variables and continuous-time signal-processing blocks (Crawford, Cheng, and Menon 1999; Dewell and Menon 1999; Menon, Yousefpor, Lam, and Steinberg 1995; and Sweriduk, Menon, and Steinberg 1998, 1999). Neural programming and PADO (Teller 1996a, 1996b, 1998, 1999; Teller and Veloso 1995a, 1995b, 1995c, 1995d, 1995e, 1996, 1997) provide a method for graphical representation of programs involving mutually interacting independent processes. In 1996, Marenbach, Bettenhausen, and Freyer (1996) used automatically defined functions and genetic programming to represent internal feedback in a system used for system identification where the overall multi-branch program represented a continuous-time system with continuously interacting processes. The essential feature of this work was aptly named "multiple interacting programs" in later work (Angeline 1997, 1998a, 1998b; Angeline and Fogel 1997).

In addition, genetic programming has been previously applied to many other specific types of control problems, including, but not limited to, discrete-time problems where the evolved program receives the system's current state as input, performs an arithmetic or conditional calculation on the inputs, and computes a value for the controller's output. These discrete-time problems include cart centering (Koza and Keane 1990a), broom balancing (Koza and Keane 1990b), wall following, box moving, backing a tractor-trailer truck to a loading dock (Koza 1992d), controlling the food foraging strategy of a lizard (Koza 1992a; Koza, Rice, and Roughgarden 1992), navigating a robot with a nonzero turning radius to a destination point (Koza, Bennett, Keane, and Andre 1997), using linear genomes for robotic control (Andersson, Svensson, Nordin, and Nordin, and Mats 1999; Banzhaf, Nordin, Keller, and Olmer 1997), and using developmental genetic programming (cellular encoding) for a pole-balancing problem involving two poles (Whitley, Gruau, and Preatt 1995).

3.4.6 Possible Approaches to Automatic Controller Synthesis Using Genetic Programming

Program trees (consisting of both result-producing branches and automatically defined functions) evolved by genetic programming may be employed in several different ways.

In the approach where genetic programming is being used to automatically create a computer program to solve a problem, the program tree is simply executed. The result of the execution is a set of returned values, a set of side effects on some other entity (e.g., an external entity such as a robot or an internal entity such as computer memory), or a combination of returned values and side effects. In this approach, the functions in the program are individually executed, in time, in accordance with a specified order of evaluation such that the result of the execution of one function is available at the time when the next function is going to be executed. The functions in a subroutine are executed at the distinct time when the subroutine is invoked by the calling program. The subroutine produces a result that is available at the time when the next function in the calling program is executed.

The second approach is a developmental approach. In this approach, a program tree is interpreted as a set of instructions for constructing a complex structure. The developmental approach has been successfully used to automatically synthesize neural networks (Kitano 1990; Gruau 1992a, 1992b), to automatically synthesize electrical circuits (Koza, Bennett, Andre, and Keane 1996a, 1996b, 1996c, 1996d, 1996e, 1996f, 1999a; Bennett, Koza, Andre, and Keane 1996; Koza, Bennett Andre, Keane, and Dunlap 1997), to automatically synthesize patterns and structures by means of Lindenmayer rewrite rules (Koza 1993), and to solve other problems (Spector and Stoffel 1996a, 1996b) by means of ontogenetic programming. The developmental approach is implemented by progressively applying the functions in the program tree to an embryonic structure (e.g., an embryonic circuit, an embryonic neural network, an embryonic pattern) so as to develop the embryo into a fully developed structure. As in the first approach, the functions in the program are executed separately, in time, in accordance with a specified order of evaluation.

Program trees are used in a third way in this chapter. In this third approach, the program trees are converted directly to a block diagram (without any intervening developmental processes). The block diagram consists of signal-processing functions linked by directed lines representing the flow of information. All the controller's signal-processing blocks and the to-be-controlled plant interact with one another as part of a closed system. There is no order of evaluation among the signal-processing blocks constituting the controller.

As will be seen in this book, genetic programming can be successfully used to automatically synthesize both the topology and tuning for a controller using this third approach. The automatically created controllers can potentially accommodate

- one or more externally supplied reference signals,
- one or more controller outputs,
- one or more plant outputs,
- external feedback of one or more plant outputs to the controller,
- comparisons (of many types) between reference signal(s) and the corresponding plant output(s),
- internal state variable(s) of the plant,
- direct feedback of the controller's output(s) back into the controller, and
- internal feedback of signals from one part of the controller to another part of the controller.

3.5 Our Approach to the Automatic Synthesis of the Topology and Tuning of Controllers

Our approach to the problem of automatically creating both the topology and tuning of a controller involves

(1) establishing a representation for controllers involving program trees that can be progressively bred by means of genetic programming, and
(2) defining a fitness measure that measures how well the behavior and characteristics of a candidate controller satisfy the problem's high-level design requirements.

The representation and fitness measure are then used during the run of genetic programming. During the run, the evaluation of the fitness of each individual in the population involves

(1) converting each individual program tree in the population into an analog electrical circuit representing the controller,
(2) converting each analog electrical circuit into a netlist of the type accepted by the circuit simulator,
(3) obtaining the behavior and characteristics of the individual controller by simulating the corresponding analog electrical circuit, and
(4) using the controller's behavior and characteristics to calculate its fitness.

There are seven representations for a controller that are used to some degree in the work described in this book.

- *Block Diagram:* Control engineers ordinarily represent a controller by means of a block diagram. In this representation, the individual blocks in the diagram represent signal-processing functions and the directed lines represent time-domain signals flowing between the blocks. This representation is discussed in section 3.3 and an example of a block diagram is shown in figure 3.2.
- *Transfer Function:* A linear controller can also be represented as a transfer function involving the Laplace transform variable s (section 3.6.1).
- *Symbolic Expression:* A controller can also be represented as a symbolic expression (S-expression) in the style of the LISP programming language (section 3.6.2). This representation is used internally by genetic programming.
- *Program Tree:* A controller can also be represented as a program tree whose internal points (nodes) are functions and whose external points (leaves) are terminals (section 3.6.3). This representation enables genetic programming to breed a population of programs in a search for a controller that satisfies pre-specified design objectives.
- *Mathematica:* A controller can also be represented in the style of the Mathematica programming language (section 3.6.4). Mathematica was used in analyzing and simplifying the results of certain runs in this book.
- *Connection List:* A controller can also be represented as a connection list (section 3.6.5). This representation is sometimes useful for explanatory purposes and is closely related to the SPICE netlist (mentioned next).

- *SPICE Netlist:* In a run of genetic programming on a computer, the entire system (consisting of the controller and the plant) is typically simulated using a simulator. The SPICE (Simulation Program with Integrated Circuit Emphasis) simulator (Quarles, Newton, Pederson, and Sangiovanni-Vincentelli 1994) is used throughout this book for simulating controllers, circuits, and networks of chemical reactions. In order to invoke the SPICE simulator, the controller and the plant are represented in the form of a netlist (described in section 3.6.6).

3.5.1 Repertoire of Functions

The repertoire of functions includes the following time-domain functions.

The two-argument GAIN function multiplies the time-domain signal represented by its first argument by a scalar numerical value represented by its second argument. The numerical parameter value for this function (and other functions described later that possess one or more numerical parameters) may be represented using one of three different approaches (described in sections 3.5.5.1, 3.5.5.2, and 3.5.5.3).

The one-argument INTEGRATOR function integrates the time-domain signal represented by its one argument. That is, this function applies the transfer function $1/s$, where s is the Laplace transform variable.

The one-argument DIFFERENTIATOR function differentiates the time-domain signal represented by its argument. The transfer function for a differentiator is s. Because differentiation can produce an unbounded output for a bounded input, we implement the DIFFERENTIATOR function in this book with a transfer function that is substantially equivalent to s for frequencies of interest in this book, but that never produces an unbounded output. In particular, we implement differentiation with $s/(1 + \tau s)$, where $\tau = 2$ microseconds. For simplicity, we use only s as the transfer function for the DIFFERENTIATOR function when presenting figures and transfer functions for genetically evolved results in this book.

The two-argument ADD_SIGNAL, SUB_SIGNAL, and MULT_SIGNAL functions perform addition, subtraction, and multiplication, respectively, on the two time-domain signals represented by their two arguments. Note that the ADD_SIGNAL, SUB_SIGNAL, and MULT_SIGNAL functions differ from the ordinary arithmetic +, −, and * functions in that both arguments of the former three functions are time-domain signals whereas both arguments of the latter three functions are scalar numerical values.

Note also that the MULT_SIGNAL function differs from the GAIN function (described earlier) in that both arguments of the MULT_SIGNAL function are time-domain signals whereas one argument of the GAIN function is a time-domain signal and its other argument is a scalar numerical value.

The three-argument ADD_3_SIGNAL function adds the time-domain signals represented by its three arguments.

The two-argument DIFFERENTIAL_INPUT_INTEGRATOR function integrates the time-domain signal representing the difference between its two arguments.

The one-argument INVERTER function negates the time-domain signal represented by its argument.

The two-argument LEAD function applies the transfer function $1 + \tau s$, where τ is a real-valued numerical parameter. The first argument is the time-domain input signal. The second argument, τ, is a numerical parameter representing the time constant (usually expressed in seconds).

The two-argument LAG function applies the transfer function $1/(1 + \tau s)$, where τ is a numerical parameter. This function's first argument is the time-domain input signal. The second argument, τ, is the time constant.

The three-argument LAG2 (second-order lag) function applies the transfer function

$$\frac{\omega_0^2}{s^2 + 2\zeta\omega_0 s + \omega_0^2},$$

where ζ is the damping ratio, and ω_0 is the corner frequency.

The one-argument ABS_SIGNAL function performs the absolute value function on the time-domain signal represented by its argument.

The three-argument LIMITER function limits a signal by constraining it between an upper and lower bound. This function returns the value of its first argument (the incoming signal) when its first argument lies between its second and third arguments (the two bounds). If the first argument is greater than its third argument (the upper bound), the function returns its third argument. If its first argument is less than its second argument (the lower bound), the function returns its second argument. Note that, in many systems, the output of the controller is often passed through a LIMITER function before it is passed on to the plant.

The four-argument DIV_SIGNAL function divides the time-domain signals represented by its two arguments and constrains the resulting output by passing the quotient through a built-in LIMITER function with a specified upper and lower bound.

The one-argument DELAY function has one numerical parameter, its time delay, and applies the transfer function

$$e^{-sT},$$

where T is the time delay.

The three-argument IF_POSITIVE function is a switching function that operates on three time-domain signals and produces a particular time-domain signal depending on whether its first argument is positive. If, at a given time, the value of the time-domain function in the first argument of the IF_POSITIVE function is positive, the value of the IF_POSITIVE function is the value of the time-domain function in the second argument of the IF_POSITIVE function. Otherwise, the value of the IF_POSITIVE function is the value of the time-domain function in the third argument of the IF_POSITIVE function.

Automatically defined functions (e.g., ADF0, ADF1) possessing one or more arguments may also be included in the function set (as described in sections 2.3.2 and 3.5.4).

The two-argument $+$, $-$, $*$, and $\div$ functions perform addition, subtraction, multiplication, and division, respectively. These functions are typically used herein to establish numerical parameter values for signal-processing functions that possess numerical parameters.

3.5.2 Repertoire of Terminals

The repertoire of terminals includes the following:

The REFERENCE_SIGNAL terminal is the time-domain signal representing the desired plant response. If there are multiple reference signals, they are named REFERENCE_SIGNAL_0, REFERENCE_SIGNAL_1, and so forth. The reference signal is sometimes also called the *setpoint* or *command signal*.

The PLANT_OUTPUT terminal is the time-domain signal representing the plant output. If the plant has multiple outputs, the plant outputs are named PLANT_OUTPUT_0, PLANT_OUTPUT_1, and so forth. Notice that the inclusion of PLANT_OUTPUT in the terminal set merely makes the plant output available to genetic programming. There is no human-imposed requirement that genetic programming actually use the plant output in synthesizing a controller. In particular, the existence of a terminal in the repertoire of available terminals does not mandate that genetic programming incorporate feedback into the genetically evolved controller. If the genetically evolved controller ultimately does not use PLANT_OUTPUT, the result is an open loop controller. If the genetically evolved controller uses PLANT_OUTPUT, the result is a closed-loop controller. Feedback of the plant's output to the controller is useful—indeed essential—for constructing most practical controllers. However, this domain knowledge about the field of control is not provided to genetic programming in advance. Instead, genetic programming must discover the usefulness of feedback (well known to control engineers) *on its own* during the run.

The CONTROLLER_OUTPUT terminal is the time-domain signal representing the output of the controller (sometimes called the *control variable*). If the controller has multiple outputs, the outputs are named CONTROLLER_OUTPUT_0, CONTROLLER_OUTPUT_1, and so forth. Inclusion of the terminal CONTROLLER_OUTPUT in the problem's terminal set provides the potential means to implement direct feedback of the controller's own output back into the controller. There is, however, no human-imposed requirement that genetic programming actually use the controller's own output in synthesizing the controller.

If the plant has internal state(s) that are available to the controller, then the terminals STATE_0, STATE_1, etc. are the plant's internal state(s).

The CONSTANT_0 terminal is the constant time-domain signal whose value is always 0. Similar terminals may be defined, if desired, for other particular constant-valued time-domain signals.

Constant numerical terminals (called $\Re$) are included in the terminal set.

Zero-argument automatically defined functions may also be included in the terminal set of a particular problem (as described in section 3.5.4).

In problems involving parameterized topologies (chapters 9 and 13), there are additional terminals representing the problem's free variables.

3.5.3 Representing the Plant

Use of genetic programming to automatically synthesize controllers requires ascertaining the behavior of the overall system (composed of both the controller and plant). Thus, the plant's behavior must be available to the genetic programming system in some way.

If a model of the plant is available, the controller's behavior and characteristics can be obtained by simulating the controller and plant together. In practice, the plant can usually be modeled with the same repertoire of functions and terminals listed in sections 3.5.1 and 3.5.2.

If a model of the plant is not initially available, genetic programming can be used to create a model (block diagram) for the plant. This can be accomplishing by making a run of genetic programming to perform system identification on data representing the plant outputs corresponding to particular plant inputs. This type of run of genetic programming could employ the same repertoire of functions and terminals listed in sections 3.5.1 and 3.5.2. See also Witczak 2003.

3.5.4 Automatically Defined Functions

Automatically defined functions resemble subroutines (section 2.3.2). An automatically defined function (ADF) is a function whose body is dynamically evolved during the run and which may be invoked by the main result-producing branch(es), by other automatically defined function(s), and, in some cases, other types of branches (section 2.3.2). When automatically defined functions are being used, an individual program tree consists of one or more reusable automatically defined functions (function-defining branches) as well as the main result-producing branch(es). Automatically defined functions may possess zero, one, or more dummy arguments (formal parameters) and may be reused with different instantiations of their dummy arguments.

Automatically defined functions are typically used in genetic programming to provide a means for dynamically evolving subroutines that reuse code. However, reuse is not the major reason for using automatically defined functions in the context of control problems.

In control problems, automatically defined functions provide a convenient mechanism for representing takeoff points. Once an automatically defined function is defined, it may be referenced repeatedly by other parts of the overall program tree representing the controller. Thus, an automatically defined function may be used to disseminate the output of a particular processing block to one or more points in the block diagram. A variation in the just-described method for handling takeoff points is described in section 13.1.1.

In addition, automatically defined functions provide a convenient mechanism for implementing internal feedback within a controller. In control problems, the function set for each automatically defined function and each result-producing branch includes each existing automatically defined function (including itself). Specifically, each automatically defined function is a composition of the functions and terminals listed in sections 3.5.1 and 3.5.2, all existing automatically defined functions, and (possibly) dummy variables (formal parameters) that parameterize the automatically defined functions. Note that in the style of ordinary computer programming, a reference to `ADF0` from inside the function definition for `ADF0` would be considered to be a recursive reference. However, in the context of control problems, an automatically defined function that references itself represents a loop in the block diagram (i.e., internal feedback within the controller).

The program trees in the initial random generation (generation 0) may consist only of result-producing branches. Automatically defined functions are then introduced sparingly on subsequent generations of the run by means of the architecture-altering operations. Alternatively, a certain predetermined number of automatically defined functions may be present in all the individuals of generation 0.

The to-be-evolved controller can accommodate one or more externally supplied reference signals, external feedback of one or more plant outputs to the controller, computations of error between the reference signals and the corresponding external plant outputs, one or more internal state variables of the plant, and one or more outputs passed from the controller to the plant. These automatically created controllers can also accommodate internal feedback of one or more signals from one part of the controller to another part of the controller. The amount of internal feedback is automatically determined during the run.

For additional information on automatically defined functions, see Koza 1990a, 1992a, 1992c, 1994a, 1994b; Koza and Rice 1991, 1992b, 1994a; and Koza, Bennett, Andre, and Keane 1999a.

3.5.5 Three Approaches for Establishing Numerical Parameter Values

All the problems in this book (including those involving controllers, electrical circuits, antennas, metabolic pathways, and genetic networks) entail the automatic creation of both the topological features and the numerical aspects of a structure.

For example, many of the signal-processing blocks appearing in controllers (e.g., gain, lead, lag, delay) possess a numerical parameter value. In addition, some signal-processing blocks (e.g., the limiter, second-order lag) possess more than one numerical parameter value. The value of each numerical parameter for these signal-processing functions is established by one of the function's arguments. Similarly, many components (e.g., capacitors, resistors, inductors) used in electrical circuits elsewhere in this book possess a numerical parameter value. Wire segments in antennas (chapter 6) possess numerical parameter values (e.g., length, thickness). Chemical reaction functions (chapter 8) are parameterized by rate constants. Genetic switches (chapter 7) have thresholds.

The following three approaches are used in this book to establish the numerical parameter values:

(1) an arithmetic-performing subtree consisting of one or more arithmetic functions and one or more constant numerical terminals;
(2) a subtree consisting of a single perturbable numerical value; and
(3) an arithmetic-performing subtree consisting of one or more arithmetic functions, one or more perturbable numerical values, and (optionally) one or more free variables and conditional operators.

In all three approaches, a constrained syntactic structure (section 2.3.1) maintains the distinction between the subtrees that are used to establish numerical values and all other parts of the program tree.

It is often advantageous to use a nonlinear mapping (*NLM*) to interpret the value returned by the arithmetic-performing subtree (or perturbable numerical value). A logarithmic nonlinear mapping enables constants that differ by several orders of magnitude to be evolved efficiently. Unless otherwise stated, the nonlinear mapping used throughout this book converts numbers ranging between −5.0 and +5.0 into numbers ranging over 10 orders of magnitude. Specifically, the nonlinear mapping transforms its argument x to the final parameter value as follows:

$$NLM(x) = \begin{cases} 10^0 & \text{if } x < -100 \\ 10^{-\frac{100}{19}-\frac{1}{19}x} & \text{if } -100 \leqslant x < -5 \\ 10^x & \text{if } -5 \leqslant x \leqslant 5 \\ 10^{\frac{100}{19}-\frac{1}{19}x} & \text{if } 5 < x \leqslant 100 \\ 10^0 & \text{if } x > 100 \end{cases}$$

3.5.5.1 Arithmetic-Performing Subtrees In the first approach, the numerical parameter is established by an arithmetic-performing subtree. The value returned by the entire arithmetic-performing subtree is interpreted as the desired numerical parameter. The subtree consists of one or more arithmetic functions (such as addition, subtraction, multiplication, and division) and one or more constant numerical terminals. In the initial random generation (generation 0), each constant numerical terminal (called $\Re$) is set, individually and separately, to a random value in a chosen range (e.g., between −5.0 and +5.0). This constant numerical terminal is (usually) subsequently interpreted as a component value lying in a range of positive values between 10^{-5} and 10^5 using a nonlinear mapping (as just described in section 3.5.5). Each such constant numerical terminal remains unchanged during all subsequent generations of the run of genetic programming. This first approach was extensively employed in connection with our early work on the automatic synthesis of analog electrical circuits (Koza, Bennett, Andre, and Keane 1999a). A constrained syntactic structure maintains one function and terminal set for the arithmetic-performing subtrees and a different function and terminal set for all other parts of the program tree.

Figure 3.3 shows a value-setting subtree consisting of two arithmetic functions (addition and multiplication) and three constant numerical terminals (2.963, 1.234, and 3.292).

This approach is used in sections 3.7.1.2, 3.10.1.2, 4.2.1.4, 4.3.1.3, 4.4.1.3, and 5.2.1.4 of this book.

This first approach is now recognized to be relatively inefficient because when the crossover operation exchanges sub-subtrees of arithmetic-performing subtrees, it comes perilously close to a random mutation operation.

3.5.5.2 Perturbable Numerical Value In the second approach, the numerical parameter value is established by a single perturbable numerical value. Unlike the

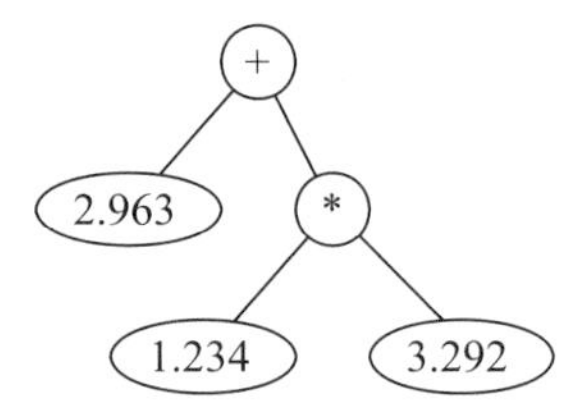

Figure 3.3 Value-setting subtree consisting of two arithmetic functions and three constant numerical terminals.

constant numerical terminals of the just-described first approach (section 3.5.5.1), the perturbable numerical values are changed during the run. In generation 0, each perturbable numerical value (called $\Re_p$) is set, individually and separately, to a random value in a chosen range (e.g., between −5.0 and +5.0). After generation 0, the perturbable numerical value may be perturbed. The perturbation is usually relatively small. The to-be-perturbed value is considered to be the mean of a Gaussian distribution. A relatively small preset parameter establishes the standard deviation of the Gaussian distribution. For example, the standard deviation of the Gaussian perturbation might be 1.0 (corresponding to one order of magnitude, if the number between −5.0 and +5.0 is later interpreted using the nonlinear mapping described in section 3.5.5). The to-be-perturbed value is then perturbed by an amount determined by the Gaussian distribution. These perturbations are implemented by a special genetic operation for mutating the perturbable numerical values. A constrained syntactic structure maintains one function and terminal set for the arithmetic-performing subtrees and a different function and terminal set for all other parts of the program tree.

Figure 3.4 shows a value-setting subtree consisting of a single perturbable numerical value.

This second approach is patterned after the Gaussian mutation operation used in evolution strategies (Rechenberg 1965, 1973). This second approach has the advantage (over the first approach described in section 3.5.5.1) of changing numerical parameter values by a relatively small amount. Therefore, the search for a numerical value in the space of possible parameter values is conducted most thoroughly in the immediate neighborhood of the existing value (which is, by virtue of Darwinian selection, necessarily part of a relatively fit individual). This approach retains the usual advantage of the genetic algorithm with crossover, namely the ability to exploit co-adapted sets of functions and numerical values. It is also possible to perform an ordinary crossover operation in which a perturbable numerical value is inserted in lieu of a chosen other perturbable numerical value.

Our experience is that the perturbable numerical values work considerably better than the first approach. Thus, we use perturbable numerical values (either in the form of this second approach or the third approach described below) for the vast majority of the problems in this book.

3.5.5.3 Arithmetic-Performing Subtree Containing Perturbable Numerical Values
The third approach is more general than the second approach and employs arithmetic-performing subtrees in conjunction with perturbable numerical values. This approach differs from the second approach in that a full subtree is used (instead of just the single perturbable numerical value).

A constrained syntactic structure maintains one function and terminal set for the arithmetic-performing subtrees and a different function and terminal set for all other parts of the program tree.

4.809

Figure 3.4 Value-setting subtree consisting of a single perturbable numerical value.

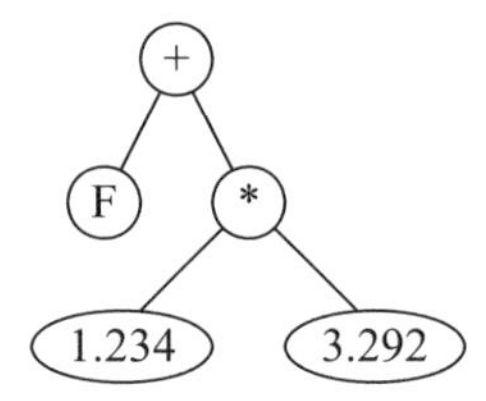

Figure 3.5 Value-setting subtree consisting of two functions, two perturbable numerical values, and one free variable F.

This third approach is especially advantageous when externally supplied free variables are being used to establish numerical parameter values for components within the to-be-evolved structure. This approach is also useful when a dummy variable (formal parameter) of an automatically defined function is used to establish the numerical parameter values for components that are defined in the body of the automatically defined function (or perhaps in some other branch that is called hierarchically from the automatically defined function). In both cases, the free variables confer generality on the structure that they parameterize.

Figure 3.5 shows a value-setting subtree consisting of two functions (addition and multiplication), two perturbable numerical values (1.234 and 3.292), and one free variable, F.

An arithmetic-performing subtree may (optionally) include conditional operators.

The percentages of genetic operations (as well as the maximum sizes of the various branches, minor control parameters, and the various default parameters) are shown for each problem in appendix B.

When the first approach (section 3.5.5.1) is used, the genetic operations (described in section 2.2.1) are typically performed at rates such as

- 89% one-offspring crossover,
- 10% reproduction, and
- 1% ordinary mutation (i.e., replacement of a subtree with a randomly grown subtree).

When architecture-altering operations and the first approach are used, the allocation of percentages are altered slightly to reflect the presence of the architecture-altering operations. The genetic operations are typically performed at rates such as

- 86.5% one-offspring crossover,
- 10% reproduction,
- 1% ordinary mutation,
- 1% subroutine duplication,
- 1% subroutine creation, and
- 0.5% subroutine deletion.

When the second approach (section 3.5.5.2) is used, these percentages must be adjusted to reflect the presence of the perturbable numerical values. In this event, the genetic operations are typically performed at rates such as

- 20% Gaussian mutation on perturbable numerical values,
- 1% ordinary mutation,
- 48.5% one-offspring crossover on internal points of the program tree other than perturbable numerical values,
- 9% one-offspring crossover on terminals other than perturbable numerical values,
- 9% one-offspring crossover on perturbable numerical values,
- 10% reproduction,
- 1% subroutine creation,
- 1% subroutine duplication, and
- 0.5% subroutine deletion.

Note that ordinary mutation remains at 1%.

3.5.6 Constrained Syntactic Structure for Program Trees

The program trees in the initial generation 0 as well as any trees created in later generations of the run by the mutation, crossover, and architecture-altering operations are created in compliance with a constrained syntactic structure (section 2.3.1) that limits the particular functions and terminals that may appear at particular points in each branch of the overall program tree.

An individual tree consists of main result-producing branch(es) and zero, one, or more automatically defined functions (function-defining branches). Each of these branches is composed of various functions and terminals.

The functions and terminals in the trees are divided into four categories:

(1) signal-processing block functions,
(2) automatically defined functions that appear in the function-defining branches and that enable both internal feedback within a controller and the dissemination of the output from a particular signal-processing block within a controller to other points in the block diagram,
(3) terminals and functions that may appear in arithmetic-performing subtrees for the purpose of establishing the numerical parameter value for signal-processing blocks possessing numerical parameters, and
(4) terminal representing time-domain signals (e.g., the reference signal).

3.6 Additional Representations of Controllers

3.6.1 Representation of a Controller by a Transfer Function

The controller of figure 3.2 can also be represented as a transfer function, $G_c(S)$, as follows.

$$G_c(s) = 214.0 + \frac{1000.0}{s} + 15.5s = \frac{214.0s + 1000.0 + 15.5s^2}{s}.$$

In this representation, the 214.0 is the proportional (P) element of the controller. The 1,000.0/s is the integrative (I) term, where s is the Laplace transform variable. The 15.5s is the differentiating (D) term.

The entire transfer function is conventionally expressed as a quotient of two polynomials in the Laplace transform variable s (as is done on the right).

3.6.2 Representation of a Controller as a LISP Symbolic Expression

A controller (such as the PID controller of figure 3.2) may be represented by a composition of functions and terminals in the LISP programming language in the form of S-expressions (symbolic expressions). LISP S-expressions are not to be confused, of course, with the expressions (such as those in section 3.6.1) containing the Laplace transform variable s. The block diagram for the PID controller of figure 3.2 can be represented as the following LISP S-expression:

```
1 (PROGN
2   (DEFUN ADF0 ()
3     (VALUES
4       (- REFERENCE_SIGNAL PLANT_OUTPUT)))
5     (VALUES
6       (+
7         (GAIN 214.0 ADF0)
8         (DERIVATIVE (GAIN 15.5 ADF0))))
9         (INTEGRATOR (GAIN 1000.0 ADF0))
10 )
```

Notice that the automatically defined function ADF0 provides the mechanism for disseminating a particular signal (the difference taken on line 4) to three places (lines 7, 8, and 9) and corresponds to the takeoff point 520 in figure 3.2. An alternative approach to takeoff points is discussed in section 13.1.1.

Although the work described in this book was done using software written in C or Java, we often present the results in the form of LISP S-expressions.

3.6.3 Representation of a Controller as a Program Tree

A controller may also be represented as a point-labeled tree with ordered branches (i.e., a program tree) that corresponds directly with the LISP S-expression representation (as just shown in section 3.6.2). The terminals (leaves) of such a program tree may represent inputs to the controller, constant numerical values, perturbable numerical values, or externally supplied free variables. The functions in such program trees correspond to the signal-processing blocks in a block diagram representing the controller. The value returned by an entire result-producing branch of the program tree corresponds to the controller output that is to be passed on to the plant. If the controller has more than one output, the program tree has one result-producing branch for each such output. Figure 3.6 presents the block diagram for the PID controller of figure 3.2 as a program tree. All block diagrams for controllers may be represented in this manner.

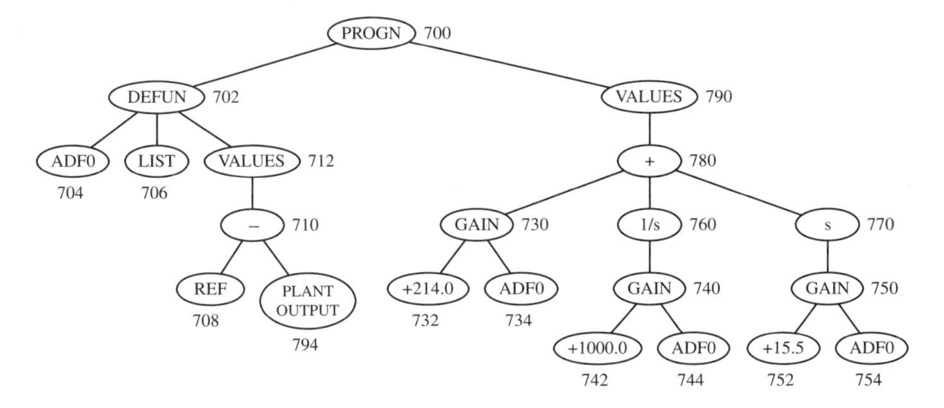

Figure 3.6 Program tree representation of the PID controller of figure 3.2.

3.6.4 Representation of a Controller in Mathematica

The LISP S-expression for the PID controller of figure 3.2 may be converted to the following two equations in Mathematica:

```
RPB0 == ((214.0*ADF0)+((15.5*ADF0)*S)+((1000.0*ADF0)/S))
ADF0 == (REFERENCE-PLANT_OUTPUT)
```

The SOLVE command of the symbolic algebra package in Mathematica can be used to solve this system of two equations for `RPB0` and `ADF0`. It yields the following:

```
RPB0 = (214.0+1000.0/S+15.5*S)(REFERENCE-PLANT_OUTPUT)
ADF0 = REFERENCE-PLANT_OUTPUT
```

This solution represents the transfer functions at the points labeled `RPB0` and `ADF0`.

The transfer function of the controller is

```
(214.0+1000.0/S+15.5*S)(REFERENCE-PLANT_OUTPUT)
```

3.6.5 Representation of a Controller and Plant as a Connection List

A controller (such as the PID controller of figure 3.2) may be represented by a connection list.

A connection list for a block diagram is a data structure that defines both the topology and the parameter values of each element of a block diagram. Each line of a connection list for a block diagram corresponds to one signal-processing block of the block diagram. Each line of a connection list of a block diagram contains the name of a processing block, the points to which each input and output of that processing block is connected, and the parameter(s), if any, of that signal-processing block.

The connection list includes both the controller and the plant. Figure 3.7 shows a two-lag plant consisting of a series composition of a `LIMITER` block (with a range −40.0 Volts to +40.0 Volts) and two `LAG` blocks (each with a lag of +1.0). In the

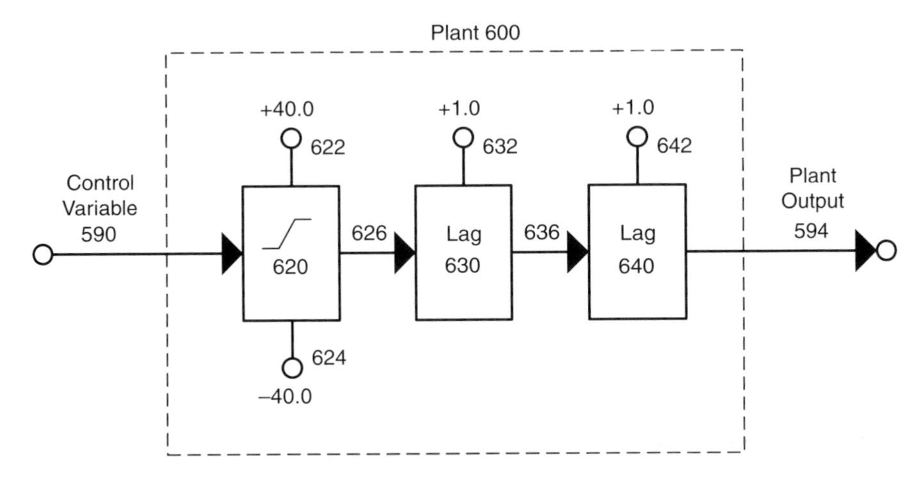

Figure 3.7 Block diagram of a two-lag plant with a LIMITER block that constrains the plant's input to the range between −40 and +40 Volts.

figure, the input to the plant 600 is the output of some controller not shown in the figure. This signal is first passed into LIMITER function block 620 whose upper limit is established by the numerical parameter +40.0 (at 622 of the figure) and whose lower limit is established by the numerical parameter −40.0 (at 624). The output of LIMITER function block 620 is signal 626. This signal then passes into first LAG block 630. The time constant of this first lag is established by the numerical parameter +1.0 (at 632). The output of first LAG block 630 is signal 636. This signal then passes into second LAG block 640. The time constant of this second lag is established by the numerical parameter +1.0 (at 642). The output of this second LAG block 640 is plant output 680.

The connection list for the block diagram of the PID controller of figure 3.2 and the two-lag plant of figure 3.7 is as follows:

```
 1 508 596 512 SUBTRACT
 2 512 538 GAIN 214.0
 3 512 548 GAIN 1000.0
 4 548 568 INTEGRATOR
 5 512 558 GAIN 15.5
 6 558 578 DERIVATIVE
 7 538 568 578 590 ADDITION
 8 590 626 LIMITER −40.0 40.0
 9 626 636 LAG 1.0
10 636 596 LAG 1.0
```

Each line of a connection list begins with a group of two or more numbers, each number representing an input or output signal of one signal-processing block in the block diagram. The connection list for a block diagram reflects the directionality of all connections between signal-processing blocks. In this regard, the last number of each group represents the single output signal of the signal-processing block. The other numbers of each group represent the input(s) to the signal-processing block. If the signal-processing function possesses parameter values, then the parameter value(s) follow the name of the function.

Lines 1 through 7 of this connection list represent the block diagram of the controller of figure 3.2. Lines 8 through 10 represent the block diagram of the two-lag plant of figure 3.7.

Line 1 of this connection list for a block diagram indicates that the two-argument `SUBTRACT` function (corresponding to 510 in figure 3.2), has inputs 508 and 596, and has output signal 512. In this example, the inputs to this function are ordered. The first input 508 represents the positive input to the `SUBTRACT` function and the second input 596 represents the negative input.

Line 2 indicates that the one-argument `GAIN` function (corresponding to 530 in figure 3.2) has input signal 512 and output signal 538. In addition, line 2 indicates that the function has a parameter value (amplification factor) for this signal-processing block of +214.0 (corresponding to 532 in figure 3.2). Output 512 of the `SUBTRACT` function of line 1 is the input to this `GAIN` function of line 2, and the input to the `GAIN` functions in lines 3 and 5. Line 2 corresponds to the proportional part of the PID controller of figure 3.2.

Line 3 indicates that the `GAIN` function (corresponding to 540 of figure 3.2) has input signal 512 and output signal 548. In addition, line 3 indicates that the parameter value (amplification factor) for this signal-processing block is +1,000.0 (corresponding to 542 in figure 3.2). Line 4 indicates that the `INTEGRATOR` function (corresponding to 560 in figure 3.2) has input signal 548 and output signal 568. Lines 3 and 4 together correspond to the integrative part of the PID controller of figure 3.2.

Line 5 indicates that the `GAIN` function (corresponding to 550 in figure 3.2) has input signal 512 and output signal 558. In addition, line 5 indicates that the parameter value (amplification factor) for this signal-processing block is +15.5 (corresponding to 552 in figure 3.2). Line 6 indicates that the `DERIVATIVE` function (corresponding to 570 in figure 3.2) has input signal 558 and output signal 578. Lines 5 and 6 together correspond to the derivative part of the PID controller of figure 3.2.

Line 7 indicates that the three-argument `ADDITION` function (corresponding to 580 in figure 3.2) has positive inputs 538, 568, and 578 and that this `ADDITION` function has output signal 590, corresponding to the output of the PID controller (590 of figure 3.2).

Line 8 indicates that the three-argument `LIMITER` function (corresponding to 620 in figure 3.7) has input 590, has output signal 626, and that the two parameter values (upper and lower limits) for this `LIMITER` block are −40.0 and +40.0 (corresponding to 624 and 622, respectively, in figure 3.7).

Line 9 indicates that the two-argument `LAG` function (corresponding to 630 in figure 3.7) has input 626, has output signal 636, and that the parameter value for this `LAG` function is +1.0 (corresponding to 632 of figure 3.7).

Line 10 indicates that the two-argument `LAG` function (corresponding to 640 in figure 3.7) has input 636, has output signal 596 (i.e., the plant output), and that the parameter value for this `LAG` function is +1.0 (corresponding to 642 in figure 3.7).

3.6.6 Representation of a Controller and Plant as a SPICE Netlist

SPICE (Simulation Program with Integrated Circuit Emphasis) is a large family of programs written over several decades at the University of California at Berkeley for the simulation of analog, digital, and mixed analog/digital electrical circuits. SPICE3 is a version of Berkeley SPICE that consists of about 217,000 lines of C source code residing in 878 separate files (Quarles, Newton, Pederson, and Sangiovanni-Vincentelli 1994). Our work in this book on controllers, circuits, and networks of chemical reactions is based on the SPICE3 version of Berkeley SPICE.

In order to simulate a circuit with the SPICE simulator, it is necessary to provide SPICE with a netlist describing the circuit to be analyzed. It is also necessary to supply SPICE with commands specifying the type of analysis to be performed and the nature of the output to be produced.

Specifically, the input to the SPICE simulator consists of

(1) a command line that names the circuit to be simulated (always first),
(2) the netlist of the circuit to be simulated,
(3) commands that instruct SPICE about the type of analysis to be performed and the nature of the output to be produced,
(4) (optionally) subcircuit definitions, and
(5) an `END` command (always last).

Although the SPICE simulator was originally designed primarily for simulating electrical circuits, it can be adapted for simulating controllers and plants. In fact, SPICE may also be used to simulate a wide variety of other types of systems, including mechanical systems and networks of chemical reactions (chapter 8).

One of the issues that must be addressed when SPICE is being used to simulate controllers and plants is the absence of many signal-processing blocks that are commonly found in controllers and plants (e.g., lead, lag) from the basic version of SPICE. This deficiency can be easily remedied using SPICE's facility to create subcircuit definitions and SPICE's facility to implement user-defined continuous-time mathematical calculations. Of course, basic signal-processing blocks, such as integrators and differentiators, can be represented by capacitors and inductors in conjunction with appropriate voltage-to-current or current-to-voltage converters.

Table 3.1 shows a SPICE input file for simulating a system consisting of a PID controller (figure 3.2), a two-lag plant (figure 3.7), and additional built-in circuitry to calculate the integral of the time-weighted absolute error (ITAE). The lines beginning with an asterisk are comments that are ignored by SPICE.

The input file of table 3.1 contains seven major types of items.

First, the group of lines labeled X1, B2, B3, X4, B5, X6, and X7 in table 3.1 define the PID controller 500 of figure 3.2.

Second, the group of lines labeled X8, X9, X10, V11, V12, V13, and V14 define the two-lag plant 600 of figure 3.7.

Third, the line labeled X15 invokes a calculation of the integral of time-weighted absolute error.

Fourth, the line labeled VP16 provides the pulse reference signal 508 (figure 3.2).

Fifth, the line starting with "`.TRAN`" is a SPICE command that instructs SPICE to perform a transient analysis in the time domain.

Table 3.1 SPICE input file for PID controller, a two-lag plant, and an in-line calculation of ITAE

```
.PID CONTROLLER, TWO-LAG PLANT, AND ITAE CALCULATOR
*
* THE PID CONTROLLER
* THE OUTPUT OF CONTROLLER (CONTROL VARIABLE) IS AT 590
* THE REFERENCE SIGNAL IS AT 508
* THE PLANT OUTPUT (FEEDBACK TO CONTROLLER) IS AT 596
X1    508    596    512    SUBV_SUBCKT
B2    538    0      V=V(512)*214.0
B3    548    0      V=V(512)*1000.0
X4    548    0      568    DII_SUBCKT
B5    558    0      V=V(512)*15.5
X6    558    578    DIFFB_SUBCKT
X7    538    568    578    590    ADD3_SUBCKT
*
* THE TWO-LAG PLANT
* THE PLANT INPUT IS AT 590
* THE PLANT OUTPUT (FEEDBACK TO CONTROLLER) IS AT 594
X8     590   626   622   624     LIMIT_SUBCKT
X9     626    632    636    LAG_SUBCKT
X10    636     642    594    LAG_SUBCKT
V11    622   0     DC    40
V12    624   0     DC    -40
V13    632   0     DC    1.0
V14    642   0     DC    1.0
*
* CALCULATION OF INTEGRAL OF TIME-WEIGHTED ABSOLUTE ERROR (ITAE)
X15     508     594     508     7       ITAE_SUBCKT
*
* REFERENCE SIGNAL 508
VP16   508  0   PULSE(0.0    1    0.1   0.001  0.001   10   15)
*
* COMMANDS TO SPICE FOR TRANSIENT ANALYSIS AND PLOT
.TRAN  0.08    9.6     0.0     0.04    UIC
.PLOT TRAN V(7)
*
* SUBCIRCUIT DEFINITION FOR SUBV
.SUBCKT SUBV 1 2 3
B1      3 0          V=V(1)-V(2)
.ENDS SUBV
*
* SUBCIRCUIT DEFINITION FOR TWO-ARGUMENT ADDV
.SUBCKT ADDV 1 2 3
B1      3 0          V=V(1)+V(2)
.ENDS ADDV
*
* SUBCIRCUIT DEFINITION FOR THREE-ARGUMENT ADD3
.SUBCKT ADD3 1 2 3 4
B1         4 0                V=V(1)+V(2)+V(3)
.ENDS ADD3
*
```

(Continued)

Table 3.1 *(Continued)*

```
 * SUBCIRCUIT DEFINITION FOR INVERTER
.SUBCKT INVERTER_SUBCKT 1 2
B1         2 0               V=-V(1)
.ENDS INVERTER_SUBCKT
*
* SUBCIRCUIT DEFINITION FOR MULV
.SUBCKT MULV 1 2 3
B1         3 0               V=V(1)*V(2)
.ENDS MULV
*
* SUBCIRCUIT DEFINITION FOR DIVV
.SUBCKT DIVV 1 2 3
B1         3 0               V=V(1)/V(2)
.ENDS DIVV
*
* SUBCIRCUIT DEFINITION FOR ABSV
.SUBCKT ABSV 1 2
B1         2 0               V=ABS(V(1))
.ENDS ABSV
*
* SUBCIRCUIT DEFINITION FOR DIFFB (DERIVATIVE)
.SUBCKT DIFFB_SUBCKT 1 2
G1         4 0 1 0           1.0
L1         4 0               1.0
B1         2 0               V=-V(4)
.ENDS DIFFB_SUBCKT
*
* SUBCIRCUIT DEFINITION FOR DII
.SUBCKT DII_SUBCKT 1 2 3
G1         4 0 1 2           1.0
R1         4 0               1000MEG
C1         4 0               1.0 IC=0V
X1         4 3               INVERTER_SUBCKT
.ENDS DII_SUBCKT
*
* SUBCIRCUIT DEFINITION FOR LAG
.SUBCKT LAG_SUBCKT 1 2 3
X1         1 3 4             DII_SUBCKT
X2         4 5 3             MULV
X3         6 2 5             DIVV
V1         6 0               DC 1.0
.ENDS LAG_SUBCKT
*
* SUBCIRCUIT DEFINITION FOR LEAD
.SUBCKT LEAD_SUBCKT 1 2 3
X1         1 4               DIFFB_SUBCKT
X2         2 4 5             MULV
X3         1 5 3             ADDV
.ENDS LEAD_SUBCKT
*
```

(Continued)

Table 3.1 *(Continued)*

```
 * SUBCIRCUIT DEFINITION FOR LAG2
.SUBCKT LAG2_SUBCKT 1 2 3 4
X1          1 4 5               DII_SUBCKT
B2          6 0
+ V=0.5*V(5)*V(2)/V(3)
X3          6 7 4               LAG_SUBCKT
B1          7 0                 V=1/(2*V(2)*V(5))
.ENDS LAG2_SUBCKT
*
* SUBCIRCUIT DEFINITION FOR LIMIT
.SUBCKT LIMIT_SUBCKT 1 2 3 4
B1          2 0
+ V=URAMP(V(1)-V(4))+V(4)-URAMP(V(1)-V(3))
.ENDS LIMIT_SUBCKT
*
* MODEL FOR SSW
.MODEL SSW SW()
*
* SUBCIRCUIT DEFINITION FOR ITAE
.SUBCKT ITAE_SUBCKT 31 32 34 33
VOSPCT      3 0                 DC 0.02V
VOSPEN      11 0                DC 10V
X1          6 34 4              DIVV
V2          10 0                DC 1.0
X3          9 12 7              MULV
S4          11 9 4 3            SSW
S5          14 13 31 0          SSW
V6          14 0                DC 1.0
X7          15 34 33            DIVV
X8          7 18 17             MULV
X9          6 12                ABSV
X10         32 31 6             SUBV
X11         13 0 18             DII_SUBCKT
X12         17 0 15             DII_SUBCKT
R13         9 10                1K
R14         0 13                1K
R15         0 33                1K
.ENDS ITAE_SUBCKT
*
* END COMMAND FOR SPICE INPUT FILE
.END
```

Sixth, the line starting "`.PLOT`" is a SPICE command that instructs SPICE to produce a tabular sequence of values for a particular signal (probe point).

Seventh, this file contains 13 subcircuit definitions (many of which are not necessary for the illustrative system at hand, but are used elsewhere in this book), including

- subtraction (`SUBV`),
- two-argument addition (`ADDV`),

- three-argument addition (ADD3),
- inversion (INVERTER),
- multiplication (MULV),
- division (DIVV),
- absolute value (ABSV),
- differentiation (DIFF),
- differential input integrator (DII),
- lag (LAG),
- lead (LEAD),
- the second-order LAG function (LAG2),
- limiter (LIMIT), and
- integral of the time-weighted absolute error (ITAE).

In explaining table 3.1, we start with the overall inputs and outputs of the overall system, then discuss the subcircuit definitions, and finally discuss the controller and plant.

The line beginning VP16 defines the reference signal.

```
VP16  508  0   PULSE(0.0  1  0.1  0.001  0.001  10  15)
```

The reference signal here is created by the seven-argument PULSE function. In particular, the reference signal is a step function that starts at 0 Volts, rises to 1 Volt at 0.1 seconds, has a rise time of 0.001 seconds, has a fall time of 0.001 seconds, has a pulse width of 10 seconds, and a period of 15 seconds.

The TRAN command instructs SPICE to perform a transient analysis in the time domain. Specifically, the line

```
.TRAN 0.08 9.6 0.000 0.04 UIC
```

instructs SPICE to perform a transient analysis in the time domain in step sizes of 0.08 seconds between an ending time of 9.6 seconds and a starting time of 0.000 seconds. The fourth parameter (0.04 seconds) is a limit on the step size used by the SPICE simulator. The "UIC" instructs SPICE to "use initial conditions" (in this case, for the differential input integrator subcircuit described below).

The PLOT command causes SPICE to provide output in the form of a tabular sequence of values for a specified signal (probe point). Specifically, the line

```
.PLOT V(7)
```

instructs SPICE to capture the circuit's behavior in terms of the voltage V(7) at node 7. The signal at node 7 is the integral of time-weighted absolute error that is calculated by the ITAE subcircuit (invoked by the line beginning with X15). That is, the output of the SPICE simulation is a tabular sequence of values of ITAE between 0 and 9.6 seconds. The value of ITAE for the final time (after 9.6 seconds) may then be used as part of the calculation of the fitness of the individual controller.

It should be noted that the actual time steps that SPICE uses internally in performing its simulation of a system are not the same as the set of tabular values that SPICE provides as external output in response to the user's TRAN and PLOT commands. The actual time steps that SPICE uses internally in performing its simulation are dynamically determined during the simulation by algorithms within SPICE. They are often much finer than the reporting intervals requested by the user.

SPICE permits the mixing of mathematical functions and electrical components. Electrical components may be used to perform mathematical functions in SPICE. For example, differentiation may be performed with an inductor. It is also possible to write a formula in SPICE that implements a mathematical function that does not correspond to any single electrical component. This facility is useful, for example, to implement multiplication. Although it is possible to perform multiplication with a computational circuit, the required circuit contains numerous electrical components and would consume a considerable amount of computer time to simulate. It is much simpler (and more efficient) to use SPICE's ability to define multiplication by writing a simple formula.

The `SUBCKT` command enables a particular topological combination of components (each with associated component values) to be defined once and be invoked thereafter in the netlist as if it were a single primitive component. Each `SUBCKT` definition in SPICE consists of the name of the subcircuit, one or more formal parameters that identify its leads, a netlist that defines the subcircuit, and the `ENDS` command (which must include the name of the subcircuit if nested subcircuit definitions are being used).

A subcircuit definition in SPICE resembles a definition for a subroutine in several ways. First, a subcircuit may be invoked multiple times. Second, a subcircuit may be invoked with different instantiations of their dummy arguments (formal parameters). Third, the numbering of nodes in a subcircuit definition is entirely local to the subcircuit definition (thus permitting numbers such as 1, 2, and 3 to be used in other subcircuit definitions).

An "X" at the beginning of a line of a SPICE input file indicates that a subcircuit definition is being invoked.

As an example of a subcircuit definition, consider the two-argument subtraction function `SUBV`.

```
.SUBCKT  SUBV  1  2  3
B1           3  0            V=V(1)-V(2)
.ENDS  SUBV
```

The first line of the subcircuit definition for `SUBV` indicates that this subcircuit is connected into the invoking circuit by means of three nodes (1, 2, and 3).

The second line of the subcircuit definition for `SUBV` contains the body of a subcircuit definition. The body is only one line here, but may, in general, be many lines. Nodes 1, 2, and 3 are the dummy variables (formal parameters) of this subcircuit. Node 2 represents the voltage that is to be subtracted. Node 1 represents the voltage from which node 2 is to be subtracted. Node 3 represents the voltage that is the output of the subtraction. The "B" at the beginning of the second line indicates that the component B1 is an ideal current or voltage source whose value is specified by an arbitrary mathematical equation. Because the left side of this equation is the letter "V," the component B1 is a voltage source (as opposed to a current source). Specifically, the value of time-varying voltage source B1 is equal to the difference between the voltage at node 1 and the voltage at node 2, as represented by the equation

V = V(1) – V(2). The voltage source B1 has its positive end at node 3 and its negative end at 0 (ground). The voltage at ground (node 0) is defined to be zero at all times during the simulation. Thus, the voltage at node 3 is simply equal to V(1) – V(2) at all times during the simulation. That is, the voltage at node 3 represents the result of the subtraction of the voltage at node 2 from that at node 1.

The third line of the subcircuit definition contains the `ENDS` command that ends the subcircuit definition.

The subcircuit definitions for two-argument addition (`ADDV`), three-argument addition (`ADD3`), inversion (`INVERTER`), multiplication (`MULV`), division (`DIVV`), and absolute value (`ABSV`) follow the same principles as the subcircuit definition for `SUBV`.

As another example of a subcircuit definition, consider the definition of differentiation `DIFFB`. This subcircuit definition employs an electrical component (an inductor) to simulate the mathematical function of differentiation. The voltage across an inductor is equal to its inductance times the derivative of current with respect to time.

```
.SUBCKT DIFFB_SUBCKT 1 2
G1        4 0 1 0            1
L1        4 0                1
B1        2 0                V=-V(4)
.ENDS DIFFB_SUBCKT
```

The first line of the subcircuit definition for `DIFFB` indicates that this subcircuit is connected into the invoking circuit by means of two nodes (1 and 2). The overall effect of this subcircuit definition will be that the subcircuit differentiates the incoming voltage at dummy node 1 and places the output voltage at dummy node 2.

The second line of the subcircuit definition for `DIFFB` begins with “G.” The “G” defines a voltage-controlled current source (VCCS) that converts the voltage between local nodes 1 and 0 (ground) to a current flowing from local nodes 4 to 0.

The third line of the subcircuit definition for `DIFFB` begins with L and defines an inductor (with component value of 1 Henry) that is positioned between local node 4 and ground (node 0).

The fourth line of the subcircuit definition for `DIFFB` begins with B. The “B” at the beginning of this line indicates that this line contains a source (a voltage source, in this case) whose value is specified by a mathematical formula. This fourth line specifies that the voltage at node 2 (the subcircuit's output) is to be equal to the negation of the voltage at node 4.

As yet another example of a subcircuit definition, consider the definition of a differential input integrator. It employs a capacitor. Integration is simulated with a combination of a 1 farad capacitor (C), a 1-giga-Ohm resistor (R), and a voltage-controlled current source. The capacitor has an initial charge (voltage difference) of 0 Volts. The voltage across a capacitor is proportional to the integral of current with respect to time.

Refer now to the group of seven lines (beginning X1, B2, B3, X4, B5, X6, and X7) in table 3.1. These seven lines implement the PID controller 500 of figure 3.2.

```
X1     508    596      512    SUBV_SUBCKT
B2     538    0      V=V(512)*214.0
```

```
B3      548    0       V=V(512)*1000.0
X4      548    0       568    DII_SUBCKT
B5      558    0       V=V(512)*15.5
X6      558    578       DIFFB_SUBCKT
X7      538    568       578    590    ADD3_SUBCKT
```

The first line (beginning X1) computes the difference between the reference signal and the plant output. It accomplishes this by invoking the subcircuit definition for subtraction (SUBV). The voltage produced at node 512 is equal to the difference between the voltage at node 508 (the reference signal of the PID controller of figure 3.2) and the voltage at node 596 (the plant output).

The second line (beginning B2) simulates the proportional (P) part of the PID controller of figure 3.2 and corresponds to the GAIN function block 530 of figure 3.2. This line performs the mathematical function of multiplying the constant +214.0 and the voltage at node 512 (the just-computed difference between the reference signal and the plant output). The output (i.e., the voltage at node 538) is equal to the product between node 538 and ground (node 0).

The third and fourth lines (beginning B3 and X4) simulate the integrative (I) part of the PID controller (corresponding to the GAIN function block 540 and the integrative function block 560 of figure 3.2).

The third line (beginning B3) corresponds to the GAIN function block 540 of figure 3.2. This line performs the mathematical function of multiplying the voltage at node 512 (the difference between the reference signal and the plant output) by the numerical constant +1,000.0 and puts out a voltage equal to the product (the voltage at node 548).

The fourth line (beginning X4) corresponds to the integrative function block 560 of figure 3.2. This line invokes the subcircuit definition for the differential input integrator (DII) found in table 3.1.

The fifth and sixth lines (beginning B5 and X6) simulate the derivative (D) part of the PID controller (corresponding to the GAIN function block 550 and the derivative function block 570 of figure 3.2).

The fifth line (beginning B5) corresponds to the GAIN function block 550 of figure 3.2. The line performs the mathematical function of multiplying the voltage at node 512 (the difference between the reference signal and the plant output) by the numerical constant +15.5 and produces as output a voltage (at node 538) equal to the product.

The sixth line (beginning X6) corresponds to the derivative function block 570 of figure 3.2. This line invokes the subcircuit definition for the derivative (DIFFB).

The seventh line (beginning X7) of this group of seven lines invokes the three-argument addition subcircuit (ADD3). This line simulates the three-argument addition block 580 of figure 3.2. This line sums the proportional (P) part 538, the integrative (I) part 568, and the derivative (D) part 578 of the PID controller. The signal coming out of the ADD3 signal-processing block is the output 590 of the controller.

Refer now to the group of seven lines (beginning X8, X9, X10, V11, V12, V13, and V14) in table 3.1. This group of lines simulates the two-lag plant (labeled 600 in figure 3.7).

```
X8    590    626    622    624    LIMIT_SUBCKT
X9      626     632          636   LAG_SUBCKT
X10     636       642        594    LAG_SUBCKT
V11     622  0       DC        40
V12     624  0       DC        -40
V13     632  0       DC        1.0
V14     642  0       DC        1.0
```

The first line (beginning X8) simulates the limiter block 620 of the plant in figure 3.7. This line invokes the three-argument `LIMIT` subcircuit (defined in table 3.1). The input to the limiter is the output signal 590 of the controller of figure 3.2. The output of the limiter is signal 626 of figure 3.7. The limiter function possesses two arguments that specify its lower and upper limits.

The fourth line (beginning V11) references node 622. The limiter's upper limit of +40 Volts is specified on the line beginning with V11.

The fifth line (beginning V12) references node 624. The limiter's lower limit of −40 Volts is specified on the line labeled V12.

The second line (beginning X9) of this group of seven lines corresponds to the plant's first lag block (namely block 630 of figure 3.7). This line invokes the `LAG` subcircuit (defined in table 3.1). The input to the plant's first `LAG` is signal 626 (coming from the limiter). The output is signal 636. The plant's first `LAG` possesses one numerical argument (i.e., 1.0) as established for node 632 on the sixth line (beginning V13) of this group of seven lines.

Similarly, the third line (beginning X10) of this group of seven lines corresponds to the plant's second lag block (namely block 640 of figure 3.7). This line invokes the `LAG` subcircuit. The input to the plant's second `LAG` is signal 636 (coming from the first `LAG`). Its output is the plant output (labeled 594 in figures 3.2 and 3.7). The plant's second `LAG` possesses one numerical argument (i.e., 1.0) as established for node 632 on the seventh line (beginning V14) of this group of seven lines.

The fitness of a controller may be evaluated in numerous ways. Various signals (including the plant output) may be captured and used internally or externally to evaluate the controller's fitness.

The integral of the time-weighted absolute error is frequently a part of the evaluation of the fitness of a controller. It may be advantageous to evaluate ITAE internally as part of an overall circuit that also includes the controller and the plant. The group of 15 lines (X1, V2, X3, S4, S5, V6, X7, X8, X9, X10, X11, X12, R13, R14, R15) in table 3.1 calculates the integral of time-weighted absolute error between the reference signal 508 and the plant output 594 and makes the resulting integrated value available at node 7. This node may then be probed in order to compute the fitness of the individual controller. Of course, the calculation of ITAE can alternatively be performed using external software operating on the data produced by the SPICE simulation.

There are many additional SPICE commands that can be used to specify other types of analyses to be performed on the signals at the probe points. SPICE may also be instructed to perform more than one type of analysis or to plot more than one probed value.

3.7 Two-Lag Plant

We illustrate how genetic programming can automatically synthesize both the topology and tuning of a controller with a problem calling for the design of a robust controller for a two-lag plant.

The transfer function of a two-lag plant is

$$G(s) = \frac{K}{(1 + \tau s)^2},$$

where K is the plant's internal gain and τ is the plant's time constant.

As with most practical problems of controller design, there are multiple (conflicting) objectives that must be satisfied.

The high-level requirements of this problem are:

- The plant output is to reach the level of the reference signal while minimizing the integral of the time-weighted absolute error.
- The overshoot in response to a step input (reference signal) is to be less than 2%.
- The controller is to be robust in the face of significant variation in the plant's internal gain and the plant's time constant. In this problem, the nominal value of the plant's internal gain, K, is 1.0 and the nominal value of the plant's time constant, τ, is 1.0.
- The input to the plant is limited. This constraint is implemented by inserting a `LIMITER` block between the controller's output and the plant in order to limit the input to the plant to the range between -40 and $+40$ Volts.
- The effect, on the plant output, of high frequency noise in the reference signal is to be limited. This element of the fitness measure reflects an often-unspoken constraint that is typically imposed in the real world in order to prevent potential damage to the plant.

Dorf and Bishop (1998, page 707) developed a PID compensator preceded by a lowpass pre-filter that delivers credible performance on this problem. Dorf and Bishop say that this design has "the optimum ITAE transfer function."

Of course, when Dorf and Bishop say that they achieved the "optimum," they mean that they achieved it given that they had already decided to employ the PID topology. When we applied genetic programming to this problem, we did not pre-specify that the controller must be in the form of a PID compensator preceded by a lowpass pre-filter. Instead, we started with the high-level design requirements of the problem and gave genetic programming a free hand to create any topology that satisfies the problem's requirements.

As will be seen (section 3.7.2), the controller created by genetic programming does not employ the conventional PID topology. The genetically evolved controller employs a second derivative block. The genetically evolved controller is 2.42 times better than the Dorf and Bishop controller as measured by the criterion used by Dorf and Bishop (namely, the integral of the time-weighted absolute error). The genetically evolved controller has only 56% of the rise time and only 32% of the settling time of the Dorf and Bishop controller and is 8.97 times better in terms of suppressing the effects of a step disturbance at the plant input.

Certain constraints are often taken for granted in books written by engineers working in specialized areas. As a result, the above statement of the problem differs from the problem found in Dorf and Bishop 1998 in two respects. First, the last two elements in our statement of the problem are not found in Dorf and Bishop 1998. They reflect our interpretation of the implicit constraints appropriate for this problem. Second, our 2% overshoot requirement is more stringent than the 4% overshoot requirement used by Dorf and Bishop. Note that the Dorf and Bishop controller satisfies our statement of the problem.

3.7.1 Preparatory Steps for the Two-Lag Plant

The preparatory steps are the means of communicating what needs to be done to the genetic programming system. Six major preparatory steps are required before applying genetic programming to a problem involving the automatic synthesis of the topology and tuning for a controller:

(1) determine the architecture of the program trees,
(2) identify the terminals,
(3) identify the functions,
(4) define the fitness measure,
(5) choose control parameters for the run, and
(6) choose the termination criterion and method of result designation.

3.7.1.1 Program Architecture Because the problem involves a controller with one output, each program tree in the population necessarily has one result-producing branch.

All program trees in the initial random population (generation 0) consist of one result-producing branch and no automatically defined functions. In subsequent generations, the architecture-altering operations may add automatically defined function(s) to individual program trees (selected probabilistically based on fitness). Also, in subsequent generations, the architecture-altering operations may delete automatically defined functions from individual program trees (selected probabilistically based on fitness). The result is that, after generation 0, the population becomes architecturally diverse.

We established a generous maximum of five automatically defined functions for this problem. Similarly, we established a generous maximum of 150 for the number of points (i.e., functions or terminals) in the result-producing branch and a generous maximum of 100 points for each automatically defined function. We do not list minor parameters such as these in the body of this book for subsequent problems. Instead, they are listed in appendix B.

3.7.1.2 Terminal Set Taken together, the terminal set and function set (section 3.7.1.3) provide genetic programming with the set of primitive ingredients from which to try to construct a solution to the problem at hand.

There are two types of terminals (external points) in the program trees for problems involving controllers, namely

- terminals representing numerical values, and
- terminals representing time-domain signals.

In this problem, the numerical parameter value for each signal-processing block possessing a parameter is established by an arithmetic-performing subtree. These arithmetic-performing subtrees consist of one or more arithmetic functions and one or more constant numerical terminals. This approach for establishing numerical parameter values is described in section 3.5.5.1.

A constrained syntactic structure enforces the use of one terminal set for the arithmetic-performing subtrees and another for all other parts of the program tree.

The terminal set, T_{aps}, for the arithmetic-performing subtrees is

$$T_{aps} = \{\Re\},$$

where $\Re$ denotes constant numerical terminals in the range from -1.0 to $+1.0$.

The terminal set, T, for all other parts of the result-producing branch and automatically defined functions (if any) contains time-domain signals.

T = {REFERENCE_SIGNAL, CONTROLLER_OUTPUT, PLANT_OUTPUT, CONSTANT_0}.

3.7.1.3 Function Set There are two types of functions in the program trees, namely

- functions used in the arithmetic-performing subtrees for establishing the numerical values (tuning) of the signal-processing blocks possessing numerical parameters, and
- functions used in all other parts of the result-producing branch and automatically defined functions (if any).

A constrained syntactic structure enforces the use of one function set for the arithmetic-performing subtrees and another for all other parts of the program tree.

The function set, F_{aps}, for the arithmetic-performing subtrees is

$$F_{aps} = \{+, -\}.$$

The function set, F, for all other parts of the result-producing branch and automatically defined functions (if any) is

F = {GAIN, INVERTER, LEAD, LAG, LAG2, DIFFERENTIAL_INPUT_INTEGRATOR, DIFFERENTIATOR, ADD_SIGNAL, SUB_SIGNAL, ADD_3_SIGNAL, ADF0, ADF1, ADF2, ADF3, ADF4}.

Here ADF0, ADF1, ADF2, ADF3, and ADF4 denote automatically defined functions. Automatically defined functions are added by the architecture-altering operations.

The function set, F_{adf}, for each automatically defined function consists of F_{rpb} plus whatever automatically defined functions may be called.

3.7.1.4 Fitness Measure Genetic programming is a probabilistic algorithm that searches the space of compositions of the available functions and terminals. The search is guided by a fitness measure. The fitness measure is a mathematical implementation of the high-level requirements of the problem. The fitness measure is couched in terms of what needs to be done—not how to do it. The fitness measure is the main way for communicating the problem's high-level requirements to the genetic programming system.

The fitness measure may incorporate any measurable behavior or characteristic or combination of behaviors or characteristics. Construction of a fitness measure requires translating the problem's high-level requirements into an explicit computation.

All the fitness measures in this book assign a single numerical value to a candidate individual in the population.

The fitness measure for the two-lag plant problem (and, indeed, virtually every control problem) is multiobjective in the sense that it incorporates more than one consideration.

The process of constructing the fitness measure is especially straightforward for many of the design problems that practicing engineers encounter in their daily activity. The reasons are that design problems are usually couched in terms of measurable behaviors or characteristics of the to-be-designed entity and that design problems usually explicitly address the tradeoffs that the engineer is willing to accept between the problem's (often competing) considerations.

The high-level requirements for the two-lag plant problem are stated in terms of several different considerations, including

- minimizing the integral of the time-weighted absolute error while the plant output rises to the level of the reference signal,
- keeping overshoot below 2%,
- ensuring that the controller is robust in the face of significant variation in the plant's internal gain and the plant's time constant,
- limiting the plant's input to non-extreme values, and
- limiting the system bandwidth of the overall system (i.e., the controller and the plant).

The process of evaluating the fitness of each individual in the population of individuals at each generation of the run of genetic programming begins by executing the program tree (i.e., the result-producing branch and any existing automatically defined functions). This execution yields an interconnected sequence of signal-processing blocks—that is, the block diagram for the individual controller.

The plant's input is then connected to the time-domain output signal produced by the controller.

Note that no other connection is automatically made in our approach to the synthesis of controllers. In particular, the reference signal is not automatically connected to the controller even though the controller indeed must be connected to, and use, the reference signal in order to achieve any reasonable value of fitness for the problem at hand. Likewise, the plant output is not automatically connected to the controller. Terminals representing the reference signal and plant output are available in the terminal set. The evolutionary process is required to discover the utility of these available terminals on its own. Similarly, we do not build in any explicit comparison of the reference signal and the plant output. Subtraction is available in the function set. Other means for comparing two signals (such as division) are also available. The evolutionary process is required to discover the usefulness of comparing these two signals on its own.

A netlist for the resulting system (i.e., the to-be-evolved controller in combination with the to-be-controlled plant) is then generated.

The netlist is then wrapped inside an appropriate set of SPICE commands instructing SPICE to carry out various types of analyses.

SPICE is also provided with subcircuit definitions to implement all the signal-processing functions that are contained in the function set or the plant.

The netlist, the SPICE commands, and the subcircuit definitions together constitute the input to the SPICE simulator (as shown in table 3.1).

The system composed of the controller and the plant is then simulated using our modified version of the SPICE simulator. For details on our modifications to SPICE, see section 25.18.4 of *Genetic Programming III* (Koza, Bennett, Andre, and Keane 1999a).

SPICE is instructed to provide the results of the requested analyses in the form of tabular output. An interface communicates this information to code that parses the tabular output and computes the individual's fitness.

For the two-lag plant problem, the fitness of a controller is measured using 10 elements, including

- eight time-domain-based elements measuring the integral of time-weighted absolute error, robustness, and overshoot, and
- one time-domain-based element measuring the controller's response when it is faced with a spiked reference signal, and
- one frequency-domain-based element measuring the reasonableness of the frequency response (bandwidth) of the entire system (i.e., the controller and plant).

The fitness of an individual controller is the sum of the detrimental contributions of these 10 elements. The smaller the sum, the better.

Note that the genetic programming system receives a single numerical value as a result of the computation of fitness. It does not receive the 10 separate elements of the fitness measure.

The first eight elements of the fitness measure together evaluate (i) how quickly the controller causes the plant to reach the reference signal, (ii) the robustness of the controller in the face of significant variations in the plant's internal gain and the plant's time constant, and (iii) the success of the controller in avoiding overshoot.

These first eight elements of the fitness measure together represent

- two choices of values of the height of the reference signal, in conjunction with
- two choices of values of the plant's internal gain, K, in conjunction with
- two choices of values of the plant's time constant τ.

The first reference signal is a step function that rises from 0 to 1 Volt at $t = 100$ milliseconds. The second reference signal is a step function that rises from 0 to 1 microvolt at $t = 100$ milliseconds.

The nominal value of the plant's internal gain, K, is 1.0. Two values of K (1.0 and 2.0) are considered.

The nominal value of the plant's time constant, τ, is 1.0. Two values of τ (0.5 and 1.0) are considered.

Exposing candidate controllers to different combinations of values of K and τ helps to produce a robust controller.

It is sometimes possible to construct an open loop controller that exactly cancels out the effect of the plant for a particular combination of values of K and τ. This process is called *pole cancellation* or *pole elimination*. Such a controller is not what is desired in practice. Multiple combinations of values of K and τ help prevent the evolution of a controller that engages in such pole cancellation.

The differing heights of the two step-functions are used to address the nonlinearity caused by the limiter.

For each of these eight fitness cases, SPICE is instructed to perform a transient analysis in the time domain. The contribution to fitness for each of the eight elements of the fitness measure is based on the integral of time-weighted absolute error

$$\int_{t=0}^{9.6} t|e(t)|A(e(t))Bdt.$$

Here $e(t)$ is the difference (error) at time t between the plant output and the reference signal. Integration from $t=0$ to $t=9.6$ seconds is sufficient to capture all interesting behavior of any reasonable controller for this problem. B is a factor that is used to normalize the contributions associated with the two step-functions. The multiplication of each value of $e(t)$ by B makes both reference signals equally influential. Specifically, B multiplies the difference $e(t)$ associated with the 1-Volt step function by 1 and multiplies the difference $e(t)$ associated with the 1-microvolt step function by 10^6. The integral also contains an additional weighting function, A, that heavily penalizes non-compliant amounts of overshoot. Specifically, the function A depends on $e(t)$ and weights all variations up to 2% above the reference signal by a factor of 1.0. The function A weights overshoots above 2% by a factor of 10.0.

Fitness may be computed in two ways. It may be computed in software after the controller and plant are simulated by SPICE. Alternatively, fitness may be computed by SPICE at the same time that the controller and plant are simulated. That is, a SPICE subcircuit can be constructed for computing ITAE and overshoot. For example, a SPICE subcircuit can calculate ITAE by employing the `IF_POSITIVE` function (to detect overshoot) along with appropriate other signal processing functions (e.g., integrators, subtractors, adders, multipliers) from the same repertoire used for the controller and plant.

The ninth element of the fitness measure evaluates the response of the controller when it is faced with a spiked reference signal. The spiked reference signal rises to 10^{-9} Volts at time $t=0$ and persists for 10 nanoseconds. The reference signal is 0 for all other times. A transient analysis is performed using the SPICE simulator for 121 time values representing times $t=0$ to $t=120$ microseconds. If the plant output never exceeds a fixed limit of 10^{-8} Volts (i.e., an order of magnitude greater than the pulse's magnitude) for any of these 121 time values, then this element of the fitness measure is zero. However, if the absolute value of plant output goes above 10^{-8} Volts for any time t, then the contribution to fitness is $500(0.000120-t)$, where t is the first time (in seconds) at which the absolute value of plant output goes above 10^{-8} Volts. This penalty is a ramp starting at the point (0, 0.06) and ending at the point (1.2, 0) so 0.06 seconds is the maximum penalty and 0 is the minimum penalty.

The 10th element of the fitness measure is designed to constrain the effect, on the plant output, of high frequency noise in the reference signal. The constraint imposed here is that the closed-loop frequency response of the entire system must lie below a 40 decibel (dB) per decade lowpass curve whose corner is at 100 Hz. A *decibel* is a unitless measure of relative voltage that is defined as 20 times the common (base 10) logarithm of the ratio between two voltages. Thus, for example, 40 decibels corresponds to a ratio of 100-to-1 and 60 decibels corresponds to a ratio of 1,000-to-1. In other words, the entire system (i.e., the controller plus the plant) must (ideally) be fully responsive to all low-frequency inputs below 100 Hz and (ideally) progressively less responsive to inputs above 100 Hz. Specifically, the response at 1,000 Hz should be no more than 1/100th of that at 100 Hz; the response at 10,000 Hz should be no more than 1/10,000 of that at 100 Hz; and so forth. If the closed-loop frequency response is acceptable, this element of the fitness measure will be zero.

This element of the fitness measure is based on 121 frequencies. Specifically, SPICE is instructed to perform an AC sweep of the reference signal over 20 sampled frequencies (equally spaced on a logarithmic scale) in each of six decades of frequency between 0.01 Hz and 10,000 Hz. A gain of 0 dB is ideal for the 80 frequency values in the first four decades of frequency between 0.01 Hz and 100 Hz; however, a gain of up to +3 dB is acceptable. The contribution to fitness for each of these 80 frequency values is zero if the gain is ideal or acceptable, but 18/121 per fitness case otherwise. The maximum acceptable gain for the 41 frequency values in the two decades between 100 Hz and 10,000 Hz is given by the straight line connecting (100 Hz, −3 dB) and (10,000 Hz, −83 dB) on a graph with a logarithmic horizontal axis for the frequency and a linear vertical axis for the gain (measured in decibels). The contribution to fitness for each of these frequency values is zero if the gain is on or below this straight line, but otherwise 18/121 per frequency.

Note that many real-world considerations in designing controllers (notably stability) are not explicitly incorporated into the fitness measure for this first illustrative problem. However, later problems in this book will incorporate such considerations.

For this problem, the 10 elements of the fitness measure are combined into a single numerical value by simply adding them together. This approach is the simplest way to create a multiobjective fitness measure. Another relatively simple approach is to use a lexical fitness measure in which a secondary consideration comes into play only after the problem's primary consideration is satisfied.

Numerous alternative approaches are available for multiobjective optimization in the field of genetic and evolutionary computation (Osyczka 1984; Bagchi 1999; Deb 2001; Coello Coello, Van Veldhuizen, and Lamont 2002). There are two informative tutorials and extensive bibliographies in the proceedings of the First International Conference on Evolutionary Multi-Criterion Optimization (Zitzler, Deb, Thiele, Coello Coello, and Corne 2001). Of course, the issues surrounding multiobjective optimization are not unique to genetic and evolutionary computation. They are, for example, extensively addressed in the literature of various other fields (e.g., operations research) and in connection with other search methods (e.g., simulated annealing, gradient methods).

The SPICE simulator is remarkably robust; however, it cannot simulate every conceivable controller. In particular, many controllers that are randomly created for the initial random generation of a run of genetic programming as well as many controllers that are created in later generations by the crossover and mutation operations are so pathological that SPICE cannot simulate them. A controller that cannot be simulated by SPICE is assigned a high penalty value of fitness (10^8). Such controllers become worst-of-generation individuals for their generation. The practice of assigning a high penalty value to unsimulatable entities (controllers, circuits, antennas, and networks of chemical reactions) is used throughout this book.

Past experience with many different types of problems indicates that a small percentage of individuals in the population often consume an inordinate percentage of the computer time during a run of genetic programming. We set a maximum time of 20 seconds for each SPICE simulation. This same time limit is used for all other problems in this book involving SPICE. If a simulation takes more than a specified maximum amount of time, the simulation is terminated and the individual is assigned a high penalty value of fitness (again, 10^8). Preliminary testing indicated that the average time for evaluating the fitness of an individual in this problem is about 5 seconds.

This time limit was originally implemented in order to efficiently allocate available computer time. However, it should be noted that the simulation of an individual with a complex structure often consumes an inordinately large amount of computer time. Thus, the time limit indirectly and imperfectly exerts a certain amount of pressure in favor of parsimony. Of course, if parsimony is an important objective in a particular problem, it should simply be included explicitly as an element of the fitness measure.

The reader may feel that the fitness measure for the problem of synthesizing a controller for a two-lag plant is complex. This perceived complexity is especially stark when this fitness measure is compared to the relatively simple fitness measure for the tutorial problem (section 2.2.1) involving discovery of a computer program whose output is equal to the values of the quadratic polynomial x^2+x+1. In that tutorial problem, fitness was simply the sum of the deviations between a particular program's output and the target values.

The reason for the perceived complexity of the fitness measure for the present problem is that the specifications of a controller *are* complex. Controller specifications are inherently complex because they almost always entail multiple considerations, each of which is inherently technical and many of which directly conflict with one another.

Moreover, it is simply not possible to avoid addressing this multiplicity of conflicting technical considerations in designing a practical controller. In particular, one cannot simplify the problem of designing controllers by separately considering each element of the fitness measure.

It is easy, for example, to construct a controller that forces the plant's output up to a desired target value quickly (i.e., with a very low ITAE) if there is no limit on the magnitude of the stimulus to the plant and if overshoot is not a consideration. However, no one is interested in such a controller. In any practical situation, there are almost always significant constraints on the magnitude of the stimulus and on overshoot.

Similarly, it is relatively easy to construct a controller that operates only under ideal conditions (i.e., a non-robust controller). However, no one is interested in a controller that, like a sphere balanced on the apex of a pyramid, performs correctly only for an isolated set of ideal conditions, but whose performance degrades catastrophically under slightly different conditions. Thus, robustness is almost always an important consideration in the design of controllers.

Having said this, it is important to recognize what is, *and is not*, the origin of the complexity of the fitness measure. The fitness measure for controller problems are not complex because of any burden imposed by genetic programming. Fitness measures for controllers are complex because of the multiplicity of the highly technical and conflicting considerations involved in controller design. If a general automated method other than genetic programming were used to synthesize a controller's topology and sizing, the same highly technical and conflicting considerations would have to be taken into account.

In passing, note that many alternative approaches could have been used in constructing the fitness measure for this problem (and other problems of controller synthesis in this book). Different optimization metrics might have been used. For example, the rise time and settling time might have been used in lieu of the integral of the time-weighted absolute error. There are numerous alternative time-domain and frequency-domain constraints that might have been incorporated into the fitness measure. The fitness measure might have additionally considered robustness with respect to any aspect of the system that may potentially deviate from its ideal value. Also, the fitness measure might have been constructed so as to impose constraints on the plant's internal states by, for example, penalizing extreme values. In addition, the fitness measure may be constructed to include elements measuring the plant's behavior in the face of sensor noise added to the plant output (or, if applicable, the plant's internal states). The fitness measure might also have been be constructed to include elements measuring the robustness of the plant's behavior with respect to changes of some external variable that affects the plant's operation (e.g., temperature, the plant's production rate, line speed, or flow rate).

3.7.1.5 Control Parameters The population size is 66,000.

The minor control parameters (e.g., the percentages of genetic operations) are detailed in appendix B for this problem (and all other problems in this book).

All the runs in this book (except for the genetic network problem in chapter 7) were run on one of two home-built Beowulf-style parallel cluster computer systems. This particular problem was run on a system consisting of 66 533-MHz DEC Alpha® units arranged in a two-dimensional 6×11 toroidal mesh. Chapter 17 and appendix B contain information about the computer system, the migration rates, and other details of each problem in this book.

3.7.1.6 Termination Criterion and Results Designation For this problem (and all the other problems in this book), we have no way of knowing, in advance, whether genetic programming can produce a satisfactory solution. Also, if genetic programming can produce a satisfactory solution, we do not know, in advance, the quality of the best solution. Thus, our practice for all problems in this book is to manually monitor and manually terminate the run. This is accomplished by setting a very large maximum for the number of generations that are to be run and then monitoring the progress of the run from generation to generation. Each run is typically terminated when the values of fitness for numerous successive best-of-generation individuals appear to have reached a plateau. The single best-so-far individual is then harvested and designated as the result of the run. There is, of course, no guarantee that the best-so-far individual will be a satisfactory solution to the problem. And, there is no guarantee that a run that appears to have reached a plateau will not produce a significantly improved result if it were allowed to continue. We do not further explicitly mention this preparatory step in the remainder of this book.

3.7.1.7 Knowledge Incorporated into the Preparatory Steps Our pursuit of automated problem-solving proceeds with two distinct (and, to some extent, conflicting) orientations.

The first orientation is concerned with solving problems while minimizing the amount of knowledge, information, analysis, and intelligence that is prespecified by the human user in the form of the preparatory steps prior to launching the run of genetic programming (sections 3.7.1.1 through 3.7.1.6). (Note that we use the terms "knowledge" and "domain knowledge" in this book in the same sense that these terms are ordinarily used in the fields of artificial intelligence, expert systems, and other knowledge-based methods.) In this first orientation, we are generally interested in evaluating the ratio of that which is delivered by the automated operation of the *artificial* method to the amount of *intelligence* that is supplied by the human applying the method to a particular problem (i.e., the AI ratio discussed in section 1.1.2).

The second orientation is that of a practicing engineer. The engineer generally wants to solve the problem at hand as quickly and efficiently as possible. Practicing engineers have little or no interest in minimizing the amount of human input that goes into solving a problem. Indeed, the reason that they have jobs is to supply knowledge, analysis, intuition, and creativity.

The first orientation is the dominant consideration in this particular section of this book. This section analyzes the amount and nature of the knowledge, information, analysis, and intelligence that is, *and is not*, supplied to the genetic programming system by the human user in connection with the two-lag plant problem.

First, notice that the terminal set (established in section 3.7.1.2) does not reflect any deep knowledge about the field of control. Genetic programming is a problem-independent algorithm. Moreover, genetic programming is not a knowledge-based algorithm. As a result, the genetic programming system does not contain any built-in knowledge that the reference signal and the plant output are useful inputs to a controller, but that, say, the price of gold and the time of day are not. Thus, the human user is obligated to include the reference signal and the plant output in the terminal set in order to provide genetic programming with a sufficient set of ingredients from which to construct a solution to the problem at hand. Notice that the inclusion of the reference signal and plant output in the terminal set does not mandate that genetic programming actually use either of these terminals in solving a problem of controller synthesis. These terminals are merely made available to the genetic programming system for possible use in solving the problem. Additional extraneous terminals (such as the price of gold or the time of day) could be made available to genetic programming provided the terminal set is still sufficient to solve the problem.

In a similar vein, the genetic programming system does not contain any built-in information about the syntax of the various ingredients of controllers. For example, the genetic programming system does not have built-in syntactic knowledge that, say, a gain block in a controller possesses a numerical parameter. Thus, the human user must inform genetic programming about this syntactic requirement and must include numerical terminals in the terminal set for the signal-processing blocks that require them.

The foregoing information about the reference signal, the plant output, and the numerical terminals does not constitute deep knowledge about the field of control. Quite the contrary. This information is entirely platitudinous. Indeed, it is difficult to imagine how any automated problem-solving method could solve a problem in the field of control without starting with information (in some form) about the problem's primitive elements.

Having said all this, notice that the terminal set used in the two-lag plant problem is quite general. Once the terminal set is created for the two-lag plant problem, it can be used (and, indeed, is used) on numerous additional problems of controller synthesis throughout this book. The policy (in this book) of substantial consistency in the terminal set for control problems allows the reader to eliminate superficial concerns that the success of genetic programming depends on a shrewd or fortuitous problem-specific choice for the terminal set.

Second, note that the function set for the two-lag plant problem (established in section 3.7.1.3) does not reflect any deep knowledge about the field of control. The function set informs the genetic programming system that the to-be-created controller may contain integrators, differentiators, gains, adders, subtractors, leads, lags, second-order lags, and differential input integrators. Because genetic programming is a problem-independent algorithm, this elementary domain-specific knowledge about the field of control is not built into the genetic programming system. Instead, this platitudinous information about controllers must be supplied by the human user. Note that when the human user puts functions such as integrators and differentiators and the like into the function set, it is only the symbolic name of the function that is being put into the function set. No knowledge about the appropriate way to use an integrator or a differentiator is supplied to the genetic programming system. Notice that the inclusion of a function in the function set does not mandate that genetic programming actually use the function in solving the problem. Also, notice that, once the function set is created for the present problem, it can be used (and, indeed, is used) on numerous additional problems of controller synthesis throughout this book.

Third, genetic programming does not have access to a description of the plant for the present problem (or, indeed, for any subsequent problem in this book). In particular, the genetic programming system does not have access to a structural description of the plant in the form of, say, a block diagram. Genetic programming does not have access to the plant's transfer function (or the poles or zeroes of the transfer function). Having said that, the plant is not, of course, absent from the overall process. The plant's behavior plays a role in the measurement of the fitness of a particular candidate controller. It does so in the following way: First, the controller (in response to the available reference signal and available plant output) generates a signal that becomes the plant's input. The plant responds to this signal from the controller by producing a plant output. The overall behavior of the entire system (composed of the controller and plant) is measured by the problem's fitness measure (described in section 3.7.1.4). Finally, the genetic programming system receives a single numerical value (the controller's fitness).

Fourth, note that the preparatory steps do not mandate that genetic programming employ feedback. No domain knowledge about the usefulness of feedback is provided to genetic programming. The inclusion of PLANT_OUTPUT in the terminal set makes the plant output *available* to genetic programming; however, there is no human-imposed requirement that genetic programming actually use it in synthesizing a controller. If the genetically evolved controller happens to use PLANT_OUTPUT, the result is a closed-loop controller. If the genetically evolved controller does not use PLANT_OUTPUT, the result is an open loop controller. The reality is that an open loop controller is unlikely to do a good job in satisfying the problem's requirements (notably the robustness requirement). However, genetic programming is left to discover the usefulness of feedback (well known to control engineers) *on its own* during the run.

Fifth, notice that there is no human-imposed requirement that genetic programming employ *negative* feedback (i.e., subtraction of the plant's output from the reference signal) in order to synthesize a controller to satisfy the problem's requirements. Instead, genetic programming must discover the usefulness of negative feedback *on its own* during the run. Negative feedback is indeed very useful for constructing most practical controllers and for satisfying the high-level requirements of the present problem. However, no domain knowledge about the usefulness of negative feedback or any information about the way to implement negative feedback is provided *a priori* to genetic programming.

Sixth, notice that the CONSTANT_0 terminal in the terminal set for this problem is useless. The terminal sets in this book often include extraneous terminals. Such extraneous terminals make the point that genetic programming can solve a problem even in the face of inefficient or inept choices by the user concerning the preparatory steps. If, for example, the radio signal from AM station 740 in San Francisco were included in the terminal set for this problem, genetic programming would still be able to solve the problem. However, the efficiency of the run would be reduced because genetic programming would have to learn to prevent this irrelevantly fluctuating signal from influencing the controller's output.

Seventh, note that the function set contains numerous signal-processing functions (e.g., LEAD, LAG, LAG2) that will prove to be extraneous to the solution of the problem involving the two-lag plant. In addition, many of the other functions in the function set are duplicative in the sense that they can be realized by another function or a combination of other functions (e.g., ADD_3_SIGNAL, INVERTER). Moreover, far more automatically defined functions are available than are needed to solve the present problem. Most of the function sets in this book include extraneous functions.

The inclusion of extraneous and duplicative functions makes the point that genetic programming can usually solve a problem (albeit with reduced efficiency) even in the face of inefficient or inept choices by the human user. The fact that genetic programming can overcome ill-advised choices by the human user concerning the preparatory steps was observed in our 1992 book (Koza 1992a, section 24.3) and has been subsequently noted by numerous other researchers in the field of genetic programming.

Another reason for including extraneous functions is that our choice of an overly large function set for this introductory problem enables us to use the same function set for numerous different problems throughout this book. This policy of substantial consistency in the function set helps the reader eliminate superficial concerns that the success of the run depends on a shrewd or fortuitous choice of the function set.

However, the most important reason for including extraneous functions is that many potential inventions are never discovered (or their discovery is greatly delayed) because the thought processes of scientists and engineers are channeled along well-traveled paths. Sometimes a seemingly extraneous function may enable genetic programming to solve a problem in a novel way. One of the virtues of an open-ended evolutionary method such as genetic programming is that it often unearths solutions that that might never have occurred to a human scientist or engineer. Thus, for many problems (particularly if the user is interested in novelty), it may be "too clever by half" to exclude seemingly extraneous functions in the illusory pursuit of efficiency.

Having said all that, from a practical point of view, if a practicing engineer has good reason to believe that the inclusion of a particular function in the function set will only hinder the efficiency of the run of genetic programming, there is no reason not to exclude that function and take advantage of the resulting increased efficiency.

Eighth, it is important to notice what knowledge is, *and is not*, associated with the fitness measure. The fitness measure specifies what the human user desires. The fitness measure translates the human user's desires into a partial order among candidate individuals (i.e., an indication that one candidate is better than another). The human user necessarily employs knowledge about the field of control in order to translate the desired behavior into a well-defined calculation. However, it is important to distinguish between domain knowledge employed by the human user in constructing a fitness measure and domain knowledge that is actually available to the genetic programming system during the run.

Knowledge about the field of control is assuredly required to translate each part of this problem's 10-element fitness measure into a mathematically precise calculation. In constructing the fitness measure for the two-lag plant problem, the human user first has to draw on knowledge about the field of control in order to translate the English imperative phrase, "Get the plant output to match the reference signal quickly," into a calculation involving the integral of time-weighted absolute error. Similarly, the human user has to employ additional knowledge about the field of control in order to translate the English words "overshoot" and "robustness" into mathematically precise calculations. However, the genetic programming system does not have access to any of the domain knowledge used to perform this translation. The only thing that is communicated to the genetic programming system concerning each candidate controller is a *single numerical value* representing the candidate's fitness. Distilling a candidate controller's behavior during 10 separate time-domain and frequency-domain experiments into a single number does not transfer any usable domain knowledge about the field of control to the genetic programming system.

Ninth, notice that our use of the SPICE simulator does not confer any domain knowledge about the field of control on the genetic programming system. In this connection, it is important to distinguish among knowledge possessed by the authors of the SPICE simulator, knowledge embedded inside the SPICE simulator, and knowledge that is actually available to the genetic programming system during the run as a consequence of use of the SPICE simulator. No usable domain knowledge about the field of control is transferred from SPICE to the genetic programming system for the following reasons.

One reason why SPICE does not confer any domain knowledge about the field of control on the genetic programming system is that the SPICE simulator simply does not contain any built-in knowledge about the field of control in the first place. This point becomes obvious when one realizes that SPICE was designed with electrical circuits, and not controllers, in mind.

A second reason is that SPICE does not do *synthesis*. The SPICE simulator only does *analysis* of the behavior of the already existing entity that is presented to it in the form of a netlist. SPICE does not synthesize either the topology or the numerical parameter values of the entity. Instead, SPICE simply ascertains the behavior of the entity it is instructed to analyze. In fact, SPICE does not even attempt to optimize the numerical parameter values or topology of the entity that it is given.

Yet another reason becomes clear when one examines what kind of knowledge SPICE actually does contain. Assuredly, writing the SPICE simulator required considerable human knowledge and expertise. However, the human knowledge embodied in SPICE concerns how to automatically and efficiently converge on the solution to a system of equations (with particular emphasis on the nonlinear equations and the integro-differential equations that represent electrical circuits containing transistors).

A final reason why genetic programming does not acquire any domain knowledge about the field of control from the SPICE simulator is made clear by realizing that no simulator of any kind is required in order to make a run of genetic programming in the first place. As an alternative to simulation, we could interrupt the run of genetic programming at each moment during each generation when the fitness of a candidate controller is required. We could then go to a lab bench and construct a physical embodiment of the to-be-evaluated controller. We could then measure the actual operational behavior of this physical controller as it attempts to control the given plant. We could then ascertain the numerical value of fitness for each physical controller. We could then communicate the controller's fitness to the run of genetic programming by going to the keyboard and entering this single numerical value into the computer. This single numerical value is the only information that the genetic programming system ever receives about a candidate controller. We would then restart the run of genetic programming on the computer. In practice, the difference between building thousands of physical controllers on a lab bench and using a software simulator is a matter of accuracy, precision, elapsed time, and cost (in terms of effort, time, and money).

No one would argue that merely observing, measuring, and recording the position and velocity of a falling apple is equivalent to acquiring knowledge (in the sense that this term is used in the field of artificial intelligence) of Newton's laws. There is simply no knowledge contained in such measurements. Similarly, observing, measuring, and recording the behavior and characteristics of a controller (whether physically building it or by simulating it with software) does not convey any knowledge about the field of control engineering to the observer or genetic programming.

Tenth, notice that no domain knowledge about the field of control is incorporated into the choice of control parameters for the two-lag plant problem. This book continues the policy of *Genetic Programming* (Koza 1992a), *Genetic Programming II* (Koza 1994a), and *Genetic Programming III* (Koza, Bennett, Andre, and Keane 1999a) of using a substantially fixed set of default values for most of the minor control parameters on all problems. Appendix B provides details on the control parameters for each problem in this book. The most common reason for the occasional minor differences in control parameters is that the probability of performing each of the genetic operations must be slightly adjusted to reflect the presence or absence of the architecture-altering operations. Although we do make occasional other changes in control parameters, these changes are made on a time scale measured in *years* and are, in any case, not related to the exigencies of any particular problem. Although certain particular problems in this book could possibly be solved more efficiently with different choices of the control parameters, the current published literature on the topic of optimizing control parameters is, alas, replete with conflicting advice on how to choose control parameters. We believe that our policy of substantial consistency in the choice of control parameters helps the reader eliminate superficial concerns that the demonstrated success of genetic programming depends on shrewd or fortuitous choices of the control parameters.

In summary, the information contained in the preparatory steps in this book required to launch runs of genetic programming is usually elementary, platitudinous, and *de minimus* information about the field involved. In fact, the information contained in the preparatory steps hardly warrants being called "knowledge" or "domain knowledge" in the sense that those terms are ordinarily used in the fields of artificial intelligence, expert systems, and other knowledge-based methods.

Of course, a small amount of knowledge is not the same as none. Neither the authors of this book (nor any other researchers in the field of genetic programming of whom we are aware) have ever suggested that genetic programming operates without any domain knowledge whatsoever. What we have said is that the amount of human-supplied information contained in the preparatory steps for a run of genetic programming is very small. We have also made the point that the amount of domain knowledge required by the preparatory steps for a run of genetic programming is usually considerably less (and, in any case, certainly not greater) than that required by any other method of artificial intelligence and machine learning of which we are aware.

We have also said that we know of no automated problem-solving method that does not start with human-specified primitive elements of some kind, does not incorporate some method for specifying what needs to be done to guide the method's operation, does not require administrative parameters, and does not contain a termination criterion of some kind.

Moreover, when we assert that the amount of human-supplied domain knowledge required by the preparatory steps prior to launching a run of genetic programming is small, we are not implying that there would be anything wrong if it were large. There is nothing wrong with a problem-solving method that employs substantial amounts of domain knowledge.

What we have said is that *it is important to soberly evaluate the value added by an automated problem-solving method.* This is accomplished by carefully comparing that which is supplied by the human employing the method and that which is actually delivered by the automated operation of the method. High added value can be realized in many different ways. An automated method may start with a small amount of information and deliver a significant result. Or, an automated method may start with a substantial amount of domain knowledge and produce a correspondingly greater result.

Given our dual orientation (mentioned at the beginning of this section), the practicing engineer who is primarily interested in solving problems should bear with us while we belabor seemingly hair-splitting questions such as the amount of domain knowledge contained in a particular function and terminal set; whether negative feedback is, or is not, handed to genetic programming on a silver platter; whether the receipt by genetic programming of a single numerical value (fitness) is somehow equivalent to transferring domain knowledge about the field of control engineering to the genetic programming system; whether use of simulation software constitutes a transfer of domain knowledge to the genetic programming system; or whether some minor change in the choice of control parameters or minor change in technique indirectly provides useful knowledge to a run of genetic programming. These issues (largely irrelevant to the practicing engineer interested in solving a particular problem) are important in sharpening the boundary between that which is delivered by genetic programming and that which is supplied by the human user.

3.7.2 Results for the Two-Lag Plant

A run of genetic programming starts with the creation of an initial population of individual program trees. Each program tree in the population is composed of the functions and terminals established by the preparatory steps. Each program tree has the architecture established by the preparatory steps. In addition, each program tree (in this book) is constructed in accordance with a constrained syntactic structure.

The initial random population of a run of genetic programming is a blind random parallel search of the space of computer programs defined by the preparatory steps. For non-trivial problems, the individuals in the population at generation 0 are invariably poor in terms of satisfying the problem's requirements.

Some of the randomly created controllers in the initial population pay no attention to what is happening (i.e., the plant output) or what is desired (i.e., the reference signal). For example, the only time-domain signal processed by the following individual is the useless CONSTANT_0 time-domain signal (in bold).

```
(LEAD (LAG (LAG2 (CONSTANT_0) 0.991705 0.333394)
(+ 0.403004 -0.438887)) (+ (+ 0.720283 0.273918)
(+ 0.482394 -0.075069)))
```

The controller below considers the plant output, but not the reference signal.

```
(INVERTER PLANT_OUTPUT)
```

The controller below considers the reference signal, but not the plant output.

```
(DIFFERENTIATOR (DIFFERENTIATOR REFERENCE_SIGNAL))
```

The controller below contains an internal feedback loop in which the controller's own output (in bold) is an input to the controller itself. However, this controller ignores both the plant output and reference signal.

```
(LAG2 (DIFFERENTIATOR (DIFFERENTIATOR CONTROLLER_
OUTPUT))(- (+ 0.335119 0.720933)(- 0.724019 0.773270))
-0.194707)
```

Other controllers from generation 0 consider the plant output and the fed-back controller output (but not the reference signal) or the reference signal and the fed-back controller output (but not the plant output).

Some of the program trees contain hundreds of points whereas others are small.

Although the results of blind random search are invariably poor for non-trivial problems, some individuals are better than others.

The best individual from generation 0 of our one run of this problem has a fitness of 8.26. The LISP S-expression for this individual is shown below. For brevity, the (equivalent) value 62.8637 replaces the 29-point arithmetic-performing subtree that establishes the amplification factor for the GAIN function.

```
(GAIN
  (DIFFERENTIATOR
    (DIFFERENTIAL_INPUT_INTEGRATOR
      (LAG REFERENCE_SIGNAL 0.708707)
     PLANT_OUTPUT
     )
  )
62.8637)
```

One of the noteworthy features of the best-of-generation individual from generation 0 is that it (unlike many of its cohorts in generation 0) considers both what is happening (i.e., the plant output) and what is desired (i.e., the reference signal).

Because the differentiation offsets the differential input integration, the net effect is that the plant output is subtracted from a lagged reference signal and then multiplied by 62.8637.

Although this controller does not employ conventional negative feedback, it bears some resemblance to negative feedback. If the LAG function were not present, this individual would be equivalent to the most rudimentary part of a PID controller, namely the proportionate (P) part.

In spite of its shortcomings, this rudimentary and flawed controller outperforms the 65,999 other individuals in generation 0 on the problem at hand.

This individual (along with other individuals from generation 0) provides the starting point for the evolutionary process that will unfold during subsequent generations of the run. In particular, the subtraction of the plant's output from the lagged reference signal provides a small toehold from which true negative feedback may later evolve. Moreover, rudimentary P-type control provides a foundation on which more sophisticated types of control (e.g., D-type control, I-type control) may be added later during the run.

Generation 1 (and each subsequent generation of a run of genetic programming) is created from the population of the preceding generation by performing the reproduction, crossover, mutation, and architecture-altering operations on individuals selected probabilistically from the population on the basis of fitness.

The evolutionary process exploits the small differential advantages possessed by the more fit individuals in the population. Over successive generations, both the average fitness of all individuals in the population and the fitness of the best individual in the population tend to improve.

Sixty percent of the controllers in generation 0 cannot be simulated by SPICE. These unsimulatable controllers receive a high penalty value of fitness (10^8) and thereby immediately become the worst-of-generation controllers for their generation. Significantly, the percentage of unsimulatable controllers drops to 14% by generation 1 and 8% by generation 10. That is, the vast majority of the offspring created by genetic programming are simulatable after just a few generations. This phenomenon (reflective of the cumulative effect of the pressure exerted by the selection of individuals based on fitness) is typical of all other runs of problems of controller synthesis in this book. In fact, this same phenomenon (of increased simulatability) has been repeatedly observed in connection with problems of automatic synthesis of electrical circuits (Koza, Bennett, Andre, and Keane 1999a and chapter 4 of this book).

The level of fitness for the best and average individuals of generation 0 provides a useful baseline for comparing the progress that occurs in subsequent generations.

The best-of-run individual emerges in generation 32. It has a near-zero fitness of 0.1639.

Table 3.2 shows the contribution of each of the 10 elements of the fitness measure to the total fitness (0.1639) of the best-of-run controller from generation 32 for the two-lag plant problem. The step size is in Volts. The time constant, τ, is in seconds.

Figure 3.8 shows the best-of-run controller in the form of a block diagram. In this figure, $R(s)$ is the reference signal; $Y(s)$ is the plant output; and $U(s)$ is the controller's output.

Figure 3.9 compares the time-domain response of the best-of-run controller (triangles in the figure) to a 1-Volt unit step with the time-domain response (circles) of the Dorf and Bishop controller. This comparison is made with $K = 1$ and $\tau = 1$. The best-of-run controller causes the plant output to reach a maximum value of 1.0106 at 369 milliseconds (1.06% overshoot). The Dorf and Bishop controller causes the plant to reach a maximum value of 1.020054 at 577 milliseconds (2% overshoot).

Table 3.2 Contributions of the 10 elements of the fitness measure for the best-of-run controller for the two-lag plant problem

	Step size	Internal gain, K	Time constant, τ	Fitness
1	1	1	1.0	0.0220
2	1	1	0.5	0.0205
3	1	2	1.0	0.0201
4	1	2	0.5	0.0206
5	10^{-6}	1	1.0	0.0196
6	10^{-6}	1	0.5	0.0204
7	10^{-6}	2	1.0	0.0210
8	10^{-6}	2	0.5	0.0206
9	Spiked reference signal			0.0
10	AC sweep			0.0
Total fitness				0.1639

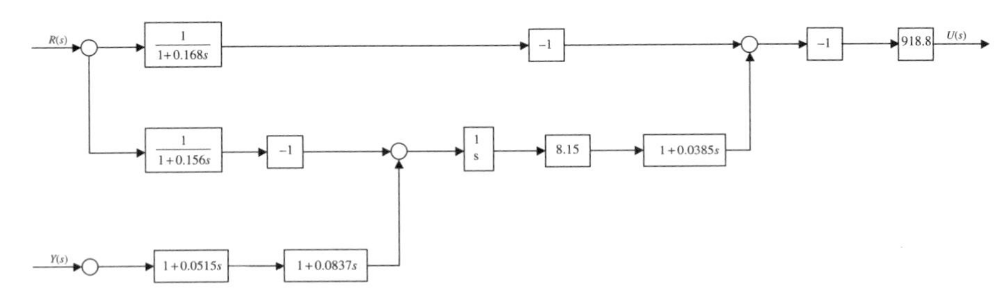

Figure 3.8 Best-of-run controller for the two-lag plant problem.

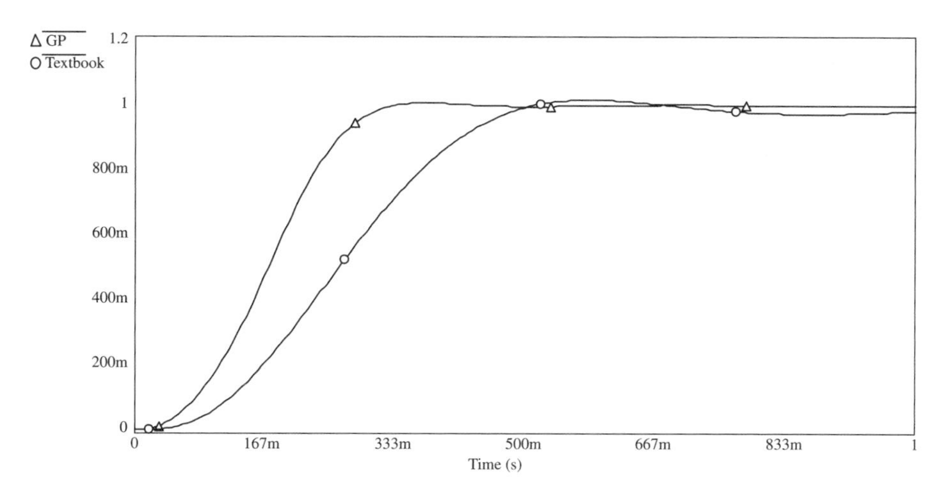

Figure 3.9 Comparison of the time-domain response to a 1-Volt step input for the best-of-run controller (triangles) and the Dorf and Bishop controller (circles) for the two-lag plant problem.

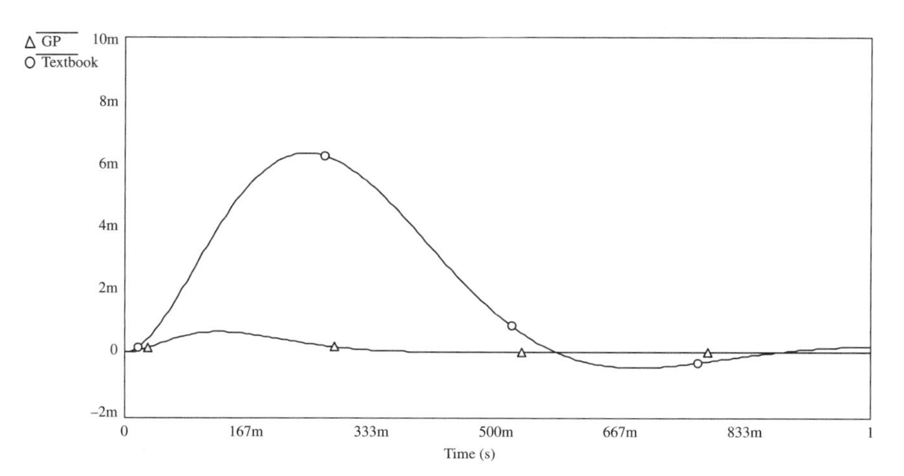

Figure 3.10 Comparison of the time-domain response to a 1-Volt disturbance signal of the genetically evolved controller (triangles) and the Dorf and Bishop controller (circles) for the two-lag plant.

The curves for all other combinations of values of K and τ are similar to figure 3.9 in that they favor the genetically evolved controller.

A practical controller must have certain behaviors and characteristics above and beyond those explicitly incorporated into the fitness measure for this particular problem involving a two-lag plant. For example, a controller should be reasonably oblivious to disturbances that may be added to the controller output. That is, the system should reject disturbance.

Figure 3.10 compares the effect of disturbance on the best-of-run controller (triangles) and the controller (circles) presented in Dorf and Bishop. This comparison is made with $K = 1$ and $\tau = 1$. The upper curve in the figure is the time-domain response to a 1-Volt disturbance signal for the Dorf and Bishop controller. The upper curve has

Table 3.3 Comparison of performance of the best-of-run controller and the Dorf and Bishop controller for the two-lag plant problem

	Units	Genetically evolved controller	Dorf and Bishop controller
ITAE	millivolt sec^2	19	46
Disturbance sensitivity	microvolts/Volt	644	5,775
Rise time	milliseconds	296	465
Settling time	milliseconds	304	944
System bandwidth (3 dB)	Hertz	1.5	1

a peak value of response to a 1-Volt disturbance signal of 5,775 microvolts. The lower curve (whose peak is only 644 microvolts) applies to the best-of-run controller from generation 32. That is, the genetically evolved controller rejects disturbance more effectively than the Dorf and Bishop controller.

Table 3.3 compares the performance (averaged over the eight combinations of values for K, τ, and the step size of the reference signal) of the best-of-run controller and the Dorf and Bishop controller.

As can be seen in table 3.3, the best-of-run controller is 2.42 times better than the Dorf and Bishop controller as measured by the integral of the time-weighted absolute error (the criterion used by Dorf and Bishop) and is 8.97 times better in terms of suppressing the effects of disturbance at the plant input.

The rise time for the best-of-run controller is 296 milliseconds. This is 64% of the rise time for the Dorf and Bishop controller (465 milliseconds).

The settling time for the best-of-run controller is 304 milliseconds. The Dorf and Bishop controller first reaches the 98% level at 477 milliseconds; however, it rings and subsequently falls below the 98% level. In fact, it does not settle until 944 milliseconds. The settling time for the genetically evolved controller is 32% of that for the Dorf and Bishop controller.

The *system bandwidth* is the frequency of the reference signal above which the plant output is attenuated by at least a specified degree (3 dB here) in comparison to a specified low frequency (e.g., DC or a very low frequency). The bandwidth of both controllers (approximately 1 Hz) is well below the acceptable level.

All the program trees in generation 0 consist of one result-producing branch (but no automatically defined functions). During the run, the architecture-altering operations may add automatically defined functions into particular program trees in the population.

The best-of-run individual from generation 32 has one result-producing branch and three automatically defined functions. The result-producing branch has 145 points (i.e., functions and terminals). The result-producing branch contains one occurrence of the `REFERENCE_SIGNAL`. In fact, this occurrence of `REFERENCE_SIGNAL` is the sole means of making the reference signal available to the genetically evolved controller. Although the result-producing branch contains two occurrences of `PLANT_OUTPUT` (in bold), these two occurrences of the plant's output are semantically irrelevant because they are subtracted from each other. The result-producing branch calls on automatically defined function `ADF0` once (underlined).

```
(GAIN (SUB_SIGNAL (ADD_SIGNAL (INVERTER (ADD_SIGNAL
(INVERTER (LAG REFERENCE_SIGNAL -0.774818)) (ADF0))) (LEAD
(DIFFERENTIATOR (LEAD (DIFFERENTIAL_INPUT_INTEGRATOR
PLANT_OUTPUT PLANT_OUTPUT) (- 0.553103 (- -0.979905
0.159026)))) (- (- (+ 0.338199 0.958669) (+ -0.719664
-0.357372)) (- 0.068355 0.059952)))) (INVERTER (LAG (GAIN
(LAG2 (LAG2 (CONSTANT_0) -0.743520 0.480632) (+ -0.130435
0.628778) (+ -0.627210 -0.206744)) (- (- 0.359068 (- (-
0.127450 0.661750) (+ 0.799922 0.546706))) (+ 0.136226
0.841644))) (+ (+ (- -0.047550 -0.930215) (- -0.381294
-0.218464)) (+ (- 0.185214 0.661750) (- (+ -0.376335
(+ -0.695582 -0.774818)) 0.287808)))))) (- (+ (+ (- (-
(+ 0.338199 0.958669) (+ -0.719664 -0.357372)) (- (+ (+
-0.047550 (+ -0.078200 (- (+ (+ 0.356728 -0.569439)
0.223178) (+ 0.287808 (+ -0.545314 -0.357372)))))
0.099732) (+ (- -0.271322 0.197696) -0.774818))) (+ (+
(+ 0.579470 0.727545) (+ -0.058393 -0.058393)) (+ (+
0.059952 0.139516) (+ 0.127450 0.672953)))) (+ (+ (-
(+ 0.002680 0.779938) (+ 0.808460 0.786740)) (+ (+ 0.836821
0.036504) (- -0.039999 -0.381294))) 0.127450)) 0.287106))
```

Automatically defined function ADF0 consists of only one point. It is merely a call to ADF2.

```
(ADF2)
```

Automatically defined function ADF2 has 41 points. It contains a reference to the PLANT_OUTPUT (in bold). ADF2 is, in fact, the sole means of making the plant output available to the genetically evolved controller.

```
(LEAD (ADD_SIGNAL (GAIN (DIFFERENTIAL_INPUT_INTEGRATOR (LEAD
(LEAD PLANT_OUTPUT (+ (+ 0.331753 -0.813045) (- -0.448149
0.359068))) (+ -0.719664 -0.357372)) (ADF1)) (- (-0.773578
0.059952) -0.197462)) (CONSTANT_0)) (+ -0.961429 (- (+
-0.629573 0.801392) (+ -0.495284 (- -0.130435 (- (+
-0.310777 -0.988344) (+ 0.601379 -0.649952)))))))
```

Automatically defined function ADF1 has 13 points. However, ADF1 (below) is not called by any other branch and therefore contributes nothing to the genetically evolved controller.

```
(LAG REFERENCE_SIGNAL (+ (+ -0.390529 (+ (- -0.382748
-0.853353) (- -0.558397 -0.021241))) -0.349639))
```

As previously mentioned, the LISP S-expression for each individual in the population at each generation of the run is converted into a netlist that is submitted to the SPICE simulator.

The input to the SPICE simulator is a text file. Table 3.4 shows the SPICE input file (containing the netlist and certain commands) for the best-of-run individual for the two-lag plant problem. This table does not include the subcircuit definitions previously shown in table 3.1.

Table 3.4 SPICE input file for the best-of-run controller for the two-lag plant problem

```
X1      6 2000 16             LAG_SUBCKT
V2      2000 0                0.167951V
B3      17 0                  V=-V(16)
B4      18 0                  V=V(17)+V(11)
B5      19 0                  V=-V(18)
X6      2 2 20                DII_SUBCKT
X7      20 2001 21            LEAD_SUBCKT
V8      2001 0                49.2078V
X9      21 22                 DIFFB
X10     22 2002 23            LEAD_SUBCKT
V11     2002 0                232.007V
B12     24 0                  V=V(19)+V(23)
X13     0 2003 2004 25        LAG2_SUBCKT
V14     2003 0                0.180501V
V15     2004 0                3.02435V
X16     25 2005 2006 26       LAG2_SUBCKT
V17     2005 0                3.15024V
V18     2006 0                0.14657V
B19     27 0                  V=V(26)*18.2863
X20     27 2007 28            LAG_SUBCKT
V21     2007 0                0.0128456V
B22     29 0                  V=-V(28)
B23     30 0                  V=V(24)-V(29)
B24     1 0                   V=V(30)*918.803
X25     6 2008 12             LAG_SUBCKT
V26     2008 0                0.156057V
X27     2 2009 33             LEAD_SUBCKT
V28     2009 0                0.0514625V
X29     33 2010 34            LEAD_SUBCKT
V30     2010 0                0.0837459V
X31     34 12 35              DII_SUBCKT
B32     36 0                  V=V(35)*8.14868
B33     37 0                  V=V(36)+V(0)
X34     37 2011 11            LEAD_SUBCKT
V35     2011 0                0.0385089V
X36     1 2012 2013 2014      LIMIT_SUBCKT
B37     2015 0                V=V(2012)* 1
X38     2015 2016 3           LAG_SUBCKT
X39     3 2016 2              LAG_SUBCKT
V40     2016 0                DC 1
V41     2013 0                DC 40
V42     2014 0                DC -40
X43     6 2 6 7               ITAE_SUBCKT
VP44    6 0                   PULSE(0.0 1
+ 0.1 0.001 0.001 10 15)
.TRAN 0.08 9.6 0.0 0.04 UIC
.PLOT TRAN V(7)
.END
```

The genetically evolved solution to this problem was evolved on the first and only run of the problem (after debugging runs).

As in most problems in this book, most of the computer time was consumed by the evaluation of fitness of candidate individuals in the population—not the genetic operations or other executional steps of genetic programming. The fitness evaluation (involving 10 separate SPICE simulations) averaged 2.57×10^9 computer cycles (4.8 seconds) per individual (on a 533-MHz computer).

In summary, we have demonstrated that genetic programming can be used to automatically create both the parameter values (tuning) and the topology for a controller for the two-lag plant problem.

The genetically evolved controller can be most directly compared to the controller developed by Dorf and Bishop by viewing the controller as a pre-filter and a compensator. Figure 3.11 presents a model for the entire system that is helpful in making this comparison. In this figure, the reference signal, $R(s)$, is fed through pre-filter $G_p(s)$. The plant has one output, namely $Y(s)$. The plant output, $Y(s)$, is passed through $H(s)$ and then subtracted, in continuous time, from the pre-filtered reference signal emanating from $G_p(s)$. The difference (error) is fed into the compensator $G_c(s)$. $G_c(s)$ has one input (the difference) and one output $U(s)$. Disturbance, $D(s)$, is added to the output, $U(s)$, of compensator $G_c(s)$. The resulting sum is subjected to a limiter (in the range between -40 and $+40$ Volts) prior to entering the plant.

For the Dorf and Bishop controller, $H(s) = 1$ and the transfer function for the pre-filter, $G_{p-dorf}(s)$, is

$$G_{p\text{-}dorf}(s) = \frac{42.67}{42.67 + 11.38s + s^2}.$$

The transfer function for the PID compensator, $G_{c-dorf}(s)$, of the Dorf and Bishop controller is

$$G_{c\text{-}dorf}(s) = \frac{12(42.67 + 11.38s + s^2)}{s}.$$

After applying standard block diagram manipulations, the transfer function for the genetically evolved best-of-run controller from generation 32 can also be expressed as a transfer function for a pre-filter and a transfer function for a compensator. The transfer function for the pre-filter, $G_{p32}(s)$, for the best-of-run controller is

$$G_{p32}(s) = \frac{1(1+0.1262s)(1+0.2029s)}{(1+.03851s)(1+.05146s)(1+.08375)(1+.1561s)(1+.1680s)}.$$

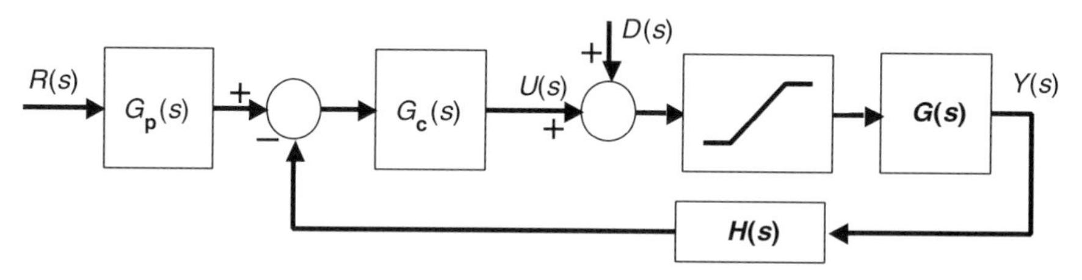

Figure 3.11 Overall model with a pre-filter $G_p(s)$ and a compensator $G_c(s)$.

For the best-of-run controller, $H(s) = 1$ and the transfer function for the compensator, $G_{c32}(s)$, is

$$G_{c32}(s) = \frac{7487(1 + .03851s)(1 + .05146s)(1 + .08375s)}{s}$$
$$= \frac{7487.05 + 1300.63s + 71.2511s^2 + 1.2426s^3}{s}.$$

The s^3 term (in conjunction with the s in the denominator) indicates a second derivative. In fact, the evolved compensator is a PID-D2 (proportional, integrative, derivative, and second derivative) controller.

Although derivatives are not helpful in controlling certain plants (because they may amplify high frequency effects such as noise), their use is appropriate here because there are no such possibly disadvantageous effects in this particular problem.

For additional information on this problem, see Koza, Keane, Yu, Bennett, Mydlowec 2000 and Koza, Keane, Bennett, Yu, Mydlowec, and Stiffelman 1999.

Dorf and Bishop solved this problem by seeking the parameter values (tuning) for a PID-type controller. A PID controller is a good type of controller for a two-lag plant (and, indeed, many other plants).

In contrast, the run of genetic programming was not limited to considering only PID-type controllers. The consequence of this more open-ended approach was that genetic programming produced a PID-D2 controller.

Genetic programming is a search that is guided by the necessities articulated by the problem's fitness measure. When performing the preparatory steps for the run of genetic programming of this problem, the human user did not preordain that the first derivative, much less a second derivative, should be incorporated into the to-be-created controller. The evolutionary process discovered that a second derivative was helpful in achieving a better value of fitness.

Similarly, the human user did not preordain any particular topological arrangement of proportional, integrative, derivative, second derivative, or other functions within the to-be-created controller. Instead, genetic programming automatically created a controller for the problem without the benefit of any user-supplied information concerning the total number of processing blocks to be employed in the controller, the type of each processing block, the topological interconnections between the blocks, the values of the numerical parameters for the blocks, or the existence of internal feedback (none in this instance) within the controller.

Genetic programming starts each run as a new adventure and is free to innovate in any manner that may satisfy the problem's high-level requirements. Genetic programming is not encumbered by the preconceptions that often channel human thinking along well-trodden paths. Human engineers tend to look at problems in particular ways—often based on ideal mathematical models (e.g., the PID controller topology). In contrast, genetic programming doesn't know anything about the problem's underlying mathematics. It simply tries to produce a sequence of incrementally improved results. Thus, we frequently see creative things come out of the evolutionary process. In this particular run, genetic programming rediscovered the PID-D2 controller. As will be seen in numerous later places in this book, genetic programming often discovers solutions that might never occur to human designers.

3.7.3 Human-Competitiveness of the Result for the Two-Lag Plant Problem

Harry Jones of The Brown Instrument Company of Philadelphia patented the PID-D2 controller topology in 1942. The PID-D2 controller was an improvement over the PID controller patented by Callender and Stevenson in 1939 (discussed in section 3.3).

As Jones (1942) states,

> "A ... specific object of the invention is to provide electrical control apparatus ... wherein the rate of application of the controlling medium may be effected in accordance with or in response to the first, second, and high derivatives of the magnitude of the condition with respect to time, as desired."

Claim 38 of the Jones 1942 patent states,

> "In a control system, an electrical network, means to adjust said network in response to changes in a variable condition to be controlled, control means responsive to network adjustments to control said condition, reset means including a reactance in said network adapted following an adjustment of said network by said first means to initiate an additional network adjustment in the same sense, and rate control means included in said network adapted to control the effect of the first mentioned adjustment in accordance with the second or higher derivative of the magnitude of the condition with respect to time."

Because the best-of-run controller from generation 32 has proportional, integrative, derivative, and second derivative blocks, it infringes the 1942 Jones patent. That is, the genetically evolved controller reads on all elements of claim 38.

The legal criteria for obtaining a U.S. patent are that the proposed invention be "new," "useful," "improved," and that

> "... the differences between the subject matter sought to be patented and the prior art are such that the subject matter as a whole would [not] have been obvious at the time the invention was made to a person having ordinary skill in the art to which said subject matter pertains." (35 *United States Code* 103a)

Referring to the eight criteria in table 1.2 for establishing that an automatically created result is competitive with a human-produced result, the rediscovery by genetic programming of the PID-D2 controller satisfies the following two of the eight criteria:

(A) The result was patented as an invention in the past, is an improvement over a patented invention, or would qualify today as a patentable new invention.

(F) The result is equal to or better than a result that was considered an achievement in its field at the time it was first discovered.

The rediscovery by genetic programming of the PID-D2 controller came about six decades after Jones received a patent for his invention. Nonetheless, the fact that the original human-designed version satisfied the Patent Office's criteria for patent-worthiness means that the genetically evolved duplicate would also have satisfied the Patent Office's criteria for patent-worthiness (if only it had arrived earlier than Jones' patent application).

The fact that genetic programming rediscovered both the topology and sizing of a controller that was unobvious "to a person having ordinary skill in the art" establishes that this evolved result satisfies Arthur Samuel's criterion (1983) for artificial intelligence and machine learning:

> "[T]he aim [is]... to get machines to exhibit behavior, which if done by humans, would be assumed to involve the use of intelligence."

Note that a PID-D2 controller is also evolved in section 9.1.2.

3.7.4 AI Ratio for the Two-Lag Plant Problem

In section 1.1.2, we mentioned that an automated problem-solving method may be measured by the ratio of that which is delivered by the automated operation of the *artificial* method to the amount of *intelligence* that is supplied by the human applying the method to a particular problem.

What is the AI ratio for the solution produced by genetic programming to the two-lag plant problem?

Ascertaining the AI ratio for this problem requires measuring the amount of "A" that is delivered by the automated problem-solving method (i.e., genetic programming) in relation to the amount of "I" that is supplied by the human user (i.e., in the form of the preparatory steps prior to the launch of the run of genetic programming). In this section (and other similar sections throughout this book), "A" and "I" are evaluated in a qualitative way.

Regarding the amount of "A," the result produced by genetic programming is considered to be human-competitive for the two reasons mentioned in the previous section (section 3.7.3). Because the result is human-competitive, there is a high amount of "A" in the numerator of the AI ratio.

Measuring the amount of "I" requires the discipline of drawing a bright line between that which is delivered by the artificial system and that which is supplied by the intelligent human user. We imposed this discipline in section 3.7.1 where we detailed the preparatory steps for this problem.

As discussed in section 3.7.1.7, we did not employ any deep knowledge about controllers in selecting the terminals and functions to be used by genetic programming in synthesizing the design of the required controller.

The terminal set (established in section 3.7.1.2) permits the construction of any controller whose input consists of a reference signal, the plant output, and the controller's own output.

The function set (established in section 3.7.1.3) permits the construction of any controller composed of the signal-processing blocks ordinarily found in controllers (e.g., integrators, differentiators, gains, adders, subtractors, leads, lags, and differential input integrators).

Certainly no knowledge of 1939 Callender and Stevenson patent on the PID controller, the 1942 Jones patent on the PID-D2 controller, or any other significant domain knowledge from the field of control was employed in these two preparatory steps. In fact, the terminal set and the function set incorporate only platitudinous information about controllers.

The fitness measure (section 3.7.1.4) is constructed directly from the high-level statement of the problem's requirements. The statement of the problem calls for minimizing the integral of the time-weighted absolute error, keeping overshoot small, achieving robustness in the face of significant variation in the plant's internal gain and time constant, limiting the plant's input to non-extreme values, and limiting the system bandwidth.

The remaining preparatory steps (concerning control parameters, the termination criterion, and the method of result designation) were unremarkable administrative steps.

In summary, the preparatory steps were straightforward and uncomplicated translations of a high-level statement of the problem. That is, the human user supplied only *de minimus* knowledge about controllers. Hence, there is only a small amount of "I" in the denominator of the AI ratio.

The high amount of "A" in the numerator of the AI ratio in conjunction with the small amount of "I" in the denominator means that the AI ratio for the genetically evolved solution to the two-lag plant problem is high.

3.8 Three-Lag Plant

A three-lag plant is similar to the plant of figure 3.7, except that there is one additional lag block. The transfer function of the three-lag plant is

$$G(s) = \frac{K}{(1 + \tau s)^3},$$

where K is the plant's internal gain and$\tau$is the plant's time constant.

The problem in this section is to create both the topology and tuning for a robust controller for a three-lag plant that satisfies the following high-level requirements:

- The plant output is to reach the level of the reference signal while minimizing the integral of the time-weighted absolute error.
- The overshoot in response to a step input is to be less than 2%.
- The controller is to be robust in the face of significant variation in the plant's internal gain and the plant's time constant. In this problem, the nominal value of the plant's internal gain, K, is 1.0 and the nominal value of the plant's time constant, τ, is 1.0.
- The input to the plant is limited to the range between -10 and $+10$ Volts.

The reader may recall that the fitness measure for the two-lag plant problem (section 3.7.1.4) did not explicitly include measuring disturbance rejection. Nonetheless, the genetically evolved controller had a good value of disturbance rejection (section 3.7.2). In order to demonstrate the ease of combining optimization requirements, time-domain constraints, frequency-domain constraints, and robustness requirements into a fitness measure, the fitness measure for this problem is slightly different from that used on the two-lag plant in that it explicitly includes disturbance rejection.

Åström and Hägglund (1995, page 225) developed a PID compensator preceded by a lowpass pre-filter that delivers credible performance on this problem.

As will be seen (section 3.8.2), the controller produced by genetic programming is better than 7.2 times as effective as the Åström and Hägglund controller for this problem as measured by the integral of the time-weighted absolute error. Relative to the Åström and Hägglund controller, the genetically evolved controller has only 50% of the rise time, has only 35% of the settling time, and is 92.7 dB better in terms of suppressing the effects of a step disturbance at the plant input.

3.8.1 Preparatory Steps for the Three-Lag Plant

3.8.1.1 Program Architecture The architecture of the to-be-evolved program is automatically determined during the run of genetic programming using the architecture-altering operations in the manner described in section 3.7.1.1 for the two-lag plant problem.

3.8.1.2 Terminal Set Except for the value-setting subtrees, the terminal set for every part of the result-producing branch and the automatically defined functions is the same as that used in the two-lag plant described in section 3.7.1.2.

The numerical parameter value for each signal-processing function possessing a parameter is established by a value-setting subtree containing a single perturbable numerical value. This approach for establishing numerical parameter values is described in section 3.5.5.2. That is, the terminal set, $\mathsf{T}_{\mathrm{vss}}$, for the value-setting subtrees is

$$\mathsf{T}_{\mathrm{vss}} = \{\Re_{\mathrm{p}}\},$$

where $\Re_{\mathrm{p}}$ denotes a perturbable numerical value.

Our use of this approach (which differs from the approach used for the two-lag plant problem in section 3.7.1.2) reflects the chronology of our work and the evolution of our thinking about the efficiency of establishing numerical parameter values (as opposed to any special requirement or exigency of the present problem).

3.8.1.3 Function Set The function set for every part of the result-producing branch and the automatically defined functions for this problem is the same as that used in the two-lag plant problem (section 3.7.1.3), except that the function set for the value-setting subtrees is empty (because each value-setting subtree contains only a single perturbable numerical value).

3.8.1.4 Fitness Measure Fitness is measured using 10 elements, including

(1) eight time-domain-based elements based on the integral of time-weighted absolute error, robustness, and overshoot,
(2) one time-domain-based element measuring the controller's response when it is faced with a spiked reference signal, and
(3) one time-domain-based element measuring disturbance rejection.

The fitness of an individual controller is the sum of the (detrimental) contributions of these 10 elements. The smaller the sum, the better.

The first nine elements are the same as for the two-lag plant problem (section 3.7.1.4).

The 10th element of the fitness measure for the present problem is based on disturbance rejection. This element is computed based on a time-domain analysis that runs for 9.6 seconds. With the reference signal being held at a value of 0, a disturbance signal is added to the controller's output at time $t = 0$. The disturbance signal is a unit step function. The resulting disturbed signal then becomes the input to the plant. The detrimental contribution to fitness is the absolute value of the largest single difference between the plant output and the reference signal (which is invariant at 0 throughout).

The number of hits is defined as the number of elements (0 to 6) for which the individual controller has a smaller (better) contribution to fitness than the controller designed using the techniques of Åström and Hägglund (1995). Note that Åström and Hägglund start by taking measurements of the plant in the field (as opposed to mathematically analyzing the plant's transfer function).

3.8.1.5 Control Parameters The population size is 66,000.

As shown in appendix B, the use of perturbable numerical values in the value-setting subtrees (section 3.8.1.2) slightly changes the percentages of the various genetic operations.

3.8.2 Results for the Three-Lag Plant

The best individual from generation 0 of our one (and only) run of this problem has a fitness of 14.35.

The best-of-run controller emerges in generation 31 and has a near-zero fitness of 1.14.

Figure 3.12 shows the best-of-run controller from generation 31 in the form of a block diagram. $R(s)$ is the reference signal, $Y(s)$ is the plant output, and $U(s)$ is the controller's output.

Table 3.5 shows, for each of the eight combinations of step size, internal gain, K, and the time constant, τ, the values of disturbance rejection (in microvolts output from the plant per disturbance Volt), the integral of time-weighted error (in Volt-second^2), the closed-loop system bandwidth (in Hertz), the rise time (in seconds) required to reach 90% of the reference signal, and the settling time in seconds (2%) for the best-of-run controller from generation 31 for the three-lag plant.

Table 3.6 shows (for each of the eight combinations of step size, internal gain, K, and time constant, τ) the corresponding values for the PID solution designed using the techniques of Åström and Hägglund (1995) for the three-lag plant.

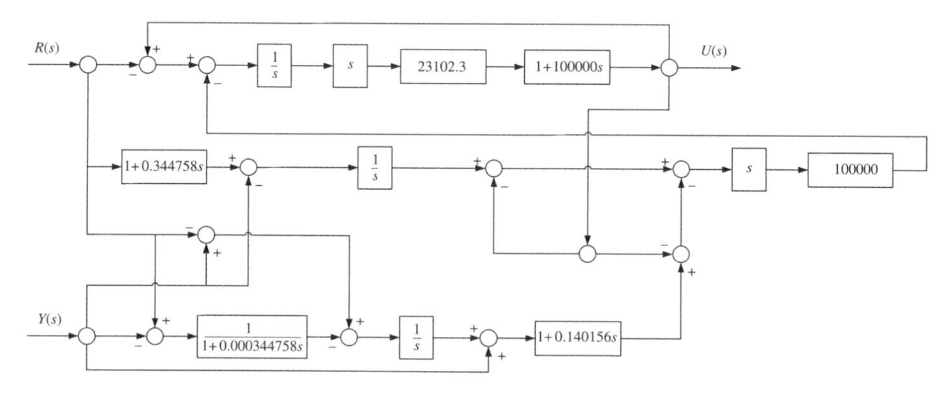

Figure 3.12 Best-of-run controller for the three-lag plant.

Table 3.5 Characteristics of the best-of-run controller for the three-lag plant problem

Step size	K	τ	Disturbance	ITAE	Bandwidth	Rise time	Settling time
1	1	1.0	4.3	0.360	0.72	1.25	1.87
1	1	0.5	4.3	0.190	0.72	0.97	1.50
1	2	1.0	4.3	0.240	0.72	0.98	1.39
1	2	0.5	4.3	0.160	0.72	0.90	1.44
10^{-6}	1	1.0	4.3	0.069	0.72	0.64	1.15
10^{-6}	1	0.5	4.3	0.046	0.72	0.53	0.97
10^{-6}	2	1.0	4.3	0.024	0.72	0.34	0.52
10^{-6}	2	0.5	4.3	0.046	0.72	0.52	0.98
Average			4.3	0.142	0.72	0.77	1.23

Table 3.6 Characteristics of the Åström and Hägglund PID solution to the three-lag plant problem

	Step size	K	τ	Disturbance	ITAE	Bandwidth	Rise time	Settling time
1	1	1	1.0	186,000	2.6	0.248	2.49	6.46
2	1	1	0.5	156,000	2.3	0.112	3.46	5.36
3	1	2	1.0	217,000	2.0	0.341	2.06	5.64
4	1	2	0.5	164,000	1.9	0.123	3.17	4.53
5	10^{-6}	1	1.0	186,000	2.6	0.248	2.49	6.46
6	10^{-6}	1	0.5	156,000	2.3	0.112	3.46	5.36
7	10^{-6}	2	1.0	217,000	2.0	0.341	2.06	5.64
8	10^{-6}	2	0.5	164,000	1.9	0.123	3.17	4.53
Average				180,750	2.2	0.21	2.8	5.5

Table 3.7 Comparison of average performance of the best-of-run controller and the Åström and Hägglund controller for the three-lag plant problem

	Units	Genetically evolved controller	PID controller
Disturbance sensitivity	μVolts/Volts	4.3	180,750
ITAE	millivolt seconds^2	0.142	2.2
System bandwidth (3 dB)	Hertz	0.72	0.21
Rise time	milliseconds	0.77	2.8
Settling time	milliseconds	1.23	5.5

Table 3.7 compares the average values of five metrics of the best-of-run controller from generation 31 for the three-lag plant and the PID controller designed using the techniques of Åström and Hägglund (1995). This average is taken over the eight combinations of values for K, τ, and the step size of the reference signal. As can be seen, the best-of-run controller has superior average values for ITAE, disturbance sensitivity, rise time, and settling time (and has an acceptable value for system bandwidth).

Åström and Hägglund (Åström and Hägglund 1995) did not consider seven of the eight combinations of values for K, τ, and the step size used in computing the averages, whereas we used all eight combinations of values in our run. Accordingly,

Table 3.8 Comparison of performance of the best-of-run controller and the Åström and Hägglund controller for the three-lag plant problem for $K = 1.0$, $\tau = 1.0$, and step size of 1.0

	Units	Genetically evolved controller	PID controller
Disturbance sensitivity	μVolts/Volts	4.3	186,000
ITAE	Volt seconds^2	0.360	2.6
System bandwidth (3 dB)	Hertz	0.72	0.248
Rise time	seconds	1.25	2.49
Settling time	seconds	1.87	6.46

table 3.8 compares the performance of the best-of-run controller for the three-lag plant and the Åström and Hägglund PID controller using the values of the plant's internal gain, the plant's time constant, and the step size of the reference signal specified by Åström and Hägglund. Specifically, the plant's internal gain, K, is 1.0; the plant's time constant, τ, is 1.0; and the step size of the reference signal is 1 Volt.

As can be seen in table 3.8, the best-of-run controller is 7.2 times better than the textbook controller as measured by the integral of the time-weighted absolute error. It has only 50% of the rise time and only 35% of the settling time. It is 92.7 dB better in terms of suppressing the effects of disturbance at the plant input. The genetically evolved controller has 2.9 times the system bandwidth of the PID controller.

For all eight combinations of values for K, τ, and the step size of the reference signal, the genetically evolved controller has superior values for ITAE, disturbance sensitivity, rise time, and settling time.

Figure 3.13 compares the time-domain response of the best-of-run controller (squares) to a 1-Volt unit step with the response of the Åström-Hägglund controller (diamonds). This comparison is made with $K = 1$ and $\tau = 1$. The faster rising curve is the response of the genetically evolved controller.

As can be seen in figure 3.13, the rise time for the best-of-run controller is 1.25 seconds. This is 50% of the 2.49-second rise time for the Åström and Hägglund (Åström and Hägglund 1995) controller.

The settling time for the best-of-run controller is 1.87 seconds. That is, the best-of-run controller settles in 35% of the 6.46-second settling time for the Åström and Hägglund controller.

The maximum value of the plant output produced by the best-of-run controller is 1.023 (achieved at 1.48 seconds). That is, its overshoot is 2.3%. The maximum value of the plant output produced by the Åström and Hägglund controller is 1.074 (achieved at 3.47 seconds). That is, its overshoot is 7.4%.

After applying standard block diagram manipulations, the transfer function for the best-of-run controller from generation 31 can be expressed as a transfer function for a pre-filter and a transfer function for a compensator. The transfer function for the pre-filter, $G_{p31}(s)$, for the best-of-run controller for the three-lag plant is

$$G_{p31(s)} = \frac{(1 + 0.2083s)(1 + 0.0002677s)}{(1 + 0.000345s)}.$$

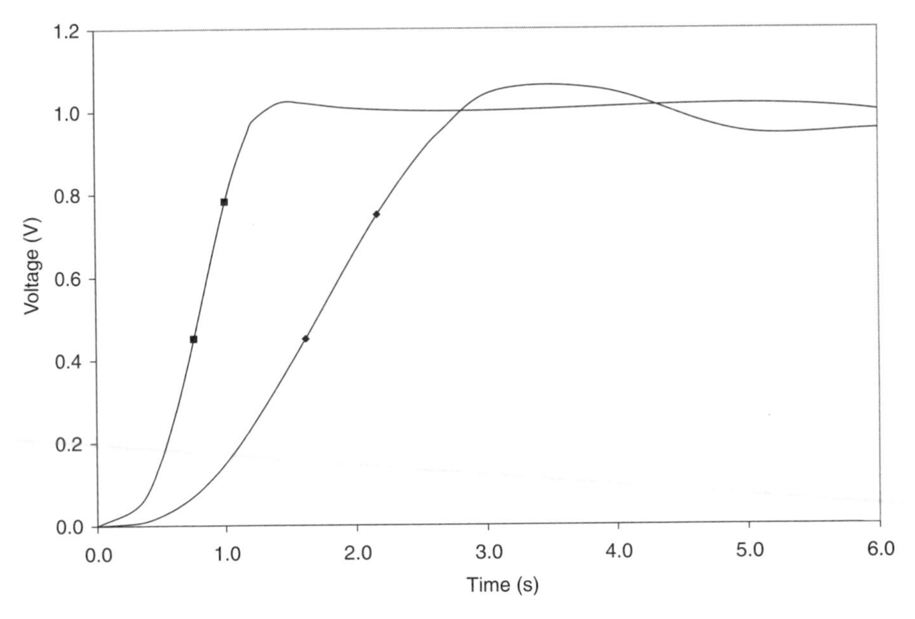

Figure 3.13 Comparison of the time-domain response of the best-of-run controller (squares) and the Åström and Hägglund controller (diamonds) for a 1-Volt unit step for the three-lag plant problem.

The transfer function for the compensator, $G_{c31}(s)$, for the best-of-run controller for the three-lag plant is

$$G_{c31}(s) = 300{,}000.$$

The feedback transfer function, $H_{31}(s)$, for the best-of-run controller for the three-lag plant is

$$H_{31}(s) = 1 + 0.42666s + 0.046703s^2.$$

The best-of-run controller from generation 31 consists of a three-point result-producing branch, a 43-point automatically defined function `ADF0`, an 18-point automatically defined function `ADF1`, a 73-point automatically defined function `ADF2`, and an 18-point automatically defined function `ADF3`. Only `ADF2` is actually called by the result-producing branch. The three-point result-producing branch applies a lead function to the signal produced by `ADF2`. Automatically defined function `ADF2` contains occurrences of the plant's output, the reference signal, and the controller's output.

For additional information, see Koza, Keane, Yu, Bennett, Mydlowec 2000; Koza, Keane, Yu, Bennett, Mydlowec, and Stiffelman 1999; and Koza, Keane, Yu, Mydlowec, and Bennett 2000a.

In summary, we have demonstrated that genetic programming can be used to automatically create both the parameter values (tuning) and the topology for controllers for a three-lag plant. The genetically evolved controller is better than the controller designed and published by experts in the field of control.

3.8.3 Routineness for the Three-Lag Plant Problem

Section 1.1.3 mentions that a problem-solving method may be measured by its routineness. As also mentioned there, a problem-solving method is "routine" if it is general and if relatively little human effort is required to get the method to successfully handle new problems within a particular domain and to successfully handle new problems from a different domain.

The three-lag plant problem in this section and the two-lag plant problem (section 3.7) are, of course, in the same domain—controller synthesis. So, how much effort was required to make the transition from one problem to the other?

The preparatory steps for the three-lag plant problem are, with two exceptions, the same as the preparatory steps for the two-lag plant problem.

First, the fitness measure for the three-lag plant problem implements Åström and Hägglund's statement of the three-lag plant problem whereas the fitness measure for the two-lag plant problem implements Dorf and Bishop's statement of the two-lag plant problem. The change in fitness measures from problem to problem is an inherent and necessary part of making the transition from one problem to another. It does not entail an enormous amount of effort.

Second, we used a different approach for establishing the numerical parameter values for the three-lag plant problem than we did for the two-lag plant problem. This change has a minor effect on the percentages of genetic operations (shown in appendix B). As already mentioned in section 3.8.1.2, this change was entirely optional. This change reflects the chronology of our work and the evolution of our thinking on how to maximize the efficiency of runs of genetic programming (as opposed to any special requirement or exigency of the three-lag plant problem).

Once the preparatory steps are performed, the executional steps of genetic programming for the two-lag plant problem are the same as for the three-lag plant problem.

Thus, relatively little effort is required to make the (intra-domain) transition from the two-lag plant problem to the three-lag plant problem and it is reasonable to say that the transition is routine.

3.8.4 AI Ratio for the Three-Lag Plant Problem

What is the AI ratio for the solution to the three-lag plant problem produced by genetic programming?

Genetic programming produced a controller that was better than the controller designed by Åström and Hägglund (1995).

Although the solution produced by genetic programming for this problem is, in fact, better than a human-produced solution, that fact alone does not qualify the result as "human-competitive" under the intentionally stringent criteria for human-competitiveness enumerated in table 1.2. The fact that a problem appears in a college textbook is not alone sufficient to establish the problem's difficulty or importance. A textbook problem may occasionally be "a problem of indisputable difficulty in its field" (criterion G). It may, or may not, satisfy one or more of the other criteria in table 1.2. However, the three-lag plant problem does not satisfy any of the eight criteria in table 1.2. On the other hand, because the problem is moderately difficult, we assign a moderately high amount of "A" to the solution produced by genetic programming.

The preparatory steps for the three-lag plant problem are (except for the two differences noted in the previous section) the same as the preparatory steps for the two-lag plant problem. Thus, for the same reasons stated in section 3.7.4, there is only a small amount of "I" in the denominator of the AI ratio.

The moderately high amount of "A" represented by the result in conjunction with the small amount of "I" provided by the human user means that the AI ratio for the solution produced by genetic programming is moderately high.

3.9 Three-Lag Plant with a Five-Second Delay

In this section, genetic programming is used to discover both the topology and parameter values for a controller for a three-lag plant with a significant (five-second) time delay in the external feedback from the plant output to the controller.

The transfer function of the plant is

$$G(s) = \frac{Ke^{-5s}}{(1+\tau s)^3},$$

where K is the plant's internal gain and τ is the plant's time constant.

The five-second delay adds to the difficulty of designing an effective controller.

The requirements of the problem are that

- the plant output reaches the level of the reference signal in minimal time (as measured by the integral of the time-weighted absolute error),
- the overshoot in response to a step input is less than 2%,
- the controller is robust in the face of significant variation in the plant's internal gain and the plant's time constant,
- the controller is robust in the face of disturbance (added into the controller's output), and
- the input to the plant is limited to the range between -40 and $+40$ Volts.

A controller is presented in Åström and Hägglund 1995 (page 225) that delivers credible performance on this problem for a value of the plant's internal gain, K, of 1.0 and a value of plant's time constant, τ, of 1.0.

The above statement of the problem differs from that presented by Åström and Hägglund in that the plant here is more general. The plant here operates over several different combinations of values for K and τ.

3.9.1 Preparatory Steps for the Three-Lag Plant with a Five-Second Delay

3.9.1.1 Program Architecture The architecture of the to-be-evolved program is automatically determined during the run of genetic programming using the architecture-altering operations in the manner described in section 3.7.1.1 for the two-lag plant problem.

3.9.1.2 Terminal Set The terminal sets are the same as that used in the three-lag plant described in section 3.8.1.2.

3.9.1.3 Function Set The function set for every part of the result-producing branch and automatically defined functions for this problem is the same as that used in the two-lag plant (section 3.7.1.3), except that the `DELAY` function is included in the function set for this problem.

3.9.1.4 Fitness Measure Fitness is measured using 13 elements, including

(1) 12 time-domain-based elements based on the integral of time-weighted absolute error, overshoot, and robustness, and
(2) one time-domain-based element measuring disturbance rejection.

The fitness of an individual controller is the sum of the (detrimental) contributions of these 13 elements. The smaller the sum, the better.

The first 12 elements of the fitness measure together evaluate (i) how quickly the controller causes the plant to reach the reference signal, (ii) the robustness of the controller in the face of significant variations in the plant's internal gain and the plant's time constant, and (iii) the success of the controller in avoiding overshoot.

These first 12 elements of the fitness measure together represent

- three choices of values of the plant's internal gain, K, in conjunction with
- two choices of values of the height of the reference signal, in conjunction with
- two choices of values of the plant's time constant τ.

The three values of K are 0.9, 1.0, and 1.1.

The first reference signal is a step function that rises from 0 to 1 Volt at $t = 100$ milliseconds. The second reference signal is a step function that rises from 0 to 1 microvolts at $t = 100$ milliseconds.

The two values of τ are 0.5 and 1.0.

For each of these 12 fitness cases, a transient analysis is performed in the time domain using the SPICE simulator.

The contribution to fitness for each of these 12 elements of the fitness measure is based on the integral of time-weighted absolute error

$$\int_{t=5}^{36} (t-5)\,|e(t)|\,A(e(t))BC\,dt.$$

Because of the built-in five-second time delay, the integration runs from time $t = 5$ seconds to $t = 36$ seconds. Here $e(t)$ is the difference (error) at time t between the delayed plant output and the reference signal. In implementing the integral of time-weighted absolute error, a discrete approximation to the integral was used employing 120 300-millisecond time steps between $t = 5$ to $t = 36$ seconds.

In addition, the penalty associated with each fitness case is multiplied by the reciprocal of the amplitude of the reference signal associated with that fitness case so both reference signals (1 microvolt and 1 Volt) are equally influential. Specifically, B is a factor that is used to normalize the contributions associated with the two step-functions.

B multiplies the difference $e(t)$ associated with the 1-Volt step function by 1 and multiplies the difference $e(t)$ associated with the 1-microvolt step function by 10^6.

Also, the integral contains an additional weight, A, that varies with $e(t)$. The function A weights all variation up to 102% of the reference signal by a factor of 1.0, and heavily penalizes overshoots over 2% by a factor of 10.0.

Finally, the integral contains a special weight, C, which is 5.0 for the two fitness cases for which $K = 1$ and $\tau = 1$, and 1.0 otherwise.

The 13th element of the fitness measure is based on disturbance rejection. The penalty is computed based on a time-domain analysis for 36.0 seconds. In this analysis, the reference signal is held at a value of 0. A disturbance signal consisting of a unit step is added to the `CONTROLLER_OUTPUT` at time $t = 0$ and the resulting disturbed signal is provided as input to the plant. The detrimental contribution to fitness is 500/36 times the time required to bring the plant output to within 20 millivolts of the 0-Volt reference signal (i.e., to reduce the effect to within 2% of the 1-Volt disturbance signal) assuming that the plant settles to within this range within 36 seconds. If the plant does not settle to within this range within 36 seconds, the detrimental contribution to fitness is 500 plus 500 times the absolute value of the plant's output in Volts. For example, if the effect of the disturbance is never reduced below 1 Volt, the detrimental contribution to fitness would be 1,000.

3.9.1.5 Control Parameters The population size is 500,000.

3.9.2 *Results for the Three-Lag Plant with a Five-Second Delay*

The best individual in generation 0 has a fitness of 1926.498.

The best-of-run controller emerged in generation 129. This best-of-run controller has a fitness of 522.605. The result-producing branch of this best-of-run individual has 119 points. There are 95, 93, and 70 points, respectively, in the three automatically defined functions.

Figure 3.14 shows the best-of-run controller from generation 129. Note that this best-of-run controller employs a 4.8-second delay (as reflected by $e^{-4.80531s}$ in the transfer function of the evolved pre-filter). This delay is reasonable in light of the problem's five-second plant delay.

When compared to a controller designed with the techniques in Åström and Hägglund 1995, the best-of-run controller has a better value of fitness for a 1-Volt step size, an internal gain, K, of 1.0; and a time-constant, τ, of 1.0 (the specific case considered by Åström and Hägglund 1995).

Figure 3.15 compares the time-domain response to step input of 1 Volt of the best-of-run controller (triangles) with the response of the Åström and Hägglund controller (squares).

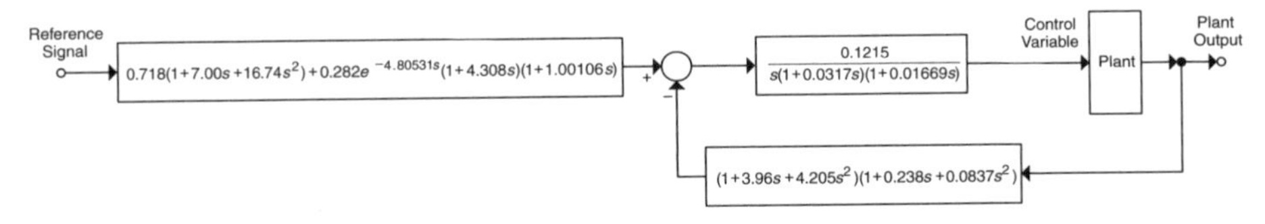

Figure 3.14 Best-of-run controller for the three-lag plant problem with a five-second delay.

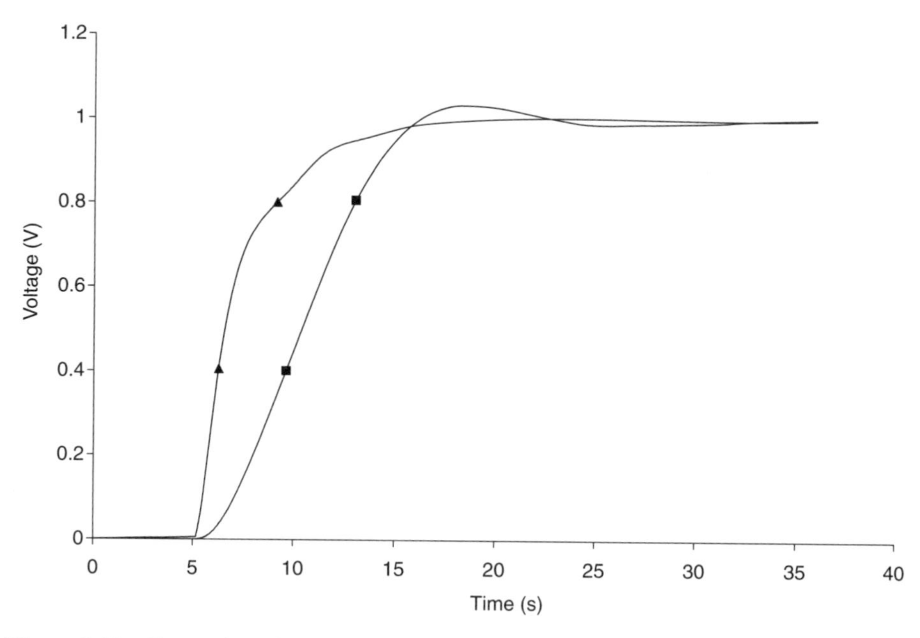

Figure 3.15 Comparison for step input for the best-of-run controller for the three-lag plant with a five-second delay.

Figure 3.16 compares the disturbance rejection of the best-of-run controller (triangles) with that of the Åström and Hägglund controller (squares). The disturbance signal has a step size of 1 Volt. The reference signal is held constant at 0 Volts. The comparison is made with an internal gain, K, of 1.0 and a time-constant, τ, of 1.0.

Table 3.9 compares the fitness of the best-of-run controller with the fitness of the Åström and Hägglund controller (1995) for this problem. The 13 elements of the fitness measure are shown in its left-most four columns. Two of the entries are divided by the special weight $C = 5.0$. All 13 entries are better for the genetically evolved controller than for the Åström and Hägglund 1995 controller.

For additional details, see Koza, Keane, Yu, Mydlowec, and Bennett 2000a.

3.9.3 Routineness for the Three-Lag Plant with a Five-Second Delay

The preparatory steps for the problem of the three-lag plant with a five-second delay are substantially the same as the preparatory steps for the three-lag plant problem in section 3.8 with two exceptions. First, this problem has a different goal (and hence a different fitness measure). Second, this problem has the `DELAY` function in the function set. Therefore, very little effort was required to make the transition from the two previously described problems of controller synthesis to the present problem (i.e., the transition was routine).

3.9.4 AI Ratio for the Three-Lag Plant with a Five-Second Delay

The solution produced by genetic programming to this problem has a moderately high amount of "A." The preparatory steps for this problem are substantially the same as

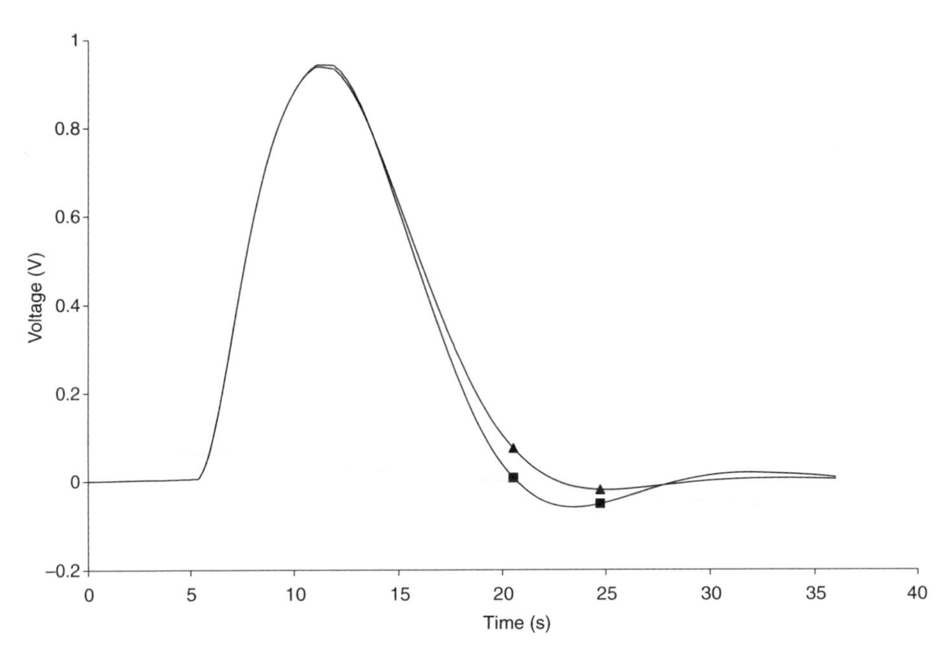

Figure 3.16 Comparison for disturbance rejection for the best-of-run controller for the three-lag plant with a five-second delay.

Table 3.9 Comparison of fitness of the best-of-run controller and the Åström and Hägglund controller for the three-lag plant with a five-second delay

Element	Step size (Volts)	Plant internal gain, K	Time constant, τ	Fitness of the best-of-run controller	Fitness of Åström-Hägglund controller
0	1	0.9	1.0	13.7	27.4
1	1	0.9	0.5	25.6	38.2
2	1	1.0	1.0	34.0/5 = 6.8	22.9
3	1	1.0	0.5	18.6	29.3
4	1	1.1	1.0	4.4	25.4
5	1	1.1	0.5	16.3	22.7
6	10^{-6}	0.9	1.0	13.2	27.4
7	10^{-6}	0.9	0.5	25.5	38.2
8	10^{-6}	1.0	1.0	30.7/5 = 6.1	22.9
9	10^{-6}	1.0	0.5	18.5	29.3
10	10^{-6}	1.1	1.0	4.3	25.4
11	10^{-6}	1.1	0.5	16.2	22.7
Disturbance	1	1	1	302	373

those for the two previously described problems of controller synthesis. For the same reasons as stated before, there is only a small amount of "I" in the denominator of the AI ratio. Therefore, the AI ratio for the solution produced by genetic programming is moderately high.

3.10 Non-Minimal-Phase Plant

In this section, genetic programming is used to discover both the topology and parameter values for a controller for a non-minimal-phase plant.

The transfer function of the non-minimal-phase plant is

$$G(s) = \frac{K(1-0.5s)}{(1+\tau s)^2},$$

where K is the plant's internal gain and τ is the plant's time constant.

The five requirements at the beginning of section 3.9 apply to this problem.

A controller is presented in Villagran and Sbarbaro (1998) that delivers credible performance on this problem for a value of the plant's internal gain, K, of 1.0 and a value of plant's time constant, τ, of 1.0.

The statement of the problem above differs from that presented in Villagran and Sbarbaro 1998 in that the plant here is more general. The plant here operates over several different combinations of values for K and τ.

3.10.1 Preparatory Steps for the Non-Minimal-Phase Plant

3.10.1.1 Program Architecture The architecture of the to-be-evolved program is automatically determined during the run of genetic programming using the architecture-altering operations in the manner described in section 3.7.1.1 for the two-lag plant problem.

3.10.1.2 Terminal Set The terminal set (including the method of establishing the numerical parameter value for each signal-processing block) is the same as that used in the two-lag plant described in section 3.7.1.2.

3.10.1.3 Function Set The function set is the same as that used in the two-lag plant described in section 3.7.1.3, except that the LAG2 function is not used for this problem.

3.10.1.4 Fitness Measure The fitness measure for the non-minimal phase plant problem consists of 11 elements.

The first ten elements are identical to those in section 3.7.1.4 for the two-lag plant problem.

The eleventh element of the fitness measure is concerned with disturbance rejection and is identical to that of section 3.8.1.4 for the three-lag plant problem, except that the test is performed for 2.4 seconds.

3.10.1.5 Control Parameters The population size is 66,000.

3.10.2 Results for the Non-Minimal Phase Plant

The best individual from the initial random generation (generation 0) for the non-minimal phase plant problem has a fitness of 94.99.

The best-of-run controller from generation 38 has a fitness of 21.72.

Table 3.10 shows the contribution of each of the 11 elements of the fitness measure to the fitness of the best-of-run controller.

Table 3.10 Contribution of the 11 elements of the fitness measure to the fitness of the best-of-run controller for the non-minimal-phase plant problem

	Step size	Internal gain	Time constant	Fitness
1	1	1	1.0	3.00
2	1	1	0.5	3.94
3	1	2	1.0	1.54
4	1	2	0.5	2.12
5	10^{-6}	1	1.0	3.00
6	10^{-6}	1	0.5	3.94
7	10^{-6}	2	1.0	1.54
8	10^{-6}	2	0.5	2.14
9	Spiked reference signal			0.00
10	AC sweep			0.00
11	Disturbance sensitivity			0.49
Total				21.72

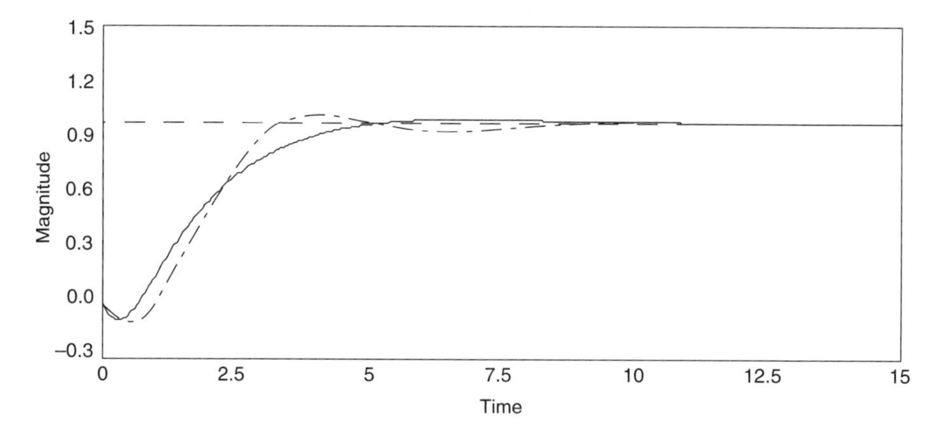

Figure 3.17 Comparison of the best-of-run individual (dashed curve) and Villagran and Sbarbaro result for the non-minimal-phase plant (solid curve).

Figure 3.17 compares the plant response in the time domain for the best-of-run individual from generation 38 (dashed curve) and the controller presented by Villagran and Sbarbaro (solid curve).

The transfer functions for the best-of-run individual from generation 38 (figure 3.18) are

$$H(s) = \frac{1.13(1.67 + 1.78s + s^2)}{s(2.51 + s)}$$

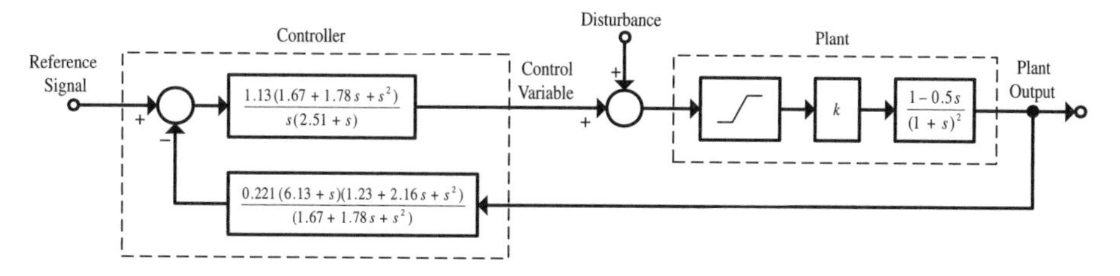

Figure 3.18 Best-of-run controller for the non-minimal-phase plant.

and

$$F(s) = \frac{0.221(6.13+s)\,(1.23+2.16s+s^2)}{(1.67+1.78s+s^2)}.$$

The evolved controller performs robustly. A Bode plot for the entire system confirms that the evolved controller conforms to the desired 40 dB per decade roll-off at frequencies higher than 100 Hz.

For additional details, see Koza, Keane, Yu, Bennett, Mydlowec, and Stiffelman 1999.

3.10.3 Routineness for the Non-Minimal Phase Plant Problem

The preparatory steps for this problem are substantially the same as the preparatory steps for the three previous problems of controller synthesis in this chapter except that the present problem has a different goal and hence a different fitness measure. Therefore, very little effort was required to make the transition from the three previously described problems of controller synthesis to the present problem (i.e., the transition was routine).

3.10.4 AI Ratio for the Non-Minimal Phase Plant Problem

The solution produced by genetic programming to this problem has a moderately high amount of "A." The preparatory steps for this problem are substantially the same as those for the three previously described problems of controller synthesis. For the same reasons as stated before, there is only a small amount of "I" in the denominator of the AI ratio. Therefore, the AI ratio for the solution produced by genetic programming is moderately high.

4

Automatic Synthesis of Circuits

The design process for electrical circuits begins with a high-level description of the circuit's desired behavior and characteristics. The process entails creation of both the topology and the sizing of a satisfactory circuit.

The *topology* of a circuit comprises

- the total number of components in the circuit,
- the type of each component (e.g., resistor, capacitor, transistor) at each location in the circuit,
- a list of all the connections that may exist between the leads of the circuit's components, input ports, output ports, power sources (if any), and ground.

The *sizing* of a circuit consists of the component value(s), if any, associated with each component. The sizing of a component is usually a numerical value (e.g., the capacitance of a capacitor).

Considerable progress has been made in recent years in automating the design of purely digital circuits; however, the design of analog circuits and mixed analog-digital circuits has not proved as amenable to automation. As Getreu (2002) noted,

> "[A]nalog designers—and tools to enhance their productivity—are in very short supply. Development of the analog segments of sophisticated designs may take longer than the digital portion, even though the analog portion is often smaller. Meanwhile, digital designers have synthesis, formal verification, and test pattern-generation tools, among others, at their disposal. They also can create large circuits from high-level languages."

As Moretti (2002) observed,

> "Analog design is more difficult and time-consuming than digital design. There are far fewer analog designers than digital designers, so as the number of mixed-signal designs increases, a productivity bottleneck forms. Engineers design analog

circuits by entering the schematic diagram and sizing and biasing the various elements using the properties and notational capabilities of the schematic editor. To simulate the design to observe its operational characteristics, engineers must develop testbenches, set operational goals, run simulations, modify the circuit if necessary, and repeat the process until they meet their design goals. Because analog simulation is computationally intensive, this process takes longer than does digital simulation."

As Aaserud and Nielsen (1995) noted:

"[M]ost ... analog circuits are still handcrafted by the experts or so-called 'zahs' of analog design. The design process is characterized by a combination of experience and intuition and requires a thorough knowledge of the process characteristics and the detailed specifications of the actual product.

"Analog circuit design is known to be a knowledge-intensive, multiphase, iterative task, which usually stretches over a significant period of time and is performed by designers with a large portfolio of skills. It is therefore considered by many to be a form of art rather than a science."

Genetic programming was first used to automatically create (synthesize) both the topology and sizing of analog electrical circuits in 1995 (Koza, Bennett, Andre, Keane 1996a, 1996b, 1996c, 1996d, 1996e, 1996f, 1999b; Bennett, Koza, Andre, and Keane 1996; Koza, Bennett Andre, Keane, and Dunlap 1997). Numerous examples of the automatic synthesis of analog electrical circuits composed of transistors, capacitors, resistors, inductors, and other components are presented in the 1999 book *Genetic Programming III: Darwinian Invention and Problem Solving* (Koza, Bennett, Andre, and Keane 1999a) and the accompanying videotape *Genetic Programming III Videotape: Human-Competitive Machine Intelligence* (Koza, Bennett, Andre, Keane, and Brave 1999).

Section 4.1 describes our approach to the automatic creation of both the topology and sizing of electrical circuits by means of genetic programming.

Later sections of this chapter demonstrate:

- the use of genetic programming to search for the solution to a seemingly impossible problem (section 4.2),
- the reinvention of the Philbrick circuit that was patented in 1956 (section 4.3),
- the automatic synthesis of a NAND circuit (section 4.4),
- the automatic synthesis of an arithmetic logic unit (ALU) circuit (section 4.5),
- the automatic synthesis of a square root computational circuit (section 4.6), and
- the ability to automatically create, during a run of genetic programming, surrogates for the source and load resistors that are normally hard-wired into the problem's test fixture, as illustrated by a problem involving the automatic synthesis of a lowpass filter (section 4.7).

Later, chapter 5 demonstrates the simultaneous synthesis of the topology, sizing, placement, and routing of both a lowpass filter and an amplifier.

Chapter 10 demonstrates the automatic synthesis of parameterized topologies representing electrical circuits.

Chapter 11 demonstrates the automatic synthesis of parameterized topologies with conditional developmental operators for circuits.

Chapter 14 describes the reinvention of negative feedback.

Chapter 15 discusses the automatic creation of circuits that duplicate the functionality of six post-2000 patented inventions.

4.1 Our Approach to the Automatic Synthesis of the Topology and Sizing of Circuits

Our approach to the problem of automatically creating both the topology and sizing of an electrical circuit involves

(1) establishing a representation for electrical circuits involving LISP symbolic expressions (S-expressions) and program trees, and
(2) defining a fitness measure that measures how well the behavior and characteristics of a candidate circuit satisfy the problem's high-level design requirements.

The representation and fitness measure are then used during the run of genetic programming. During the run, the evaluation of the fitness of each individual in the population involves

(1) converting each individual program tree in the population into a netlist of the type accepted by a circuit simulator,
(2) obtaining the behavior of the individual circuit by simulating it, and
(3) using the circuit's behavior and characteristics to calculate its fitness.

The implementation of our approach entails working with four different representations of an electrical circuit (each described in detail later):

- *Circuit Diagram*: Electrical engineers ordinarily represent a circuit by means of a diagram (a labeled graphical structure with cycles).
- *Program Tree*: A circuit can also be represented as a program tree whose internal points (nodes) represent circuit-constructing functions and whose external points (leaves) represent terminals. This representation enables genetic programming to breed a population of programs in a search for a circuit that satisfies the pre-specified design objectives.
- *Symbolic Expression*: A circuit can also be represented as a symbolic expression (S-expression) in the style of the LISP programming language. This representation is used internally by genetic programming.
- *SPICE Netlist*: The fully developed circuit (embedded in a test fixture) is simulated using a simulator. The SPICE (Simulation Program with Integrated Circuit Emphasis) simulator is used throughout this book for simulating circuits, controllers, and networks of chemical reactions. In order to invoke the SPICE simulator, the

fully developed circuit and test fixture are presented to SPICE in the form of a netlist.

Electrical circuits are ordinarily represented as labeled *cyclic* graphical structure with cycles (circuit diagrams). When genetic programming is used to automatically create computer programs, the programs are usually represented as program trees—that is, *acyclic* graphs. Thus, there is a representational obstacle that must be overcome before genetic programming can be applied to the problem of automatic circuit synthesis.

Our approach to the automatic synthesis of circuits using genetic programming employs a developmental process to overcome this representational obstacle. This approach is inspired by the principles of developmental biology, the innovative work of Kitano (1990) on using developmental genetic algorithms to evolve neural networks, and the creative work of Gruau (1992a, 1992b) on using developmental genetic programming (cellular encoding) to evolve neural networks. The reader is also referred to early work on ontogenetic programming by Spector and Stoffel (1996a, 1996b) and on evolving Lindenmayer rules for creating structures (Koza 1993).

The developmental process transforms a program tree (an acyclic graph) into a fully developed electrical circuit (a graphical structure with cycles). The developmental process entails the execution of functions in a circuit-constructing program tree. The circuit-constructing program tree may contain component-creating functions, topology-modifying functions, development-controlling functions, arithmetic-performing functions, and automatically defined functions.

The starting point for our developmental process consists of a simple initial circuit. The initial circuit consists of

- an embryo, and
- a test fixture.

The embryo contains at least one modifiable wire. All development originates from the embryo's modifiable wire(s). If the original modifiable wire(s) are not modified by the developmental process, the circuit produces only trivial output.

An electrical circuit is developed by progressively applying the functions in a circuit-constructing program tree (in the population being bred by genetic programming) to the modifiable wires of the original embryo and, as the circuit grows, to the modifiable wires and modifiable components that sprout from it. The execution of the functions in the program tree transforms the initial circuit into a fully developed circuit. That is, the functions in the circuit-constructing program tree progressively side-effect the embryo and its successors until a fully developed circuit eventually emerges.

Test fixtures are commonly used in electrical engineering to measure the behavior of a circuit. When genetic programming is being used to automatically synthesize an electrical circuit, the test fixture is the entity (external to the entity that is being automatically created) that facilitates measurement of the fully developed circuit. The test fixture feeds external input(s) into the circuit of interest. It also enables the circuit's output(s) to be probed. The test fixture supplies the measurements that enable the fitness measure to assign a single numerical value of fitness to the behavior and characteristics of the fully developed circuit. The test fixture is a hard-wired structure composed of nonmodifiable wires and nonmodifiable electrical components. The test fixture has one or more ports that enable the embryo (and, later, the fully developed circuit) to be embedded into it. In turn, the embryo (and, later, the fully developed circuit) has one or more ports that enable it to communicate with the test fixture in which it is embedded. The hard-wired components of the test fixture often include a source resistor and a load resistor. Occasionally, a test fixture contains a voltage-to-current converter or a current-to-voltage converter that enables a signal to be provided or measured in a certain preferred way (e.g., voltage or current). After completion of the run of genetic programming, the test fixture plays no further role. At that time, the genetically evolved circuit is attached to whatever input(s) it is intended to receive and to whatever points its output(s) are intended to go.

The functions in the circuit-constructing program trees are divided into five categories:

- topology-modifying functions (e.g., series division, parallel division, cut, via) that modify the topology of the developing circuit,
- component-creating functions that insert components (e.g., resistors, capacitors, transistors) into the developing circuit,
- development-controlling functions that control the developmental process by which the embryo and its successor circuits are converted into a fully developed circuit (e.g., the no-operation function),
- arithmetic-performing functions (e.g., addition, subtraction) that may appear in a value-setting subtree that is an argument to a component-creating function and that specifies the numerical value of the component, and
- automatically defined functions that enable certain substructures to be reused (including parameterized reuse).

The component-creating functions generally have value-setting subtree(s), whereas topology-modifying functions and development-controlling functions do not.

Many of the functions possess one or more construction-continuing subtrees.

The terminals in the circuit-constructing program trees may include

- constant numerical values,
- perturbable numerical values,
- externally supplied free variables,
- symbolic values (e.g., discrete alternative types for certain components), and
- zero-argument functions (e.g., the development-ending function).

In a run of genetic programming, all the individual program trees created in generation 0 of the population are syntactically valid executable programs. All the genetic operations of genetic programming (i.e., crossover, mutation, reproduction, and the architecture-altering operations) operate so as to create syntactically valid executable programs. Thus, all the individuals encountered during the run (including, in particular, the best-of-run individual) are syntactically valid executable programs.

Each circuit-constructing program tree is created in accordance with a constrained syntactic structure that imposes limits on how the available functions and terminals may be combined (section 2.3.1). All the individuals in the initial random population (generation 0) of a run of genetic programming for automatic circuit synthesis comply with the constrained syntactic structure. All the genetic operations that are performed during the run (i.e., crossover, mutation, reproduction, and the architecture-altering operations) are designed to preserve the constrained syntactic structure. In particular, the crossover operation uses *point typing* to ensure preservation of the constrained syntactic structure.

Part 5 of *Genetic Programming III: Darwinian Invention and Problem Solving* (Koza, Bennett, Andre, and Keane 1999a) contains additional detailed information about automatic circuit synthesis using genetic programming. In particular, section 25.11 of that book contains the grammar governing the construction of circuit-constructing program trees in problems of automatic circuit synthesis.

Our developmental approach is more than just a mechanism for mapping an acyclic graph (the circuit-constructing program tree) into a graphical structure with cycles (the fully developed circuit).

For one thing, the developmental approach has the specific advantage of preserving locality. Because most of the component-creating, topology-modifying, and development-controlling functions operate on a small local area of the circuit, the subtrees that are transplanted by the crossover operation generally operate locally. The developmental process works in conjunction with the crossover operation in preserving locality. When a crossover replaces a subtree in one individual with a subtree from another individual, it (usually) replaces a local structure in the circuit created by the first individual with a local structure in the circuit created by the second individual. The developmental process similarly works in conjunction with the mutation operation in preserving locality.

The developmental process has the additional advantage of preserving electrical connectivity. There are no unconnected leads in the initial circuit. Each component-creating, topology-modifying, and development-controlling function preserves connectivity at each stage of the developmental process. The result is that there are no unconnected leads in the fully developed circuit.

Also, the developmental approach enables useful parts of a circuit-constructing program tree to be reused (as explained in section 16.2.2). Reuse eliminates the need to "reinvent the wheel" on each occasion when a particular sequence of steps may be useful. Reuse makes it possible to exploit a problem's modularities, symmetries, and regularities (and thereby potentially accelerate the problem-solving process).

The preservation of syntactic validity and executablity of the circuit-constructing program trees, the preservation of the constrained syntactic structure, the preservation of electrical connectivity, the preservation of locality during crossover, and the facilitation of reuse together contribute to the efficiency by which developmental genetic programming is able to automatically synthesize circuits.

More detailed information about the developmental process and our approach to automatic circuit synthesis may be found in *Genetic Programming III: Darwinian Invention and Problem Solving* (Koza, Bennett, Andre, and Keane 1999a).

4.1.1 Evolvable Hardware

Work in the field of automatic synthesis of analog electrical circuits is part of a growing field called *evolvable hardware*. The growing literature on in this field includes

- the proceedings of the International Conference on Evolvable Systems (Higuchi, Iwata, and Lui 1997; Sipper, Mange, and Perez-Uribe 1998; Miller, Thompson, Thomson, and Fogarty 2000; Liu, Tanaka, Iwata, Higuchi, and Yasunaga 2001),
- the proceedings of the NASA/DoD Conference on Evolvable Hardware (Stoica, Keymeulen, and Lohn 1999; Lohn, Stoica, Keymeulen, and Colombano 2000; Keymeulen, Stoica, Lohn, and Zebulum 2001; Stoica, Lohn, Katz, Keymeulen, and Zebulum 2002),
- edited collections of papers (Sanchez and Tomassini 1996; Mazumder and Rudnick 1999), and
- books (Kruiskamp 1996; Thompson 1998; Drechsler 1998; Zebulum, Pacheco, and Vellasco 2002).

Early papers in the field include those by Higuchi, Niwa, Tanaka, Iba, de Garis, and Furuya (1993a, 1993b); Kruiskamp and Leenaerts (1995); Grimbleby (1995); Thompson (1996); and Koza, Bennett, Andre, and Keane (1996a, 1996b, 1996c, 1996d, 1996e, 1996f; Bennett, Koza, Andre, and Keane 1996; Koza, Bennett Andre, Keane, and Dunlap 1997).

4.2 Searching for the Impossible

One of the themes of this book is that many potential inventions are never discovered (or their discovery is long delayed) because the thought processes of scientists and engineers are channeled along well-traveled paths. It is thus appropriate that this book's first illustrative example of automatic circuit synthesis by means of genetic programming be a circuit that is seemingly impossible to create.

In his regular column in *Electronic Design*, Robert Pease, the legendary analog-design engineer and chief scientist at National Semiconductor, posed the question of whether it is possible to design an electrical circuit composed only of resistors and capacitors that delivers a gain that is greater than one (Pease 1996).

Pease (1996) observed that most electrical engineers would instinctively react by saying that it is "absurd" to try to build such a circuit using only resistors and capacitors. As Pease said,

> "Resistors are passive. How can you take a network of Rs and Cs and generate a gain of greater than one? That's impossible!"

Indeed, when we posed Pease's challenge informally to a number of electrical engineers, they all initially reacted by saying that the design of such a circuit was impossible. Interestingly, once these engineers were told of the existence of such a circuit (but *before* being told how the circuit is constructed), all were quickly able to come up with a reasonable and correct explanation of how such a circuit would work. Why is this? Years of thinking along traditional lines often prevents scientists and engineers from seeing certain things. In this case, the engineers were initially unable to see something that they were readily able to explain. That is, they had the knowledge needed to explain how an RC circuit can generate a gain that is greater than one. However, until they were given the information that such a circuit was indeed possible, they could not bring their knowledge to bear on the challenge.

Pease's explanation was based on a circuit that was patented in 1956 by George Philbrick, one of the early pioneers of analog circuit design. The Philbrick circuit, consisting of only three resistors and three capacitors, was used to preprocess the input to cathode-ray oscilloscopes (Philbrick 1956). Philbrick's RC circuit has a surprising characteristic unrelated to the circuit's intended purpose. As Pease observed in his 1996 column in *Electronic Design*, if the output of the Philbrick circuit is attached to a unity-gain follower (such as an op amp with a gain of 1) and the output of the follower is, in turn, fed back to the input of the Philbrick circuit, the resulting circuit is a phase-shift oscillator. According to the Nyquist criterion, a circuit can only oscillate if the gain around the loop is greater than 1. However, because the unity-gain op amp has a gain of only 1, then, by inference, Philbrick's RC circuit must have a gain that is greater than 1. In fact, it does.

Figure 4.1 shows the circuit composed of only resistors and capacitors that Philbrick patented in 1956. In the figure, an incoming signal VSRC passes into the network of resistors and capacitors. The voltage is probed at probe point V.

Figure 4.2 shows the output voltage measured at point V of the circuit of figure 4.1 for a sine wave input signal with 1-Volt amplitude. The horizontal axis represents six decades of frequencies between 1 millihertz and 1,000 Hertz (Hz) on a logarithmic scale. As can be seen, Philbrick's RC circuit delivers a gain that is greater than 1.0 for frequencies between approximately 2.0 Hz and 20.0 Hz. Specifically, the voltage peaks at 1.19 Volts at 4.64 Hz for Philbrick's circuit.

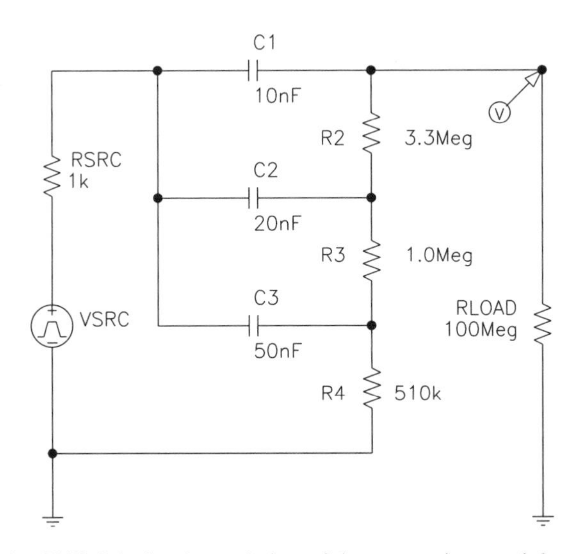

Figure 4.1 Philbrick circuit consisting of three capacitors and three resistors.

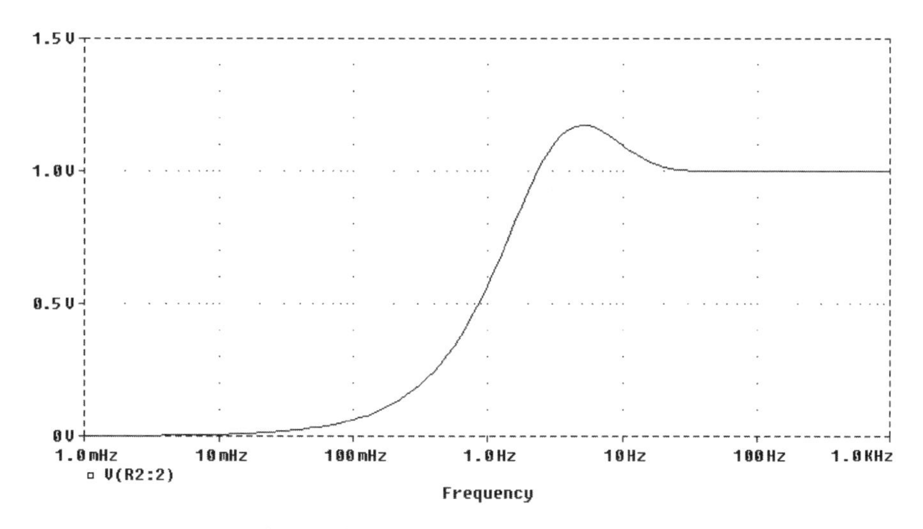

Figure 4.2 Output of the Philbrick circuit.

The question arises as to whether it is possible to automate the process of exploring challenges, such as the one posed by Pease, concerning the existence of seemingly impossible constructions. This challenge lies at the heart of the invention process.

This section demonstrates how genetic programming can be used to automate the process of exploring queries, conjectures, and challenges concerning the existence or design of seemingly impossible entities.

The evolutionary process opportunistically solves problems without considering whether its solution comports with human preconceptions about whether the goal is impossible. Because evolution is a probabilistic process that is not encumbered by the preconceptions that often channel human thinking down familiar paths, it often creates novel designs.

The observed counter-intuitive voltage multiplication of the Philbrick circuit comes about as a result of a phase shift. In particular, the output of the Philbrick circuit is the difference of a 1-Volt (peak amplitude) input signal of the form $sin\ 2\pi ft$ and another signal of the form $sin\ (2\pi ft+\Delta)$, where Δ represents a phase shift. As previously mentioned, the particular circuit in Philbrick's 1956 patent delivers a gain of 1.19.

Each of the electrical engineers to whom we posed Pease's query mentioned that this line of reasoning indicated that it would clearly be possible to construct an RC circuit that delivers a gain approaching two in a similar way.

Of course, this observation immediately raises yet another query. Is it possible to construct an RC circuit that can deliver a gain that is greater than two? Again, several electrical engineers (and the authors here) thought that such a circuit would be impossible to construct.

The counter-intuitive result is that an electrical circuit composed only of resistors and capacitors can deliver a gain that is greater than two.

4.2.1 Preparatory Steps for the RC Circuit with Gain Greater than Two

Seven major preparatory steps are required to apply genetic programming to a problem of circuit synthesis:

(1) identify the initial circuit (i.e., the test fixture and embryo),
(2) determine the architecture of the circuit-constructing program trees,
(3) identify the primitive functions of the program trees,
(4) identify the terminals of the program trees,
(5) create the fitness measure,
(6) choose control parameters, and
(7) determine the termination criterion and method of result designation.

4.2.1.1 Initial Circuit Figure 4.3 shows a one-input, one-output initial circuit consisting of an embryo embedded in a test fixture.

The embryo (highlighted in figure 4.3 by an outline box) consists of two modifiable wires (called Z0 and Z1).

The test fixture in figure 4.3 has an incoming signal source VSOURCE, a source resistor RSOURCE, a nonmodifiable wire ZOUT, a voltage probe point VOUT (the output of the overall circuit), a variable output load (called LOAD), and a nonmodifiable wire ZGND providing a connection to ground. The test fixture has three ports (nodes 2, 3, and 4) that are connected to the embryo (and, after completion of the developmental process, to the fully developed circuit). One port (node 2) makes the input signal VSOURCE available to the embryo. A second port (node 4) provides access to ground. A third port (node 3) is connected to the voltage probe point VOUT. The initial circuit shown in figure 4.3 produces only trivial (i.e., always 0) output at its probe point.

It is important to establish the amount of knowledge that is, or is not, be incorporated into the specification of the test fixture. The test fixture is the means to implement the problem's fitness measure. Specifically, the test fixture provides the means to feed incoming signals (voltage of current) into the genetically evolved circuit and to probe output signals (voltage or current) in order to compute fitness.

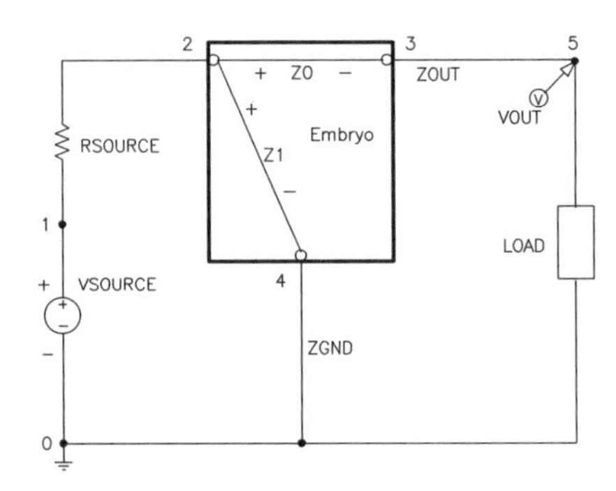

Figure 4.3 One-input, one-output initial circuit with two modifiable wires.

In this book, we always specifically identify the test fixture. However, test fixtures are often not explicitly mentioned in the literature of machine learning and artificial intelligence because they are simply taken for granted. Test fixtures are widely used for evaluating the performance of entities that are created by other automated problem-solving methods (including neural networks, decision trees, reinforcement learning, genetic classifier systems, and pattern recognizers). The "input interface" and "output interface" often mentioned in the literature of machine learning and artificial intelligence corresponds to a test fixture.

For example, consider the intertwined spiral problem. This problem was solved using a neural network in the 1980s and was subsequently solved by genetic programming (as described in section 17.3 of Koza 1992a). The goal in the intertwined spiral problem is to classify a point, represented by its Cartesian *x-y* coordinate, as belonging to the red or blue spiral. When a neural network is applied to this problem, the fitness of the neural network is computed by a fixed hard-wired entity that is external to the neural network. This fixed hard-wired external entity does several things. First, it presents the neural network with a sequence of Cartesian *x-y* coordinates of points that lie on either the red or blue spiral. These pairs of points are called "training cases" in the field of neural networks (and "fitness cases" in the field of genetic programming). The presentation of training cases to the neural network is accomplished by connecting the signals representing the values of x or y to the input points of each neuron in the input layer of the neural network. Second, the fixed hard-wired external entity captures the numerical value emitted by the network's final output neuron. This capture of the neural network's output is accomplished by making a connection to the signal(s) emanating from the neural network's output neuron(s). Third, the fixed hard-wired external entity converts the numerical output signal(s) into one of two discrete symbolic values (red or blue). Fourth, the fixed hard-wired external entity compares the resulting symbolic value (red or blue) with the correct color for the fitness case (stored in some database) to see if they match. Fifth, the fixed hard-wired external entity computes the fitness of the neural network (measured in terms of, say, correlation, sum of squared errors, accuracy, error rate).

Notice that the process (chapter 3) of attaching the plant to a candidate controller for the purpose of simulating the overall system and calculating the controller's fitness can be viewed as embedding the controller into a test fixture.

4.2.1.2 Program Architecture There is one result-producing branch in the program tree for each modifiable wire in the embryo. Because there are two modifiable wires in the embryo in figure 4.3, the architecture of each individual circuit-constructing program tree in the population necessarily has two result-producing branches. Automatically defined functions and the architecture-altering operations are not used on this problem.

4.2.1.3 Function Set There are two types of internal points in the circuit-constructing program trees for this problem, namely

- arithmetic-performing functions for establishing the numerical values (sizing) of components, and
- circuit-constructing functions.

A constrained syntactic structure enforces the use of one function set for the arithmetic-performing subtrees and another function set for all other parts of the program tree.

In this problem, the numerical value for each electrical component possessing a parameter (i.e., resistors and capacitors) is established by an arithmetic-performing subtree. Each arithmetic-performing subtree consists of one or more arithmetic functions and one or more constant numerical terminals. This approach for establishing numerical parameter values is described in section 3.5.5.1. The function set, F_{aps}, for each arithmetic-performing subtree is

$$F_{aps}=\{+, -\}.$$

The circuit-constructing functions include component-inserting functions, topology-modifying functions, and development-controlling functions.

Because the specific challenge of this problem involves constructing a circuit using only resistors and capacitors, only resistor-creating and capacitor-creating functions are included in the function set.

The function set, F_{ccs}, for each construction-continuing subtree is

F_{ccs}={R, C, SERIES, PARALLEL0, PARALLEL1, FLIP, NOP,
PAIR_CONNECT_0, PAIR_CONNECT_1}.

Briefly, the component-creating R function inserts a resistor into a developing circuit and establishes the numerical value of the resistor. Likewise, the C function inserts a capacitor and establishes the numerical value of the capacitor.

The topology-modifying SERIES function divides a modifiable wire or modifiable component into a series composition of modifiable wires or components. Likewise, the two versions of the PARALLEL function perform a parallel division.

The topology-modifying FLIP function reverses the polarity of a modifiable component or wire.

The NOP ("no operation") function is a development-controlling function.

The two versions of the topology-modifying PAIR_CONNECT function provide a way to connect two (usually distant) points in the developing circuit.

For a more detailed explanation of each of these functions, see Koza, Bennett, Andre, and Keane 1999a.

4.2.1.4 Terminal Set There are two types of terminals (external points) in the circuit-constructing program trees for this problem, namely

- terminals representing numerical values (which appear in arithmetic-performing subtrees), and
- terminals pertaining to circuit construction (which appear in construction-continuing subtrees).

A constrained syntactic structure enforces the use of one terminal set for the arithmetic-performing subtrees and another terminal set for all other parts of the program tree.

The terminal set, T_{aps}, for each arithmetic-performing subtree is

$$T_{aps}=\{\Re\},$$

where R represents the set of floating-point constants from -1.0 to $+1.0$.

The terminal set pertaining to circuit construction for this problem contains a zero-argument development-controlling function and a zero-argument topology-modifying function. Specifically, the terminal set, T_{ccs}, for each construction-continuing subtree is

$$T_{ccs} = \{\texttt{END, SAFE\_CUT}\}.$$

Briefly, the zero-argument `END` function is a development-controlling function that makes the modifiable wire or modifiable component with which it is associated into a nonmodifiable wire or component (thereby ending a particular developmental path).

The zero-argument `SAFE_CUT` function is a topology-modifying function that deletes a modifiable wire or component from the developing circuit while preserving circuit validity.

For a more detailed explanation of each of these terminals, see Koza, Bennett, Andre, and Keane 1999a.

4.2.1.5 Fitness Measure The evaluation of each individual circuit-constructing program tree in the population begins with its execution. The execution progressively applies the functions in the program tree to the embryo (and its successor circuits), thereby creating a fully developed circuit.

A netlist is then created that identifies each component of the fully developed circuit, the nodes to which each component is connected, and the numerical values associated with each component. The netlist includes the test fixture's components and connections (which, for this problem, include the source resistor and the variable load resistor as well as the connections to the input port, the output port, and ground).

The netlist becomes the input to our modified version of the SPICE simulator. SPICE then determines the circuit's behavior.

Because this problem concerns the circuit's frequency-domain behavior, the output voltage VOUT is measured in the frequency domain. Specifically, SPICE is instructed to perform two AC analyses on each circuit using different output loads.

In the first simulation, LOAD has infinite resistance (i.e., represents an open circuit).

In the second simulation, LOAD consists of a 10 mega-Ohm resistor in parallel with a 6-pico-farad capacitor (i.e., a load resembling that of an oscilloscope probe).

In each case, only the voltage associated with 1,000 Hz is used in computing fitness.

The fitness of a circuit is the sum of two numbers:

$$1/(1+v_{\text{out-infinite}}) + 1/(1+v_{\text{out-oscilloscope}}),$$

where $v_{out\text{-}infinite}$ is the voltage at 1,000 Hz for the open circuit and $v_{out\text{-}oscilloscope}$ is the voltage at 1,000 Hz for the load resembling that of an oscilloscope probe.

The SPICE simulator is remarkably robust; however, it cannot simulate every conceivable circuit. In particular, many circuits that are randomly created for the initial population of a run of genetic programming and many circuits that are created by the crossover and mutation operations in later generations are so pathological that SPICE cannot simulate them. These circuits receive a high penalty value of fitness (10^8) and become the worst-of-generation programs for each generation. The technique of assigning a high penalty value to unsimulatable circuits is used throughout this book.

As an aside, we should mention that there is an automated equivalent to a breadboard. A field-programmable analog array (FPAA) (Ohr 2002; Macbeth 2002) or a field-programmable transistor array (FPTA) (Stoica, Zebulum, and Keymeulen 2001) may be used to rapidly create a physical embodiment of an analog circuit. A rapidly reconfigurable field-programmable gate array (FPGA) may be used to rapidly create a physical embodiment of a digital circuit. For example, a FPGA was used to automatically synthesize a mathematical algorithm for a sorting network (Koza, Bennett, Hutchings, Bade, Keane, and Andre 1997, 1998). Instead of analyzing or simulating each sorting network that was encountered during the run of genetic programming, each network was embodied electronically, for a tiny fraction of a second, in the form of a digital circuit on the FPGA. The behavior and characteristics of this ephemeral physical embodiment of the sorting network were then observed, measured, and recorded (by a different part of the same FPGA chip). Thompson (1996) used an FPGA as a programmable analog device. It is similarly possible to embody a controller (chapter 3) for a fraction of a second (using a field-programmable transistor array) while observing, measuring, and recording its behavior and characteristics.

4.2.1.6 Control Parameters The population size is 660,000.

4.2.2 Results for the RC Circuit with Gain Greater than Two

The best circuit-constructing program of the 660,000 programs of generation 0 has a fitness of 0.956.

In generation 15, a circuit (figure 4.4) consisting of three resistors and three capacitors emerged with a fitness of 0.913. It produces a gain of 1.19. This circuit is topologically different from the circuit in Philbrick's 1956 patent (figure 4.1); however, it delivers approximately the same gain (figure 4.2).

Figure 4.5 shows the behavior in the frequency domain of the best circuit from generation 15.

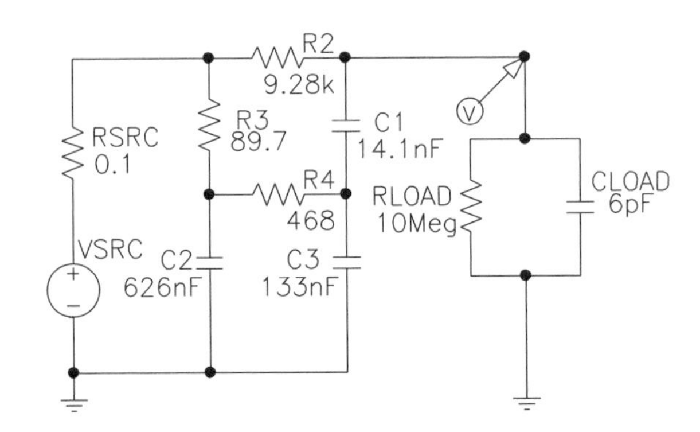

Figure 4.4 Best-of-generation circuit from generation 15.

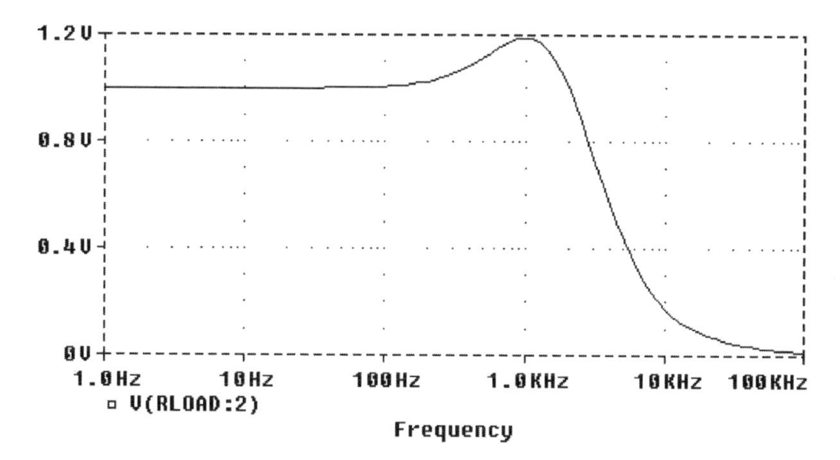

Figure 4.5 Behavior in frequency domain of the best-of-generation circuit from generation 15.

In generation 927, a circuit (figure 4.6) consisting of 38 resistors and 35 capacitors emerged. This best-of-run circuit has a fitness of 0.622. It produces a maximum gain of 2.24.

Figure 4.7 shows the behavior in the frequency domain of this best-of-run circuit from generation 927.

For additional information, see Koza, Bennett, Keane, Yu, Mydlowec, and Stiffelman 1999.

In summary, narrowly, we demonstrated the automatic synthesis, using genetic programming, of both the topology and sizing of a circuit composed of only resistors and capacitors that delivers a gain that is greater than two. More broadly, we demonstrated that genetic programming can be used to automate the process of exploring queries, conjectures, and challenges concerning the existence of seemingly impossible entities. This, in turn, suggests that genetic programming can be used to automate the invention process.

4.2.3 Routineness of the Transition from a Problem of Controller Synthesis to a Problem of Circuit Synthesis

Section 1.1.3 mentions that a problem-solving method may be measured by its routineness. As mentioned there, a problem-solving method is "routine" if it is general and if relatively little human effort is required to get the method to successfully handle new problems within a particular domain and to successfully handle new problems from a different domain.

In three places in chapter 3 (sections 3.8.3, 3.9.3, and 3.10.3), we discussed the effort required to make the transition from one problem of controller synthesis to another. That is, we discussed the *intra-domain* aspect of routineness. In this section, we consider the effort required to make the transition from a problem of controller synthesis (from chapter 3) to a problem of circuit synthesis—that is, the *inter-domain* aspect of routineness.

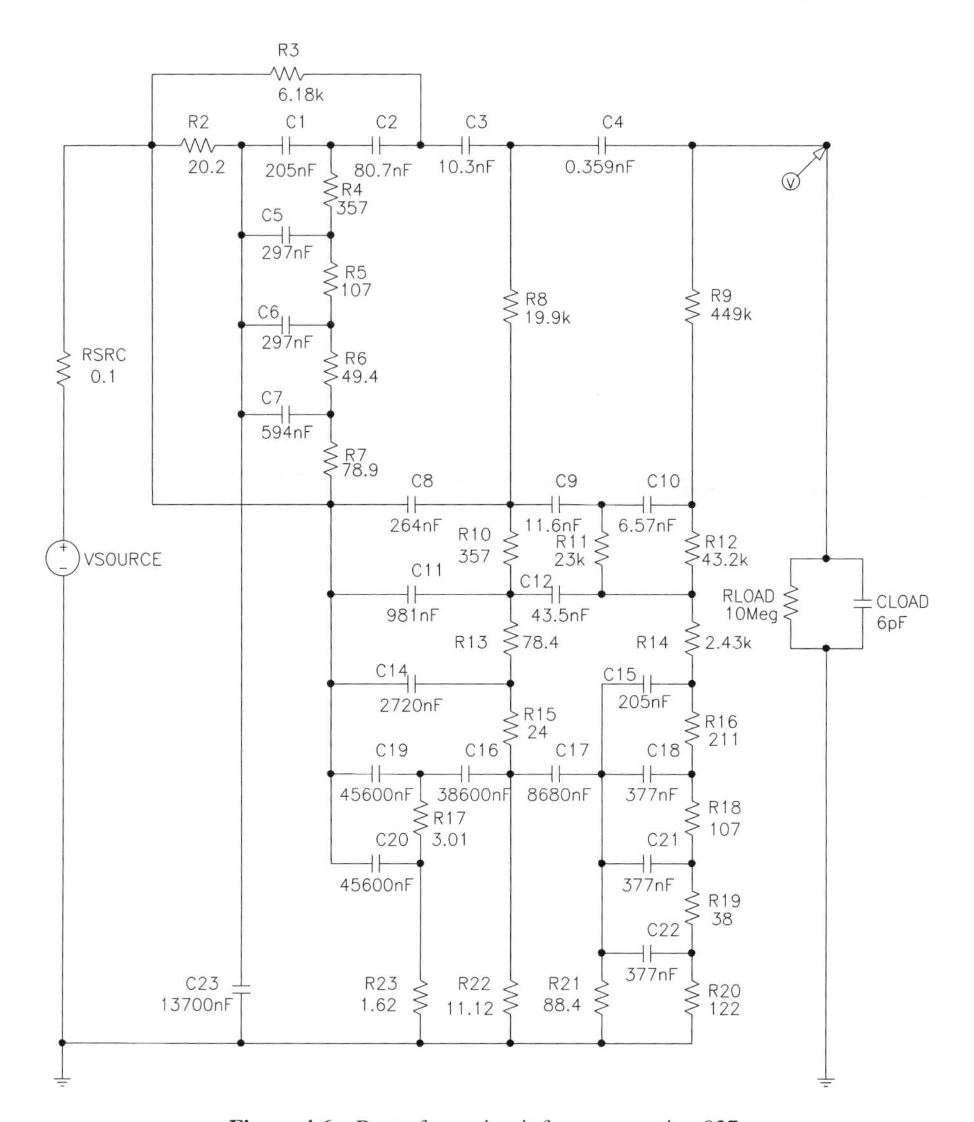

Figure 4.6 Best-of-run circuit from generation 927.

First, there are, of course, differences in the function and terminal sets. These differences reflect the fact that controllers and circuits are composed of different ingredients. For problems involving the synthesis of controllers, the function and terminal sets permit the construction of entities composed of ingredients such as integrators, differentiators, gains, adders, subtractors, reference signals, and plant outputs. For a problem of circuit synthesis, the function and terminal sets permit the construction of entities composed of ingredients such as resistors, capacitors, input points, and output points. The required change in the function and terminal sets does not entail an enormous amount of effort. The transition entails exchanging a set of platitudes about controllers with a set of platitudes about circuits.

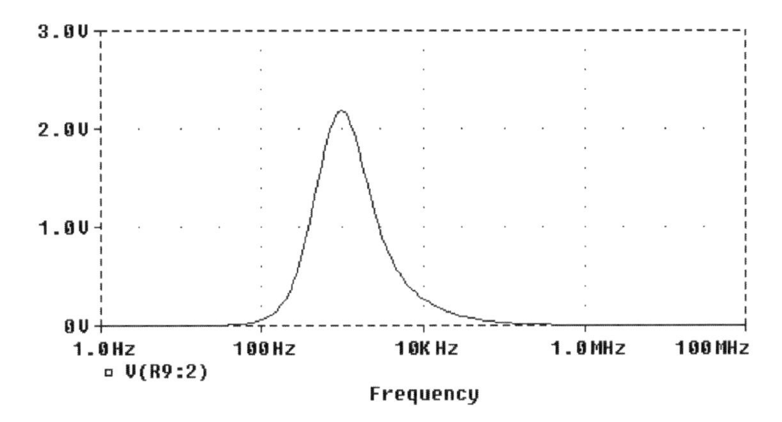

Figure 4.7 Behavior in frequency domain of the best-of-run circuit from generation 927.

Second, there is a difference in fitness measures. The fitness measure for a problem of controller synthesis is necessarily couched in terms of characteristics and performance relevant to the field of control (e.g., the integral of the time-weighted absolute error, overshoot, and robustness). On the other hand, the fitness measure for a problem of circuit synthesis is couched in terms of characteristics and performance relevant to circuits (e.g., the circuit's frequency-domain and time-domain response). The change in fitness measures is an inherent and necessary part of making the transition from one problem to another. It does not entail an enormous amount of effort.

As always, once the preparatory steps are performed, the executional steps of genetic programming are the same, so no effort at all is required to make the transition from one domain to another in this regard.

Thus, relatively little effort is required to make the transition from the domain of controllers to the domain of automatic synthesis of circuits and it is therefore reasonable to say that the transition is routine.

As will be seen later in this book, the transition from problems involving circuits to problems in yet other domains (such as antennae and networks of chemical reactions) will be similarly routine.

4.2.4 AI Ratio for the RC Circuit with Gain Greater than Two

In section 1.1.2, we mentioned that a method for solving problems may be measured by the ratio (i.e., the AI ratio) of that which is delivered by the automated operation of the *artificial* method to the amount of *intelligence* that is supplied by the human applying the method to a particular problem.

What is the AI ratio for the solution produced by genetic programming to the challenge of designing an RC circuit with a gain that is greater than two?

Because many (if not most) electrical engineers would initially opine that construction of the desired RC circuit is impossible, the surprising outcome produced by the automated approach to Pease's challenge has a moderately high amount of "A."

Ascertaining the amount of "I" requires the discipline of drawing a bright line between that which is delivered by the artificial system and that which is supplied by the intelligent human user. We imposed this discipline when we clearly identified the preparatory steps performed by the human user prior to the launch of the run of genetic programming.

The function and terminal sets (established in the preparatory steps) do not employ any deep knowledge about the design of electrical circuits. They merely reflect the statement of the problem and provide the ingredients from which to construct candidate circuits.

The component-creating functions in the function set reflect the fact that the circuit must, according to the statement of the problem, be composed only of resistors and capacitors. The topology-modifying functions in the function set reflect the well-known fact that one way to construct electrical circuits is to build them from a sequence of series divisions, parallel divisions, connections to distant points, and cuts.

The numerical constants in the terminal set reflect the fact that resistors and capacitors possess numerical parameters. The development-controlling terminals provide a way to end the developmental process. The former is essentially syntactic information and the latter is essentially a mechanical necessity of the developmental process.

The knowledge needed to create the function set and the terminal set is either intuitively obvious or platitudinous.

The function and the terminal sets are general and could be used for the automatic synthesis of *any* RC circuit. Moreover, by simply adding the transistor-creating function to the function set, the function set can also be used for the automatic synthesis of *any* circuit containing transistors, resistors, and capacitors (which, as a practical matter, includes the vast majority of present-day circuits).

The distinguishing feature of the problem at hand is its fitness measure. The fitness measure reflects the statement of the problem. We are seeking an RC circuit whose gain is greater than two.

The remaining preparatory steps (concerning control parameters, the termination criterion, and the method of result designation) are not noteworthy.

Before concluding, we should mention the role of the SPICE simulator. The fitness measure is based on the circuit's desired behavior. Each candidate circuit encountered during the run is analyzed and simulated using the SPICE simulator. The use of the SPICE simulator does not supply any knowledge to the run of genetic programming for the following reasons.

First, the SPICE simulator only does *analysis* for a circuit that it is given. The SPICE simulator does not have any capability to synthesize a circuit.

Second, although writing the SPICE simulator assuredly required considerable human expertise and knowledge, the human knowledge embodied in the simulator is not knowledge about the synthesis of electrical circuits, but instead, knowledge about how to automatically and efficiently converge on the solution to a system of equations (with particular emphasis on the integro-differential equations representing electrical components and the nonlinear equations representing transistors).

Third, when genetic programming calls on a simulator to ascertain the behavior and characteristics of a candidate individual in the population, genetic programming does not acquire any domain knowledge about the field involved. This important point is made clear by recognizing that no simulator is required in order to make a run of genetic programming in the first place. Instead, we could construct a breadboard for each circuit that is generated by genetic programming during the run. We could then observe and measure the behavior and characteristics of each breadboard. We could then calculate the fitness of each breadboard and communicate that single numerical value to the run of genetic programming. This single numerical value is the only information that the genetic programming system receives about the individual. The difference between building thousands of breadboards and using a simulator is a matter of accuracy, precision, elapsed time, and cost (in terms of effort, time, and money). Alternatively, we could use a field-programmable analog array (FPAA) (Ohr 2002; Macbeth 2002) or a field-programmable transistor array (FPTA) (Stoica, Zebulum, and Keymeulen 2001) in lieu of the breadboard.

No one would argue that merely observing, measuring, and recording the position and velocity of a falling apple is equivalent to acquiring knowledge (in the sense that this term is used in the field of artificial intelligence) of Newton's laws. There is simply no knowledge contained in such measurements. Similarly, observing, measuring, and recording the behavior and characteristics of a circuit (whether by simulating it or physically building it) does not convey any domain knowledge about the field of electrical engineering to either genetic programming or the observer because there is no knowledge contained in the measurements in the first place.

In summary, the moderately high amount of "A" represented by the result in conjunction with the small amount of "I" provided by the human user means that the AI ratio for the solution produced by genetic programming is moderately high.

4.3 Reinvention of the Philbrick Circuit

In this section, we evolve the Philbrick circuit (cited by Robert Pease in section 4.2). The circuit patented by George Philbrick was intended to preprocess an analog signal that was to be fed into an oscilloscope. As Philbrick (1956) stated,

> "This invention relates to an electric filter network, and more particularly to a delayed-recovery, high-pass filter network, which transmits the early portion of a transient voltage signal substantially without distortion but the output of which thereafter relatively rapidly recovers to a quiescent value of zero voltage when the impressed signal becomes quiescent at any voltage.
>
> "Filter networks of this type frequently are required in electronic equipment for A. C. coupling, pulse forming and shaping, differentiating, etc. For example, in cathode-ray oscilloscopes it often is desirable to display high-frequency transients without distortion and thereafter to return the horizontal trace relatively rapidly to zero value when the input signal returns to a steady state value. This necessitates a special input circuit.
>
> "Prior to the present invention no simple filter was available to accomplish this result. In the past it has been customary to use a simple series-capacitor, shunt-resistor, high-pass filter network for this purpose. When this conventional type of filter network is used in such circuits, it is subject to certain disadvantages.

If high-frequency transients are to be transmitted with negligible distortion its recovery time usually is entirely too long for practical purposes or, if made to have a recovery time of practical length, considerable distortion is introduced into the transient being transmitted.

"Accordingly, it is an object of the present invention to provide a filter network having a transmission characteristic such that initial high-frequency transients of the input signal are transmitted substantially without distortion followed after a predetermined time-delay by a relatively rapid return of the output voltage to zero irrespective of the actual magnitude of the quiescent value of the input signal."

4.3.1 Preparatory Steps for the Philbrick Circuit

4.3.1.1 Initial Circuit Figure 4.8 shows the one-input, one-output initial circuit for this problem. The embryo consists of one modifiable wire Z0. The test fixture consists of an incoming signal source VSOURCE, a 1,000-Ohm source resistor, a nonmodifiable wire ZOUT, a voltage probe point VOUT, a 100 mega-Ohm load resistor, and a nonmodifiable wire ZGND connecting to ground. The test fixture has two ports to which the embryo is connected and one port to which the developing circuit may connect. The two ports to which the embryo is connected are those at nodes 2 (that makes the input signal VSOURCE available to the circuit) and node 3 (that provides access to the voltage probe point VOUT). A third port provides optional access to ground.

Note that our use of an embryo with one modifiable wire for this problem (in contrast to the two modifiable wires used in section 4.2) is not motivated by any special requirement or exigency of the present problem. Problems are presented in this book in an order that seems appropriate to the overall organization and theme of this book

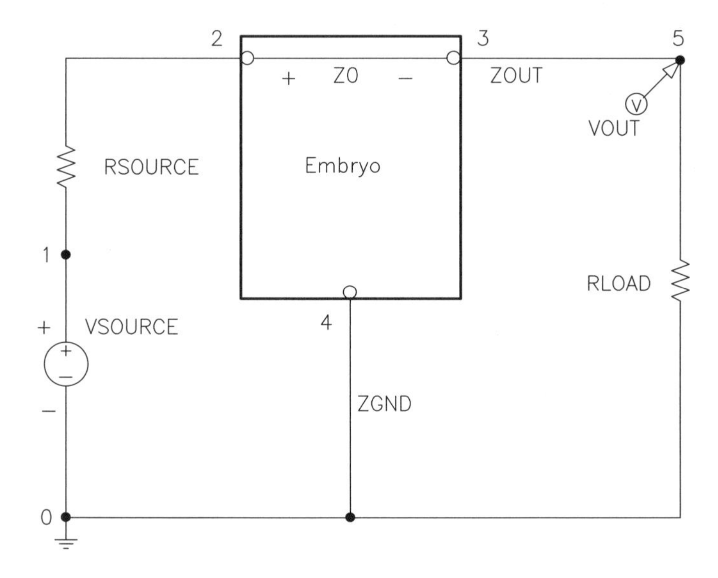

Figure 4.8 One-input, one-output initial circuit with one modifiable wire.

(and not in terms of the chronology of the work). Our use of an embryo with one modifiable wire for this problem reflects the chronology of our work and the evolution of our thinking (in particular, a recently adopted preference in favor of using the simplest possible embryo).

4.3.1.2 Program Architecture Because there must necessarily be one result-producing branch in the program tree for each modifiable wire in the embryo and there is one modifiable wire in the embryo in figure 4.8, the architecture of each circuit-constructing program tree has one result-producing branch. Neither automatically defined functions nor architecture-altering operations are used.

4.3.1.3 Terminal Set The terminal set is the same as that used in section 4.2.1.4.

4.3.1.4 Function Set A constrained syntactic structure enforces the use of one function set for the arithmetic-performing subtrees and another function set for all other parts of the program tree.

The function set, F_{aps}, for each arithmetic-performing subtree is

$$F_{aps}=\{+, -\}.$$

As in section 4.2, the challenge of this particular problem involves constructing a circuit using only resistors and capacitors. Thus, only resistor-creating and capacitor-creating functions are used for this particular problem. The function set, F_{ccs}, for each construction-continuing subtree is

F_{ccs}={R, C, SERIES, PARALLEL0, PARALLEL1, FLIP, NOP, PAIR_CONNECT_0, PAIR_CONNECT_1, RETAINING_THREE_GROUND0, RETAINING_THREE_GROUND1}.

Briefly, the two versions of the RETAINING_THREE_GROUND functions provide a way to connect a point in the developing circuit to ground. For additional details, see Koza, Bennett, Andre, and Keane 1999a.

4.3.1.5 Fitness Measure The fitness measure is based on the closeness of the behavior of a candidate circuit to the behavior of the circuit in Philbrick's patent. The defining characteristics of the Philbrick circuit are expressed in terms of its time-domain behavior and its frequency-domain behavior.

Fitness is the sum of two elements. The first element of the fitness measure is a weighted sum of the discrepancies between the behavior of the circuit in Philbrick's patent and the candidate circuit's behavior in the frequency domain (determined using an AC analysis in SPICE). The candidate circuit is simulated at 121 frequency values in an interval of six decades between 1 millihertz and 1,000 Hz. This element of the fitness measure is the sum, over the 121frequencies, of the absolute weighted deviation between the actual value of the voltage that is produced by the circuit at the probe point VOUT and the target value for voltage. For the 61 points between 1 Hz and 1,000 Hz, the absolute difference between the desired voltage and the actual output voltage is weighted by 1.0 if the voltage is above 970 millivolts, but otherwise weighted by 10. For the single point at 1 millihertz, the absolute difference between the desired voltage and the actual output voltage is weighted by 60.0. Discrepancies for the remaining 59 points are ignored.

The second element of the fitness measure is a weighted sum of the discrepancies between the behavior of the circuit in Philbrick's patent and the candidate circuit's behavior in the time domain (determined using a transient analysis in SPICE). The circuit is simulated over 121 values in an interval between 0 and 120 milliseconds. This element of the fitness measure is the sum, over the 121time values, of the absolute weighted deviation between the actual value of the voltage that is produced by the circuit at the probe point VOUT and the target value for voltage. The absolute difference between 1 Volt (the desired voltage between 0 and 10 milliseconds) and the actual output voltage for the 11 points between 0 and 10 milliseconds is weighted by 60 if the absolute difference is 30 millivolts or less, but otherwise weighted by 600. The absolute differences between 0 Volts (the desired voltage between 60 and 120 milliseconds) and the actual output voltage for the 61 points are weighted by 1 if the absolute difference is 300 millivolts or less, but otherwise weighted by 10. The discrepancies for the remaining 49 points are ignored.

4.3.1.6 Control Parameters The population size is 660,000.

4.3.2 Results for the Philbrick Circuit

The best-of-generation circuit from generation 39 (figure 4.9) consists of six resistors and six capacitors (not counting the components in the test fixture). This circuit has a fitness of 665.55. The element of the fitness measure pertaining to the frequency domain is 49.93 and the element pertaining to the time domain is 615.62.

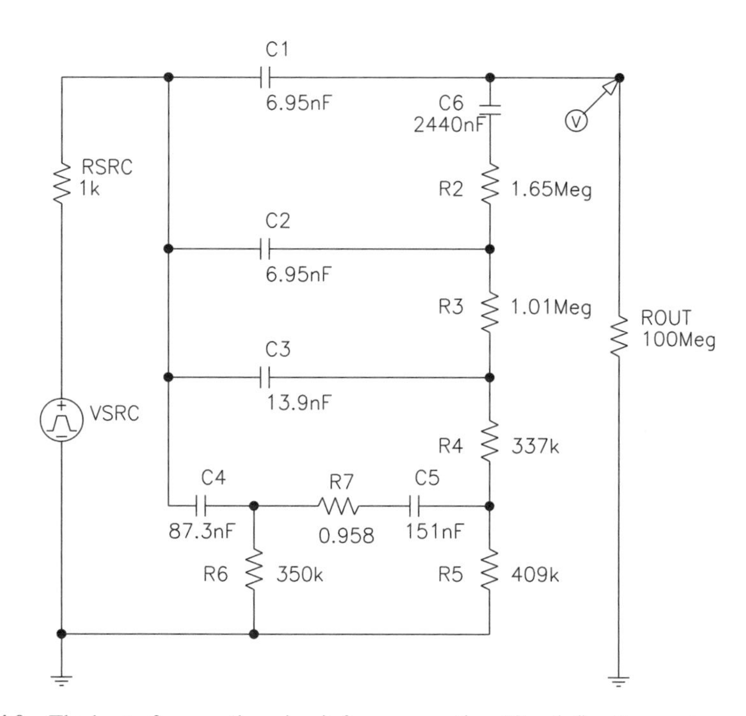

Figure 4.9 The best-of-generation circuit from generation 39 satisfies the requirements of the Philbrick circuit problem.

This run was continued beyond generation 39. A somewhat better circuit appeared in generation 214. This circuit is considerably larger than the best-of-generation circuit from generation 39. The circuit from generation 214 consists of 12 resistors and 15 capacitors. The fitness of this circuit is 663.03 (47.64 for the frequency domain and 615.39 for the time domain). Because the fitness of the considerably larger circuit from generation 214 is only 0.4% better than that of the best-of-generation circuit from generation 39, we focus now on the circuit from generation 39.

Figure 4.10 shows the behavior in the frequency domain of the best-of-generation circuit from generation 39.

The requirement that Philbrick (1956) established for (and that is satisfied by) his patented circuit was to transmit "the early portion of a transient voltage signal substantially without distortion" for approximately 10 milliseconds and to recover "to a quiescent value of zero voltage" after 100 milliseconds. Figure 4.11 compares the time-domain behavior of the Philbrick circuit (top curve with diamonds) and the best-of-generation circuit from generation 39 (bottom curve with squares). As can be seen in the figure, the first 10 milliseconds of the transient voltage signal of the evolved best circuit from generation 39 (bottom curve with boxes) is virtually the same as that for Philbrick's circuit (top curve with diamonds). After 100 milliseconds, the transient voltage signal of the evolved best circuit from generation 39 (with squares) is suppressed to a greater degree than that of Philbrick's circuit (without boxes). In other words, the genetically evolved circuit from generation 39 satisfies Philbrick's own requirement to a greater degree than Philbrick's patented circuit.

For additional information, see Koza, Bennett, Keane, Yu, Mydlowec, and Stiffelman 1999.

4.3.3 Human-Competitiveness of the Result for the Philbrick Circuit Problem

As just mentioned, the genetically evolved circuit from generation 39 satisfies the requirements stated in Philbrick's patent.

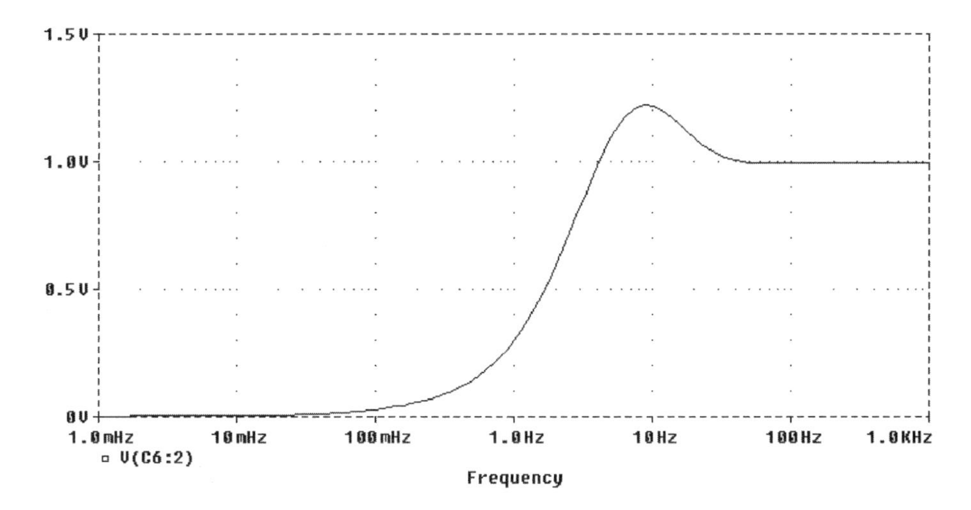

Figure 4.10 Frequency domain behavior of the best-of-generation circuit from generation 39.

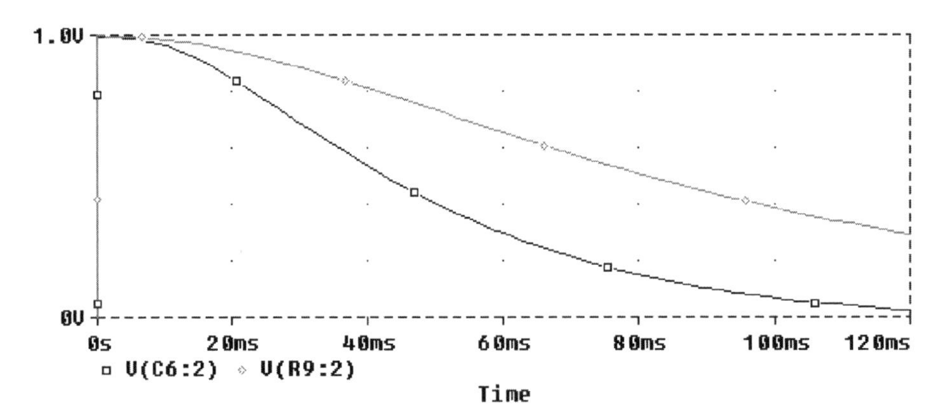

Figure 4.11 Comparison of the Philbrick circuit and the best-of-generation circuit from generation 39.

Referring to the eight criteria in table 1.2 for establishing that an automatically created result is competitive with a human-produced result, the rediscovery by genetic programming of the Philbrick circuit satisfies the following two of the eight criteria:

(A) The result was patented as an invention in the past, is an improvement over a patented invention, or would qualify today as a patentable new invention.
(F) The result is equal to or better than a result that was considered an achievement in its field at the time it was first discovered.

The rediscovery by genetic programming of the Philbrick circuit came over 40 years after Philbrick received a patent for his invention. Nonetheless, the fact that the original human-designed version satisfied the Patent Office's criteria for patent-worthiness means that the genetically evolved duplicate would also have satisfied the Patent Office's criteria for patent-worthiness (if only it had arrived earlier than Philbrick's patent application).

The fact that genetic programming rediscovered both the topology and sizing of an electrical circuit that was unobvious "to a person having ordinary skill in the art" (35 *United States Code* 103a) establishes that this evolved result satisfies Arthur Samuel's criterion (1983) for artificial intelligence and machine learning:

> "The aim [is] ... to get machines to exhibit behavior, which if done by humans, would be assumed to involve the use of intelligence."

4.3.4 Routineness for the Philbrick Circuit Problem

As mentioned in section 1.1.3, an automated problem-solving method (such as genetic programming) may be measured by its routineness. As mentioned there, a problem-solving method is "routine" if it is general and if relatively little human effort is required to get the method to successfully handle new problems within a particular domain and to successfully handle new problems from a different domain.

In section 4.2.3, we discussed the effort required to make the transition from a synthesis problem from the domain of controllers to a synthesis problem from the domain of electrical circuits. That is, we discussed the *inter-domain* aspect of routineness. The Philbrick circuit problem in this section and the problem (in section 4.2) of evolving an RC circuit with a gain that is greater than two are, of course, in the same domain—circuit synthesis. Thus, in this section, we consider the effort required to make the transition from one problem of circuit synthesis to another—that is, the *intra-domain* aspect of routineness.

The preparatory steps for the Philbrick circuit problem are substantially the same as the preparatory steps for the problem of evolving an RC circuit with a gain greater than two except that the problem involving the Philbrick circuit necessarily uses a different fitness measure. The new fitness measure reflects the requirements of the Philbrick circuit (as opposed to the requirements of the earlier problem). The change in fitness measures from problem to problem is an inherent and necessary part of making the transition from one problem to another. It does not entail an enormous amount of effort.

Once the preparatory steps are performed, the executional steps of genetic programming for the Philbrick problem are the same as for the problem of evolving an RC circuit with a gain that is greater than two (section 4.2).

Thus, relatively little effort is required to make the transition from the problem of evolving an RC circuit with a gain that is greater than two to the Philbrick problem. That is, the transition is routine.

4.3.5 AI Ratio for the Philbrick Circuit Problem

What is the AI ratio for the solution produced by genetic programming to the Philbrick problem?

As previously mentioned (section 4.3.3), the result produced by genetic programming on this problem is human-competitive because the result was a previously patented invention. Thus, there is a high amount of "A" in the numerator of the AI ratio.

The preparatory steps for the Philbrick circuit problem are (except for the fitness measure) substantially the same as the preparatory steps for the problem of evolving an RC circuit with a gain greater than two. Thus, for the same reasons that we stated in section 4.2.4, the solution produced by genetic programming for the Philbrick problem incorporates a small amount of "I."

The high amount of "A" represented by the human-competitive result in conjunction with the small amount of "I" provided by the human user means that the AI ratio for the solution produced by genetic programming is high.

4.4 Circuit for the NAND Function

Our early work on circuit synthesis using genetic programming neglected several categories of analog circuits, including analog circuits that perform digital functions and mixed analog-digital circuits.

The practical reason for this initial neglect is that the evaluation of the behavior of many purely analog circuits can be performed in the frequency domain. As a general rule, frequency-domain simulations usually consume considerably less computer time than time-domain simulations. When we were doing our early work on circuit synthesis, we did not have computational resources capable of performing a large number of computationally expensive time-domain simulations (within any reasonable amount of time).

The evaluation of the behavior of digital circuits and mixed analog-digital circuits typically requires simulations in the time domain. Moreover, for many problems, the evaluation of the behavior of digital circuits and mixed analog-digital circuits requires simulation of each candidate circuit over an enormous number of fitness cases. The required number of fitness cases is potentially $c2^k$, where k is the number of input bits and c is some constant.

A NAND circuit performs the elementary two-input Boolean NAND ("Not AND") function. NAND circuits are useful because any Boolean function can be constructed from a composition of NAND functions and because NAND functions can be efficiently manufactured using present-day silicon chip technology.

The problem in this section is to create an analog circuit composed of transistors, capacitors, and resistors that performs the NAND digital function with a response time of less than 100 nanoseconds.

4.4.1 Preparatory Steps for the NAND Circuit

4.4.1.1 Initial Circuit Figure 4.12 shows the two-input, one-output initial circuit consisting of an embryo embedded in a test fixture for this problem. The embryo in this initial circuit consists of three modifiable wires Z1, Z2, and Z3. The test fixture receives two incoming signals VSOURCE1 and VSOURCE2, each with a 1-Ohm source resistor (RSOURCE1 and RSOURCE2, respectively). The test fixture also has a nonmodifiable wire ZOUT1 between nodes 3 and 5, an output probe point VOUT1 at node 5, and a 1,000-Ohm load resistor RLOAD1 situated between nodes 5 and 0. The test fixture has three ports to which the embryo is connected. Two ports make the input signals (VSOURCE1 and VSOURCE2) available to the developing circuit. A third port is connected to the voltage probe point VOUT1.

4.4.1.2 Program Architecture Because there must be one result-producing branch in the program tree for each modifiable wire in the embryo and there are three modifiable wires in the embryo of figure 4.12, the program tree in the population has three result-producing branches. Neither automatically defined functions nor architecture-altering operations are used.

4.4.1.3 Terminal Set The terminal set is the same as that used in section 4.2.1.4.

4.4.1.4 Function Set A constrained syntactic structure enforces the use of one function set for the arithmetic-performing subtrees and another function set for all other parts of the program tree.

The function set, F_{aps}, for each arithmetic-performing subtree is

$$F_{aps} = \{+, -\}.$$

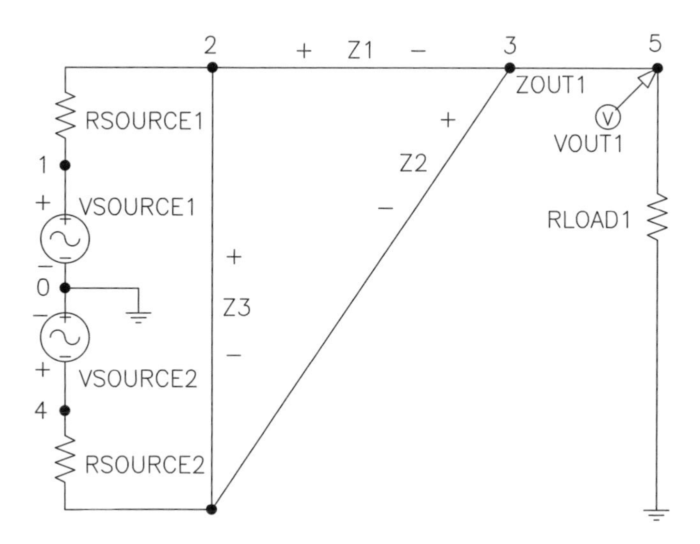

Figure 4.12 Two-input, one-output initial circuit.

The function set, F_{ccs}, for each construction-continuing subtree is

F_{ccs} = {R, SERIES, PARALLEL0, PARALLEL1, FLIP, NOP, RETAINING_THREE_GROUND_0, RETAINING_THREE_GROUND_1, RETAINING_THREE_ POS5V_0, RETAINING_THREE_POS5V_1, PAIR_CONNECT_0, PAIR_CONNECT_1, Q_DIODE_NPN, Q_DIODE_PNP, Q_THREE_NPN0,..., Q_THREE_ NPN11, Q_THREE_PNP0,..., Q_THREE_PNP11, Q_POS5V_COLL_NPN, Q_POS5V_EMIT_PNP, Q_GND_EMIT_NPN, Q_GND_EMIT_PNP}.

Briefly, the 12 versions of the transistor-creating Q_THREE_NPN function each act by inserting an NPN transistor into a developing circuit. Likewise, the 12 versions of the Q_THREE_PNP function each act by inserting a PNP transistor. We use the commercially common 2N3904 (*npn*) and 2N3906 (*pnp*) transistor models for all problems throughout this book (unless otherwise indicated).

The two versions of the component-creating Q_DIODE_NPN function insert a diode into a developing circuit.

The two versions of the transistor-creating Q_POS5V function each insert a transistor whose designated lead (emitter or collector) makes a connection to the +5-Volt power supply.

The two versions of the transistor-creating Q_GND_EMIT function each insert a transistor whose emitter makes a connection to ground.

The two versions of the RETAINING_THREE_GROUND function each make a connection to ground.

The two versions of the RETAINING_THREE_POS5V function each make a connection to the +5-Volt power supply.

For additional details about these functions, see Koza, Bennett, Andre, and Keane 1999a.

Note the following implementation detail: Whenever a connection is made to a power source, a uniquely numbered, separate power source is placed in the netlist. That is, if two points in the circuit are connected to the +5-Volt power supply, two different power supplies are used. This approach avoids certain errors that sometimes arise in the simulator.

4.4.1.5 Fitness Measure In this problem, the output voltage VOUT is measured in the time domain. Hence, the SPICE simulator is instructed to perform a transient (time domain) analysis.

A binary zero is represented by a 0-Volt signal and a binary 1 is represented by a 5-Volt signal.

Both of the input ports (called V0 and V1 in figure 4.13) are exposed to 18 100-nanosecond signals over a 1.8-microsecond period. Each 100-nanosecond signal is sampled every 20 nanoseconds (i.e., five sample points per 100 nanoseconds). Thus, there are 91 fitness cases over the full 1.8-microsecond period. Both inputs are zero during the first 100 nanoseconds. The next 17 100-nanosecond periods represent the 16 possible transitions between each of the four possible combinations of the two input signals.

The fitness of a circuit is the sum, over the 91 fitness cases, of the weighted absolute value of the difference between the actual output voltage at the probe point VOUT and the desired output voltage.

The weighting used in the fitness measure is designed not to penalize ideal voltage values, to slightly penalize every acceptable voltage, and to heavily penalize every unacceptable voltage.

Concerning the two ideal outcomes, if the desired digital signal is 1 and the actual voltage is 5.0 Volts or the desired digital signal is 0 and the actual voltage is 0, the deviation is 0.

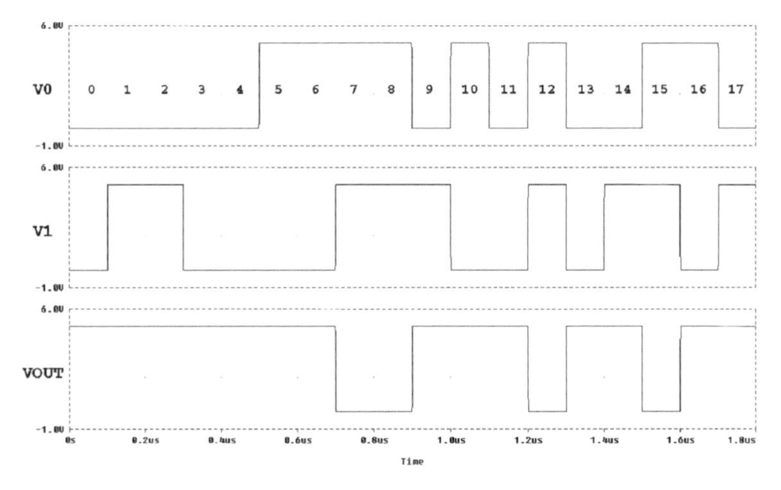

Figure 4.13 The input signals in the time domain for the NAND circuit.

Concerning the acceptable outcomes, if the desired digital signal is 1 and the actual voltage is within 0.3 Volts of 5.0 Volts or the desired digital signal is 0 and the actual voltage is within 0.4 Volts of 0.0 Volts, the absolute value of the deviation from the desired output voltage (5 Volts or 0 Volts, respectively) is weighted by a factor of 1.0.

If the actual voltage is outside this range, the absolute value of the deviation is weighted by a factor of 10.0.

The smaller the overall value of fitness, the better.

The number of hits is defined as the number of fitness cases for which the voltage is acceptable.

4.4.1.6 Control Parameters The population size is 132,000.

4.4.1.7 Termination Criterion Because the goal in this particular problem is to generate a variety of 100%-compliant circuits for examination, the run was not automatically terminated upon creation of the first 100%-compliant individual. Instead, the maximum number of generations, G, is set to an arbitrary large number (e.g., 501); numerous 100%-compliant circuits are harvested; and the run is manually monitored and manually terminated.

4.4.2 *Results for the NAND Circuit*

The best-of-run circuit (figure 4.14) from generation 17 has five transistors and five resistors. This 100%-compliant circuit scores 91 (out of 91) hits and has a fitness of 7.85.

Figure 4.15 shows the behavior of the best-of-run circuit from generation 17 in the time domain for the 91 fitness cases (as shown in figure 4.13).

Figure 4.16 shows a textbook TTL (Transistor-Transistor Logic) NAND circuit (Wakerly 1990) consisting of five transistors, four resistors, and three diodes.

Notice that the two inputs to the genetically evolved circuit (figure 4.14) are tied together (after the two source resistors) and the circuit thus resembles RTL ("Resistor-Transistor Logic"), a predecessor of TTL logic.

There have been numerous patents issued for NAND circuits, including the 1971 patent to David H. Chung and Bill H. Terrell of Texas Instruments Inc. (Chung and Terrell 1971).

For additional information, see Bennett, Koza, Keane, Yu, and Mydlowec 1999.

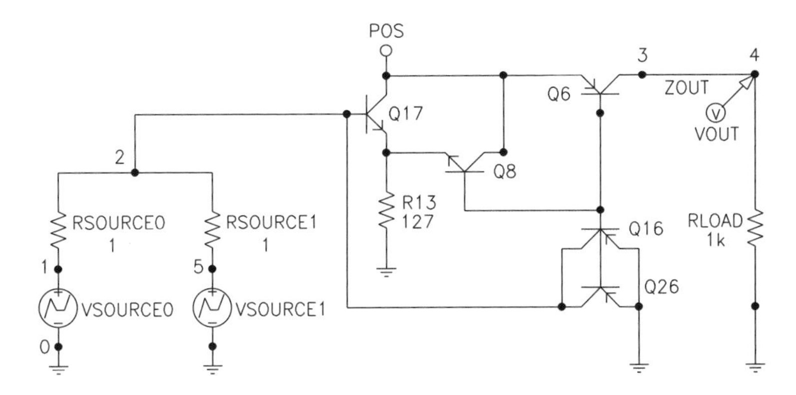

Figure 4.14 Best-of-run NAND circuit.

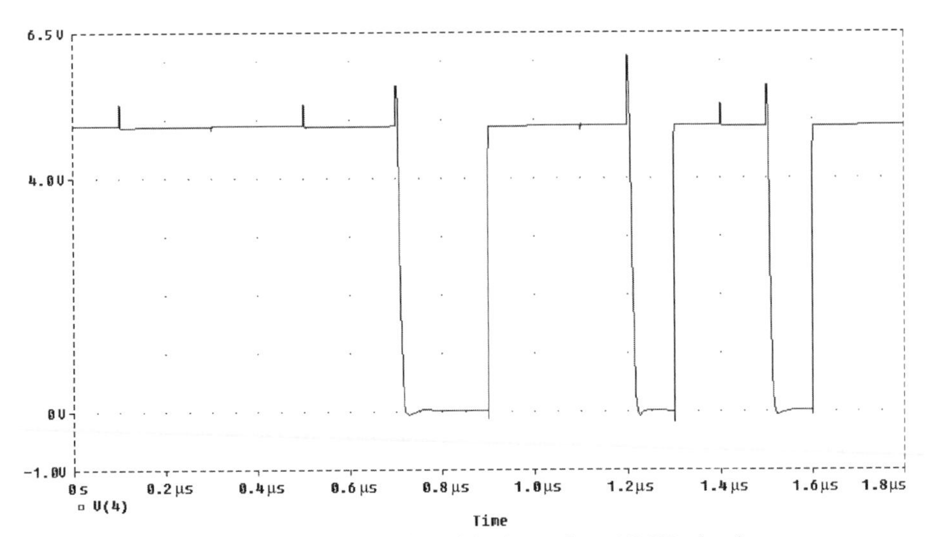

Figure 4.15 Behavior of the best-of-run NAND circuit.

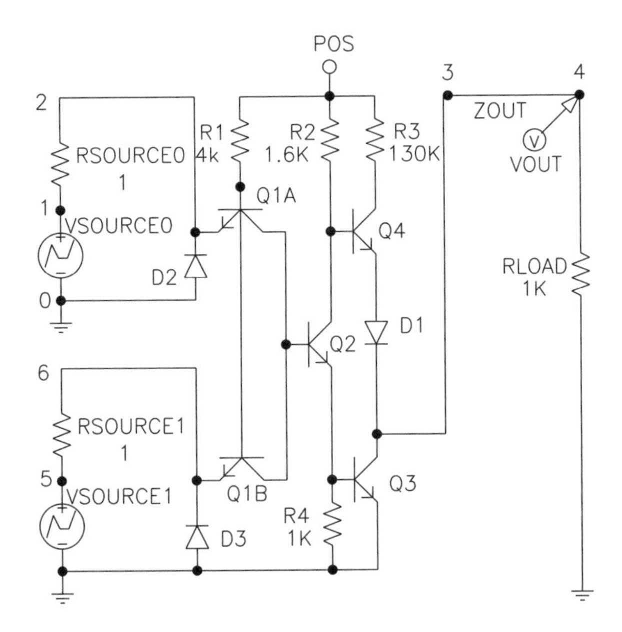

Figure 4.16 Textbook TTL NAND circuit.

4.4.3 Human-Competitiveness of the Result for the NAND Circuit Problem

The evolved NAND circuit is another example of a human-competitive result produced by genetic programming.

Referring to the eight criteria in table 1.2 for establishing that an automatically created result is competitive with a human-produced result, the automatic synthesis of a NAND circuit satisfies the following two criteria:

(A) The result was patented as an invention in the past, is an improvement over a patented invention, or would qualify today as a patentable new invention.
(F) The result is equal to or better than a result that was considered an achievement in its field at the time it was first discovered.

This evolved result satisfies Arthur Samuel's criterion (1983) for artificial intelligence and machine learning:

> "The aim [is] ... to get machines to exhibit behavior, which if done by humans, would be assumed to involve the use of intelligence."

4.4.4 Routineness for the NAND Circuit Problem

The preparatory steps for the NAND circuit problem are substantially the same as the preparatory steps for the RC circuit with a gain that is greater than two (section 4.2) and the Philbrick circuit (section 4.3) with two exceptions.

First, the NAND circuit problem has a different goal (and hence a different fitness measure).

Second, the function set for the NAND circuit problem includes the transistor-creating function. This minor change in the function set does not entail an enormous amount of effort. The vast majority of present-day electrical circuits are composed of resistors, capacitors, and transistors. Having once expanded the repertoire of functions to include transistors, the expanded function set can hereafter be used for *any* problem involving the automatic synthesis of *any* circuit composed of resistors, capacitors, and transistors.

In summary, relatively little effort is required to make the transition from the previous two problems of analog circuit synthesis to the present problem. That is, the transition is routine.

4.4.5 AI Ratio for the NAND Circuit Problem

The solution produced by genetic programming to this problem is human-competitive and therefore has a high amount of "A." The preparatory steps for this problem are substantially the same as the preparatory steps for earlier problems of circuit synthesis (except that this problem has a different goal and hence a different fitness measure). Thus, the solution produced by genetic programming to this problem incorporates a small amount of "I." Thus, the AI ratio is high for the solution produced by genetic programming.

4.5 Evolution of a Computer

This section presents the automatic synthesis of a two-instruction arithmetic logic unit (ALU) circuit that has two one-bit data inputs and one one-bit instruction input. The possible instructions are NAND (represented by an input of 1) and NOR (represented by an input of 0). The circuit executes one of these instructions on the two input bits and produces one bit as its output.

4.5.1 Preparatory Steps for the Arithmetic Logic Unit

The preparatory steps for the ALU problem are the same as those for the NAND circuit (section 4.4), except as mentioned below.

4.5.1.1 Initial Circuit Figure 4.17 shows the three-input, one-output initial circuit for the arithmetic logic unit problem. The embryo consists of three modifiable wires Z0, Z1, and Z2. The test fixture receives three incoming signals (VSOURCE0, VSOURCE1, and VSOURCE2). There is a 1-Ohm source resistor associated with each of these three incoming signals. The circuit also has a voltage probe point VOUT and a 1,000-Ohm load resistor RLOAD. The test fixture has four ports to which the embryo is connected. Three ports make the input signals available to the developing circuit. The fourth port is connected to the voltage probe point VOUT.

4.5.1.2 Fitness Measure In this problem, the output voltage V(1) is measured in the time domain. Hence, the SPICE simulator is instructed to perform a transient (time domain) analysis.

A total of 64 combinations of the three digital input signals are presented to the circuit. The 64 input triples represent all possible transitions between each of the eight possible combinations of the three input signals. Each combination of inputs persists for a total of 10 microseconds. Each combination of inputs is sampled every 2 microseconds (i.e., five sample points per 10 microseconds). Thus, there are a total of 321 fitness cases. SPICE is instructed to perform a transient (time domain) analysis.

In figure 4.18, the signals labeled V(3), V(4), and V(5) together represent various combinations of inputs. In particular, V(4) and V(5) are the two input bits and V(3) is the bit that specifies the instruction (NAND or NOR) that the ALU circuit is to execute

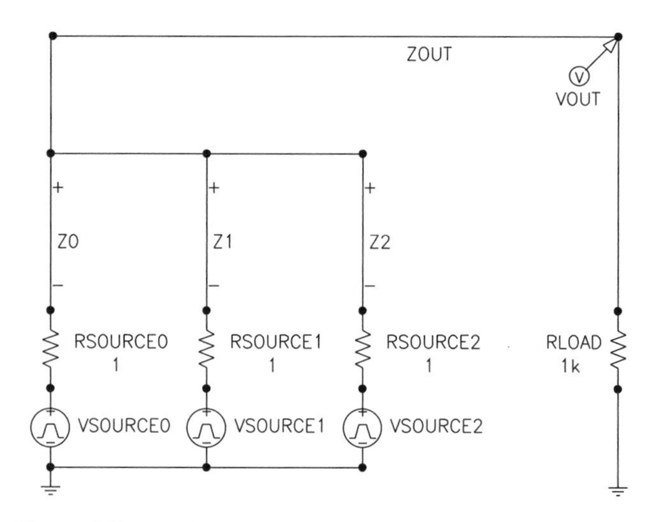

Figure 4.17 Initial circuit for the arithmetic logic unit problem.

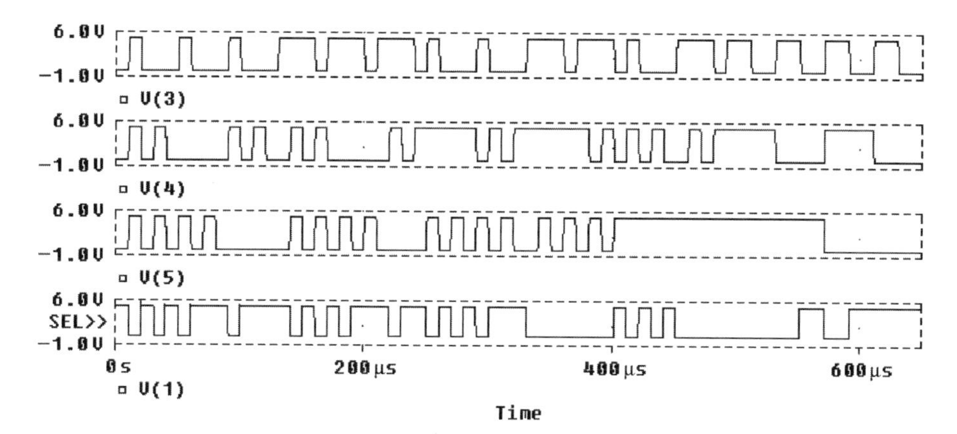

Figure 4.18 Fitness cases for the arithmetic logic unit problem.

on the two input bits. Signal V(1) shows the correct output for the ALU circuit. A binary zero is represented by a 0-Volt signal and a binary 1 is represented by a 5-Volt signal.

The fitness of a circuit is the sum, over the 321 fitness cases, of the weighted absolute value of the difference between the actual output voltage at the probe point VOUT and the desired output voltage.

If the voltage is exactly equal to the desired voltage, the deviation is 0. If the voltage is within 0.40 Volts of the desired voltage, the absolute value of the deviation from the desired output voltage is weighted by a factor of 1.0. If the voltage is outside this range, the absolute value of the deviation is weighted by a factor of 10.0.

The number of hits is defined as the number of fitness cases for which the voltage is acceptable.

4.5.1.3 Control Parameters The population size is 1,320,000.

4.5.2 Results for the Arithmetic Logic Unit

The 100%-compliant best-of-run circuit (figure 4.19) from generation 33 scores 321 (out of 321) hits and has a fitness of 215.6.

Figure 4.20 shows (in the time domain) the 100%-correct output V(1) of the best-of-run circuit from generation 33 for the 321 fitness cases.

For additional information, see Koza, Bennett, Keane, Yu, Mydlowec, and Stiffelman 1999.

4.5.3 Routineness for the Arithmetic Logic Unit Circuit Problem

The preparatory steps for this problem are substantially the same as the preparatory steps for previous problems of analog circuit synthesis, except that the present problem has a different goal (and hence a different fitness measure). The transition from the previous problems of analog circuit synthesis to the present problem is routine.

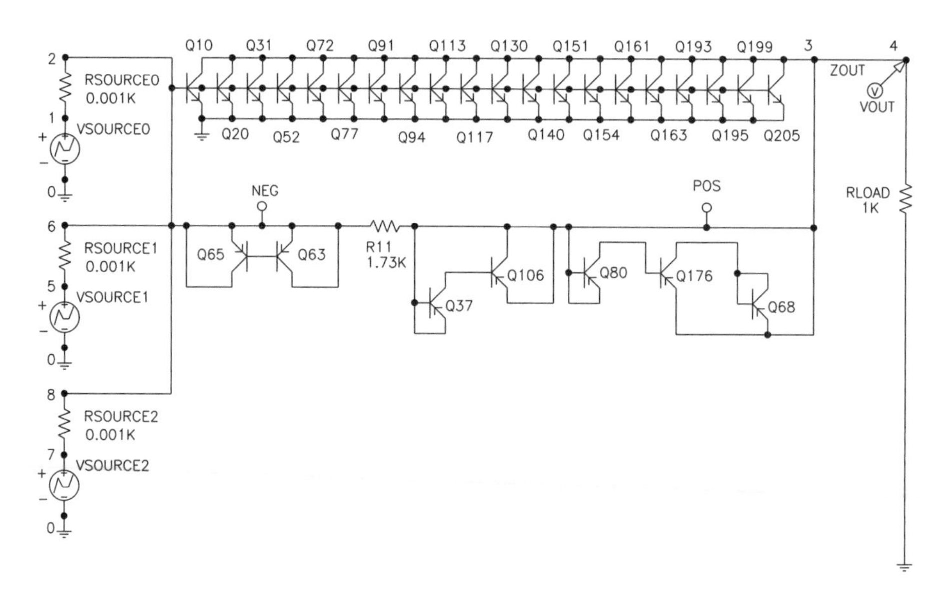

Figure 4.19 Best-of-run circuit for the arithmetic logic unit problem.

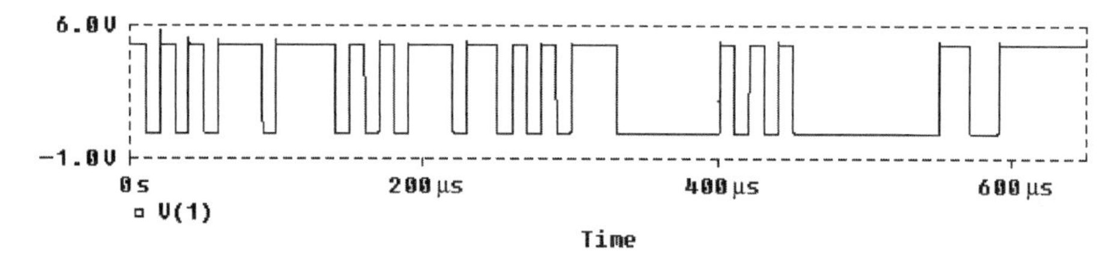

Figure 4.20 Behavior of the best-of-run circuit for the arithmetic logic unit problem.

4.5.4 AI Ratio for the Arithmetic Logic Unit Circuit Problem

The solution produced by genetic programming to this problem has a moderately high amount of "A." The preparatory steps for this problem incorporate a small amount of "I." Thus, the AI ratio for the solution produced by genetic programming is moderately high.

4.6 Square Root Circuit

An analog electrical circuit whose output is a mathematical function (e.g., square, square root) is called a *computational circuit.*

Genetic programming has been previously used to automatically synthesize analog computational circuits, including squaring, cubing, square root, cube root, logarithm, and Gaussian circuits (Koza, Bennett, Andre, and Keane 1999a). However, these previous results have limited applicability because they were created using DC sweeps. DC sweeps are considerably less computationally expensive than time-domain simulations. However, this type of computationally inexpensive simulation may not yield a robust circuit that performs the desired mathematical computation correctly over time. This section addresses this shortcoming by using multiple time-domain simulations to automatically synthesize a one-input computational circuit for the square root function.

4.6.1 Preparatory Steps for Square Root Circuit

4.6.1.1 Initial Circuit A one-input, one-output initial circuit with one modifiable wire (figure 4.8) is used. The initial circuit has an incoming signal source, a 1,000-Ohm source resistor, a voltage probe point VOUT, and a 1-Ohm load resistor.

4.6.1.2 Program Architecture The circuit-constructing program tree has one result-producing branch for each modifiable wire in the embryo of the initial circuit. Thus, each program tree has one result-producing branch. Automatically defined functions and architecture-altering operations are not used.

4.6.1.3 Terminal Set A constrained syntactic structure enforces the use of one terminal set for the arithmetic-performing subtrees and another terminal set for all other parts of the program tree.

The numerical parameter value for each electrical component possessing a parameter in this problem is established by a value-setting subtree containing a single perturbable numerical value. That is, the argument to each component-creating function possessing a component value consists of a single terminal representing a perturbable numerical value. This approach for establishing numerical parameter values is described in section 3.5.5.2.

The terminal set, T_{vss}, for the value-setting subtrees is

$$T_{vss} = \{\Re_p\},$$

where $\Re_p$ denotes a perturbable numerical value.

The terminal set for each construction-continuing subtree is

$$T_{rpb} = \{\texttt{END, SAFE\_CUT}\}.$$

4.6.1.4 Function Set The function set for this problem is identical to that used for the NAND circuit in section 4.4, except for the inclusion of the capacitor-inserting `C` function and the inductor-inserting `L` function in function set, F_{ccs}, for each construction-continuing subtree.

The `L` function inserts an inductor and establishes the numerical value of the inductor. Note that the inductor-inserting and capacitor-inserting functions are unnecessary for this problem (and, in fact, do not appear in the best-of-run individual). For a more detailed explanation of the `L` function, see Koza, Bennett, Andre, and Keane 1999a.

4.6.1.5 Fitness Measure For the square root problem, each individual circuit is exposed to four time-domain signals. Each of these four signals lasts for a duration of one second each, as shown in figures 4.21, 4.22, 4.23, and 4.24. These input signals are structured to provide a representative mixture of input values (all of whose correct outputs are well within the operating range of the transistors that we use).

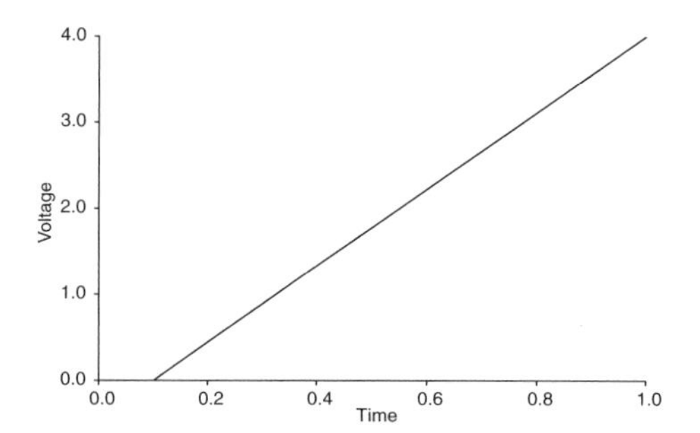

Figure 4.21 Rising ramp for the square root problem.

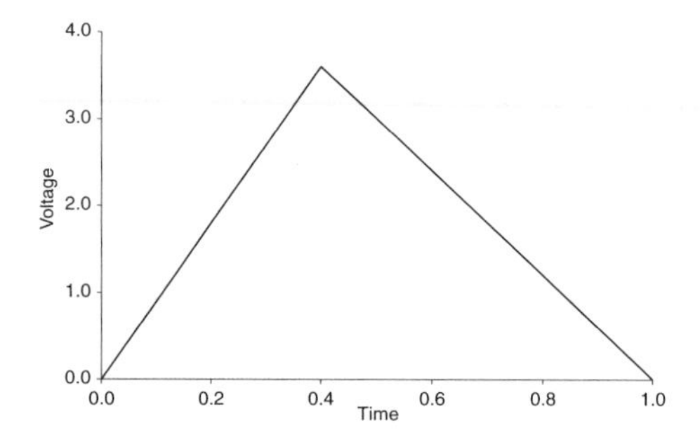

Figure 4.22 Triangle for the square root problem.

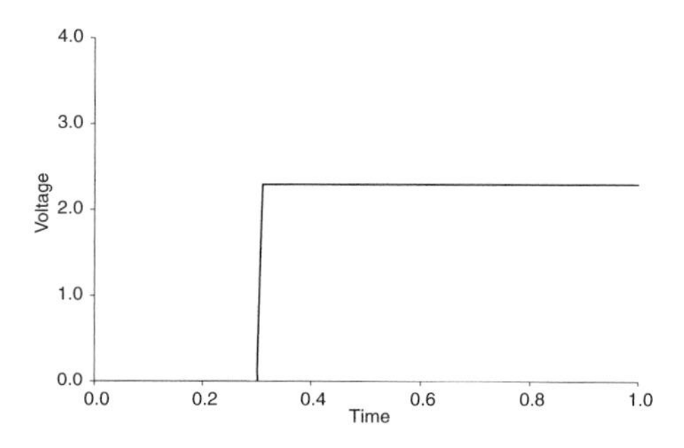

Figure 4.23 Rising step for the square root problem.

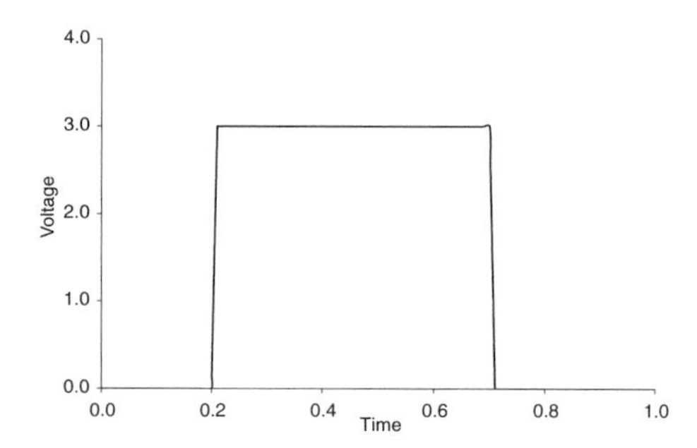

Figure 4.24 Pulse for the square root problem.

We first considered running this problem using the "obvious" fitness measure consisting of the sum, over a certain number of time steps for each of the four fitness cases, of the absolute value of the difference between the actual output voltage at probe point VOUT and the desired output voltage. However, initial tests on a single Pentium workstation indicated that this fitness measure often yielded circuits having noticeable spikes in the output. These spikes were reminiscent of the glitches that we encountered in our work involving the digital-to-analog (DAC) converter circuit (Bennett, Koza, Keane, Yu, Mydlowec, and Stiffelman 1999). Therefore, we adopted the same fitness measure that we previously used for the DAC problem. Specifically, we considered the difference only at times that correspond to SPICE's internally-created turn-defining points. Thus, the fitness measure for the square root computational circuit is the sum, for each of SPICE's turns, of the absolute value of the difference between the actual output voltage at the probe point VOUT and the desired output voltage.

4.6.1.6 Control Parameters The population size is 10,000,000.

4.6.2 Results for Square Root Circuit

The fitness of the best individual from generation 0 is 212.0.

The best-of-run individual appeared in generation 66. This best-of-run circuit has 39 components and a fitness of 6.26989 (figure 4.25). The circuit-constructing program tree has 284 points.

Figure 4.26 shows the output voltage produced by the best-of-run individual from generation 66 for the rising ramp input superimposed on the (virtually indistinguishable) correct output voltage for the square root function. As can be seen from this figure (and the three following figures), the output voltage produced by the best-of-run circuit from generation 66 closely matches the desired output for a square root circuit.

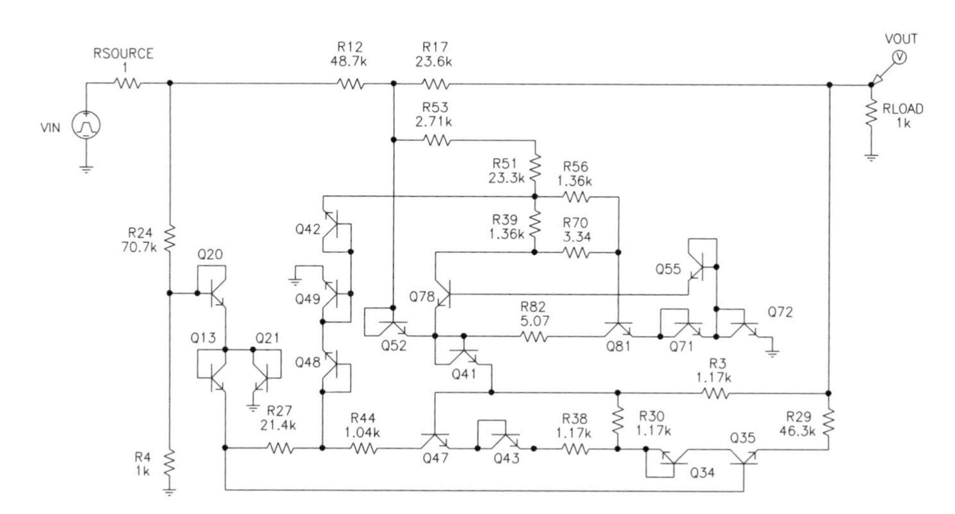

Figure 4.25 Best-of-run square root circuit.

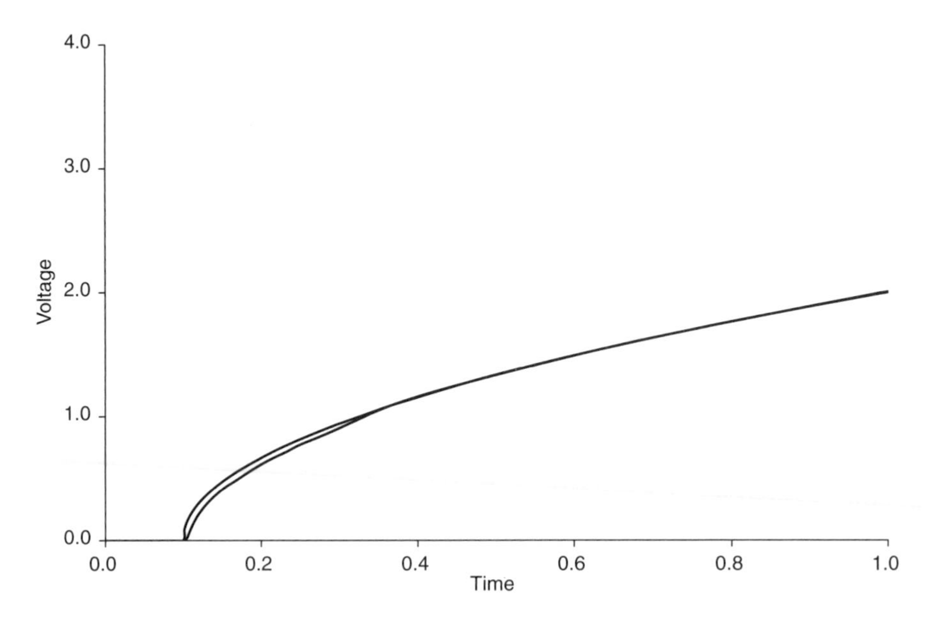

Figure 4.26 Output of the best-of-run square root circuit for the rising ramp input.

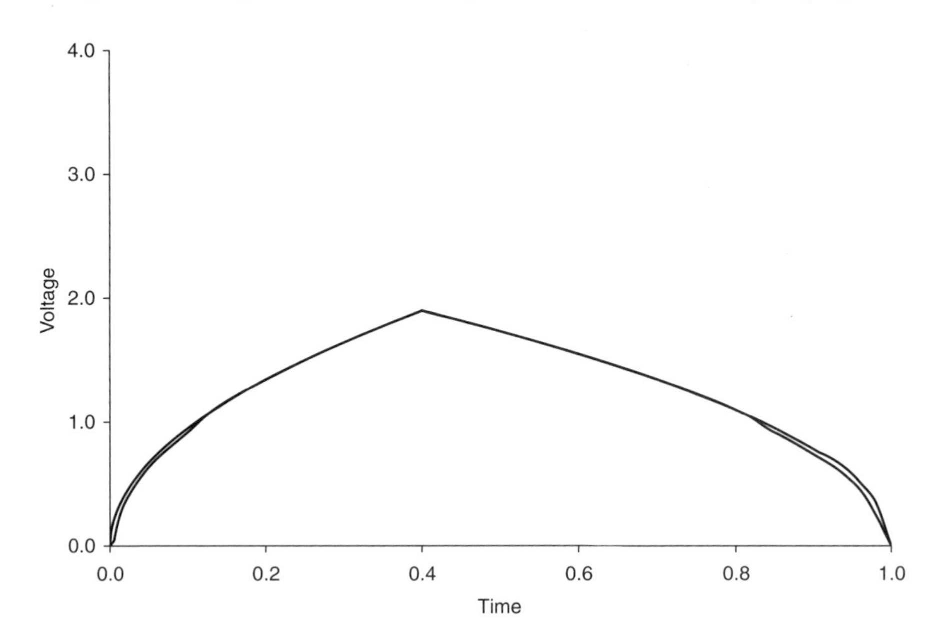

Figure 4.27 Output of the best-of-run square root circuit for the triangle input.

Figure 4.27 shows the output for the triangle input superimposed on the (virtually indistinguishable) correct output voltage for the square root function.

Figure 4.28 shows the output for the rising step input superimposed on the (virtually indistinguishable) correct output voltage for the square root function.

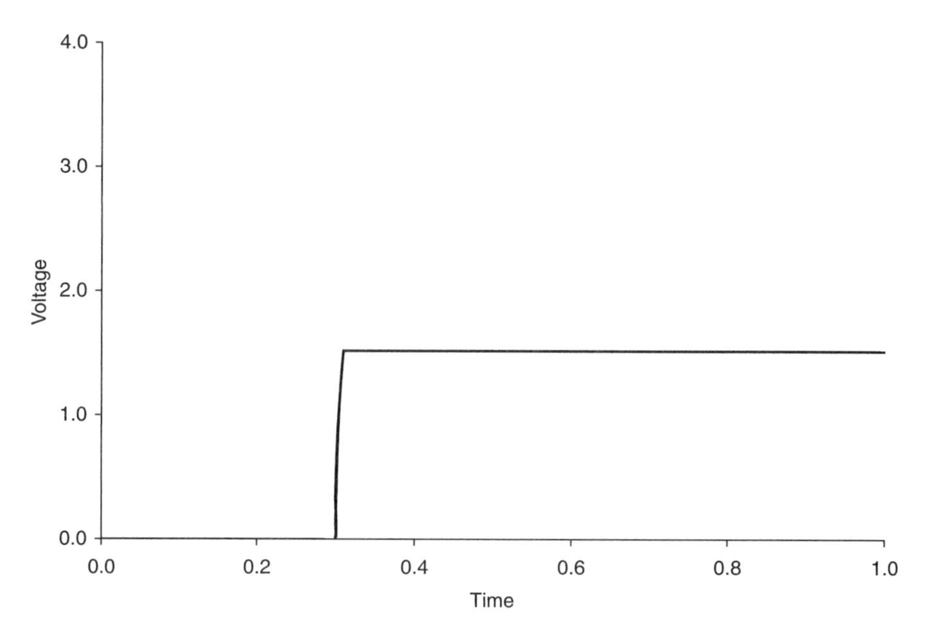

Figure 4.28 Output of the best-of-run square root circuit for the rising step input.

Figure 4.29 shows the output for the pulse input superimposed on the (virtually indistinguishable) correct output voltage for the square root function.

SPICE employs 299 points in simulating the best-of-run circuit from generation 66 for the four input signals. The average absolute error over these 299 points is 0.020 Volts.

For additional information, see Mydlowec and Koza 2000.

The authors are aware that the circuits that are genetically created in this book do not possess all the characteristics that one might reasonably require for a commercial version of the type of circuit under discussion. Commercial circuits have requirements that go considerably beyond those incorporated into the fitness measures in this book. Moreover, all the circuits described in this book were validated in simulation, but not in silicon. Also, the SPICE simulator operates on circuits composed of discrete components. In addition, when an integrated circuit is manufactured (particularly for high-frequency operation), it is necessary to consider the parasitic and timing effects of the actual physical (geographic) placement of the components on the silicon chip and the actual routing of wires on the chip. It is also necessary to consider manufacturing variations in designing practical circuits. Moreover, the available device models used in SPICE (although good in general) are generally crafted on the assumption that the components will operate in everyday regimes. These models do not necessarily accurately represent every conceivable operating regime—particularly those that an evolutionary process may devise. It is therefore possible that some of the components in some of the genetically evolved circuits in this book are operating in regimes in which their models are not representative of their actual behavior.

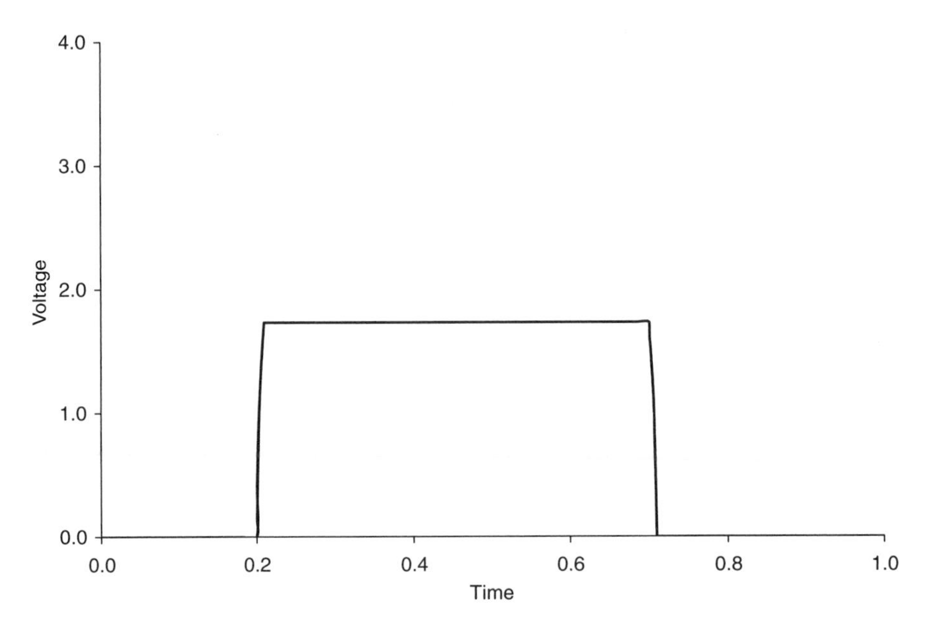

Figure 4.29 Output of the best-of-run square root circuit for the pulse input.

4.6.3 Routineness for the Square Root Circuit Problem

The preparatory steps for this problem are substantially the same as the preparatory steps for previous problems of analog circuit synthesis, except that the present problem has a different goal (and hence a different fitness measure). The transition from the previous problems of analog circuit synthesis to the present problem is routine.

4.6.4 AI Ratio for the Square Root Circuit Problem

The solution produced by genetic programming to this problem has a moderately high amount of "A." The preparatory steps for this problem incorporate a small amount of "I." Thus, the AI ratio for the solution produced by genetic programming is moderately high.

4.7 Automatic Circuit Synthesis Without an Explicit Test Fixture

This section pursues the question of the extent to which genetic programming can solve problems using a minimal amount of human-supplied information.

In this section, we demonstrate that it is possible for a run of genetic programming to solve a problem of circuit synthesis without the benefit of a built-in source resistor and built-in load resistor in the test fixture. This is accomplished at an increased cost in terms of computer time.

The demonstration involves a problem of synthesizing a lowpass filter.

A *filter* is a one-input, one-output circuit that passes the frequency components of the incoming signal that lie in a specified range (called the *passband*) while suppressing the frequency components that lie in all other frequency ranges (the *stopband*).

A *lowpass filter* passes all frequencies below a certain specified frequency, but stops all higher frequencies.

A *highpass filter* stops all frequencies below a certain specified frequency, but passes all higher frequencies.

An ideal filter would deliver full voltage in the passband (1 Volt in this problem) and a voltage of 0 in the stopband.

The *attenuation* of a filter is the ratio between the voltage at a circuit's output probe point and a reference voltage. Attenuation is often measured in terms of decibels (section 3.7.1.4).

The problem in this section is to automatically synthesize a lowpass filter circuit composed of passive components (i.e., inductors and capacitors) with a passband boundary of 1,000 Hz, a stopband boundary of 2,000 Hz, and with 1,000-to-1 (or better) attenuation of the stopband voltage relative to the passband voltage.

In practice, the ideal voltage (1 Volt) is not achieved throughout the passband. The (preferably small) variation within the passband is called the allowable *passband ripple*. A voltage in the passband that is between 970 millivolts and 1,030 millivolts (i.e., a passband ripple of 30 millivolts or less) is regarded as acceptable for purposes of this problem. Conversely, other voltages are regarded as unacceptable.

Similarly, the incoming signal is, in practice, not fully attenuated to the ideal level of 0 in the stopband. The (preferably small) variation within the stopband is called the allowable *stopband ripple*. A voltage in the stopband that is between 0 Volts and 1 millivolt (i.e., a stopband ripple of 1 millivolt or less) is regarded as acceptable for this problem. Any voltage above 1 millivolt in the stopband is regarded as unacceptable.

The frequencies between 1,000 Hz and 2,000 Hz constitute the filter's transitional "don't care" region.

These specifications are identical to those of chapter 25 of *Genetic Programming III* (Koza, Bennett, Andre, and Keane 1999a). In that previous work, the test fixture contained a built-in 1,000-Ohm source resistor and a built-in 1,000-Ohm load resistor. The test fixture in the previous work has the topology of figure 4.8 of this book. In this section, there is no built-in source or built-in load resistor.

4.7.1 Preparatory Steps for the Lowpass Filter Problem Without an Explicit Test Fixture

4.7.1.1 Initial Circuit In this problem, the initial circuit has the incoming signal source VIN and a voltage probe point VOUT; however, there is no built-in source resistor or built-in load resistor. Thus, each individual circuit must grapple with the task of either creating a suitably-valued explicit source resistor and suitably-valued explicit load resistor or figuring out a way of doing without them.

The *floating embryo* used in this problem consists of a single modifiable wire that is not initially connected to the circuit's input, output, or ground. Thus, each developing circuit must grapple with the task of discovering the circuit's input, output, or ground.

Use of the floating embryo requires functions that enable a circuit to make connection to the circuit's input(s) and output(s).

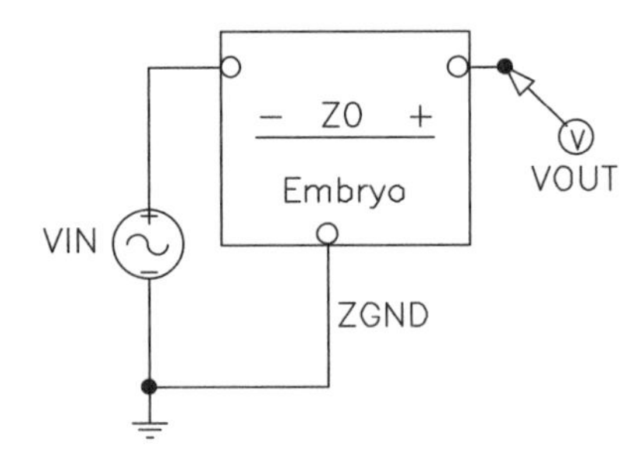

Figure 4.30 Initial circuit for the lowpass filter problem without an explicit test fixture.

The two-argument `INPUT_0` function makes a connection from a point in a developing circuit to the circuit's external input port 0. This function creates a new node and a series composition consisting of a first modifiable wire, a nonmodifiable wire, and a second modifiable wire. The nonmodifiable wire is associated with the new node. The nonmodifiable wire makes a connection to external input 0 at the new node. The `INPUT_0` function takes two arguments, one for each of the construction-continuing subtrees associated with the two modifiable wires. The `INPUT_1`, `INPUT_2`,... functions operate in a similar way with respect to external inputs 1, 2, and so forth.

Similarly, the two-argument `OUTPUT_0`, `OUTPUT_1`, `OUTPUT_2`,... functions are used to make a connection from a point in a developing circuit to the circuit's external output ports.

Figure 4.30 shows the test fixture for this problem in the same style that is used elsewhere in this book. This figure shows a source signal VIN, an output probe point VOUT, and ground.

Two implementation details should be mentioned at this time in connection with our use of the floating embryo in this problem. These two techniques are used on all problems of circuit synthesis in this book.

First, dangling wires are not allowed by the SPICE simulator. The floating embryo has dangling wires. To accommodate this requirement of the simulator, our software automatically modifies all netlists (before they are passed along to SPICE) by connecting each end of a dangling wire to a nonmodifiable, grounded 1-giga-Ohm resistor. In particular, each end of the modifiable wire in a floating embryo is connected (just prior to simulation) to a nonmodifiable, grounded 1-giga-Ohm resistor. Because this resistance is so high, these added resistors have no noticeable effect on the circuit's behavior. However, the addition of these functionless resistors enables SPICE to simulate the circuit. These added high-valued resistors should not be confused with the missing source or load resistors (which must necessarily possess non-extreme values of resistance in order to perform their desired functions).

Second, to accommodate a specific requirement of the SPICE simulator concerning short circuits, a nonmodifiable 1-pico-Ohm resistor is inserted after VIN. Because this resistance is so low, this added resistor has negligible effect on the circuit's behavior. This added resistor is included in the netlist that is supplied to the SPICE simulator.

Note that because these huge 1-giga-Ohm resistors and these tiny 1-pico-Ohm resistors are not modifiable, they cannot serve as the starting point for developing any part of the (missing) test fixture.

Of course, a circuit only containing resistors with extreme (high or low) values will receive a very poor value of fitness.

4.7.1.2 Program Architecture Because there must be a result-producing branch in the program tree for each modifiable wire in the embryo and there is one modifiable wire in the embryo, the architecture of each circuit-constructing program tree has one result-producing branch. Automatically defined functions and the architecture-altering operations are not used.

4.7.1.3 Function Set A constrained syntactic structure enforces the use of one function set for the first argument of the TWO_LEAD function and another function set for all other parts of the program tree.

The `TWO_LEAD` function is described in section 10.1.4 and provides a way to insert two-leaded components (such as resistors, capacitors, and inductors) into the developing circuit. Because resistors, inductors, and capacitors may be used in this problem, the function set, $F_{\text{two-lead}}$, for the first argument of the `TWO_LEAD` function is

$$F_{\text{two-lead}} = \{\texttt{R\_NEW}, \texttt{L\_NEW}, \texttt{C\_NEW}\}.$$

The function set, F_{ccs}, for each construction-continuing subtree is

$$F_{ccs} = \{\texttt{TWO\_LEAD}, \texttt{SERIES}, \texttt{PARALLEL\_NEW}, \texttt{NODE}, \texttt{TWO\_GROUND}, \texttt{THREE\_GROUND}, \texttt{INPUT\_0}, \texttt{OUTPUT\_0}\}.$$

The `PARALLEL_NEW` function is described in section 10.1.2.

The `NODE` function is described in section 10.1.1.

The two-argument `TWO_GROUND` function enables a part of a circuit to be connected directly to ground. The connection to ground is unconditionally made (i.e., even if there is only one `TWO_GROUND` function in the circuit-constructing program tree). After the execution of a `TWO_GROUND` function, there are two new modifiable wires.

Similarly, the three-argument `THREE_GROUND` function enables a part of a circuit to be connected directly to ground. The connection to ground is unconditionally made (i.e., even if there is only one `THREE_GROUND` function in the circuit-constructing program tree). After the execution of a `THREE_GROUND` function, there are three new modifiable wires.

4.7.1.4 Terminal Set A constrained syntactic structure enforces the use of one terminal set for the value-setting subtrees, another for the first argument of the `PARALLEL_NEW` function (section 10.1.2), and yet another for all other parts of the program tree.

The numerical parameter value for each electrical component possessing a parameter in this problem is established by a value-setting subtree containing a single perturbable numerical value. This approach for establishing numerical parameter values is described in section 3.5.5.2.

The terminal set, T_{vss}, for the value-setting subtrees is

$$T_{vss} = \{\Re_p\},$$

where $\Re_p$ denotes a perturbable numerical value.

The terminal set, $T_{parallel}$, for the first argument of the PARALLEL_NEW function is

$$T_{parallel} = \{\texttt{UP\_OR\_LEFT}, \texttt{DOWN\_OR\_RIGHT}\}.$$

The terminal set, T_{ccs}, for each construction-continuing subtree is

$$T_{ccs} = \{\texttt{END}, \texttt{SAFE\_CUT}\}.$$

4.7.1.5 Fitness Measure Because the high-level statement of the behavior for the desired circuit is expressed in terms of the passing or suppressing of signals at various frequencies, the output voltage VOUT is measured in the frequency domain. SPICE is instructed to perform an AC small signal analysis (AC sweep) and report the circuit's behavior over five decades (between 1 Hz and 100,000 Hz) with each decade being divided into 20 parts (using a logarithmic scale). Thus, there are a total of 101 sampled frequencies.

Fitness is the sum, over the 101 frequencies, of the absolute weighted deviation between the actual value of the voltage that is produced by the circuit at the probe point VOUT and the target value for voltage (0 or 1 Volt). Specifically, this term is

$$F(t) = \sum_{i=0}^{100} (W(d(f_i), f_i) d(f_i)),$$

where *fi* is the frequency of fitness case *i*; *d(x)* is the absolute value of the difference between the target and observed values at frequency *x*; and *W(y,x)* is the weighting for difference *y* at frequency *x*.

The fitness measure is designed not to penalize ideal voltage values, to slightly penalize acceptable voltage deviations, and to heavily penalize unacceptable voltage deviations.

Specifically, the procedure for computing the detrimental contribution to fitness for each of the 61 points in the three-decade interval between 1 Hz and 1,000 Hz for the intended passband is as follows: If the voltage is equal to the ideal value of 1 Volt in this interval, the deviation is 0.0. If the voltage is between 970 millivolts and 1,030 millivolts, the absolute value of the deviation from 1 Volt is weighted by a factor of 1.0. Otherwise, the absolute value of the deviation from 1 Volt is weighted by a factor of 10.0.

The acceptable and unacceptable deviations for each of the 35 points from 2,000 Hz to 100,000 Hz in the intended stopband are similarly weighted (by 1.0 or 10.0) based on the amount of deviation from the ideal voltage of 0 Volts. For the stopband, the maximum acceptable voltage is 1 millivolt. For each of the five "don't care" points between 1,000 and 2,000 Hz, the deviation is deemed to be zero.

The smaller the value of fitness, the better.

The number of hits is defined as the number of fitness cases for which the voltage is acceptable or ideal or that lie in the "don't care" band.

4.7.1.6 Control Parameters The population size is 1,000,000.

4.7.2 Results for the Lowpass Filter Problem without an Explicit Test Fixture

The best-of-run individual emerged in generation 211. Figure 4.31 shows the frequency-domain behavior of the best-of-run circuit. The circuit scores 101 hits (out of 101) and is 100% compliant with the problem's requirements.

The evolutionary process created the 100%-compliant circuit in figure 4.32 without the benefit of a human-supplied test fixture containing suitably valued source and load resistors. The evolutionary process also successfully grappled with the task of discovering the circuit's input point, output point, and ground. As can be seen in the figure, the

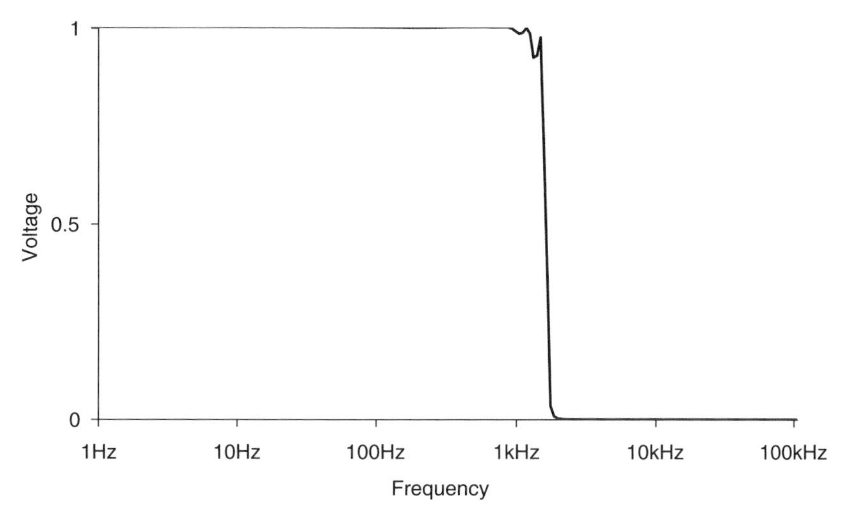

Figure 4.31 Behavior of the best-of-run circuit for the lowpass filter without an explicit test fixture.

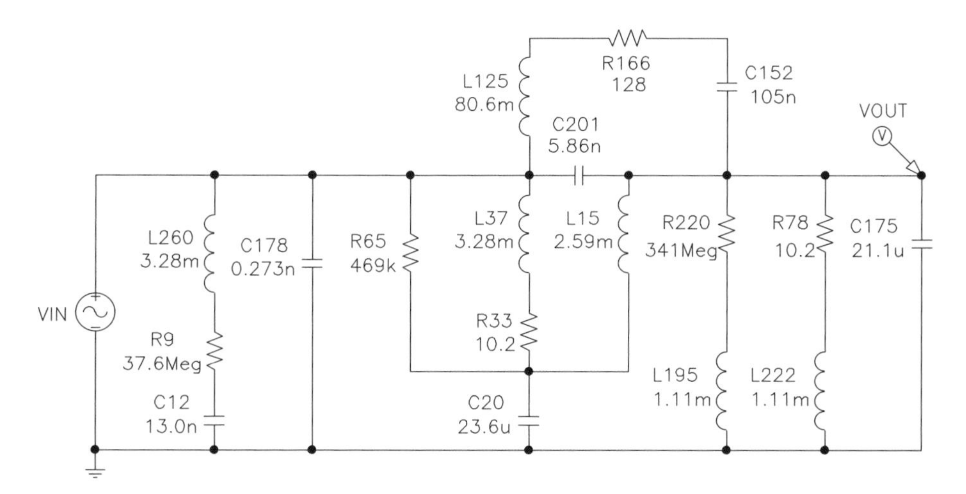

Figure 4.32 Best-of-run lowpass filter circuit for the lowpass filter without an explicit test fixture.

evolutionary process did not create a test fixture in the way that a human ordinarily would. There is no single source resistor neatly positioned on the left side of the circuit diagram. And, there is no single load resistor. Instead, a distributed set of components cooperatively work together as a substitute for the missing test fixture. Specifically, every path from the incoming signal VIN to ground contains either a resistor or capacitor and every path from the output probe point VOUT to ground also contains either a resistor or capacitor. In this regard, this highly distributed method for addressing the absence of an explicit source resistor and an explicit load resistor would probably never occur to a human engineer.

Of course, there is a cost associated with indulging in the luxury of not providing an explicit test fixture to the evolutionary process.

The most efficient way (of which we are aware) for solving the problem of synthesizing a lowpass filter (to satisfy the particular specifications herein) is the "base case" using a population size of 30,000, as described in section 54.2 of *Genetic Programming III* (Koza, Bennett, Andre, and Keane 1999a). The statistics associated with this base case were calculated after making 64 independent runs of the problem. One of the 64 runs was successful by generation 22; 72% of the runs were successful by generation 42; and 86% of the runs were successful by generation 81. The performance curve (figure 54.1 of *Genetic Programming III*) shows that a total computational effort, E, of 4,683,183 individuals (i.e., 30,000×43 generations×3.63 runs) is sufficient to yield a solution to this problem with 99% probability.

In contrast, a total of 212,000,000 individuals were processed by the time that our one (and only) run of the present problem yielded its solution in generation 211. Because the run of the present problem took nine hours of computer time, we were, of course, not inclined to make 63 additional runs of this problem merely to obtain statistics of the same quality as in *Genetic Programming III*. However, the fact that our one run here required the processing of 212,000,000 individuals (more than 45 times greater than 4,683,183) strongly suggests that there is a considerable increased cost in terms of computational resources associated with not supplying the test fixture.

Nonetheless, if one is interested in the boundary question of whether genetic programming can successfully solve a problem in the context of minimizing the amount of human-supplied information, this run demonstrates the principle that it is possible to relieve the human user of the task of specifying the test fixture.

4.7.3 Routineness for the Lowpass Filter Problem without an Explicit Test Fixture

The preparatory steps for this problem are substantially the same as the preparatory steps for previous problems of analog circuit synthesis, except that the present problem has a different goal (and hence a different fitness measure) and that there is, in effect, no test fixture. The transition from the previous problems of analog circuit synthesis to the present problem is routine.

4.7.4 AI Ratio for the Lowpass Filter Problem without an Explicit Test Fixture

The solution produced by genetic programming to this problem has a moderately high amount of "A." The preparatory steps for this problem incorporate a small amount of "I." Thus, the AI ratio for the solution produced by genetic programming is moderately high.

5

Automatic Synthesis of Circuit Topology, Sizing, Placement, and Routing

The previous chapter described how genetic programming can be used to automatically synthesize both the topology and sizing of an electrical circuit. The output of the automated process consists of the number of components in the circuit, the identity of each component (e.g., transistor, capacitor, resistor, inductor), the connections between each lead of each component, and the sizing of each component. However, the automated process in chapter 4 does not yield the physical (geographic) placement of the circuit's components on a printed circuit board or silicon wafer or the physical routing of the wires.

Placement involves the assignment of each of the circuit's components to a particular geographic location on a printed circuit board or silicon wafer.

Routing involves the assignment of a particular geographic location to the wires connecting the components.

A more complete version of the design process for electrical circuits begins with a high-level description of the circuit's desired behavior and typically entails creation of the circuit's topology, sizing, placement, and routing. As shown in figure 5.1, design engineers typically perform these four tasks sequentially.

Each of these four tasks is, by itself, either vexatious or computationally intractable for analog electrical circuits. In particular, the placement problem and the

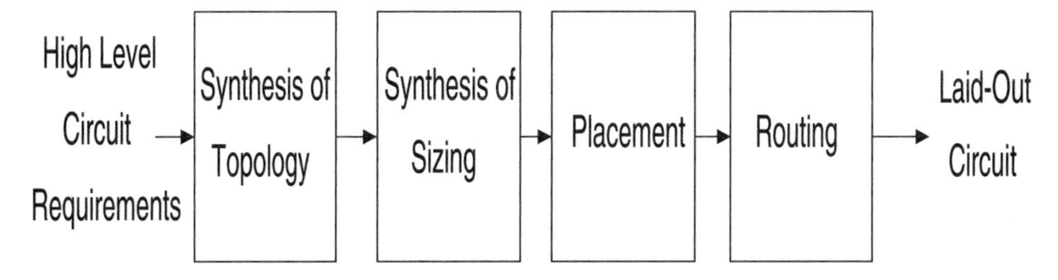

Figure 5.1 The process of designing a circuit includes creation of the circuit's topology, sizing of the circuit's components, placement of the components onto the substrate, and the routing of wires between the components.

routing problem (for both analog and digital circuits) are examples of intractable combinatorial optimization problems that require a computing effort that (in general) increases exponentially with problem size (Wong, Leong, and Liu 1988; Garey and Johnson 1979; Ullman 1984). Because it is considered unlikely that efficient algorithms can be found for such problems, considerable effort has been expended on finding efficient approximate solutions to such problems (Wong, Leong, and Liu 1988; Cohn, Garrod, Rutenbar, and Carley 1994; Maziasz and Hayes 1992; Sechen 1988).

The mandatory requirements for an acceptable scheme for placement and routing are that

- there must be a wire connected to every lead of all the circuit's components,
- wires must not cross on a particular layer of a silicon chip or on a particular side (or layer) of a printed circuit board, and
- minimum clearance distances must be maintained between wires, between components, and between wires and components.

It is possible to make connections (called *vias*) between layers of the silicon chip or printed circuit board; however, the total number of layers that are available and the type of components that can be located on each layer are strictly limited by the particular fabrication technology being used.

Once these mandatory requirements are satisfied, other considerations come into play. Minimization of area of the laid-out circuit is often the next most important consideration in the placement and routing of electrical circuits. Minimization of area has a substantial and direct economic effect on the fabrication of electronic circuitry. This objective is often expressed in terms of the area of the minimal bounding rectangle that encloses all the circuit's components and wires. In addition, it is often desirable to minimize the length of wires connecting components and to minimize the number of vias.

The question arises as to whether all four aspects of circuit design can be combined into a unified automatic process. At first glance, it would appear that the four aspects are so different in character that such a combination would be impossible. However, this chapter demonstrates that genetic programming can indeed be used to automatically create the topology, sizing, placement, and routing of analog electrical circuits.

Our method will be illustrated on the problems of creating the topology, sizing, placing, and routing for

- a lowpass filter and
- an amplifier.

The desired lowpass filter is to pass all frequencies below 1,000 Hertz (Hz), to suppress all frequencies above 2,000 Hz, and to have 1,000-to-1 (or better) attenuation in the stopband (i.e., satisfy the same requirements as the filter in section 4.7). The circuit is to be constructed on a printed circuit board whose top side contains discrete components (capacitors and inductors) that are connected by perpendicularly intersecting metallic wires and whose bottom side is devoted solely to making connections to ground.

The desired amplifier ideally has 60 dB gain, zero distortion, zero bias, and the smallest possible total area for the bounding rectangle of the fully laid-out circuit.

(See Bennett, Koza, Andre, and Keane 1996 for a more detailed statement of the 60 dB amplifier problem). The circuit is to be constructed on a two-sided printed circuit board with two internal layers. The top side contains discrete components (e.g., transistors, capacitors, and resistors) that are connected by perpendicularly intersecting metallic wires. The bottom side is devoted to connections to ground. The two internal layers are associated with `VIA0` and `VIA1`.

5.1 Our Approach to the Automatic Synthesis of Circuit Topology, Sizing, Placement, and Routing

A printed circuit board or silicon wafer has a limited number of layers that are available for wires and a limited number of layers (usually one for a wafer and one or two for a board) that are available for both wires and components. Each wire and component is located at a particular geographic location on the printed circuit board or silicon wafer.

In our approach, an electrical circuit is created using a developmental process in which the component-creating functions, topology-modifying functions, and development-controlling functions of a circuit-constructing program tree are executed. Each of these three types of functions is associated with a modifiable wire or modifiable component in the developing circuit. The starting point of the developmental process is an initial circuit consisting of an embryo and a test fixture. The embryo consists of modifiable wire(s). The embryo is embedded into a test fixture consisting of hardwired components and certain fixed wires that provide connectivity to the circuit's external inputs and outputs. Unless the modifiable wires are modified by the developmental process, the circuit produces only trivial output. An electrical circuit is developed by progressively applying the functions in a circuit-constructing program tree (in the population being bred by genetic programming) to the modifiable wires of the original embryo and to the modifiable components and modifiable wires created during the developmental process. The functions in the program tree are progressively applied (in a breadth-first order) to the initial circuit and its successors until a fully developed circuit emerges.

This section describes how all four aspects of circuit design (topology, sizing, placement, and routing) can be combined into a unified approach using genetic programming.

5.1.1 Initial Circuit

The initial circuits consists of an embryo and a test fixture. All the wires and components of the initial circuit are located at specified physical locations on a printed circuit board or silicon wafer.

The embryo is an electrical substructure consisting of at least one modifiable wire. The test fixture is a substructure composed of nonmodifiable wires and nonmodifiable electrical components. The test fixture provides access to the circuit's external input(s) and permits testing of the circuit's behavior and probing of the circuit's output. The embryo here is connected to the test fixture.

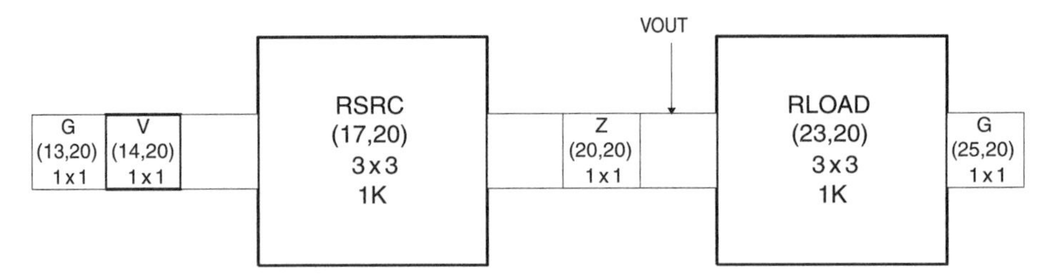

Figure 5.2 Initial circuit for synthesis of topology, sizing, placement, and routing of lowpass filter.

Figure 5.2 shows a one-input, one-output initial circuit located on one layer of a silicon wafer or printed circuit board. This initial circuit consists of the ground G (at the far left), the source V for the incoming voltage signal, a piece of nonmodifiable wire (hatched), a fixed 1,000-Ohm source resistor RSRC, another piece of nonmodifiable wire, a piece of modifiable wire Z, another piece of nonmodifiable wire, a fixed 1,000-Ohm load resistor RLOAD, and the ground G (at the far right). The output probe point VOUT is the place where the circuit's output voltage is measured.

Each element of a circuit resides at a particular physical location on the circuit's two-dimensional substrate and occupies a particular amount of space.

The location of the 1×1 ground point G at the far left of the initial circuit (figure 5.2) is (13, 20); the location of the incoming signal source V is (14,20); the location of the center of the 3×3 source resistor RSRC is (17, 20); the location of the 1×1 modifiable wire Z is (20, 20); the location of the center of the 3×3 load resistor RLOAD is (23, 20); and the location of the 1×1 ground point G at the far right is (25, 20).

All development originates from the modifiable wire Z at (20, 20). As will be seen momentarily, circuit elements typically undergo changes in geographic location during the developmental process.

Note that the initial circuit complies with the requirement that wires must not cross on a particular layer of a silicon chip or on a particular side of a printed circuit board. The initial circuit also complies with the requirement that there must be a wire connecting all the leads of all the circuit's components. In addition, the initial circuit complies with the minimum requirements for clearance between wires, between components, and between wires and components. Each of the component-creating and topology-modifying functions operates in a compliance-preserving way. Thus, every fully developed circuit will also comply with these requirements.

5.1.2 Circuit-Constructing Functions

A program tree typically contains component-creating functions, topology-modifying functions, and development-controlling functions. Each of these three types of functions operates on a modifiable wire or modifiable component in the developing circuit. The construction-continuing subtree(s), if any, of these functions point to a successor function or terminal in the circuit-constructing program tree.

A program tree typically also contains arithmetic functions and constants. The arithmetic-performing subtree of a component-creating function consists of a composition of arithmetic functions (addition and subtraction) and random numerical constants (in the range -1.0 to $+1.0$) as described in section 3.5.5.1. The arithmetic-performing subtree specifies the numerical value of a component by returning a floating-point value that is interpreted by means of the nonlinear mapping described in section 3.5.5.

5.1.3 Component-Creating Functions

The component-creating functions insert a component into the developing circuit and assign component value(s) to the new component.

The two-argument capacitor-creating `LAYOUT-C` function inserts a capacitor into a developing circuit in lieu of a modifiable wire (or modifiable component). Figure 5.3 shows a partial circuit containing four capacitors (C2, C3, C4, and C5) and a modifiable wire Z0. Each capacitor occupies a 3×3 area. Each piece of wire occupies a $1 \times n$ or an $n \times 1$ area. The modifiable wire Z0 occupies a 1×1 area and is located at (20, 20).

Figure 5.4 shows the result of applying the two-argument capacitor-creating `LAYOUT-C` function to the modifiable wire Z0 of figure 5.3. The newly created capacitor C6 occupies a 3×3 area and is centered at location (20, 20). The newly created component is larger in both directions than the 1×1 piece of modifiable wire that it replaces. Thus, its insertion affects the locations of all preexisting components in the developing circuit. In particular, preexisting capacitor C2 is pushed north and west by one unit, thereby relocating it from (18, 23) to (17, 24). Similarly, preexisting capacitor C5 is pushed south and east by one unit, thereby relocating it from (22, 17) to (23, 16). In our actual implementation, all adjustments in location are made after the completion of the entire developmental process; however, each circuit-constructing function below will be explained as if the required adjustments are made at the time that each function is executed.

The first argument of the capacitor-creating function is an arithmetic-performing subtree that specifies the capacitance of the newly created capacitor in nanofarads.

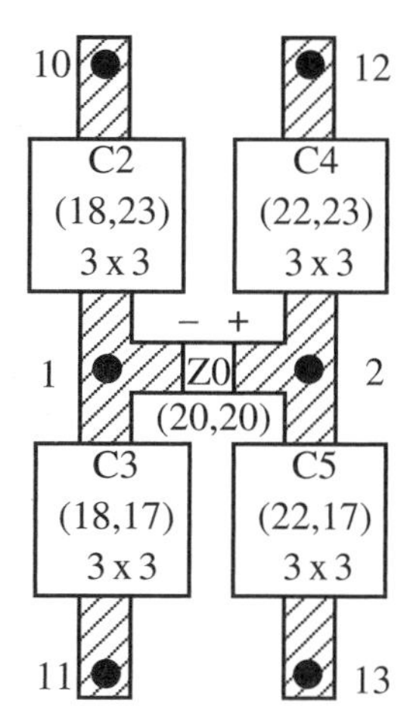

Figure 5.3 Partial circuit with a 1×1 piece of modifiable wire Z0 at location (20, 20) and four capacitors.

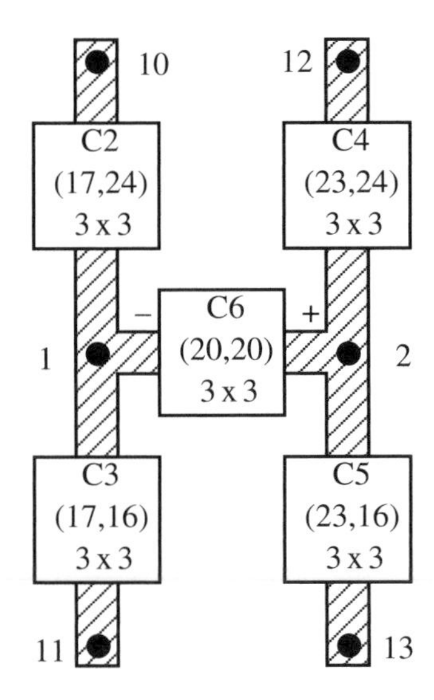

Figure 5.4 The application of the LAYOUT-C function to the modifiable wire Z0 of figure 5.3 causes a 3×3 capacitor C6 to be inserted at location (20, 20). The insertion of the new capacitor C6 forces a change in location of the other capacitors.

The second argument of the capacitor-creating function is the construction-continuing subtree. The newly created capacitor C6 remains subject to subsequent modification.

Similarly, the two-argument inductor-creating LAYOUT-L function causes an inductor to be inserted into a developing circuit in lieu of a modifiable wire (or other modifiable component). The inductors in this chapter each occupy a 3×3 area; however, different components may, in general, have different dimensions. The inductance of the new inductor (in microhenrys) is specified by the arithmetic-performing subtree (the function's first argument). The function's second argument is the construction-continuing subtree.

Three-leaded components, such as transistors, may also be inserted into a developing circuit. Figure 5.5 shows the result of applying the one-argument transistor-creating NPN-TRANSISTOR-LAYOUT function to the modifiable wire Z0 of figure 5.3. The newly created *npn* (Q2N3904 BJT) transistor Q6 occupies a 3×3 area and is located at (18, 20). The newly created component is larger than that which it replaces, thereby affecting the locations of two preexisting components (C2 and C3) in the developing circuit. Specifically, preexisting capacitor C2 is pushed north by one unit, thereby relocating it from (18, 23) to (18, 24). Similarly, preexisting capacitor C3 is pushed south by one unit, thereby relocating it from (18, 17) to (18, 16). The argument of the transistor-creating function is an arithmetic-performing subtree that specifies the transistor's width. The newly created transistor Q6 is not subject to subsequent modification (and hence this function possesses only one argument and has no construction-continuing subtree).

Similarly, the PNP-TRANSISTOR-LAYOUT function inserts a *pnp* (Q2N3906 BJT) transistor.

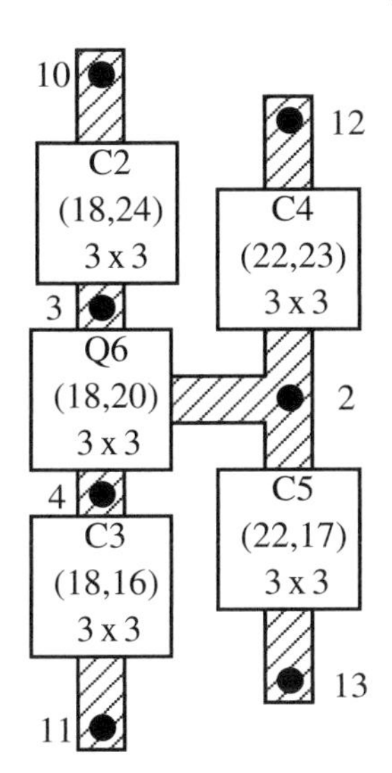

Figure 5.5 The application of NPN-TRANSISTOR-LAYOUT function to the modifiable wire Z0 of figure 5.3 causes a three-leaded transistor Q6 occupying a 3×3 area to be inserted at location (18, 20).

5.1.4 Topology-Modifying Functions

Each topology-modifying function modifies the topology of the developing circuit.

The two-argument SERIES-LAYOUT function creates a series composition consisting of the modifiable wire or modifiable component with which the function is associated and a copy of it. The function also creates one new node. Figure 5.6 shows a partial circuit containing six capacitors (C2, C3, C4, C5, C6, and C7). Figure 5.7 shows the result of applying the SERIES-LAYOUT function to modifiable capacitor C6 located at (20, 20) of figure 5.6. The SERIES-LAYOUT function creates a new capacitor C8 occupying a 3×3 area. The newly created capacitor C8 has the same values as modifiable capacitor C6. Room is made for the newly created capacitor C8 in the direction of a specified one of the two leads (the positive lead) of the preexisting component. Thus, C8 is located at (22, 20) to the east of C6 in this example. The addition of the four units horizontally to accommodate C8 affects the horizontal (but not vertical) location of the four preexisting capacitors. For example, preexisting capacitor C2 is pushed west by two units, thereby relocating it from (17, 24) to (15, 24). Similarly, preexisting capacitor C5 is pushed east by two units, thereby relocating it from (23, 16) to (25, 16). The addition of the four units horizontally to accommodate C8 also affects other parts of the developing circuit. For example, the wires to the east and west of preexisting capacitor C7 are lengthened (by two units each) to reflect the addition of the four horizontal units associated with the creation of C8. The function does not change the position of the preexisting component (C6) relative to C2 or C3. Both arguments of the SERIES-LAYOUT function are construction-continuing subtrees. Thus, both C6 and C8 remain subject to subsequent modification. New node 3 is located between preexisting capacitor C6 and new capacitor C8 at the original location (20, 20) of preexisting capacitor C6.

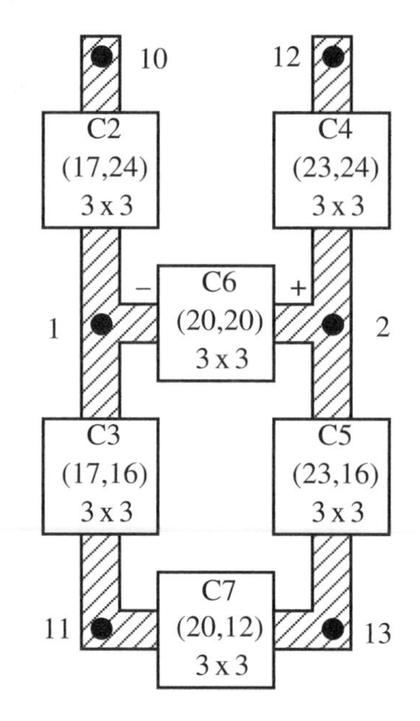

Figure 5.6 Partial circuit with a 3×3 modifiable capacitor C6 at location (20, 20) with five nearby capacitors.

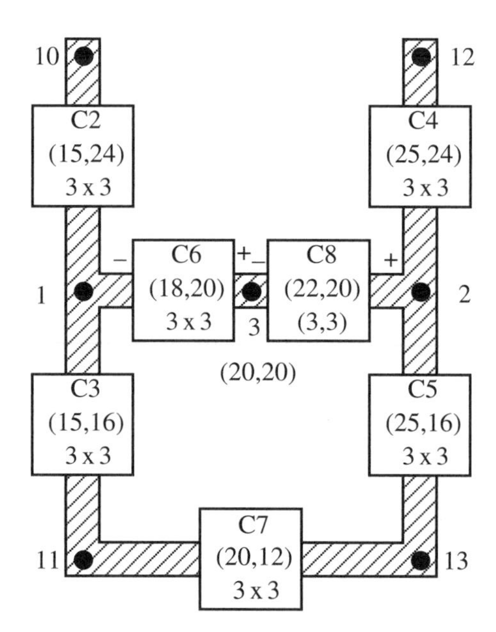

Figure 5.7 The application of the SERIES-LAYOUT function to the modifiable capacitor C6 of figure 5.6 causes a new 3×3 capacitor C8 to be inserted at location (22, 20) in series with C6.

Each of the two functions in the PARALLEL-LAYOUT family of four-argument functions creates a parallel composition consisting of two new modifiable wires, the preexisting modifiable wire or modifiable component with which the function is associated, and a copy of the modifiable wire or modifiable component. Each function also creates two new nodes. Figure 5.8 shows the result of applying the PARALLEL-LAYOUT-LEFT function to modifiable capacitor C6 located at (20, 20) of figure 5.4. The function does not change the location of the modifiable component or modifiable wire with which the function is associated. The function creates a new capacitor C7 occupying a 3×3 area with the same numerical value as C6. When looking from the negative to the positive lead of the modifiable component or modifiable wire with which the function is associated (C6 here), the PARALLEL-LAYOUT-LEFT function positions the new capacitor C7 to the left of C6 (i.e., to the north of C6). The PARALLEL-LAYOUT-LEFT function does not affect the location of preexisting circuitry to the south of C6 (i.e., C3 and C5). The function inserts a new 1×1 modifiable wire Z9 at (23, 22) to the north of C6, a new 1×1 piece of wire between preexisting node 2 and new modifiable wire Z9, and a new 1×1 piece of wire between new node 4 and Z9. The function inserts a new 1×1 modifiable wire Z8 at (17, 22) to the north of C6, a new 1×1 piece of wire between preexisting node 1 and new modifiable wire Z8, and a new 1×1 piece of wire between new node 3 and Z8. The new capacitor C7 is located at (20, 24).

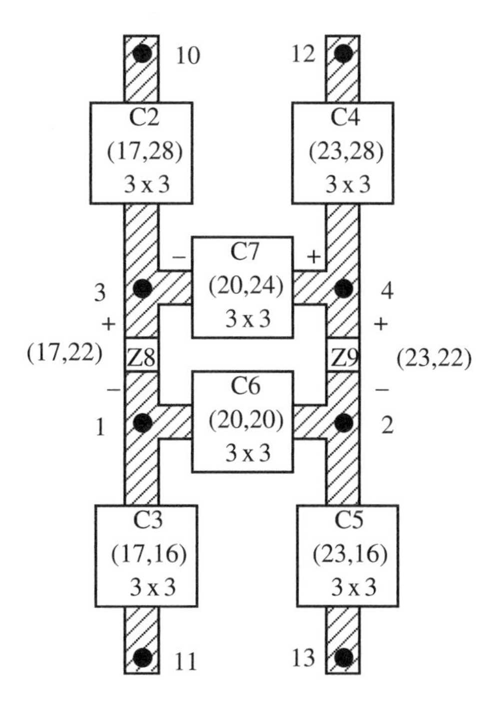

Figure 5.8 The application of the PARALLEL-LAYOUT-LEFT function to the modifiable capacitor C6 of figure 5.4 causes a new 3×3 capacitor C7 to be inserted at location (20, 24) in parallel with C6.

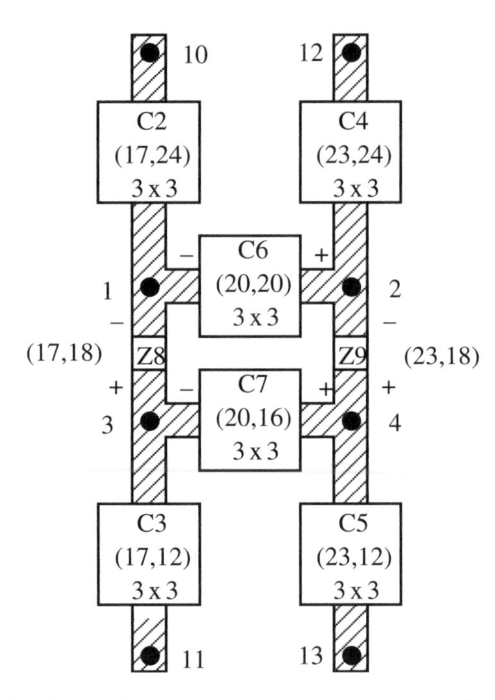

Figure 5.9 The application of the PARALLEL-LAYOUT-RIGHT function to the modifiable capacitor C6 of figure 5.4 causes a new 3×3 capacitor C7 to be inserted at location (20, 16) in parallel with C6.

The function also relocates the preexisting circuitry to the north of C6. Specifically, preexisting capacitor C2 is pushed to the north from (17, 24) to (17, 28) and preexisting capacitor C4 is pushed to the north from (23, 24) to (23, 28).

Figure 5.9 shows the result of applying the PARALLEL-LAYOUT-RIGHT function to modifiable capacitor C6 located at (20, 20) of figure 5.4. This function operates in a manner similar to the PARALLEL-LAYOUT-LEFT function, except that new capacitor C7 is located to the south of C6 and the location of preexisting circuitry to the south of C6 is pushed south.

The one-argument polarity-reversing FLIP function reverses the polarity of the modifiable component or modifiable wire with which the function is associated.

All the foregoing circuit-constructing functions operate on a plane. However, most practical circuits are not planar. Vias provide a way to connect distant points of a circuit. Each of the four functions in the VIA-TO-GROUND-LAYOUT family of three-argument functions creates a T-shaped composition consisting of the modifiable wire or modifiable component with which the function is associated, a copy of it, two new modifiable wires, and a via to ground. The function also creates two new nodes.

Figure 5.10 shows the result of applying the VIA-TO-GROUND-NEG-LEFT-LAYOUT function to modifiable capacitor C6 located at (20, 20) of figure 5.4. The function creates a new node 3 at the location (20, 20) of the modifiable component or modifiable wire with which the function is associated and also creates a new 2×1 area at the negative end of the modifiable component or modifiable wire with which

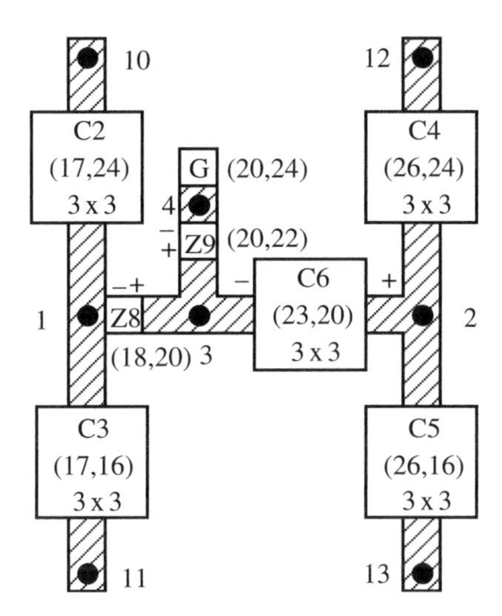

Figure 5.10 The application of the `VIA-TO-GROUND-NEG-LEFT-LAYOUT` function to the modifiable capacitor C6 of figure 5.4 causes a new 1×1 connection to ground G to be inserted at location (20, 24).

the function is associated and a new 4×1 area to the north (i.e., to the left when looking from the negative to the positive lead of C6).

The new 4×1 area consists of a new 1×1 piece of wire perpendicular to, and to the north of, the modifiable component or modifiable wire with which the function is associated (facing from the negative to positive lead of the modifiable component or modifiable wire with which the function is associated), a new 1×1 modifiable wire Z9 at location (20, 22) beyond the new 1×1 piece of wire, a new node 4 at (20, 23) beyond Z9 and the new 1×1 piece, and a via to ground G at (20, 24) beyond node 4, Z9, and the new 1×1 piece.

The new 2×1 area consists of a 1×1 piece of wire at (19, 20) at the negative lead of the modifiable component or modifiable wire with which the function is associated and a modifiable wire Z8 at (18, 20). Because the `VIA-TO-GROUND-NEG-LEFT-LAYOUT` function creates a new 2×1 area at the negative end of preexisting capacitor C6 and a new 1×1 node 3, capacitor C6 is pushed to the east by three units with the result that C6 becomes centered at (23, 20). Consequently, preexisting capacitor C4 is pushed east from (23, 24) to (26, 24) and preexisting capacitor C5 is pushed east from (23, 16) to (26, 16).

The three other members of this family of functions are named to reflect the fact that they create the new 2×1 area at the positive (instead of negative) end of the modifiable component or modifiable wire with which the function is associated and that they create the new 4×1 area to the right (instead of left).

If desired, similar families of three-argument functions can be defined to allow direct connections to a positive power supply or a negative power supply.

In addition, if desired, numbered vias can be created to provide connectivity between two different parts of a circuit. A distinct four-member family of three-argument functions is used for each via. For example, `VIA-2-NEG-LEFT-LAYOUT` makes a connection with a layer (numbered 2) of the imaginary multi-layered silicon wafer (or multi-layered printed circuit board) on which the circuit resides.

The initial circuit complies with the requirement that wires cannot cross on a particular layer of a silicon chip or on a particular side of a printed circuit board and with the requirement that there must be a wire connecting 100% of the leads. Each of the component-creating and topology-modifying functions preserves compliance with these two mandatory requirements for successful placement and routing, so any sequence of such functions yields a fully laid-out circuit that complies with these two requirements.

5.1.5 Development-Controlling Functions

The zero-argument END function makes the modifiable wire or modifiable component with which it is associated into a nonmodifiable wire or component (thereby ending a particular developmental path).

The one-argument NOOP ("no operation") function has no effect on the modifiable wire or modifiable component with which it is associated; however, it has the effect of delaying the developmental process on the particular path on which it appears.

5.1.6 Developmental Process

An electrical circuit is created by executing the functions in a circuit-constructing program tree as part of a developmental process. The functions are progressively applied to the embryonic circuit (and its successors) until all functions in the program tree are executed. That is, the functions in the circuit-constructing program tree progressively side-effect the embryonic circuit and its successors until a fully developed circuit eventually emerges. The functions are applied in a breadth-first order.

5.2 Lowpass Filter with Layout

5.2.1 Preparatory Steps for the Lowpass Filter with Layout

5.2.1.1 Initial Circuit The one-input, one-output initial circuit consisting of an embryo with one modifiable wire and a test fixture as shown in figure 5.2 is suitable for the lowpass filter problem.

5.2.1.2 Program Architecture Because there must be one result-producing branch in the program tree for each modifiable wire in the embryo and there is one modifiable wire in the embryo, the architecture of each circuit-constructing program tree has one result-producing branch. Neither automatically defined functions nor architecture-altering operations are used in this chapter.

5.2.1.3 Function Set The function set, F_{ccs}, for each construction-continuing subtree is

F_{ccs}={C-LAYOUT, L-LAYOUT, SERIES-LAYOUT, PARALLEL-LAYOUT-LEFT, PARALLEL-LAYOUT-RIGHT, FLIP, NOOP, VIA-TO-GROUND-NEG-LEFT-LAYOUT, VIA-TO-GROUND-NEG-RIGHT-LAYOUT, VIA-TO-GROUND-POS-LEFT-LAYOUT, VIA-TO-GROUND-POS-RIGHT-LAYOUT}.

These functions possess 2, 2, 1, 4, 4, 1, 1, 3, 3, 3, and 3 arguments, respectively.

5.2.1.4 Terminal Set The terminal set, T_{ccs}, for each construction-continuing subtree is

$$\mathsf{T}_{ccs} = \{\texttt{END}\}.$$

In this problem, the numerical parameter value(s) for each electrical component possessing a parameter are established by an arithmetic-performing subtree consisting of one or more arithmetic functions and one or more constant numerical terminals. This approach for establishing numerical parameter values is described in section 3.5.5.1.

A constrained syntactic structure enforces the use of one function set and one terminal set for the arithmetic-performing subtrees and another function set and terminal set for all other parts of the program tree.

The terminal set, T_{aps}, for each arithmetic-performing subtree consists of

$$\mathsf{T}_{aps} = \{\Re\},$$

where $\Re$ represents floating-point random constants from -1.0 to $+1.0$.

The function set, F_{aps}, for each arithmetic-performing subtree is

$$\mathsf{F}_{aps} = \{+, -\}.$$

5.2.1.5 Fitness Measure The fitness measure for this problem is multiobjective and is expressed in terms of

- suppression of frequencies in the stopband of the desired filter,
- passage at full power of frequencies in the passband of the desired filter, and
- minimization of area of the smallest bounding rectangle that encloses the fully laid-out circuit.

The evaluation of each individual circuit-constructing program tree in the population begins with its execution. This execution progressively applies the functions in the program tree to the embryo (and its successors), thereby creating a fully developed (and a fully laid out) circuit. A netlist is then created that identifies each component of the fully developed circuit, the nodes to which each component is connected, and the numerical value associated with each component. The netlist becomes the input to our modified version of the SPICE simulator.

Because the high-level statement of the behavior for the desired circuit is expressed (in part) in terms of the passing or suppressing of signals at various frequencies, the output voltage VOUT is measured in the frequency domain. SPICE is instructed to perform an AC small signal analysis (AC sweep) and report the circuit's behavior over five decades (between 1 Hz and 100,000 Hz) with each decade being divided into 20 parts (using a logarithmic scale). Thus, there are a total of 101 sampled frequencies.

The area of the bounding rectangle for the fully developed circuit is easily computed because the developmental process for creating the circuit is aware of the actual geographical location of all components and wires.

The desired lowpass filter is to have a passband below 1,000 Hz, a stopband above 2,000 Hz, and a 1,000-to-1 (or better) attenuation in the stopband relative to the passband. The circuit is driven by an incoming AC voltage source with an amplitude of 2 Volts. The test fixture contains a 1,000-Ohm source resistor RSRC and a 1,000-Ohm load resistor RLOAD. The desired lowpass filter should have a sharp drop-off from 1 to 0 Volts in the transitional ("don't care") region between 1,000 Hz and 2,000 Hz.

The first element is the sum, over the 101 frequencies, of the absolute weighted deviation between the actual value of the voltage that is produced by the circuit at the probe point VOUT and the target value for voltage (0 or 1 Volt) as described in section 4.7.1.5.

The second element of the fitness measure is the area of the bounding rectangle for the fully developed circuit divided by 100,000 square units of area. This element of the fitness measure is much smaller than the term involving the filter's frequency response until a circuit scores 101 (or near 101) hits.

For individuals not scoring the maximum number (101) of hits, fitness is the sum of the two terms.

For individuals scoring the maximum number of hits, fitness is only the area-based term.

The smaller the value of fitness, the better.

The number of hits is defined using the first element of the fitness measure. The number of hits is defined as the number of fitness cases for which the voltage is acceptable or ideal or that lie in the "don't care" band.

5.2.1.6 Control Parameters The population size is 1,120,000.

5.2.2 *Results for the Lowpass Filter with Layout*

We use the best-of-generation circuit-constructing program tree (figure 5.11) from generation 0 to illustrate the developmental process that converts a tree into a fully developed and fully laid-out electrical circuit. The program tree is a composition of component-creating, topology-modifying, and development-controlling functions.

The circuit-constructing program tree of figure 5.11 is shown below in the style of a LISP S-expression:

```
(L-LAYOUT
  V1
  (PARALLEL-LAYOUT-LEFT
    (FLIP (FLIP (NOOP END)))
    (C-LAYOUT
       (-(-0.656 -0.507) (+ -0.463 -0.970))
       (FLIP (FLIP END)))
    (FLIP
       (SERIES-LAYOUT
          (C-LAYOUT -0.776 END)
          (VIA-TO-GROUND-NEG-LEFT-LAYOUT END END END)))
    (FLIP
       (NOOP (L-LAYOUT -0.765 END)))))
```

The program begins with a two-argument inductor-creating `L-LAYOUT` function. The value of the new inductor is established by the first argument of this `L-LAYOUT`

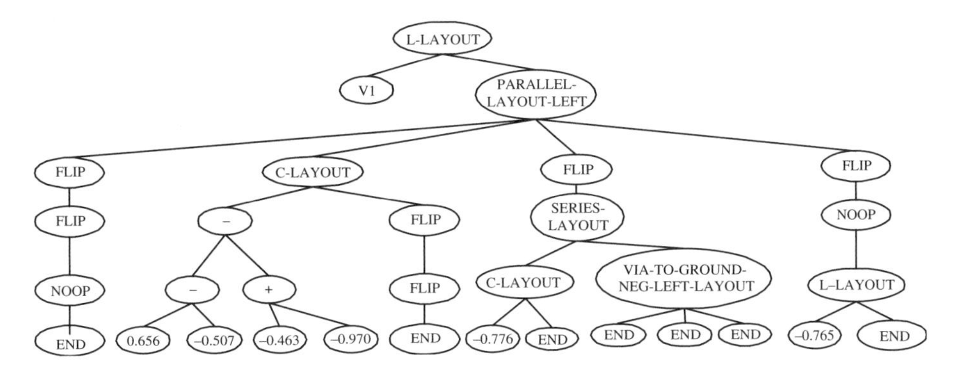

Figure 5.11 Best-of-generation circuit-constructing program tree of generation 0.

function. Because this particular argument is a large arithmetic-performing subtree composed of addition, subtraction, and floating-point random constants, it is abbreviated and labeled `V1` in figure 5.11. The second argument (construction-continuing subtree) of this `L-LAYOUT` function is a `PARALLEL-LAYOUT-LEFT` function.

The first argument of the four-argument `PARALLEL-LAYOUT-LEFT` function executes two one-argument polarity-reversing `FLIP` functions and one one-argument `NOOP` ("no operation") function before reaching a development-terminating zero-argument `END` function.

The second argument of the `PARALLEL-LAYOUT-LEFT` function executes a capacitor-creating two-argument `C-LAYOUT` function whose value is established by a seven-point arithmetic-performing subtree (shown in its entirety in the figure) and whose construction-continuing subtree contains two polarity-reversing `FLIP` functions and one `END` function.

The third argument of the `PARALLEL-LAYOUT-LEFT` function is a one-argument `FLIP` function whose construction-continuing subtree consists of a two-argument `SERIES-LAYOUT` function. The first construction-continuing subtree of the `SERIES-LAYOUT` function executes a second capacitor-creating `C-LAYOUT` function. The value of the second capacitor is established (in the manner described earlier) by the one-point arithmetic-performing subtree consisting of the floating-point random constant −0.776. The second construction-continuing subtree of the `SERIES-LAYOUT` function executes a three-argument `VIA-TO-GROUND-NEG-LEFT-LAYOUT` function.

The fourth argument of the `PARALLEL-LAYOUT-LEFT` function is a `FLIP` function whose construction-continuing subtree executes a one-argument `NOOP` function which, in turn, causes execution of a second two-argument inductor-creating `L-LAYOUT` function. The value of the second inductor is established by the one-point arithmetic-performing subtree consisting of the floating-point constant −0.765.

When this circuit-constructing program tree for the best-of-generation circuit of generation 0 is executed, it yields a fully laid-out circuit (figure 5.12) with two inductors (L2 and L11) and two capacitors (C10 and C19). The four components, the connecting wires, the ground points, and the source of the incoming signal are all assigned a precise physical location in this fully laid-out circuit. Notice that all the nonmodifiable elements of the original test fixture of the initial circuit of figure 5.2

(including, in particular, the incoming signal V, the source resistor RSRC, and the load resistor RLOAD) appear in the fully developed and fully laid-out circuit (albeit in different physical locations). In addition, notice that this fully laid-out circuit complies with the requirement that wires cannot cross on a particular layer of a silicon chip or on a particular side of a printed circuit board and with the requirement that there must be a wire connecting 100% of the leads.

The best-of-generation circuit from generation 0 (figure 5.12) scores 53 hits and has a fitness of 57.961037. The incoming signal V first passes through source resistor RSRC located at position (−7.8, 7.7) at the top left of figure 5.12. The signal passes into inductor L2 located at position (1.2, 7.7), into load resistor RLOAD located at position (11.8, 7.7), and into ground G (at the top right of the figure). In addition, capacitor C10 located at (−2.7, 4.2) is connected to ground G (in the middle left of the figure). Also, the series composition of inductor L11 and capacitor C19 is connected to the same ground point G in the middle left of the figure. The area-based term of the fitness measure is only 0.003710 at this early stage of the run.

The behavior in the frequency domain of the best circuit of generation 0 bears only a faint resemblance to a lowpass filter. The circuit passes approximately 1 Volt for frequencies up to about 50 Hz and passes approximately 0 Volts for a few frequencies near 100,000 Hz. However, none of the intermediate frequencies have the desired voltage.

Both the average fitness of all individuals in the population and the fitness of the best individual in the population improve over successive generations. The best circuit of generation 8 (figure 5.13) scores 82 hits and has a fitness of 9.731077 (of which only 0.008138 is contributed by the area-based term of the fitness measure). The result-producing branch of its circuit-constructing program tree contains 165 points. The circuit has five inductors and three capacitors. The incoming signal V passes through source resistor RSRC (at the bottom left of the figure) and is fed into a parallel-series composition of inductors L13, L2, and L12 (which are together electrically equivalent to one inductor). In a lowpass filter, the capacitors that are connected to ground are called *shunts* and the inductors positioned in series (along the bottom of this figure) between the source resistor and load resistor are called *series* inductors. When all the parallel and series compositions of like components are combined, this circuit is equivalent to a first series inductor (the L13, L2, and L12 combination), a first capacitive shunt (C18), a second series inductor (L11), a second capacitive shunt (the C16 and C19 combination), and a third series inductor (L10). The series inductors and capacitive shunts of a lowpass filter are typically drawn on paper to resemble a ladder (as they are, for example, in figure 11.7). Figure 5.13 is equivalent to a two-rung ladder. The capacitive shunts correspond to the rungs of the ladder. The series inductors correspond to one side of the ladder. The second side of the ladder is a common ground wire to which each capacitive shunt is connected. The best-of-run circuit that emerges in generation 138 is a four-rung ladder.

The behavior in the frequency domain of the best circuit of generation 8 bears some resemblance to the desired lowpass filter. The circuit passes approximately 1 Volt for frequencies up to about 900 Hz and passes approximately 0 Volts for all frequencies above 10,000 Hz. However, none of the intermediate frequencies have the desired voltage.

The first circuit scoring 101 hits (out of 101) appears in generation 25 (figure 5.14). It has a fitness of 0.01775. The result-producing branch of its circuit-constructing program tree contains 548 points. The circuit has five capacitors and 11 inductors. The incoming signal V passes through source resistor RSRC (at the bottom left of the figure) and is fed into a composition of inductors along the straight line connecting the incoming signal V to the load resistor RLOAD (at the bottom right). Along the way, there is a series composition of L2, L12, the L11/L23 parallel combination, L10, L26, the L9/L33 parallel combination, L32, and L31. There are four shunts to ground G. Three of these shunts consist of one capacitor each (C13, C29, and C40). The fourth

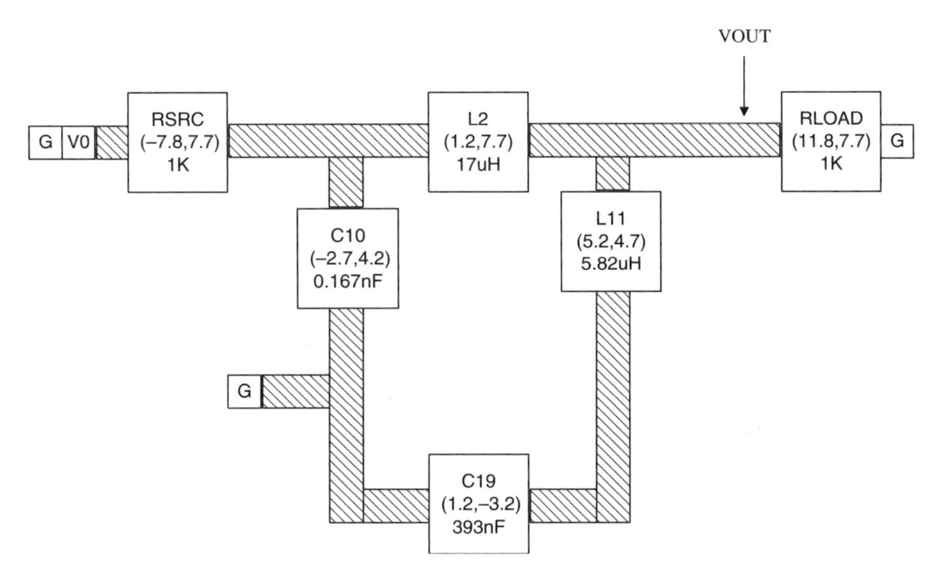

Figure 5.12 Best-of-generation circuit of generation 0 containing two inductors and two capacitors.

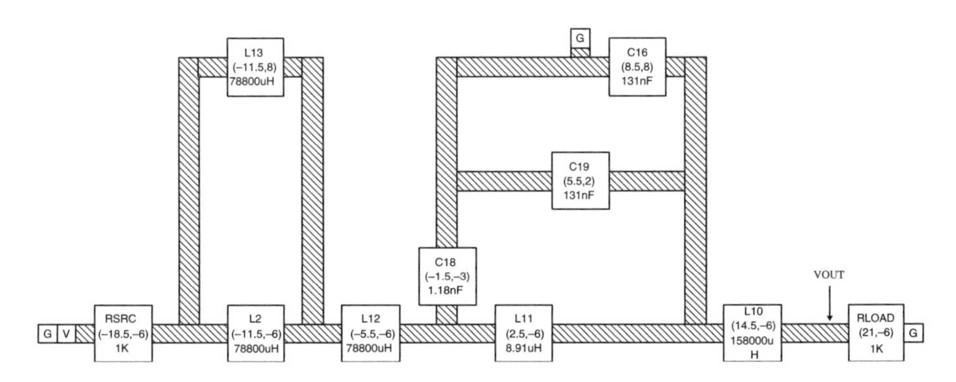

Figure 5.13 Best-of-generation circuit of generation 8 containing five inductors and three capacitors.

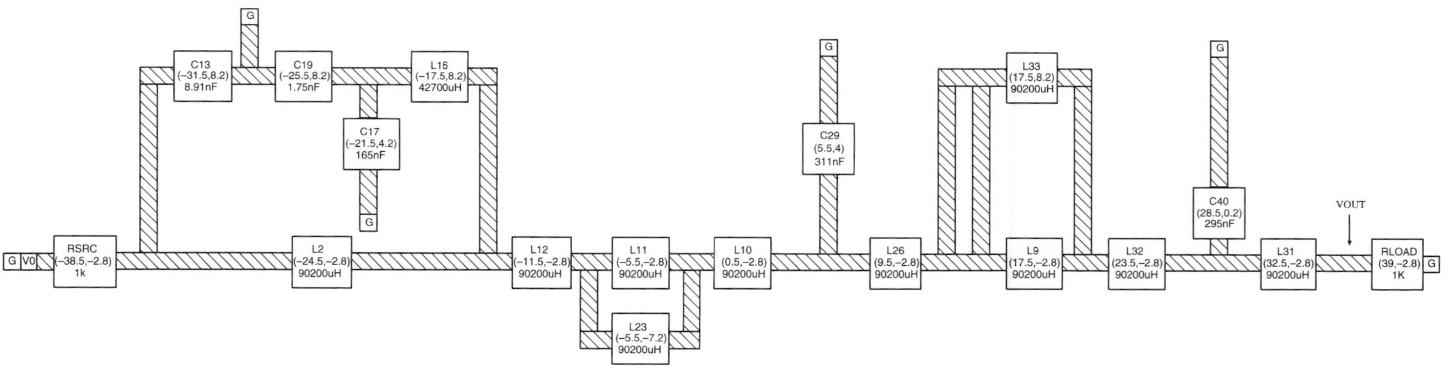

Figure 5.14 100%-compliant best-of-generation circuit of generation 25 containing five capacitors and 11 inductors (total of 16 components) occupying an area of 1775.2.

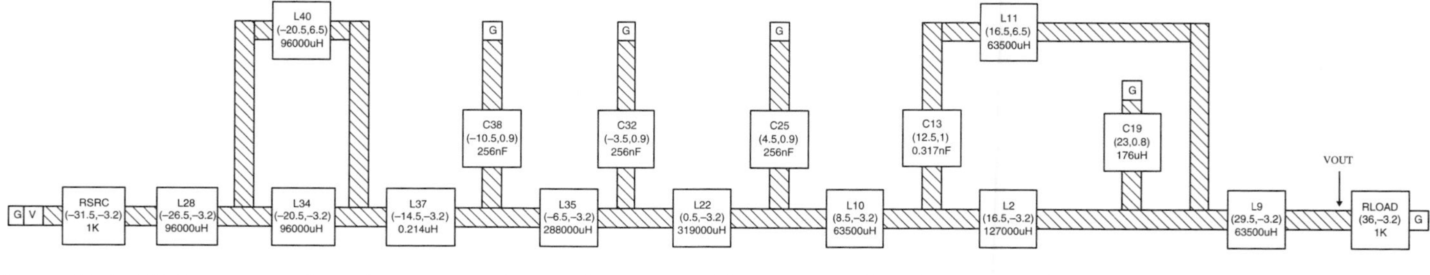

Figure 5.15 100%-compliant best-of-generation circuit of generation 30 containing 10 inductors and five capacitors (total of 15 components) occupying an area of 950.3.

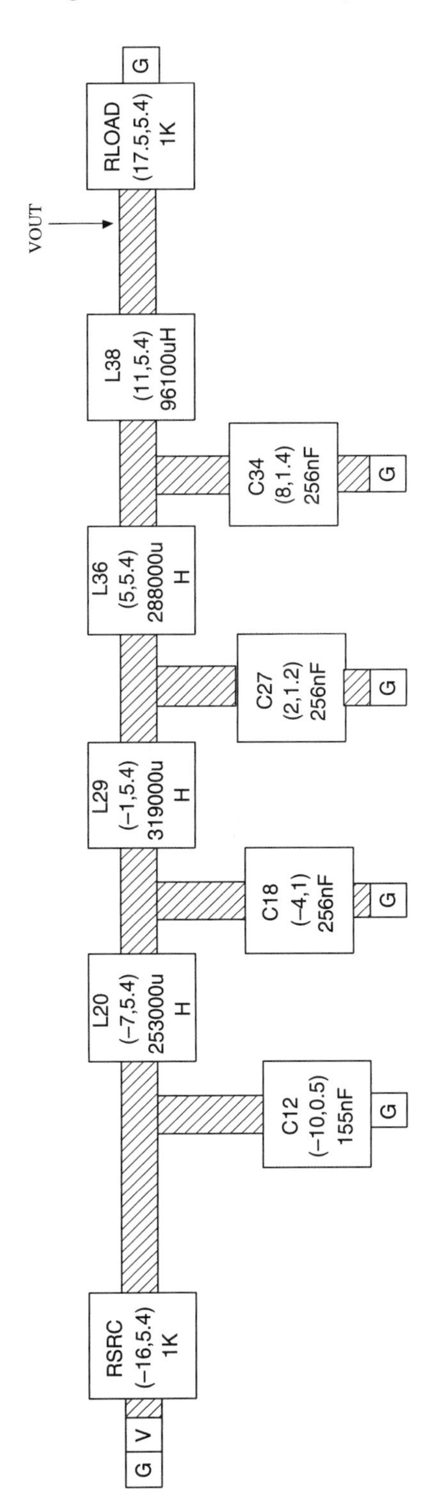

Figure 5.16 100%-compliant best-of-run circuit of generation 138 containing four inductors and four capacitors (total of only eight components) occupying an area of 359.4.

shunt consists of inductor L16 and the C17/C19 parallel combination. As can be seen, the circuit in figure 5.14 occupies a considerable geographical area (1775.2).

The behavior of the 16-component best-of-generation circuit from generation 25 is 100% compliant in terms of problem's required filter characteristics. It passes approximately a full Volt for all frequencies up to about 1,000 Hz and passes approximately 0 Volts for all frequencies above 2,000 Hz. There is a sharp drop-off in the transition region between 1,000 Hz and 2,000 Hz. However, in addition to having 16 components, it occupies an area of 1775.2.

In generation 30, the best-of-generation circuit (figure 5.15) has a fitness of 0.00950. It also scores 101 hits. This circuit has 10 inductors and five capacitors. This 100%-compliant 15-component circuit (occupying an area of 950.3) occupies only 54% of the area of the 16-component circuit from generation 25.

The best-of-run circuit (figure 5.16) appears in generation 138. It has a near-zero fitness of 0.00359 (more than five orders of magnitude better than the fitness of the best circuit of generation 0). The result-producing branch of its circuit-constructing program tree contains 463 points. This circuit has four inductors and four capacitors (half as many components as the best circuit from generation 25). In addition, the circuit has a compact layout. There are no series or parallel compositions of capacitors or inductors that redundantly connect the same two points of the circuit. The topology of this best-of-run circuit is a four-rung ladder.

This best-of-run circuit occupies an area of 359.4 (only 20% of the area of the 100%-compliant best circuit from generation 25 in figure 5.14). Note that, for reasons of space, figures 5.14, 5.15, and 5.16 are not drawn to scale.

Table 5.1 shows the number of capacitors, the number of inductors, the number of shunts to ground, the area in terms of the number of square units of the bounding rectangle, the frequency-based element of fitness, and the fitness for the three best-of-generation circuits (each scoring 101 hits) from generations 25, 30, and 138.

As can be seen in table 5.1, all three of these three best-of-generation circuits have the same number (four) of shunts to ground. That is, they solve the problem with more or less the same approach. However, the best-of-generation circuits from the two earlier generations (25 and 30) each have a total of 16 inductors and capacitors whereas the best-of-run circuit from generation 138 has a total of only eight. Moreover, the best-of-run circuit from generation 138 has only one fifth of the area of the best-of-generation circuit from generation 25. The frequency-based element of the fitness measure (column 6) is shown for reference only. Because all three of the circuits in this table are 100%-compliant, their fitness (shown in column 7) is simply the area of

Table 5.1 Comparison of characteristics of three best-of-generation circuits scoring 101 hits for the lowpass filter problem with layout

Generation	Number of capacitors	Number of inductors	Number of shunts to ground	Area	Frequency-based element	Fitness
25	5	11	4	1775.2	0.264698	0.01775
30	10	5	4	950.3	0.106199	0.00950
138	4	4	4	359.4	0.193066	0.00359

the bounding rectangle (column 5) for the fully developed circuit divided by 100,000. For additional details see Bennett, Koza, Yu, and Mydlowec (2000).

5.2.3 Human-Competitiveness of the Result for the Lowpass Filter Problem with Layout

The best-of-run circuit from generation 138 (figure 5.16) has the recognizable features of the circuit for which George Campbell of American Telephone and Telegraph received U.S. patent 1,227,113 in 1917 (Campbell 1917). Claim 2 of Campbell's patent covers:

> "An electric wave filter consisting of a connecting line of negligible attenuation composed of a plurality of sections, each section including a capacity element and an inductance element, one of said elements of each section being in series with the line and the other in shunt across the line, said capacity and inductance elements having precomputed values dependent upon the upper limiting frequency and the lower limiting frequency of a range of frequencies it is desired to transmit without attenuation, the values of said capacity and inductance elements being so proportioned that the structure transmits with practically negligible attenuation sinusoidal currents of all frequencies lying between said two limiting frequencies, while attenuating and approximately extinguishing currents of neighboring frequencies lying outside of said limiting frequencies."

An examination of the evolved circuit of figure 5.16 shows that it indeed consists of "a plurality of sections" (specifically, four). Also, as can be seen in the figure, "Each section include[s] a capacity element and an inductance element." Specifically, the first of the four sections consists of inductor L20 and capacitor C12; the second section consists of inductor L29 and capacitor C18; and so forth. Moreover, "one of said elements of each section [is] in series with the line and the other in shunt across the line." As can be seen in the figure, inductor L20 of the first section is indeed "in series with the line" and capacitor C12 is "in shunt across the line." This is also the case for the remaining three sections of the evolved circuit. In addition, the topology of the circuit in figure 5.16 herein exactly matches the topology of the circuit in figure 7 in Campbell's 1917 patent. Finally, this circuit's 100%-compliant frequency domain behavior confirms the fact that the values of the inductors and capacitors are such as to transmit "with practically negligible attenuation sinusoidal currents" of the passband frequencies "while attenuating and approximately extinguishing currents" of the stopband frequencies. In short, the circuit created by genetic programming has all the features contained in claim 2 of Campbell's 1917 patent.

Referring to the eight criteria in table 1.2 for establishing that an automatically created result is competitive with a human-produced result, the rediscovery by genetic programming of a filter that infringes Campbell's 1917 patent satisfies the following two of the eight criteria:

- **(A)** The result was patented as an invention in the past, is an improvement over a patented invention, or would qualify today as a patentable new invention.
- **(F)** The result is equal to or better than a result that was considered an achievement in its field at the time it was first discovered.

The rediscovery by genetic programming of the Campbell filter circuit came about eight decades after Campbell received a patent for his invention. Nonetheless, the fact that the original human-designed version satisfied the Patent Office's criteria for patent-worthiness means that the genetically evolved duplicate would also have satisfied the Patent Office's criteria for patent-worthiness (if only it had arrived earlier than Campbell's patent application).

The Campbell filter was, in fact, rediscovered by genetic programming in 1995 (and previously reported in Koza, Bennett, Andre, and Keane 1996a, 1996d, and 1999a). This fact is included in table 1.3 of human-competitive results. This previous rediscovery of the Campbell filter by genetic programming did not, of course, entail the circuit's placement and routing.

In this chapter, genetic programming not only automatically synthesized the topology and sizing of an analog electrical circuit, it is also simultaneously performed the tasks of placement and routing. The problem of automatic placement and routing of a circuit whose topology and sizing has already been devised is alone considered to be a formidable problem. Referring again to the eight criteria in table 1.2, the simultaneous automatic synthesis of the topology, sizing, placement, and routing of the lowpass filter in this section (and the 60 dB amplifier in section 5.3) satisfies the following additional criterion:

(G). The result solves a problem of indisputable difficulty in its field.

The automatic synthesis by genetic programming of the topology, sizing, placement, and routing of analog circuits satisfies Arthur Samuel's criterion (1983) for artificial intelligence and machine learning:

> "[T]he aim [is]...to get machines to exhibit behavior, which if done by humans, would be assumed to involve the use of intelligence."

5.2.4 Routineness of the Transition from a Problem of Circuit Synthesis without Layout to a Problem of Circuit Synthesis with Layout

What is the nature of the effort required to make the transition from a basic circuit synthesis problem involving only topology and sizing to a problem involving placement and routing (i.e., layout) in addition to topology and sizing?

First, there is the issue of the function and terminal sets. The function and terminal sets for handling a problem combining placement and routing with topology and sizing are similar to the function and terminal sets that we used in chapter 4 for handling just topology and sizing. The difference is that geographically-aware functions are used for handling a problem entailing the simultaneous synthesis of topology, sizing, placement, and routing. This difference is an inherent and necessary part of making the transition from a basic circuit synthesis problem entailing just topology and sizing to a problem additionally entailing placement and routing. The required change in function and terminal sets does not require an enormous amount of effort.

Second, the fitness measure for a problem involving topology, sizing, placement, and routing necessarily has an added element concerning layout (i.e., the area of the bounding rectangle of the laid-out circuit). The change in fitness measures from problem to problem is an inherent and necessary part of making the transition from one problem to another. The required change does not entail an enormous amount of effort.

Once the preparatory steps are performed for a problem combining both basic circuit synthesis and layout, the executional steps of genetic programming are the same as for a problem involving basic circuit synthesis.

Thus, the transition is routine from the domain of basic circuit synthesis to the domain involving topology, sizing, placement, and routing.

5.2.5 AI Ratio for the Lowpass Filter Problem with Layout

What is the AI ratio for the solution produced by genetic programming to the problem of synthesizing the topology, sizing, placement, and routing of a lowpass filter?

The result produced by genetic programming on this problem is human-competitive because the result (without any consideration of placement and routing) was patented as an invention in the past. Thus, the solution produced by genetic programming for this problem has a high amount of "A."

The preparatory steps for the present problem are (except for the fitness measure and the geographically-aware functions) substantially the same as the preparatory steps for other problems of circuit synthesis in chapter 4. Thus, for the same reasons stated in chapter 4, the solution produced by genetic programming for the present problem incorporates a small amount of "I."

The high amount of "A" represented by the human-competitive result in conjunction with the small amount of "I" provided by the human user means that the AI ratio for the solution produced by genetic programming is high.

5.3 60 dB Amplifier with Layout

5.3.1 Preparatory Steps for 60 dB Amplifier with Layout

5.3.1.1 Initial Circuit Figure 5.17 shows a one-input, one-output initial circuit (consisting of an embryo and a test fixture) located on one layer of a silicon wafer or printed circuit board. The embryo consists of the three modifiable wires, Z0, Z1, and Z2 (in the middle of the figure). All development originates from these modifiable wires. The test fixture contains three ground points G, an input point V (lower left), an output point O (upper right), nonmodifiable wires (hatched), and four nonmodifiable resistors. There is a fixed 1,000-Ohm source resistor R4, a fixed 1,000-Ohm load resistor R18, a fixed 1-giga-Ohm feedback resistor R14, and a fixed 999-Ohm balancing resistor R3.

Each element of this initial circuit (and all successor circuits created by the developmental process) resides at a particular geographic location on the circuit's two-dimensional substrate. Each element occupies a particular amount of space. For example, the resistors each occupy a 3×3 area; the source point V and the output probe point O each occupy a 1×1 area; the nonmodifiable wires each occupy a $1 \times n$ or $n \times 1$ rectangular area; the modifiable wires each (temporarily) occupy a 1×1 area.

5.3.1.2 Program Architecture There is one result-producing branch in the program tree for each modifiable wire in the embryo. Thus, the architecture of each circuit-constructing program tree has three result-producing branches. Neither automatically defined functions nor architecture-altering operations are used.

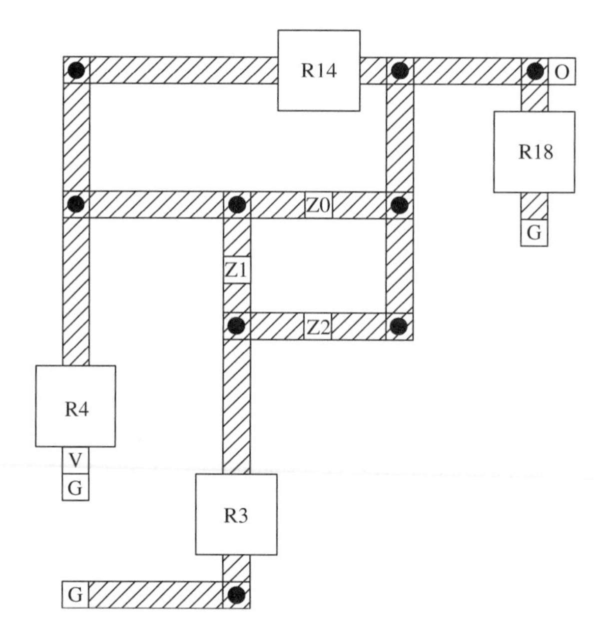

Figure 5.17 Initial circuit consisting of test fixture and embryo with three modifiable wires.

5.3.1.3 Function Set The function set, F_{ccs}, for each construction-continuing subtree includes component-creating functions for *npn* transistors, *pnp* transistors, resistors, capacitors, and inductors (a totally extraneous component for this problem); the development-controlling NOOP function; and topology-modifying functions for series, parallel, flips, and vias to ground, the positive power supply, the negative power supply, and layers 0 and 1 of the printed circuit board.

5.3.1.4 Terminal Set In this problem, the numerical parameter value(s) for each electrical component possessing a parameter are established by a value-setting subtree containing a single perturbable numerical value. This approach for establishing numerical parameter values is described in section 3.5.5.2.

The terminal set, T_{vss}, for the value-setting subtrees is

$$\mathsf{T}_{\text{vss}} = \{\Re_{\text{p}}\},$$

where $\Re_{\text{p}}$ denotes a perturbable numerical value.

5.3.1.5 Fitness Measure The fitness measure for this problem is multiobjective and is expressed in terms of

- the gain of the candidate amplifier circuit,
- bias,
- distortion, and
- minimization of the area of the smallest bounding rectangle that encloses the fully laid-out circuit.

An amplifier can be viewed in terms of its response to a DC input. An ideal inverting amplifier circuit would receive a DC input, invert it, and multiply it by the amplification factor. A circuit is flawed to the extent that it does not achieve the desired amplification, to the extent that the output signal is not centered on 0 Volts (i.e., there is a bias), and to the extent that the circuit's DC response is not linear.

The fitness measure is based on SPICE's DC sweep. The DC sweep analysis measures the circuit's DC response at several different DC input voltages. The circuits are analyzed with a 5 point DC sweep ranging from -10 millivolts to $+10$ millivolts, with points at -10, -5, 0, $+5$, and $+10$ millivolts. SPICE is used to simulate the circuit's behavior for each of these five DC voltages. Four penalties (an amplification penalty, a bias penalty, and two nonlinearity penalties) are derived from this analysis.

First, the circuit's amplification factor is measured by the slope of the straight line between the output for -10 millivolts and the output for $+10$ millivolts (i.e., between the outputs for the endpoints of the DC sweep). If the amplification factor is less than the target (60 dB), there is a penalty equal to the shortfall in amplification.

Second, the bias is computed using the DC output associated with a DC input of 0 Volts. There is a penalty equal to the bias multiplied by a weight of 0.1.

Third, the linearity is measured by the deviation between the slope of each of two shorter line segments and the circuit's overall amplification factor. The first shorter line segment connects the output value associated with an input of -10 millivolts and the output value for -5 millivolts. The second shorter line segment connects the output value for $+5$ millivolts and the output for $+10$ millivolts. There is a penalty for each of these shorter line segments equal to the absolute value of the difference in slope between the respective shorter line segment and the circuit's overall amplification factor.

Fitness is the sum of the area of the bounding rectangle for the fully developed and laid-out circuit weighted by 10^{-6}, the amplification penalty, the bias penalty, and the two nonlinearity penalties; however, if this sum is less than 0.1 (indicating achievement of a very good amplifier), the fitness becomes simply the rectangle's area multiplied by 10^{-6}. Thus, after a good amplifier design is achieved, fitness is based solely on area minimization.

5.3.1.6 Control Parameters The population size is 10,000,000.

5.3.2 Results for 60 dB Amplifier with Layout

The best-of-generation circuit from generation 0 has a fitness of 999.86890.

The first best-of-generation circuit (figure 5.18) delivering 60 dB of amplification appeared in generation 65. This 27-component circuit occupies an area of 8,234 and has an overall fitness of 33.042583.

The best-of-run circuit (figure 5.19) appears in generation 101. Note that figures 5.18 and 5.19 use different scales. The best-of-run circuit from generation 101 contains 11 transistors, 5 resistors, and 3 capacitors. The four "P" symbols indicate vias to the positive power supply. This 19-component circuit occupies an area of 4,751 and has an overall fitness of 0.004751. The best-of-run circuit from generation 101 occupies only 58% of the area of the 27-component best-of-generation circuit from generation 65.

Table 5.2 shows the number of components, the area, the four penalties comprising the non-area element of the fitness measure, and the overall fitness for the best-of-generation

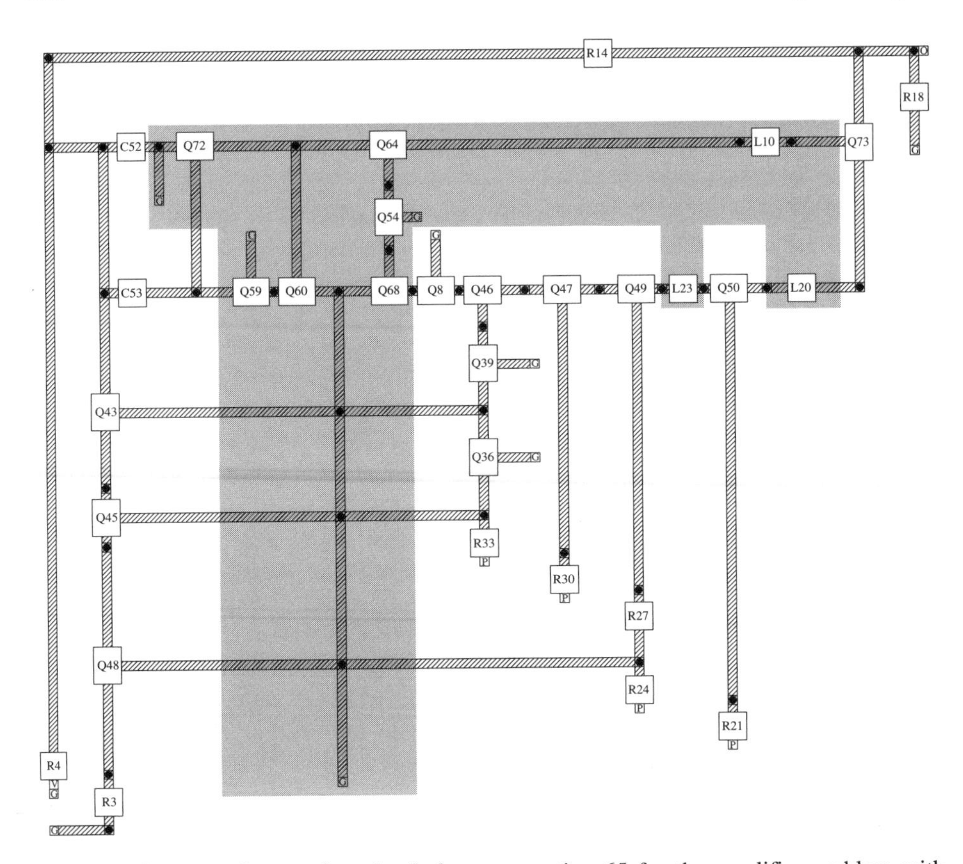

Figure 5.18 Best-of-generation circuit from generation 65 for the amplifier problem with layout.

Table 5.2 Comparison of characteristics of two best-of-generation circuits for the amplifier problem with layout

Generation	Components	Area	Four penalties	Fitness
65	27	8,234	33.034348	33.042583
101	19	4,751	0.061965	0.004751

circuit from generation 65 (figure 5.18) and the best-of-run circuit from generation 101 (figure 5.19).

The best-of-generation circuit from generations 65 has 81, 189, and 26 points, respectively, in its three branches. The best-of-run circuit from generation 101 has 65, 85, and 10 points, respectively, in its three branches. That is, the sizes of the corresponding branches of the best-of-run circuit from generation 101 are smaller than those of the best-of-generation circuit from generations 65.

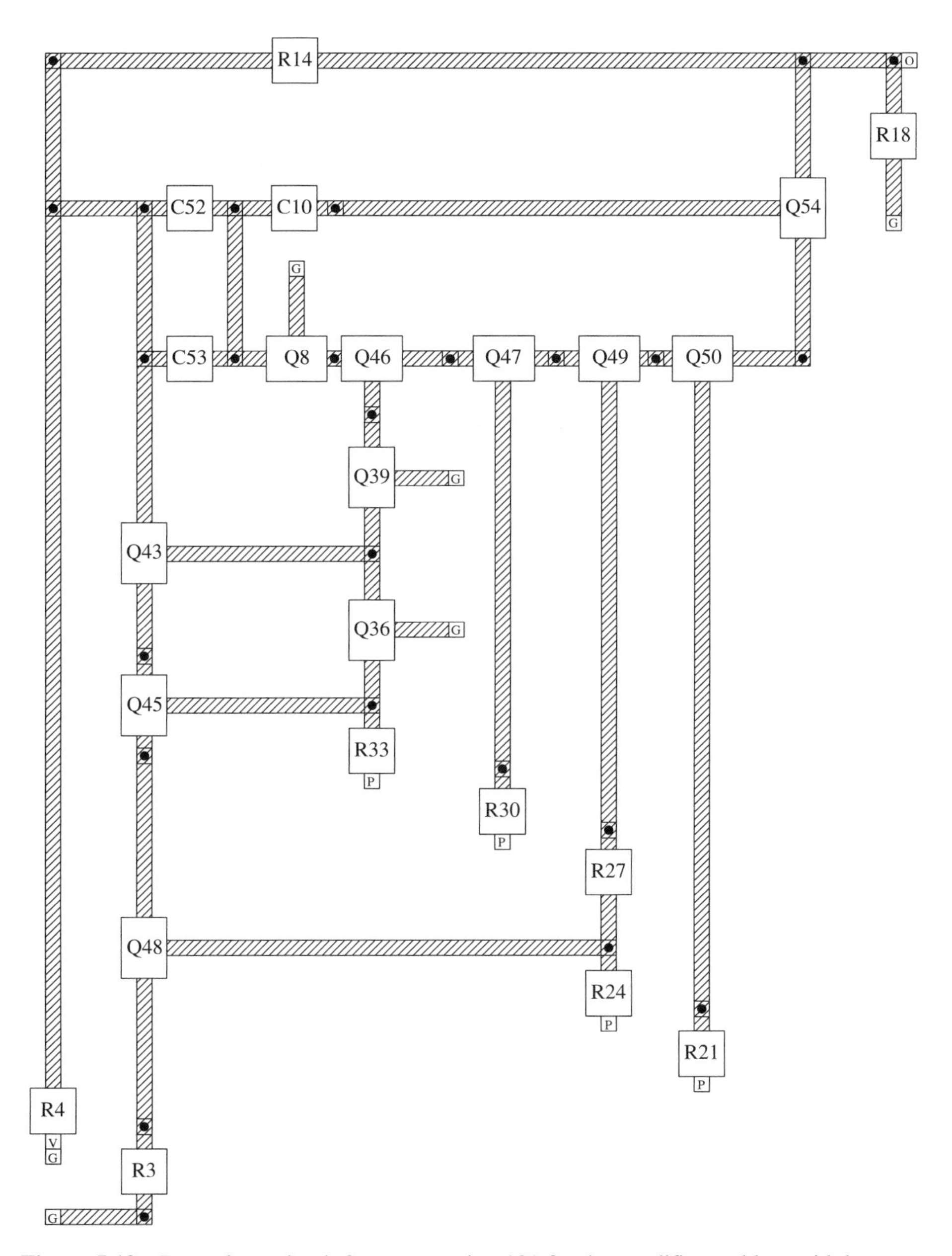

Figure 5.19 Best-of-run circuit from generation 101 for the amplifier problem with layout.

The third branches of these two individuals are both small (26 and 10 points, respectively). The only effect of these branches is to insert a single transistor (a *pnp* transistor in generation 65 and an *npn* transistor in generation 101).

The shaded portion of figure 5.18 shows the portion of the best circuit from generation 65 that is deleted in order to create the best-of-run circuit of generation 101

Table 5.3 Placement of the 19 components of the best-of-run circuit from generation 101 for the amplifier problem with layout

Component	*X*-coordinate	*Y*-coordinate	Sizing/Type
Q8	−8.398678	21.184582	Q2N3906
C10	−8.54565	31.121107	101
R21	18.245857	−24.687471	1,480
R24	12.105233	−20.687471	5,780
R27	12.105233	−12.690355	3,610
R30	5.128666	−8.690355	87.5
R33	−3.398678	−4.690355	1,160
Q36	−3.398678	3.30961	Q2N3904
Q39	−3.398678	13.309582	Q2N3904
Q43	−18.472164	8.309597	Q2N3904
Q45	−18.472164	−1.690355	Q2N3904
Q46	−3.398678	21.184582	Q2N3904
Q47	5.128666	21.184582	Q2N3904
Q48	−18.472164	−17.687471	Q2N3904
Q49	12.105233	21.184582	Q2N3904
Q50	18.245857	21.184582	Q2N3904
C52	−15.472164	31.121107	125
C53	−15.472164	21.184582	0.00778
Q54	24.873787	31.121107	Q2N3906

(figure 5.19). The first branches of these two individuals are so similar that it is clear that these two branches are genealogically related. These two first branches account for 14 components (nine transistors and five resistors) that are in common with both circuits.

The second branches of these two individuals are almost completely different. The second branch of the best-of-run circuit of generation 101 accounts for the bulk of the reduction in component count (six transistors and three capacitors). There is one added component (capacitor C10).

The difference between generations 65 and 101 caused by the first branches is that two extraneous components that are present in generation 65 are missing from the smaller 19-component circuit from generation 101.

Table 5.3 shows the *X* and *Y* coordinates for the 19 components of the best circuit of generation 101, the type (*npn* Q2N3904 or *pnp* Q2N3906) for each transistor, and the component value (sizing) for each capacitor and resistor. Capacitances are in nanofarads and resistances are in kilo-Ohms.

For additional information, see Koza and Bennett 1999 and Bennett and Koza 2002.

5.3.3 Routineness for the 60 dB Amplifier Problem with Layout

The problem involving the amplifier is in the same domain (i.e., the automatic synthesis of circuit topology, sizing, placement, and routing) as the problem concerning the lowpass filter (section 5.2). Thus, in this section, we consider the effort required to make the transition from one problem to another problem within the same domain.

The preparatory steps for the present problem are substantially the same as the preparatory steps for the problem involving the lowpass filter with two exceptions. First, the amplifier problem has a different goal (and hence a different fitness measure). Second, the lowpass filter contains only passive components (capacitors and inductors) whereas the amplifier (like most present-day active circuits) contains transistors (in addition to passive components such as capacitors and resistors). The inclusion of transistors in the parts bin, in turn, necessitates a source of external power. Neither change entails an enormous amount of effort.

Thus, relatively little effort is required to make the transition from the problem involving the lowpass filter to the problem involving the amplifier. That is, the transition is routine.

5.3.4 AI Ratio for the 60 dB Amplifier Problem with Layout

The solution produced by genetic programming to the problem involving the layout of an amplifier has a moderately high amount of "A."

The preparatory steps for this problem are the same (with the two exceptions noted in the previous section) as the preparatory steps for the previous problem involving the lowpass filter (section 5.2). Thus, the denominator of the AI ratio contains only a small amount of "I."

Thus, the AI ratio for the solution produced by genetic programming to this problem is moderately high.

6

Automatic Synthesis of Antennas

An antenna is a device for receiving or transmitting electromagnetic waves. An antenna may receive an electromagnetic wave and transform it into a signal on a transmission line. Alternately, an antenna may transform a signal from a transmission line into an electromagnetic wave that is then propagated in free space.

Some antennas are directional in the sense that they send and receive a signal primarily in a specified direction. Some operate only over a narrow band of frequencies whereas others operate over a wide band of frequencies. Antennas vary as to the impedance at their feed points and the polarization of the waves that they transmit or receive. Some antennas are two-dimensional; others are three-dimensional; and others are "two and a half" dimensional (e.g., an antenna embedded in a curved surface such an automobile windshield). The antenna's gain is usually a major design consideration.

Maxwell's equations govern the electromagnetic waves generated and received by antennas. The task of analyzing the characteristics of a given antenna is difficult. The task of synthesizing the design of an antenna with specified characteristics is even more difficult and typically calls for considerable creativity on the part of the antenna engineer (Balanis 1982; Stutzman and Thiele 1998; Linden 1997). That is, antennas resemble circuits and controllers in that analysis is difficult and synthesis is an art.

The behavior and characteristics of many antennas can be determined by simulation. For example, the *Numerical Electromagnetics Code* (NEC) is a method-of-moments (MoM) simulator for wire antennas that was developed at the Lawrence Livermore National Laboratory (Burke 1992). The NEC simulator is reasonably fast. It works from a relatively simple text input (called the *geometry table*) and produces output in the form of well-defined and easily parsed text. The NEC simulator is widely used in the antenna community and is considered to be reasonably accurate and reliable for a broad range of structures (Linden 1997). The availability of the source code (in FORTRAN) for the NEC simulator enables the simulator to be efficiently embedded inside a run of the genetic algorithm or genetic programming. In regard to these characteristics, the NEC simulator is similar to the SPICE simulator used elsewhere in this book for simulating analog electrical circuits, controllers, and metabolic pathways.

This chapter demonstrates the use of genetic programming to automatically synthesize the design of a wire antenna for an illustrative problem. This illustrative problem has been previously solved by both conventional antenna design techniques and by the genetic algorithm operating on fixed-length strings.

Genetic algorithms have been successfully applied to the design of a variety of antennas, including the design of thinned arrays (Haupt 1994), wire antennas (Linden 1997; Altshuler and Linden 1998, 1999), patch antennas (Johnson and Rahmat-Samii 1999), and linear and planar arrays (Marcano and Duran 1999). The book *Electromagnetic Optimization by Genetic Algorithms* (Rahmat-Samii and Michielssen 1999) describes numerous applications of the genetic algorithm to antenna design. Jones (1999) applied genetic programming to antenna design.

When the genetic algorithm is applied to a problem of antenna design, the human user typically prespecifies many of the key characteristics of the size and shape of the solution. In contrast, when genetic programming is applied to this problem, the human user is not required to prespecify the size and shape of the solution. Instead, these characteristics are automatically generated by the evolutionary process.

6.1 Our Approach to the Automatic Synthesis of the Geometry and Sizing of Antennas

Our approach to the problem of automatically creating both the geometry and sizing of an antenna that satisfies user-specified design requirements involves

(1) establishing a representation for antennas involving LISP symbolic expressions (S-expressions) and program trees that can be progressively bred by means of genetic programming, and
(2) defining a fitness measure that measures how well the behavior and characteristics of a candidate antenna satisfy the problem's high-level design requirements.

The representation and fitness measure are then used in a run of genetic programming. During the run, the evaluation of the fitness of each individual in the population involves

(1) converting each individual program tree in the population into an antenna,
(2) converting each antenna into a geometry table of the type accepted by the antenna simulator,
(3) obtaining the behavior of the individual antenna by simulating its behavior, and
(4) using the antenna's behavior and characteristics to calculate its fitness.

The implementation of our approach entails working with four different representations for an antenna:

- *Drawing:* An antenna can be represented as an engineering drawing in two-dimensional or three-dimensional space. In this representation, certain lines or curves represent conductive material (e.g., metal).

- *Program Tree:* An antenna can also be represented as a program tree whose internal points (nodes) are functions and external points (leaves) are terminals. This representation enables genetic programming to breed a population of programs in a search for an antenna that satisfies user-specified design requirements.
- *Symbolic Expression:* The program can also be represented as a symbolic expression (S-expression) in the style of the LISP programming language. This representation is used internally by genetic programming.
- *Geometry Table:* A wire antenna can also be represented as a geometry table that specifies the two-dimensional (or three-dimensional) coordinates of the endpoints of each wire, the radius of each wire, and certain information necessary to control the method of moment calculation performed by the NEC simulator. The geometry table used by the NEC simulator is similar in concept to the netlist used by the SPICE simulator. The geometry table is described in section 6.4.4.

6.2 Illustrative Problem of Antenna Synthesis

We illustrate the use of genetic programming for the design of antennas with a problem that was previously solved by both conventional antenna design techniques and the genetic algorithm operating on fixed-length strings (Linden 1997; Altshuler and Linden 1999).

The problem is to synthesize the design of a planar symmetric antenna composed of wires of a half-millimeter radius that

- has maximum gain in a preferred direction (specifically, along the positive X-axis) over a range of frequencies from 424 MHz to 440 MHz,
- has a reasonable value (specifically ≤ 3.0) for voltage standing wave ratio ($VSWR$) when the antenna is fed by a transmission line whose characteristic impedance, Z_0, is 50 Ω,
- fits into a bounding rectangle whose height is 0.4 meters and whose width is 2.65 meters, and
- is excited by a single voltage source. One end of the transmission line is connected to the source and the other end is connected (at the origin of a coordinate system) to the antenna's driven element. The lower left corner of the bounding rectangle is positioned at $(-250, -200)$ and the upper right corner is at (2400, 200).

These requirements can be satisfied by a Yagi-Uda antenna (Uda 1926, 1927; Yagi 1928). A Yagi-Uda antenna (figure 6.1) is a planar symmetric wire antenna consisting of a number of parallel linear elements. The Yagi-Uda antenna is widely used for home TV antennas. The *driven element* is a vertical linear element positioned along the Y-axis. The midpoint of the driven element (i.e., at position (0, 0) in the figure) is connected to a transmission line (not shown). The antenna's other elements are not connected to the transmission line. Instead, the currents in those elements are induced parasitically by mutual coupling (Balanis 1982). All the antenna's elements are symmetric (about the X-axis) in that they are all arranged in parallel (with all their midpoints lying on a straight line perpendicular to the parallel wires). The elements to the

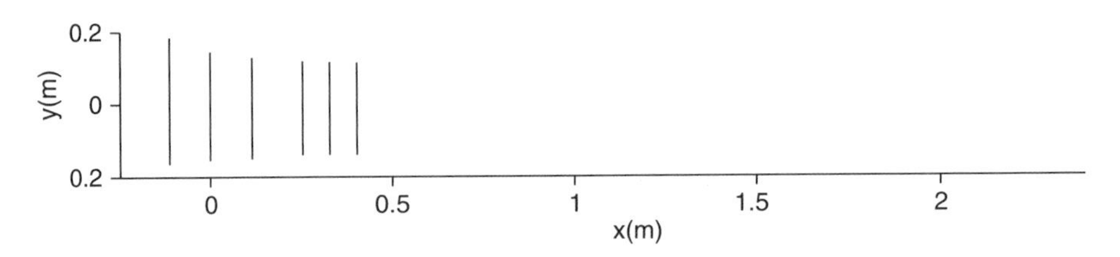

Figure 6.1 Example of a Yagi-Uda antenna.

left of the driven element act as *reflectors*. Usually (but not necessarily) there is just one reflector. The (usually numerous) elements to the right of the driven element act as *directors*. The antenna in the figure has four directors. The directors are typically spaced unequally. They are typically shorter than the driven element and of different lengths. A well-designed Yagi-Uda antenna is an *endfire* antenna in the sense that it directs most of its energy (along the positive X-axis here) toward a point in the farfield of the antenna. The driven element, directors, and reflector(s) of a Yagi-Uda antenna are physically mounted on a non-conducting physical support (not shown).

Before applying the genetic algorithm operating on fixed-length strings to a particular problem, the human user must perform four preparatory steps of:

(1) choosing the representation scheme by which the to-be-discovered parameters are mapped onto the chromosome string,
(2) defining the fitness measure,
(3) choosing control parameters for the run, and
(4) choosing the termination criterion and method of result designation.

The representation scheme for the genetic algorithm typically involves choosing the number of characters in the alphabet for the chromosome string, the number of characters that are allocated to each variable, and the location of each variable on the chromosome string (Goldberg 1989).

When Linden and Altshuler (Linden 1997; Altshuler and Linden 1999) applied the genetic algorithm operating on fixed-length strings on the present problem, they prespecified

(A) that all the antenna's wires (i.e., the directors, the reflectors, and the driven elements) would be straight,
(B) that all the antenna's wires would be arranged in parallel (with all their midpoints lying along a single straight line that is parallel to the X-axis), and
(C) that the energy source (the transmission line) would be connected to the midpoint of a single straight wire (the driven element),
(D) the number of directors, and
(E) the number of reflectors (one).

Specifications (A), (B), and (C) are key characteristics of the topology of a Yagi-Uda antenna (figure 6.1).

After specifying (A), (B), (C), (D), and (E), the only remaining issue is the discovery of certain numerical values, namely the lengths of the wires and the spacing between the wires. In applying the genetic algorithm to this problem, Linden and Altshuler constructed a chromosome string in which each part of the chromosome represented one of these to-be-discovered numerical values.

As will be seen shortly, when genetic programming is applied to this problem, the human user need not prespecify (A), (B), (C), (D), or (E).

6.3 Repertoire of Functions and Terminals

Our approach to the synthesis of a planar wire antenna involves creating a two-dimensional drawing of the antenna. The drawing is made by a turtle moving in the plane. The turtle may (or may not) deposit ink (metal) as it moves. The necessary movements can be implemented using an amalgam of features of turtle geometry (Abelson and diSessa 1980), the Logo programming language, and Lindenmayer systems (Lindenmayer 1968; Prusinkiewicz and Hanan 1980; Prusinkiewicz and Lindenmayer 1990) augmented by certain additional features of our own.

The Logo programming language is a dialect of LISP that was created in 1967 by Seymour Papert and Wallace Feurzeig, and others at Bolt, Beranek, and Newman. We use functions patterned after functions from Logo that move and turn the turtle and that iteratively execute a group of functions.

We also use a function that rubber-bands the turtle in the same manner as the brackets of Lindenmayer systems (Koza 1993).

6.3.1 Repertoire of Functions

Eight functions are used in the program trees. All are side-effecting functions that do not return any value.

The one-argument `TURN-RIGHT` function changes the facing direction of the turtle by turning the turtle clockwise by the amount specified by its argument. The argument is a floating-point number between 0.0 and 1.0. The argument is then multiplied by 2π (with the result that it represents an angle in radians).

The two-argument `DRAW` function moves the turtle in the direction that it is currently facing by an amount specified by its first argument. The turtle may or may not deposit a wire as it is moving, depending on its second argument. Drawing with `HALF-MM-WIRE` is equivalent to depositing a half-millimeter wire while moving the turtle. Drawing with `NOWIRE` is equivalent to moving the turtle without depositing metal. The first argument is a floating-point number between 0.0 and 1.0. This number is scaled between two bounds (16 to 200 millimeters for this problem). These particular bounds were chosen to avoid unreasonably long wires and unreasonably short wires (which may represent an implausible antenna geometry and may sometimes lead to simulator errors).

The two-argument `REPEAT` function causes the subtree in its second argument to be executed for the number of times specified by its first argument. The first argument is an integer modulo 100. This particular bound was chosen to give the drawing process considerable freedom while limiting the expenditure of computer time. One use of the `REPEAT` function is to permit the reuse of a useful substructure. Notice also that an iterative function such as `REPEAT` has the practical effect of enabling the creation of curved substructures (composed of a large number of very short straight wires).

The one-argument LANDMARK function causes the turtle to execute its single argument. The turtle is then restored to the position and facing direction that it had at the start of the evaluation of the LANDMARK function. The LANDMARK function operates in the same manner as the brackets of Lindenmayer systems (Lindenmayer 1968; Prusinkiewicz and Hanan 1980; Prusinkiewicz and Lindenmayer 1990).

The two-argument TRANSLATE-RIGHT function does three things. First, this function temporarily turns the turtle clockwise by the angular amount specified by its first argument (in the same manner as the argument of the TURN-RIGHT function). Second, this function moves the turtle forward (in the direction in which the turtle was temporarily turned) by the amount specified by its second argument (in the same manner as the DRAW function with NOWIRE as its second argument). Third, this function restores the turtle to its original facing direction.

The connective functions PROGN2, PROGN3, and PROGN4 sequentially execute their 2, 3, and 4 arguments (respectively).

This foregoing repertoire of functions and terminals can be easily extended in various ways. For example, three-dimensional antennas can be constructed simply by adding an additional angle to the TURN-RIGHT and TRANSLATE-RIGHT functions (thus creating a "flying turtle"). In addition, coordinate transformations can be defined to enable the turtle to move on an irregularly shaped surface (e.g., a curved automobile windshield). Variable width wires can be accommodated merely by using a numerical parameter for the wire width. The foregoing repertoire of functions and terminals assumes that there is only one connection to the transmission line. However, antennas with more than one connection to the transmission line can be easily accommodated by adding a function that creates such a connection at the turtle's current location.

Also, many antennas involve repetitions of substructures (either exactly or with slight variations). Automatically defined functions can create repetitive structures in an efficient manner. Moreover, automatically defined functions with dummy variables (formal parameters) can efficiently create structures with variation in the repeated structure. Although this degree of flexibility is not required in the present problem, this facility of genetic programming would be advantageous in synthesizing the design of more complex antennas.

6.3.2 Repertoire of Terminals

Four terminals are used in the program trees.

The terminal HALF-MM-WIRE denotes a half-millimeter wire. The terminal NOWIRE represents the absence of a wire. A constrained syntactic structure restricts the location of these two terminals to the second argument of the DRAW function.

$\Re_{real}$ denotes floating-point numbers between 0.0 and 1.0. These floating-point numbers can appear only as the first argument of a TURN-RIGHT or DRAW function. Note that we do not use the nonlinear mapping (section 3.5.5).

$\Re_{integer}$ denotes integers between 0 and 99. These integers can appear only as the first argument of a REPEAT function.

The END terminal appears at all other leaves (endpoints) of the program tree.

6.3.3 Example of the Use of the Functions and Terminals

The following composition of turtle-controlling functions and terminals illustrates how an antenna may be created:

```
1 (PROGN3
2   (TURN-RIGHT 0.125)
3   (LANDMARK
4     (REPEAT 2
5       (PROGN2
6         (DRAW 1.0 HALF-MM-WIRE)
7         (DRAW 0.5 NO-WIRE)))
8   (TRANSLATE-RIGHT 0.125 0.75))
```

As shown in figure 6.2(a), the turtle starts facing north. The PROGN3 connective function on line 1 causes the execution of the functions on lines 2, 3, and 8. As shown in figure 6.2(b), the TURN-RIGHT function on line 2 causes the turtle to turn clockwise by 1/8 of 2π (i.e., 45 degrees). The PROGN2 connective function on line 5 causes the execution of the functions on lines 6 and 7. As shown in figure 6.2(c), the DRAW function of line 6 moves the turtle in its current facing direction by length 1.0 while depositing metal (i.e., HALF-MM-WIRE). As shown in figure 6.2(d), the DRAW function of line 7 moves the turtle in its current facing direction by length 0.5 but without depositing metal (i.e., NO-WIRE). The functions on lines 5, 6, and 7 are encompassed by the iterative REPEAT function on line 4. Consequently, as shown in figure 6.2(e), this REPEAT function causes the metal of length 1.0 and the gap of length 0.5 to be repeated. The functions on lines 4, 5, 6, and 7 are encompassed by the LANDMARK function on line 3. As shown in figure 6.2(f), this LANDMARK function first causes the turtle to be restored to the position and facing direction that it had after the execution of the TURN-RIGHT function of line 2, namely the position and facing direction of figure 6.2(b). Finally, the TRANSLATE-RIGHT function of line 8 temporarily turns the turtle clockwise by 1/8 of 2π (i.e., 45 degrees) so that it is facing east and then moves the turtle forward 0.75 without depositing metal. After execution of this TRANSLATE-RIGHT function, the turtle is restored to its previous facing direction (i.e., northeast). The effect of executing the TRANSLATE-RIGHT function is shown in figure 6.2(g).

The consequence of executing lines 1 through 8 is the creation of the two wires shown in figure 6.2(g). That is, the turtle made the drawing shown in figure 6.2(g).

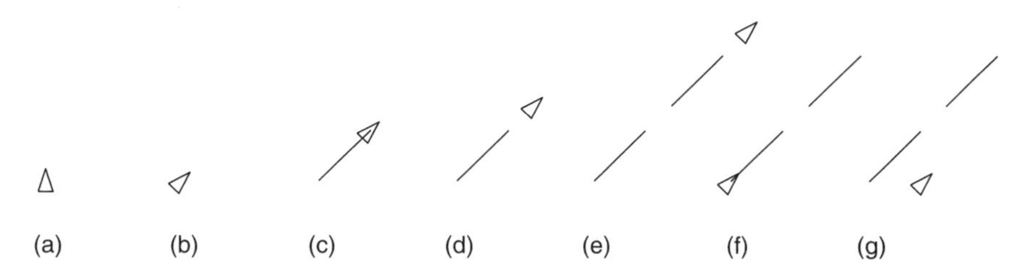

Figure 6.2 Examples of effects of turtle-controlling functions.

6.4 Preparatory Steps for the Antenna Problem

6.4.1 Program Architecture

Each individual program in the population has one result-producing branch. Automatically defined functions are not used in this problem.

6.4.2 Function Set

The function set for the result-producing branch is

F_{rpb} = {TURN-RIGHT, DRAW, REPEAT, LANDMARK, TRANSLATE-RIGHT, PROGN2, PROGN3, PROGN4}.

6.4.3 Terminal Set

In this problem, the numerical parameter value(s) for each function possessing a parameter are established by a value-setting subtree containing a single perturbable numerical value. This approach for establishing numerical parameter values is described in section 3.5.5.2.

The terminal set, T_{vss}, for the value-setting subtrees is

$$T_{vss} = \{\Re_p\},$$

where $\Re_p$ denotes a perturbable numerical value.

As previously mentioned, a constrained syntactic structure is used to restrict certain terminals (WIRE, NOWIRE, $\Re_{real}$, and $\Re_{integer}$) to be certain arguments of certain functions. For all other parts of the program tree, the terminal set is simply

$$T_{rpb} = \{\text{END}\}.$$

6.4.4 Fitness Measure

The evaluation of each individual antenna-creating program tree in the population begins with its execution. The program tree is executed in the usual depth-first order of evaluation (from left to right). The number, location, and shape of the antenna's elements are determined by the program tree. As the turtle moves, it may (or may not) deposit metal in the form of straight pieces of wire. Separate antenna elements can be created by the DRAW function (when it is executed with the NOWIRE argument) and by the TRANSLATE-RIGHT function. The result is a two-dimensional drawing of the antenna that typically consists of multiple, distinct wire segments. A geometry table is created that identifies each distinct wire of the antenna.

More specifically, the turtle starts at position (0, 7.5) with a facing direction of north (i.e., along the positive *Y*-axis). Each antenna begins with a 7.5-millimeter straight piece of wire beginning at the origin and lying along the positive *Y*-axis (i.e., facing north). This stub becomes part of the antenna's driven element. The driven element is excited by a voltage source applied at (0, 0).

The desired antenna is intended to fit inside a specified area. Thus, after the execution of the program tree, any metal that has been deposited outside the boundary of the specified clipping rectangle is deleted. The lower left corner of the clipping rectangle for this problem is positioned at (−250, 0) and its upper right corner is positioned at (2400, 200).

Because the statement of this problem calls for a symmetric antenna, all metal deposited above the X-axis is, after clipping, duplicated by reflecting it around the X-axis and duplicating it below the X-axis. The result is the desired symmetric antenna lying inside the bounding rectangle (which is twice the height, but equal in width, to the clipping rectangle).

Antennas can be simulated by means of the *Numerical Electromagnetics Code* (NEC) simulator. We embedded version 4 of the *Numerical Electromagnetics Code* (NEC) antenna simulator into our genetic programming software (in much the same way as we embedded the SPICE simulator used elsewhere in this book for simulating the behavior of controllers, circuits, and networks of chemical reactions).

Table 6.1 shows a sample input file to the NEC simulator. The input to the NEC simulator consists of textual information. This textual information includes the geometry table; information about the means of excitation, the output, and the ground plane (or lack thereof); and various commands for controlling the simulation.

In table 6.1, "CM" denotes comments. The input starts with the geometry table (denoted by "GW"). There is one line in the geometry table for each wire in the antenna.

Table 6.1 Sample input file to the *Numerical Electromagnetics Code* (NEC) antenna simulator

```
CM
CM THE GEOMETRY TABLE (GW)
CM
GW 1 3 0 0 0 0 0.015 0 5.0E-4
GW 2 3 0 0.015 0 0 0.163304 0 5.0E-4
GW 3 3 0.179902 0 0 0.179902 0.140018 0 5.0E-4
GW 4 3 0.254795 0 0 0.254795 0.139828 0 5.0E-4
GW 5 3 0.329688 0 0 0.329688 0.139638 0 5.0E-4
GW 6 3 0.404581 0 0 0.404581 0.139448 0 5.0E-4
GW 7 3 0.479474 0 0 0.479474 0.139258 0 5.0E-4
GW 8 3 0.554367 0 0 0.554367 0.139068 0 5.0E-4
GW 9 3 0.629260 0 0 0.629260 0.138878 0 5.0E-4
GW 10 3 0.704153 0 0 0.704153 0.138688 0 5.0E-4
GW 11 3 0.779046 -0 0 0.779046 0.138498 0 5.0E-4
GW 12 3 0.853939 0 0 0.853939 0.138308 0 5.0E-4
GW 13 3 0.928832 0 0 0.928832 0.138118 0 5.0E-4
GW 14 3 1.003726 -0 0 1.003726 0.137928 0 5.0E-4
GW 15 3 1.078619 0 0 1.078619 0.137738 0 5.0E-4
GW 16 3 1.153512 0 0 1.153512 0.137548 0 5.0E-4
GW 17 3 1.228405 -0 0 1.228405 0.137358 0 5.0E-4
GW 18 3 1.303298 0 0 1.303298 0.137168 0 5.0E-4
GW 19 3 1.378191 0 0 1.378191 0.136978 0 5.0E-4
GW 20 3 1.453084 0 0 1.453084 0.136788 0 5.0E-4
GW 21 3 1.527977 0 0 1.527977 0.136598 0 5.0E-4
GW 22 3 1.60287 0 0 1.60287 0.136408 0 5.0E-4
GW 23 3 1.677763 0 0 1.677763 0.136218 0 5.0E-4
GW 24 3 1.752656 -0 0 1.752656 0.136028 0 5.0E-4
GW 25 3 1.827549 0 0 1.827549 0.135838 0 5.0E-4
GW 26 3 1.902442 0 0 1.902442 0.135648 0 5.0E-4
```

(*Continued*)

Table 6.1 *(Continued)*

```
GW 27 3 1.977335 −0 0 1.977335 0.135458 0 5.0E-4
GW 28 3 2.052228 0 0 2.052228 0.135268 0 5.0E-4
GW 29 3 2.127121 0 0 2.127121 0.135078 0 5.0E-4
GW 30 3 2.202014 −0 0 2.202014 0.134888 0 5.0E-4
GW 31 3 2.276907 0 0 2.276907 0.134698 0 5.0E-4
GW 32 3 2.351800 0 0 2.351800 0.134507 0 5.0E-4
GW 33 3 0.114958 0 0 0.114958 0.151050 0 5.0E-4
GW 34 2 0.141304 0.110982 0 0.141304 0.2 0 5.0E-4
GW 35 2 −0.002003 0 0 −0.002003 0.071799 0 5.0E-4
GW 36 4 −0.118964 0 0 −0.118964 0.174407 0 5.0E-4
GW 37 3 −0.235924 0.079042 0 −0.235924 0.2 0 5.0E-4
CM
CM GEOMETRY SYMMETRY REQUEST
CM
GX 2000 010
CM
CM GROUND PLANE FLAG AND ERROR BEHAVIOR FLAG
CM
GE 0 1
CM
CM SIMULATION FREQUENCY OPTIONS
CM
FR 0 1 0 0 424.0 0
CM
CM EXCITATION DESCRIPTION AND LOCATION
CM
EX 0 1 1 0 1.00
CM
CM RADIATION PATTERN REQUEST
CM
RP 0 1 1 0011 90.0 0.0 1.0 1.0
```

Each line of the geometry table for an antenna contains

- a segment number (1 to 37 in table 6.1),
- the number of segments into which each wire is to be partitioned for purposes of simulation by the method of moment calculation performed by the *Numerical Electromagnetics Code* simulator,
- the three-dimensional coordinates of the endpoints of each wire (X1, Y1, Z1, X2, Y2, X2), with the third number of each group of coordinates being 0 here because the present problem involves a two-dimensional antenna,
- the radius of each wire (consistently one half-millimeter in this example).

The remaining lines are described in detail in the manual for version four of the *Numerical Electromagnetics Code* antenna simulator (Burke 1992).

Fitness for this problem is a linear combination of *VSWR* and gain.

The *VSWR* is a measure of how much of the input energy from the source is reflected back down the transmission line from the antenna (rather than radiated by the antenna). The NEC code calculates the complex input impedance, Z, at the voltage source.

The reflection coefficient, R, is then computed from Z and Z_0 (the characteristic impedance of the transmission line feeding the antenna) as follows:

$$R=\frac{(Z-Z_0)}{(Z+Z_0)}.$$

Finally,

$$VSWR = \frac{1+|R|}{1-|R|}.$$

The value of *VSWR* ranges from 1 (representing no reflection—that is, all energy is radiated), to infinity (i.e., all energy is reflected and there is no radiation). In this implementation, the maximum value of the *VSWR* is limited to 2×10^8.

The NEC simulator is instructed to compute the farfield radiation pattern at $\theta=90$ and $\phi=0$. θ is measured from the positive Z-axis to the X-Y-plane whereas ϕ is measured from the positive X-axis to the positive Y-axis. The value for the antenna's gain is the magnitude (in decibels relative to an isotropic radiator) of the farfield radiation pattern at $\theta=90$ and $\phi=0$.

The simulator calculates the current at the center of each segment.

The accuracy of an antenna simulation can be significantly affected by the number of segments into which each straight wire in the antenna is divided for purpose of simulation. There is no *a priori* way to compute the ideal number of segments to achieve an accurate simulation. Worse yet, contrary to intuition, the largest number of segments does not necessarily produce the most accurate simulation of an antenna. Therefore, we invoked the NEC simulation 12 times for each antenna and then used only the *worst* outcome in the fitness calculation. Specifically, we performed a simulation using both 15 and 25 segments per wavelength. Then, for each of these two cases, we performed a simulation using both 1 and 2 as the minimum number of segments per straight wire section. Finally, for each of these four cases, we performed a simulation at three different frequencies (424, 432, 440 MHz). The value of the *VSWR* used in the fitness calculation is the *maximum* over all the 12 cases. The value of the gain, G, used in the fitness calculation is the *minimum* over all the 12 cases.

Fitness is $-G+C*VSWR$. A smaller fitness is better. The constant C is 0.1 when $VSWR \leq 3.0$ and is 10 when $VSWR > 3.0$. This is the same fitness measure used by Altshuler and Linden (1999), except that it has a slightly heavier penalty for poor *VSWR*. We increased the penalty because an antenna with poor *VSWR* can produce an artificially high gain (offsetting the *VSWR* in the overall fitness). The gain, G, appears with a negative sign because greater gain is more desirable.

As required by the NEC simulator, each pair of intersecting wires is replaced with four new wires (each with an endpoint at the intersection point) prior to running the simulator. In addition, each antenna-geometry is pre-checked by various rules. If the length of each segment divided by its radius is not greater than 2.0 or the center of each segment is not at least 4 wire radii from the axis of every other wire, then the individual is either replaced (for generation 0) or assigned a high penalty value of fitness (10^8).

If more than 10,000 turtle functions are executed (due to excessive iterations), execution of the individual program tree is terminated and the individual is assigned a high penalty value of fitness (10^8).

Antennas that cannot be simulated receive a high penalty value of fitness (10^8).

6.4.5 Control Parameters

The population size is 500,000.

6.5 Results for the Antenna Problem

Figure 6.3 shows a best-of-node (i.e., the best individual from a particular processing node of the parallel computer system) antenna from generation 0 of our one and only run of this problem. It consists of only one element (i.e., the driven element). It has a fitness of −2.03. This figure (and each of the following figures) is accompanied by the antenna's radiation pattern produced by the NEC simulator.

Figure 6.4 shows another best-of-node antenna from generation 0. It consists of two elements, namely a driven element and a V-shaped element. It has a fitness of −3.82.

Figure 6.5 shows yet another best-of-node antenna from generation 0. It consists of numerous straight wires and numerous V-shaped elements. It has a fitness of −4.43.

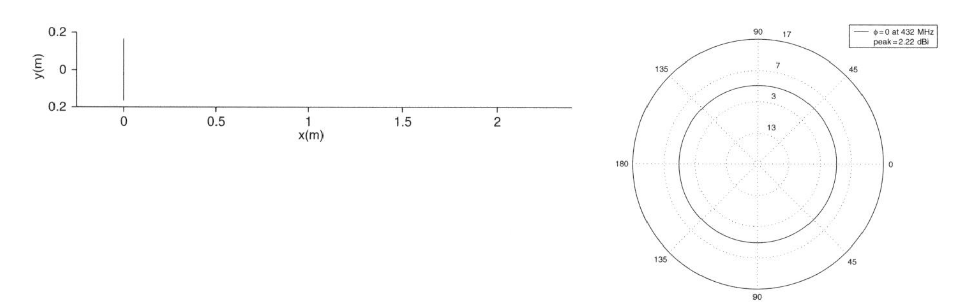

Figure 6.3 First example of a best-of-node antenna from generation 0.

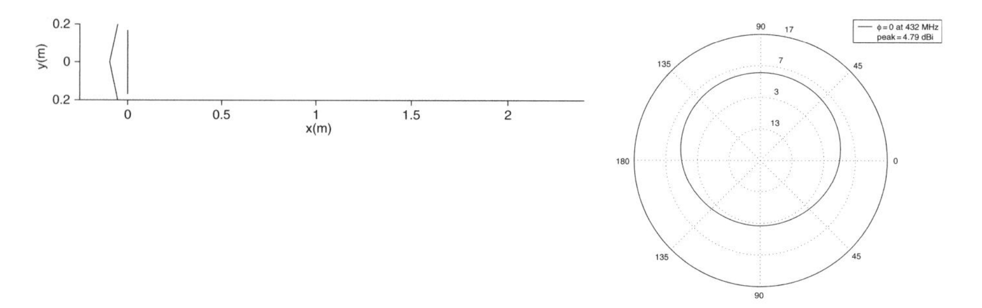

Figure 6.4 Second example of a best-of-node antenna from generation 0.

Figure 6.6 shows the best-of-generation antenna from generation 2. This antenna resembles a starburst and consists of both straight wires and V-shaped elements. It has a fitness of −5.18.

The best-of-generation antenna from generation 9 (figure 6.7) has a fitness of −7.58.

The best-of-generation antenna from generation 47 (figure 6.8) has a fitness of −14.13.

The best-of-run antenna emerged in generation 90 (figure 6.9). It has a fitness of −16.04.

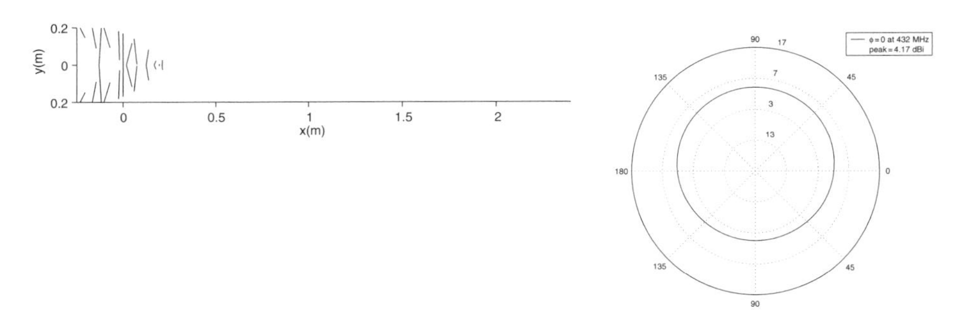

Figure 6.5 Third example of a best-of-node antenna from generation 0.

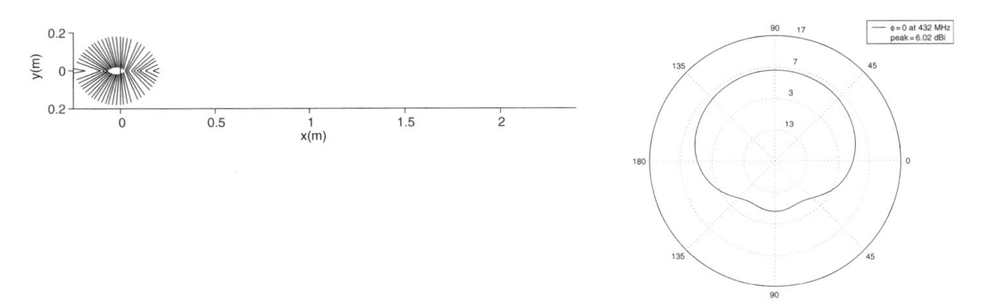

Figure 6.6 Best-of-generation antenna from generation 2.

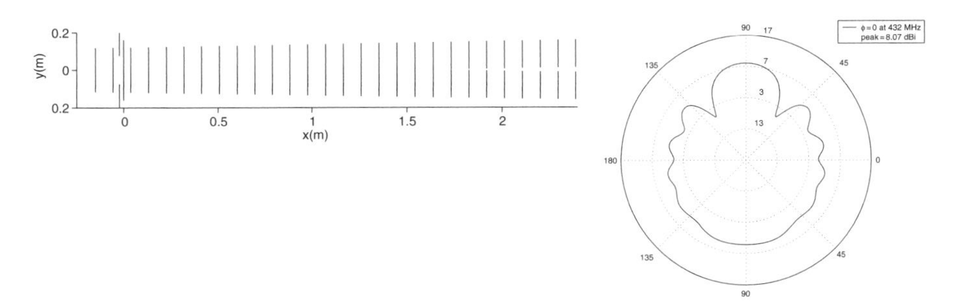

Figure 6.7 Best-of-generation antenna from generation 9.

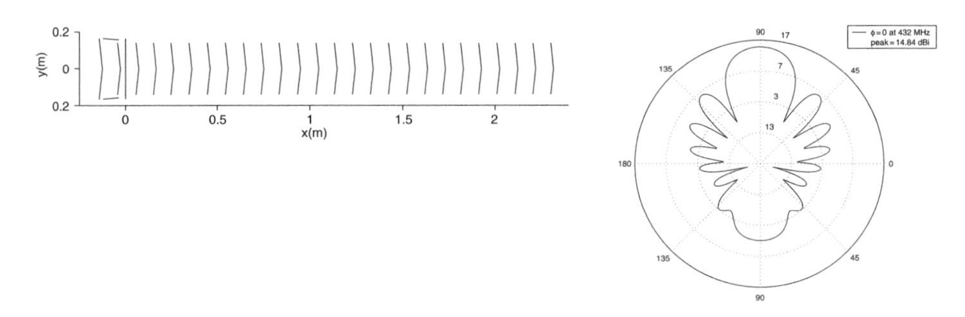

Figure 6.8 Best-of-generation antenna from generation 47.

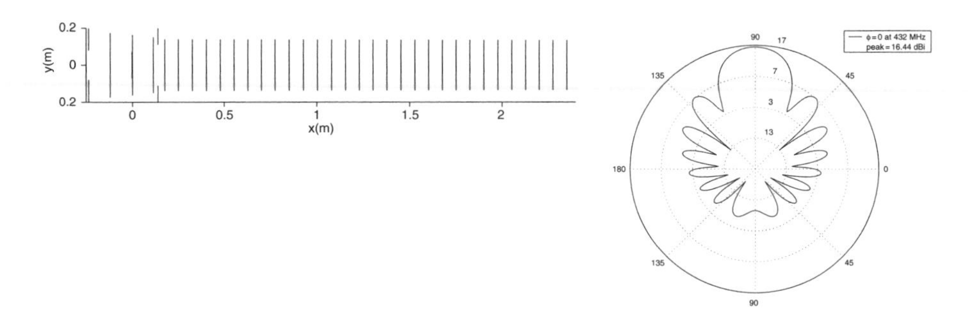

Figure 6.9 Best-of-run antenna from generation 90.

Table 6.2 Comparison of a Yagi-Uda antenna, the antenna created by Altshuler and Linden using the genetic algorithm, and the antenna created by genetic programming

Frequency	Yagi-Uda		Altshuler-Linden		GP	
(MHz)	Gain (dB)	*VSWR*	Gain (dB)	*VSWR*	Gain (dB)	*VSWR*
424	15.5	1.41	15.4	1.88	16.3	2.19
428	15.8	1.11	16.0	1.80	16.4	2.00
432	15.9	1.23	16.3	1.09	16.4	2.02
436	15.7	1.60	16.1	9.50	16.5	2.16
440	15.5	1.85	9.4	39.00	16.3	2.36

Table 6.2 compares the gain (in decibels) and *VSWR* for a conventional Yagi-Uda antenna, the antenna created by Altshuler and Linden using the genetic algorithm (Linden 1997; Altshuler and Linden 1997, 1998), and the antenna created by genetic programming. All three are satisfactory solutions to the problem.

The conventional Yagi-Uda antenna and the antennas created by the genetic algorithm and genetic programming are all approximately the same length (i.e., about 3.6 wavelengths along the *X*-axis).

Note that characteristics (A), (B), (C), (D), and (E) enumerated earlier were automatically discovered during the run of genetic programming. In contrast, all five of these characteristics were preordained by Linden and Altshuler prior to launching their run of the genetic algorithm.

Characteristics (A), (B), and (C) are key characteristics of the Yagi-Uda topology for antennas (Uda 1926, 1927, Yagi 1928).

The Yagi-Uda antenna (Uda 1926, 1927, Yagi 1928) was considered an achievement in its field at the time it was first invented. As Balanis (1982) observed,

> "Although the work of Uda and Yagi was done in the early 1920s and published in the middle 1920s, full acclaim in the United States was not received until 1928 when Yagi visited the United States and presented papers at meetings of the Institute of Radio Engineers (IRE) in New York, Washington, and Hartford. In addition, his work was published in the *Proceedings of the IRE*, June 1928, where J. H. Dellinger, Chief of Radio Division, Bureau of Standards, Washington, D. C., and himself a pioneer of radio waves, wrote 'I have never listened to a paper that I left so sure was destined to be a classic.' So true!!"

For additional information, see Comisky, Yu, and Koza 2000.

6.6 Routineness of the Transition from Problems of Synthesizing Controllers, Circuits, and Circuit Layout to a Problem of Synthesizing an Antenna

This book has previously presented the solutions to problems from three domains:

- the synthesis of controllers (chapter 3),
- the synthesis of the topology and sizing of circuits (chapter 4), and
- the synthesis of the layout (along with the topology and sizing) of circuits (chapter 5).

What is the nature of the effort required to make the transition from the previous three domains to the domain of automatic synthesis of antennas?

First, there are, of course, differences in the function and terminal sets. These differences reflect the fact that controllers, circuits, and antennas are all composed of different ingredients. For problems involving the synthesis of controllers, the function and terminal set permitted the construction of entities composed of ingredients such as integrators, differentiators, gains, adders, subtractors, reference signals, and plant outputs. For the problems involving the synthesis of the topology and sizing of circuits, the function and terminal set permitted the construction of ingredients such as transistors, capacitors, and resistors. For the problems involving the simultaneous synthesis of circuit topology, sizing, placement, and routing, a geographically-aware function set was used. For problems involving the synthesis of the design of an antenna, the function and terminal set permitted the construction of entities composed of multiple metallic wires of various lengths and orientations.

Second, there is a difference in fitness measures. The fitness measure for a problem involving the synthesis of controllers is couched in terms of behavior and characteristics relevant to the field of control, such as the integral of the time-weighted absolute error, overshoot, and robustness. The fitness measure for a problem of circuit synthesis is couched in terms of behavior and characteristics relevant to circuits, such as the circuit's frequency-domain and time-domain response. The fitness measure for a problem involving circuit layout considers factors such as the area of the bounding rectangle (in addition to the circuit's frequency-domain and time-domain response). Likewise, the fitness measure for an antenna problem is couched in terms of behavior and characteristics relevant to antennas (e.g., directionality, gain, and bandwidth).

Once the preparatory steps are performed for a problem of antenna synthesis, the executional steps of genetic programming are the same as those from the problems from the other domains.

In summary, relatively little effort is required to make the transition from the previous three domains to the domain of automatic synthesis of antennas and it is therefore reasonable to say that the transition is routine.

6.7 AI Ratio for the Antenna Problem

What is the AI ratio for the design by genetic programming of an antenna?

In 1927 and 1928, the Yagi-Uda antenna clearly solved "a problem of indisputable difficulty in its field" (criterion G of table 1.2). Genetic programming produced an antenna in this chapter that has the essential topological features of the Yagi-Uda antenna. However, we believe that the genetically produced result in this chapter falls slightly short of qualifying as human-competitive under the intentionally stringent criteria for human-competitiveness enumerated in table 1.2 because of its excessive number of wires. Because genetic programming duplicated the essential topological features of a Yagi-Uda antenna, we attribute a moderately high amount of "A" to the solution produced by genetic programming for this problem.

The function and terminal sets reflect the nature of antennas. We did not employ any deep knowledge about the design of antennas in selecting the functions and terminals. Quite the contrary. Our approach simply involves creating a two-dimensional drawing of the antenna using a turtle that may (or may not) deposit ink (metal) as it moves. These same functions and terminals could be used for constructing a wide variety of two-dimensional antennas. In fact, these same functions and terminals could be used to draw a wide variety of structures other than antennas.

Of course, a small amount of domain knowledge is embedded in certain details of the functions and terminals. For example, we choose the numerical bounds on the `DRAW` function. We also choose the half-millimeter width for the wire. Also, we choose the upper bound on the number of executions of the `REPEAT` function in order to conserve computational resources. We view all of these choices as distinctly minor.

The fitness measure specifies the performance and characteristics of the desired antenna. Each candidate antenna in each generation of the population during the run is analyzed using the *Numerical Electromagnetics Code* (NEC) simulator. Note the use of the NEC simulator does not supply any knowledge to the genetic programming system for the substantially same reasons mentioned in section 3.7.1.7 in connection with the SPICE simulator.

The moderately high amount of "A" represented by the result in conjunction with the small amount of "I" provided by the human user means that the AI ratio for the solution produced by genetic programming is moderately high.

7

Automatic Synthesis of Genetic Networks

Considerable amounts of data are becoming available concerning the concentrations of substances that participate in genetic networks (Ptashne 1992; McAdams and Shapiro 1995; Loomis and Sternberg 1995; Yuh, Bolouri, and Davidson 1998; Voit 2000).

The question arises as to whether it is possible to automatically infer the underlying logic of a genetic network from observed time-domain data for gene expression levels. The reverse engineering of a genetic network entails creating both a topological arrangement of conditional operators and comparative functions and all necessary numerical parameters such that the behavior of the genetic network matches observed time-domain data.

This chapter demonstrates that it is possible to use genetic programming to automatically create a computer program representing the underlying logic of the genetic network for the expression level of the *lac* operon as measured by its mRNA (messenger RNA). Genetic programming starts with observed time-domain expression levels of two regulatory genes (`REPRESSOR` and `CAP`) and the concentrations of two substances (`GLUCOSE` and `LACTOSE`) and automatically creates both the topology for the genetic network (including conditional operators and comparative functions) and all necessary numerical parameters associated with the network.

The work in this chapter was described in a poster paper accepted at the First International Conference on Systems Biology in Tokyo on November 14–16, 2000 (Lanza, Mydlowec, and Koza, 2000).

7.1 Statement of the Illustrative Problem

The lac operon is a basic control circuit present in many simple organisms, including *Escheria coli*. The genetic network involves two regulatory proteins (`REPRESSOR` and `CAP`) and two substances (`GLUCOSE` and `LACTOSE`). The two regulatory proteins (`REPRESSOR` and `CAP`) are involved in regulating the expression of the Z and Y genes. Figure 7.1 is a schematic representation of a genetic network for the expression

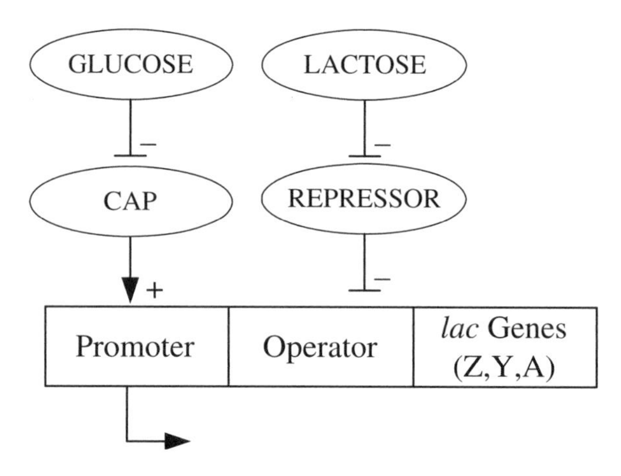

Figure 7.1 Genetic network for *lac* operon.

level of the *lac* operon (composed of the Z, Y, and A genes). The metabolism of lactose requires permease and b-galactosidase (encoded by the Z and Y genes, respectively). The permease is involved in the transport of lactose into the cell whereas the b-galactosidase is involved in cleaving the lactose molecule into glucose and galactose. This control circuit causes the expression of proteins that metabolize lactose when glucose (the preferred source of energy) is scarce and lactose is abundant.

The actual performance of the genetic network shown in figure 7.1 is determined by the expression levels of the two genes (REPRESSOR and CAP) and the concentrations of the two substances (GLUCOSE and LACTOSE) in relation to threshold values. These threshold values appear as numerical parameters of certain conditional operators and comparative functions. The logic underlying the genetic network for the *lac* operon can be succinctly written in the C-style pseudo-code shown below. The numerical value returned by this program is the expression level of the *lac* operon (LAC_MRNA_LEVEL).

```
IF(LACTOSE_LEVEL >= LACTOSE_THRESHOLD)
{
    IF(GLUCOSE_LEVEL >= GLUCOSE_THRESHOLD)
    {
        LAC_MRNA_LEVEL = LOW;
    }
    ELSE
    {
        IF (CAP_LEVEL >= CAP_THRESHOLD)
        {
            LAC_MRNA_LEVEL = HIGH;
        }
        ELSE
        {
           LAC_MRNA_LEVEL = LOW;
```

```
        }
    }
}
ELSE
{
    IF(REPRESSOR_LEVEL >= REPRESSOR_THRESHOLD)
    {
        LAC_MRNA_LEVEL = 0;
    }
    ELSE
    {
        LAC_MRNA_LEVEL = LOW;
    }
}
```

The goal in this chapter is to automatically create *both* the topology for the genetic network (including conditional operators and comparative functions) and all necessary numerical parameters representing the expression level of the *lac* operon as measured by its mRNA. In other words, the goal is to automatically create logic that is equivalent to that shown above. This will be accomplished using time-domain data for the expression levels of the two genes and the concentrations of the two substances.

7.2 Representation of Genetic Networks by Computer Programs

Each program tree represents the logic of a genetic network. Each program tree is a composition of functions from the function set and terminals from the terminal set. A program tree contains

- internal nodes representing conditional operators and comparative functions,
- external points (terminals) representing expression levels of various genes,
- external points representing concentrations of substances, and
- external points representing numerical values

The value returned by the result-producing branch of the program tree is the expression level of the *lac* operation (called "LAC_MRNA_LEVEL" in the C-style pseudo-code in section 7.1).

7.2.1 Repertoire of Functions

The three-argument IF operator returns the result of evaluating its third argument (the "else" clause) if its first argument is FALSE, but returns the result of evaluating its second argument (the "then" clause) if its first argument is TRUE.

The two-argument $<$ comparative function returns a value of TRUE if its first argument is less than its second argument, but otherwise FALSE.

The two-argument $>$ comparative function returns a value of TRUE if its first argument is greater than its second argument, but otherwise FALSE.

7.2.2 Repertoire of Terminals

The terminals GLUCOSE_LEVEL and LACTOSE_LEVEL represent substances.

The terminals REPRESSOR_LEVEL and CAP_LEVEL represent expression levels of genes.

Note that the nonlinear mapping (section 3.5.5) is not used in this chapter.

7.3 Preparatory Steps

7.3.1 Program Architecture

Each program tree has one result-producing branch.

7.3.2 Function Set

Program trees are constructed in accordance with a constrained syntactic structure. The entire program tree returns a single floating-point number. The first argument of an IF operator must be a comparative function (< or >). The two arguments of a comparative function must each be a terminal. The second and third arguments of an IF operator may be another IF operator or a perturbable numerical value.

The function set is

$$\mathsf{F}=\{\texttt{IF}, <, >\}.$$

These three functions have arity of three, two, and two, respectively.

7.3.3 Terminal Set

A constrained syntactic structure enforces the use of one terminal set for the value-setting subtrees and another terminal set for all other parts of the program tree.

In this problem, the numerical value(s) are established by a value-setting subtree containing a single perturbable numerical value. These numerical values will serve as the thresholds in the overall logic of the evolved program. This approach for establishing numerical parameter values is described in section 3.5.5.2.

The terminal set, $\mathsf{T}_{\mathrm{vss}}$, for the value-setting subtrees is

$$\mathsf{T}_{\mathrm{vss}}=\{\Re_{\mathrm{p}}\},$$

where $\Re_{\mathrm{p}}$ denotes a perturbable numerical value.

The terminal set for all other parts of the program trees is

$$T=\{\texttt{GLUCOSE\_LEVEL}, \texttt{LACTOSE\_LEVEL}, \texttt{REPRESSOR\_LEVEL}, \texttt{CAP\_LEVEL}, \Re_{\mathrm{p}}\}.$$

7.3.4 Fitness Measure

Each individual genetic network is exposed to four time-domain scenarios representing the concentrations of substances (GLUCOSE_LEVEL and LACTOSE_LEVEL) and expression values of genes (REPRESSOR_LEVEL and CAP_LEVEL) over 20 time steps (except that there are only 19 time steps in the first scenario because time

$t = 0$ is ignored). The network is exposed, during the 20 time steps, to the four combinations of high and low values of CAP_LEVEL and REPRESSOR_LEVEL.

The first of the four fitness cases is based on a high level (10) of GLUCOSE_LEVEL and a low level (0) of LACTOSE_LEVEL. Broadly speaking, CAP_LEVEL initially rises. While CAP_LEVEL is steady, REPRESSOR_LEVEL begins to rise. When REPRESSOR_LEVEL reaches it peak, CAP_LEVEL begins to fall.

The second fitness case is based on a high level (10) of GLUCOSE_LEVEL and a high level (10) of LACTOSE_LEVEL.

The third fitness case is based on a low level (0) of GLUCOSE_LEVEL and a high level (10) of LACTOSE_LEVEL.

The fourth fitness case is based on a low level (0) of GLUCOSE_LEVEL and a low level (0) of LACTOSE_LEVEL.

Fitness is the sum, over the 79 fitness cases, of the absolute weighted value of the difference between the value returned by the result-producing branch and the observed expression level of the *lac* operon (as measured by mRNA). If the value returned by the result-producing branch is within 5% of the observed expression-level data, the weight is 1.0; otherwise the weight is 10. The smaller the fitness, the better.

The number of hits is defined as the number of fitness cases (time steps 1 to 79) for which the difference is within 5% of the correct value.

7.3.5 *Control Parameters*

The population size is 10,000.

7.4 Results

The fitness of the best individual from the initial random generation (generation 0) is poor (i.e., it has a high value of 116.6). This individual scores 0 hits (out of 79).

The best-of-run individual emerges in generation 93. This individual has a fitness of 3.15 and scores 78 hits (out of 79). This program is shown below:

```
(IF (< LACTOSE_LEVEL 9.139) (IF (< REPRESSOR_LEVEL 6.270)
(IF (> GLUCOSE_LEVEL 5.491) 2.02 (IF (< CAP_LEVEL
0.639)2.033 (IF (< CAP_LEVEL 4.858) (IF (> LACTOSE_LEVEL
2.511) (IF (> CAP_LEVEL 7.807) 5.586 (IF (> LACTOSE_LEVEL
2.114) 1.978 2.137)) 0.0) (IF (> REPRESSOR_LEVEL 4.015)
0.036 (IF (< GLUCOSE_LEVEL 5.128) 10.0 (IF (< REPRESSOR_
LEVEL 4.268) 2.022 9.122)))))) (IF (> CAP_LEVEL 0.842) 0.0
5.97)) (IF (< CAP_LEVEL 1.769) 2.022 (IF (< GLUCOSE_LEVEL
2.382) (IF (> LACTOSE_LEVEL 1.256) (IF (> LACTOSE_LEVEL
1.933) (IF (> GLUCOSE_LEVEL 2.022) (IF (< GLUCOSE_LEVEL
5.183) 6.323 (IF (> CAP_LEVEL 1.208) 9.713 0.842)) 10.0)
(IF (> GLUCOSE_LEVEL 6.270) 2.109) 1.965)) 0.665) 1.982)))
```

Notice that genetic programming automatically created both the topological arrangement of conditional operators and comparative functions as well as all necessary numerical parameters of the logic for this genetic network.

When the logic of the best-of-run program from generation 93 is simplified and rewritten in C-style pseudo-code, the result is as follows:

```
IF(LACTOSE_LEVEL < 9.139)
{
    IF(REPRESSOR_LEVEL < 6.270)
    {
        LAC_MRNA_LEVEL = 2.022;
    }
    ELSE
    {
        LAC_MRNA_LEVEL = 0.0;
    }
}
ELSE
{
    IF(CAP_LEVEL < 1.769)
    {
        LAC_MRNA_LEVEL = 2.022;
    }
    ELSE
    {
        IF(GLUCOSE_LEVEL < 2.382)
        {
            LAC_MRNA_LEVEL = 10.0;
        }
        ELSE
        {
            LAC_MRNA_LEVEL = 1.982;
        }
    }
}
```

In summary, the behavior of this automatically created genetic network closely matches the observed time-domain data.

7.4.1 Routineness of the Transition from Problems of Synthesizing Controllers, Circuits, Circuits with Layout, and Antennas to a Problem of Genetic Network Synthesis

In this book, we previously solved problems involving the automatic synthesis of entities from four different domains: controllers (chapter 3), circuits (chapter 4), circuits with layout (chapter 5), and antennas (chapter 6).

What is the nature of the effort required to make the transition from these previous four domains to the domain of automatic synthesis of genetic networks?

First, there are, of course, differences in the function and terminal sets. These differences reflect the fact that controllers, circuits, antennas, and genetic networks are all composed of different ingredients.

Second, there is a difference in fitness measures. The fitness measure for a problem involving the synthesis of a genetic network is couched in terms of characteristics and performance relevant to the field of genetic networks (i.e., the difference between the single numerical value returned by the network and an observed quantity).

Once the preparatory steps are performed, the executional steps of genetic programming are the same for a problem involving genetic networks as for a problem involving the synthesis of a controller, circuit, and antenna, or anything else.

Thus, the effort required to make the transition from the previous four domains to the domain of automatic synthesis of genetic networks is relatively small. That is, the transition is routine.

7.4.2 AI Ratio for the Genetic Network Problem

What is the AI ratio for the discovery by genetic programming of a genetic network?

The function set and terminal set reflect the nature of genetic networks. We did not employ any deep knowledge about the genetic network in selecting the functions and terminals for the present problem. Quite the contrary. We simply provided conditional operators and comparative functions that create the possibility of expressing a genetic network.

The fitness measure merely reflects the statement of the problem (i.e., finding a genetic network whose behavior matches observed data).

The inference of a genetic network from data is generally considered to be a moderately difficult problem.

The moderately high amount of "A" represented by the result in conjunction with the small amount of "I" provided by the human user means that the AI ratio for the solution produced by genetic programming is moderately high.

8

Automatic Synthesis of Metabolic Pathways

A living cell can be viewed as a dynamical system in which a large number of different substances react continuously and nonlinearly with one another. In order to understand the behavior of a continuous nonlinear dynamical system with numerous interacting parts, it is usually insufficient to study behavior of each part in isolation. Instead, the behavior must usually be analyzed as a whole (Tomita, Hashimoto, Takahashi, Shimizu, Matsuzaki, Miyoshi, Saito, Tanida, Yugi, Venter, and Hutchison 1999; Voit 2000).

The concentrations of substrates, products, and intermediate substances participating in a network of chemical reactions are modeled by nonlinear continuous-time differential equations, including various first-order rate laws, second-order rate laws, power laws, and the Michaelis-Menten equations (Voit 2000). The concentrations of catalysts (e.g., enzymes) control the rates of many chemical reactions in living things.

The *topology* of a network of chemical reactions comprises

- the total number of reactions in the network,
- the number of substrate(s) consumed by each reaction,
- the number of product(s) produced by each reaction,
- the pathways supplying the substrate(s) (either from external sources or from other reactions in the network) to each reaction,
- the pathways dispersing each reaction's product(s) (either to other reactions or to external outputs), and
- an indication of which enzyme (if any) acts as a catalyst for a particular reaction.

We use the term *sizing* for a network of chemical reactions to encompass all the numerical values associated with the network (e.g., the rates of each reaction).

Biochemists have historically determined the topology and sizing of networks of chemical reactions, such as metabolic pathways, through meticulous study of particular networks of interest (Mendes and Kell 1998; Bower and Bolouri 2000).

Vast amounts of time-domain data are now becoming available concerning the concentration of biologically important chemicals in living organisms (McAdams and Shapiro 1995; Loomis and Sternberg 1995; Arkin, Shen, and Ross 1997; Yuh, Bolouri, and Davidson 1998; Laing, Fuhrman, and Somogyi 1998; Mendes and Kell 1998; D'haeseleer, Wen, Fuhrman, and Somogyi 1999; Kitano 2001; Collado-Vides and Hofestadt 2002). Such data include both gene expression data (obtained, in some cases, from microarrays) and data on the concentration of substances participating in metabolic pathways obtained from laboratory experiments.

The question arises as to whether it is possible to start with observed time-domain concentrations of final product substance(s) and automatically create both the topology and sizing of the network of chemical reactions (Arkin, Shen, and Ross 1997). In other words, is it possible to automate the process of reverse engineering a network of chemical reactions?

Although it may seem difficult or impossible to automatically infer both the topology and numerical parameters for a network of chemical reactions from observed data, the results presented in this chapter answer this question affirmatively.

As shown elsewhere in this book, genetic programming can automatically create complex networks comprising both a graphical structure and numerical values that exhibit prespecified behavior in fields such as analog electrical circuits (chapters 4, 10, 11, and 15), controllers (chapters 3 and 13), genetic networks (chapter 7), and antennas (chapter 6).

This chapter describes how genetic programming automatically created a metabolic pathway involving four chemical reactions that takes in Glycerol and fatty acid as input, uses ATP as a cofactor, and produces diacyl-glycerol as its final product. In addition, this chapter describes how genetic programming similarly created a metabolic pathway involving three chemical reactions for the synthesis and degradation of ketone bodies. Both automatically created metabolic pathways contain at least one instance of three noteworthy topological features, namely an internal feedback loop, a bifurcation point where one substance is distributed to two different reactions, and an accumulation point where one substance is accumulated from two sources.

8.1 Our Approach to the Automatic Synthesis of the Topology and Sizing of Networks of Chemical Reactions

Our approach to the problem of automatically creating both the topology and sizing of a network of chemical reactions involves

(1) establishing a representation for a network of chemical reactions involving LISP symbolic expressions (S-expressions) and program trees that can be progressively bred by means of genetic programming, and
(2) defining a fitness measure that measures how well the behavior and characteristics of a candidate network satisfy the problem's high-level design requirements.

The representation and fitness measure are then used during the run of genetic programming. During the run, the evaluation of the fitness of each individual in the population involves

(1) converting each individual program tree in the population into an analog electrical circuit representing the network of chemical reactions,

(2) converting each analog electrical circuit into a netlist (so that it can be simulated),
(3) obtaining the behavior of the network of chemical reactions by simulating the corresponding electrical circuit, and
(4) using the network's behavior and characteristics to calculate its fitness.

The implementation of our approach entails working with five different representations for a network of chemical reactions (each described in detail later):

- *Reaction Network:* Biochemists often use this representation (shown in figures 8.1 and 8.2 in section 8.2) to represent a network of chemical reactions. In this representation, the blocks represent chemical reactions and the directed lines represent flows of substances between reactions.
- *Program Tree:* A network of chemical reactions can also be represented as a program tree whose internal points (nodes) are functions and external points (leaves) are terminals (section 8.4.1.1). This representation enables genetic programming to breed a population of programs in a search for a network of chemical reactions whose time-domain behavior concerning concentrations of final product substance(s) closely matches the observed data.
- *Symbolic Expression:* A network of chemical reactions can also be represented as a symbolic expression (S-expression) in the style of the LISP programming language (section 8.4.2). This representation is used internally by genetic programming.
- *System of Nonlinear Differential Equations:* A network of chemical reactions can also be represented as a system of nonlinear differential equations (section 8.4.3).
- *Analog Electrical Circuit:* A network of chemical reactions can also be represented as an analog electrical circuit (as shown in figure 8.16 in section 8.4.4). The representation of a network of chemical reactions as a circuit facilitates simulation of the network's time-domain behavior.
- *SPICE Netlist:* The SPICE (Simulation Program with Integrated Circuit Emphasis) simulator is used in this book for simulating networks of chemical reactions, circuits, and controllers. In order to invoke the SPICE simulator, the circuit is represented in the form of a netlist (described earlier in section 3.6.6).

8.2 Statement of Two Illustrative Problems

Our technique for automatically creating (reverse engineering) both the topology and sizing of a network of chemical reactions will be demonstrated in this chapter by means of two problems.

The network for the first illustrative problem (figure 8.1) consists of four reactions that are part of the phospholipid cycle, as presented in the E-CELL cell simulation model (Tomita, Hashimoto, Takahashi, Shimizu, Matsuzaki, Miyoshi, Saito, Tanida, Yugi, Venter, and Hutchison 1999).

The first network's external inputs (shown at the top of the figure) are Glycerol (`C00116`) and fatty acid (`C00162`). The network also uses the cofactor ATP (`C00002`) (shown in the top right part of the figure).

This network's final product is diacyl-glycerol (`C00165`) (shown in the bottom left part of the figure).

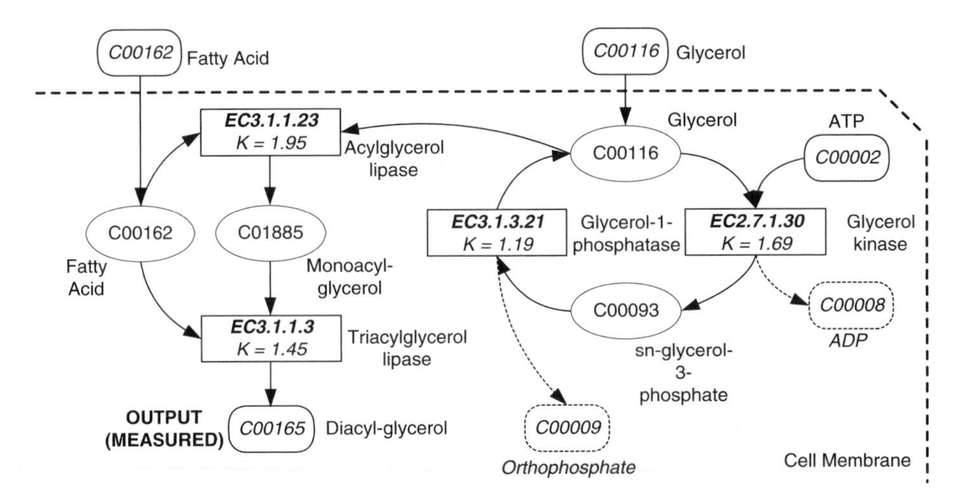

Figure 8.1 Four reactions from the phospholipid cycle.

This network has four reactions (shown as rectangles in the figure). The four reactions are catalyzed by Glycerol kinase (EC2.7.1.30), Glycerol-1-phosphatase (EC3.1.3.21), Acylglycerol lipase (EC3.1.1.23), and Triacylglycerol lipase (EC3.1.1.3). The EC numbers are the codes assigned by the Enzyme Nomenclature Commission (Webb 1992).

The reaction catalyzed by Acylglycerol lipase (EC3.1.1.23) and the reaction catalyzed by Triacylglycerol lipase (EC3.1.1.3) both have two substrates and one product. In figure 8.1, the reaction catalyzed by Glycerol kinase (EC2.7.1.30) has two substrates and two products. The reaction catalyzed by Glycerol-1-phosphatase (EC3.1.3.21) has one substrate and two products.

This network (figure 8.1) has two intermediate substances, namely sn-Glycerol-3-Phosphate (`C00093`) and Monoacyl-glycerol (`C01885`), that are produced and consumed within the network.

In this network (figure 8.1), the external supply of fatty acid (`C00162`) (shown at the top left in the figure) is distributed to two reactions (both on the left side of the figure), namely the reaction catalyzed by Acylglycerol lipase (EC3.1.1.23) and the reaction catalyzed by Triacylglycerol lipase (EC3.1.1.3). The external supply of Glycerol (`C00116`) (shown at the top right in the figure) is distributed to two reactions, namely the reaction catalyzed by Glycerol kinase (EC2.7.1.30) and the reaction catalyzed by Glycerol-1-phosphatase (EC3.1.3.21). That is, both fatty acid (`C00162`) and Glycerol (`C00116`) are consumed by two reactions. This network of chemical reactions has two instances of a bifurcation point.

In addition, the concentration of Glycerol (`C00116`) in this network (figure 8.1) is increased in two ways. First, as previously mentioned, it is externally supplied. Second, it is produced by the reaction catalyzed by Glycerol-1-phosphatase (EC3.1.3.21). That is, this network of chemical reactions has an instance of an accumulation point (where one substance is accumulated from two sources).

Also, this network (figure 8.1) has an internal feedback loop in which a substance is both consumed and produced by the reactions in the loop. Specifically, Glycerol

(C00116) is consumed (in part) by the reaction catalyzed by Glycerol kinase (EC2.7.1.30). This reaction, in turn, produces an intermediate substance, sn-Glycerol-3-Phosphate (C00093). This intermediate substance is, in turn, consumed by the reaction catalyzed by Glycerol-1-phosphatase (EC3.1.3.21). That reaction, in turn, produces Glycerol (C00116).

An oval is used to indicate a point for measuring the concentration of each intermediate substance in this figure. In addition, an oval is also used for each bifurcation point and each accumulation point. When there is both a bifurcation point and an accumulation point for one particular substance (such as Glycerol in this figure), one oval is used.

The rate, K, of each of the four reactions in figure 8.1 is specified by a numerical constant contained in the rectangle representing the reaction. For example, the rate of the reaction catalyzed by Acylglycerol lipase (EC3.1.1.23) is 1.95. The rate of reaction catalyzed by Triacylglycerol lipase (EC3.1.1.3) is 1.45. The rate of the reaction catalyzed by Glycerol kinase (EC2.7.1.30) is 1.69. The rate of the reaction catalyzed by Glycerol-1-phosphatase (EC3.1.3.21) is 1.19.

The second network (figure 8.2) consists of three reactions that are involved in the synthesis and degradation of ketone bodies.

This network's external inputs are Acetoacetyl-CoA and Acetyl-CoA.

This network's final product is Acetoacetate.

The second network (figure 8.2) has three reactions. The three reactions are catalyzed by 3-oxoacid CoA-transferase (EC 2.8.3.5), Hydroxymethylglutaryl-CoA synthase (EC 4.1.3.5), and Hydroxymethylglutaryl-CoA lyase (EC 4.1.3.4). One of this network's reactions (i.e., the reaction catalyzed by 3-oxoacid CoA-transferase (EC 2.8.3.5)) is a one-substrate, one-product reaction. The other two reactions are two-substrate, one-product reactions.

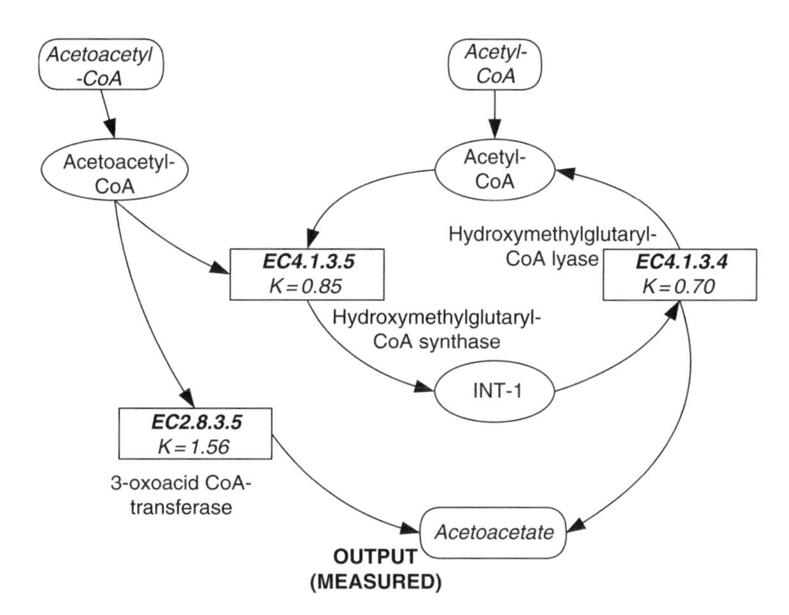

Figure 8.2 Three reactions involved in the synthesis and degradation of ketone bodies.

There is one intermediate substance, `INT-1`, in this network.

Like the first illustrative network of chemical reactions (figure 8.1), this second network (figure 8.2) incorporates three noteworthy topological features.

First, this second network of chemical reactions has one instance of a bifurcation point (where one substance is distributed to two different reactions), namely for the externally supplied substance Acetoacetyl-CoA (in the upper left part of the figure).

Second, this second network has two accumulation points. Acetyl-CoA (at the top of the figure) is an externally supplied substance and it is also produced by the reaction catalyzed by Hydroxymethylglutaryl-CoA lyase (EC 4.1.3.4). Also, the network's final product, Acetoacetate (at the bottom of the figure) is produced by the reaction catalyzed by 3-oxoacid CoA-transferase (EC 2.8.3.5) and by the reaction catalyzed by Hydroxymethylglutaryl-CoA lyase (EC 4.1.3.4).

Third, this second network has an internal feedback loop (in which a substance is both consumed and produced by the reactions in the loop). Specifically, Acetyl-CoA is consumed by the reaction catalyzed by Hydroxymethylglutaryl-CoA synthase (EC 4.1.3.5). This reaction, in turn, produces an intermediate substance (`INT-1`). This intermediate substance is, in turn, consumed by the reaction catalyzed by Hydroxymethylglutaryl-CoA lyase (EC 4.1.3.4). That reaction, in turn, produces Acetyl-CoA.

8.3 Types of Chemical Reactions

A chemical reaction typically causes a change, over time, in the concentration of the various substances participating in the reaction. The substances involved in chemical reactions include reactants (input substances), intermediate products, and products (output substances). A substance often appears as both a reactant and product in a network of reactions.

The rates of many chemical reactions are affected by a catalyst (a substance that accelerates or decelerates the rate of a reaction, but remains unchanged by the reaction). In a living cell, the catalysts are often enzymes. The reactant(s) of a catalyzed reaction are usually called *substrate(s).*

8.3.1 One-Substrate, One-Product Reaction

First consider an illustrative chemical reaction in which one chemical (the *substrate*) is transformed into another chemical (the *product*) under control of a catalyst.

Figure 8.3 shows a one-substrate, one-product chemical reaction in which pyrophosphate (the substrate) is transformed into orthophosphate (the product) under control of the catalyst pyrophosphatase (EC3.6.1.1). Herein `C00013` is an alternative designation for pyrophosphate and `C00009` is an alternative designation for orthophosphate. This reaction (and the other illustrative reactions in this section) is part of the nine reactions of a surmised phospholipid cycle that is presented in the E-CELL cell simulation model (Tomita, Hashimoto, Takahashi, Shimizu, Matsuzaki, Miyoshi, Saito, Tanida, Yugi, Venter, and Hutchison 1999).

Figure 8.4 shows a plot, over 60 seconds of time, of the concentrations of the three substances involved in the one-substrate, one-product enzymatic reaction for producing

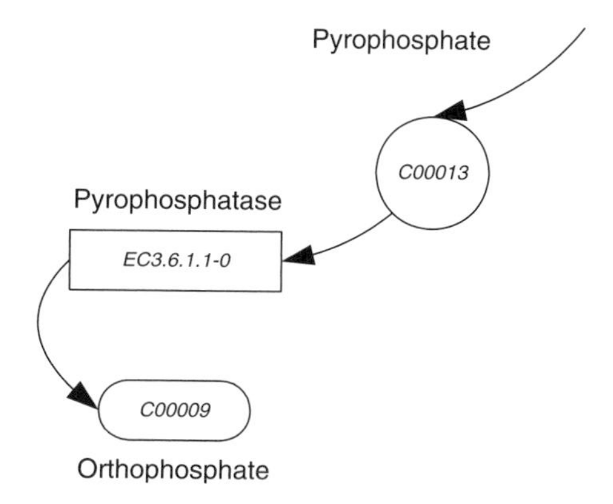

Figure 8.3 Illustrative one-substrate, one-product chemical reaction.

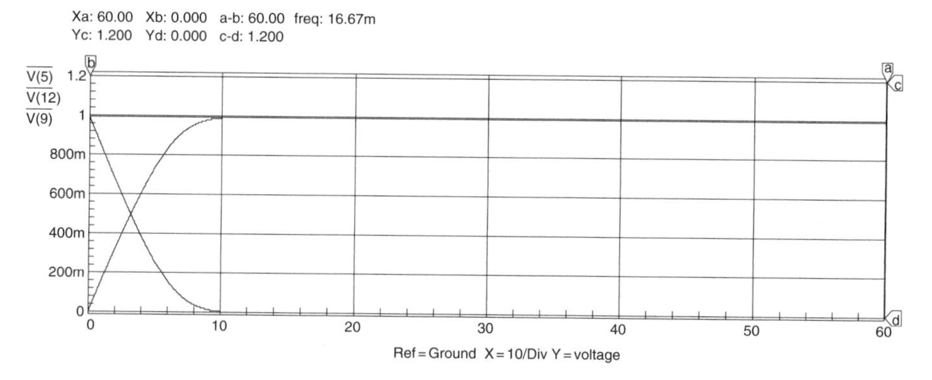

Figure 8.4 Changing concentrations of substances in the illustrative one-substrate, one-product reaction.

orthophosphate. The concentration of pyrophosphatase (the catalyst) is constant at 1.2 over the entire period involved. The concentration of pyrophosphate (the substrate) starts at 1.0 (its concentration at time $t = 0$). As the substrate (pyrophosphate) is consumed by the reaction, its concentration decreases until it reaches a level of 0.0 at about $t = 10$ (and remains at this level thereafter). The concentration of the product (orthophosphate) starts at 0.0 (its concentration at time $t = 0$). As the product (orthophosphate) is produced by the reaction, its concentration increases to a level of 1.0 at about $t = 10$ (and remains at this level thereafter).

The action of an enzyme (catalyst) in a one-substrate chemical reaction can be viewed as a two-step process in which the enzyme E first binds with the substrate S at a rate k_1 to form ES. The formation of the product P from ES then occurs at a rate k_2. The reverse reaction (for the binding of E with S) in which ES dissociates into E and S, occurs at a rate of k_{-1}, and is represented by

$$E + S \underset{k_{-1}}{\overset{k_1}{\rightleftarrows}} ES \overset{k_2}{\rightarrow} P + E.$$

The rates of chemical reactions are modeled by rate laws. The concentrations of substrates, products, intermediate substances, and catalysts participating in reactions are modeled by various rate laws, including first-order rate laws, second-order rate laws, power laws, and the Michaelis-Menten equations (Voit 2000).

For example, the Michaelis-Menten rate law for a one-substrate chemical reaction is

$$\frac{d[\mathrm{P}]}{dt} = \frac{k_2[\mathrm{E}]_0[\mathrm{S}]_t}{[\mathrm{S}]_t + K_m}.$$

In this rate law, [P] is the concentration of the reaction's product P; $[\mathrm{S}]_t$ is the concentration of substrate S at time t; $[\mathrm{E}]_0$ is the concentration of enzyme E at time 0 (and at all other times); and K_m is the Michaelis constant defined as

$$K_m = \frac{k_{-1} + k_2}{k_1}.$$

However, when the constants K_m and k_2 are considerably greater than the concentrations of substances, it is often satisfactory to use a pseudo-first-order rate law such as

$$\frac{d[\mathrm{P}]}{dt} = \frac{k_2[\mathrm{E}]_0[\mathrm{S}]_t}{K_m} = k_{new}[\mathrm{E}]_0[\mathrm{S}]_t,$$

where K_{new} is a constant defined as

$$k_{new} = \frac{k_2}{K_m}.$$

The pseudo-first-order rate law involves only multiplication.

It is both convenient and useful to represent the behavior, over time, of a chemical reaction as an analog electrical circuit. In this representation, the concentration of a substance is represented as a voltage in the time domain.

Figure 8.5 shows an electrical circuit representing the illustrative one-substrate, one-product enzymatic reaction.

This circuit is composed of the following components:

- voltage sources (represented by small circles such as V18 at the top left of the figure as well as voltage sources V20, V8, V1, and V2 elsewhere in the figure),
- voltage adders with three ports (represented by rectangles and labeled ADDV, such as M21 at the top left of the figure as well as voltage adders M23 and M8 elsewhere in the figure),
- sum-integrators (represented by a triangle-rectangle combination with three ports, such as U23 at the top left of the figure as well as sum-integrators U26 and U7 elsewhere in the figure),
- a one-substrate chemical reaction function MICH_1 (represented by the rectangle U25 with five ports at the top right of the figure),
- a ground point (at the bottom of the figure), and
- an output point (represented by the probe wire emanating from voltage adder M8 at the bottom middle of the figure and labeled as probe point 5).

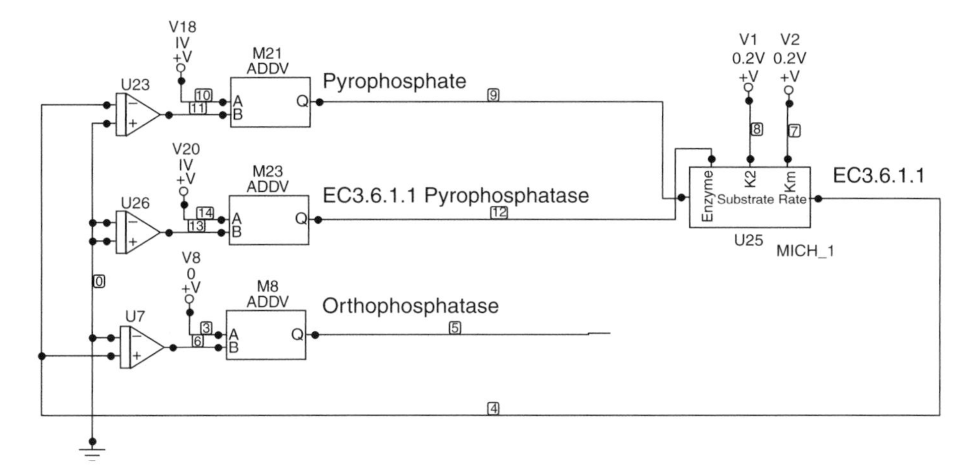

Figure 8.5 Circuit for the illustrative one-substrate, one-product chemical reaction.

The initial concentration of pyrophosphate (the substrate) is 1.0, as established by the 1-Volt voltage source V18 (in the top left part of figure 8.5).

Voltage adder M21 (in the top left part of figure 8.5) is an electrical component used for adding two voltages. V18 is one of its two inputs.

The output of voltage adder M21 represents the concentration of pyrophosphate (the substrate in this illustrative one-substrate, one-product reaction). Ignoring, for the moment, the second input port to voltage adder M21, the output of M21 goes into the port labeled "substrate" of the one-substrate reaction function MICH_1 labeled U25 (at the top right of the figure). If desired, this voltage representing the concentration of the substrate may be measured at probe point 9 (along the line at the top of the figure).

The rate law may be the Michaelis-Menten law. Alternatively, the rate law may be a first-order rate law. Then again, the rate law may be some other rate law (Voit 2000). The function MICH_1 in figure 8.5 is an analog computational subcircuit that models the Michaelis-Menten rate law.

The constant, K_m, goes into the port labeled K_m of the reaction function MICH_1 of figure 8.5. Similarly, the constant k_2 goes into the port labeled k_2 of the reaction function MICH_1.

The initial concentration of catalyst (pyrophosphatase) is 1.0 (established by the 1-Volt voltage source V20 connected to voltage adder M23). The output of voltage adder M23 goes into the port labeled "enzyme" of the one-substrate, one-product reaction function MICH_1 labeled U25. If desired, this voltage may be measured at probe point 12.

The initial concentration of product (orthophosphate) is 0.0, as established by the 0-Volt voltage source V8 connected to voltage adder M8. If desired, the output of voltage adder M8 may be measured at probe point 5.

U23, U26, and U7 are sum-integrators that integrate, over time, the sum of their respective two inputs. In each case in figure 8.5, one of the two incoming ports is negated. In other words, U23, U26, and U7 serve as differential-integrators (i.e., they sum, over time, the difference between their two inputs).

The output of the reaction function MICH_1 labeled U25 is the reaction's rate. If desired, this reaction rate may be measured at probe point 4 (along the bottom of the figure). This reaction rate is fed to

- a subtractive input to sum-integrator U23, and
- an additive input to sum-integrator U7.

These two connections are of central importance in the figure. They reflect the conservation principle inherent in a chemical reaction. Specifically, the subtractive input to sum-integrator U23 reflects the fact that the substrate substance is *consumed* by the reaction at the specified rate. The additive input to sum-integrator U7 reflects the fact that the product substance is *produced* by the reaction at the same specified rate.

Because there is nothing in the figure that consumes any of the reaction's product, the subtractive input to sum-integrator U7 is zero (i.e., grounded). Two-input voltage adder M8 adds the product's initial concentration (0.0) (at probe 3) to the concentration (at probe 6) of the product that is being produced by the reaction.

Because there is nothing in the figure that creates an additional substrate, the additive input to sum-integrator U23 is zero (i.e., grounded). Thus, two-input voltage adder M21 adds the substrate's initial concentration (1.0) (at probe 10) to the (negative) concentration (at probe 11) of the substrate that is being consumed by the reaction.

There is one voltage source, one sum-integrator, and one voltage adder associated with the enzyme that catalyzes this reaction. Because the concentration of a catalyst is unchanged by a reaction, the additive and subtractive inputs to sum-integrator U26 are both zero (i.e., grounded). This sum-integrator could, of course, be deleted. However, we employ this consistent arrangement of three components (i.e., voltage source, sum-integrator, and voltage adder) in all the figures herein to emphasize the formal procedure used to translate a chemical reaction diagram into an electrical circuit. In practice, when both inputs of a sum-integrator are connected to ground, the sum-integrator is simply deleted from the circuit that is eventually simulated.

The circuit is simulated using the SPICE simulator (Quarles, Newton, Pederson, and Sangiovanni-Vincentelli 1994) in the same manner as in the chapter on controllers (chapter 3) and the chapters on electrical circuits (chapters 4 and 5). See section 3.6.6 for additional discussion of SPICE and netlists.

A diagram representing an electrical circuit differs from a diagram representing a network of chemical reactions in several important ways. In particular, there is no directionality of wires in circuits (as there is in the block diagram representing a network of chemical reactions and in the block diagram representing a controller). In addition, because SPICE was written for the purpose of simulating the behavior of electrical circuits, it does not have built-in components that correspond to chemical rate laws (such as a first-order rate law, a second-order rate law, a power law, or the Michaelis-Menten equations). However, these rate laws can be realized by using the facility of SPICE to create subcircuit definitions (macros). Mathematical functions such as integration and differentiation can be realized by using electrical components such as capacitors and inductors. However, SPICE is not limited to electrical components that actually exist in the real world. SPICE has a facility for representing arbitrary continuous-time mathematical functions. Once a subcircuit is defined in SPICE, it operates as if it were an electrical component. Thus, it may be inserted into an electrical circuit in addition to other electrical components or other subcircuits.

For example, voltage multiplication (`XMULTV`) can be realized by a subcircuit definition (macro). The subcircuit definition entails multiplying two voltages, `V(1)` and `V(2)`, in order to produce an output voltage at node `V(3)`.

```
*NETLIST FOR SUBCIRCUIT DEFINITION OF VOLTAGE MULTIPLICATION (XMULTV)
.SUBCKT XMULTV 1 2 3
BX 3 0 V=V(1)*V(2)
.ENDS XMULTV
```

Voltage addition (`XADDV`) can be realized in a similar manner by means of the following subcircuit definition:

```
*NETLIST FOR SUBCIRCUIT DEFINITION OF VOLTAGE ADDITION (ADDV)
.SUBCKT XADDV 1 2 3
BX 3 0 V=V(1)+V(2)
.ENDS XADDV
```

Similarly, voltage division (`XDIVV`) can be realized by the following subcircuit definition:

```
*NETLIST FOR SUBCIRCUIT DEFINITION OF VOLTAGE DIVISION (DIVV)
.SUBCKT XDIVV 1 2 3
BX 3 0 V=V(1)/V(2)
.ENDS XDIVV
```

Similarly, the pseudo-first-order rate law can be realized in SPICE by a subcircuit definition (macro). The pseudo-first-order rate law `XFORL` ("First Order Rate Law") entails multiplying three voltages, `V(1)`, `V(2)`, and `V(4)` to produce an output voltage at node `V(3)`.

```
*NETLIST FOR SUBCIRCUIT DEFINITION OF PSEUDO-FIRST-ORDER RATE LAW (XFORL)
.SUBCKT XFORL 1 2 3 4
BX 3 0 V=V(1)*V(2)*V(4)
.ENDS
```

The pseudo-second-order rate law can be realized in SPICE by a subcircuit definition. The pseudo-second-order rate law `XSORL` ("Second Order Rate Law") entails multiplying four voltages, `V(1)`, `V(2)`, `V(3)`, and `V(5)` to produce an output voltage at node `V(4)`.

```
*Netlist for SUBCIRCUIT DEFINITION OF PSEUDO-SECOND-ORDER RATE LAW (XSORL)
.SUBCKT XSORL 1 2 3 4 5
BX 4 0 V=V(1)*V(2)*V(3)*V(5)
.ENDS
```

As another example, a sum-integrator (SUMINT) can be realized in SPICE by the following subcircuit definition (employing a 1-giga-Ohm resistor R2, a voltage-controlled current source G1, and a capacitor C1 with an initial condition (IC) of 0 Volts):

```
*NETLIST FOR SUMINTEGRATOR (SUMINT)
.SUBCKT SUMINT 1 2 3 4
G1 4 0 1 2 1.0
C1 4 0 1.0 IC=0.0
R2 4 0 1000.0MEG
UNARYV 3 0 V=-V(4)
.ENDS
```

Figure 8.6 shows a subcircuit for a sum-integrator SUMINT. The symbol for the sum-integrator is shown at the top of the figure in the form of a triangle-rectangle combination. The sum-integrator has two input ports (one with a positive sign and one with a negative sign). Its one output (node 3) is the integral of the difference of its two inputs. The circuit required to implement a sum-integrator consists of a voltage-controlled current source (in the lower left corner of the figure), a capacitor C1, a resistor R2, and an inverter UNARYV (the rectangle in the middle of the right side of the figure). The rounded box on the right side of the figure represents the initial condition (i.e., initial charge) on the capacitor C1. The voltage-controlled current source converts the difference in the two voltages to a current. The capacitor C1 (starting with its initial charge) integrates the current. The capacitor voltage (node 4) is equal to the total charge (i.e., the integral of the current) divided by its capacitance. Note that in this figure the input and output ports of the subcircuit are connected with wires to the

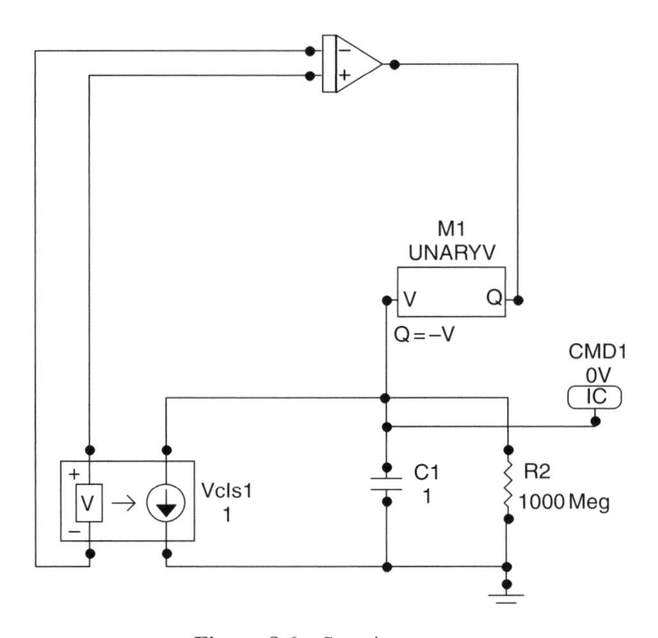

Figure 8.6 Sum-integrator.

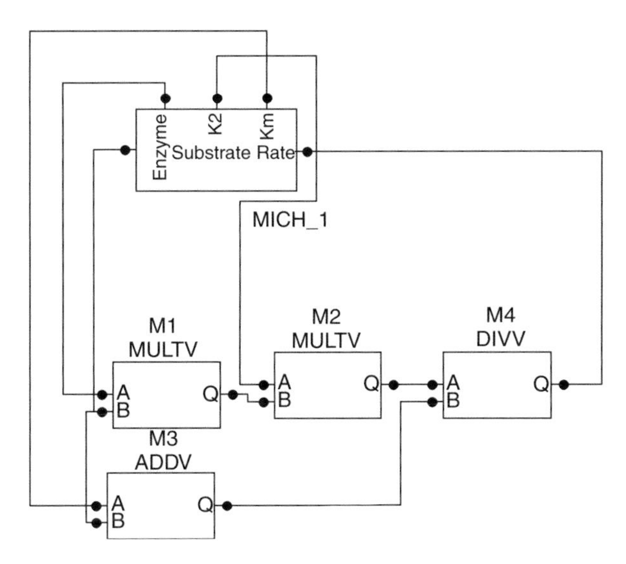

Figure 8.7 Subcircuit for the one-substrate Michaelis-Menten equation MICH_1.

corresponding ports on the schematic symbol for the sum-integrator (the triangle-rectangle combination in the figure).

The Michaelis-Menten equations can also be realized by an electrical circuit. Figure 8.7 shows a circuit for the one-substrate Michaelis-Menten equation MICH_1

$$\frac{d[\mathrm{P}]}{dt} = \frac{k_2[E]_0[S]_t}{[S]_t + K_m},$$

using two voltage multiplications (M1 and M2), a voltage addition M3, and a voltage division M4.

The subcircuit definition in SPICE for the one-substrate Michaelis-Menten equation MICH_1 is as follows:

```
*NETLIST FOR MICHAELIS-MENTEN MICH_1
.SUBCKT XMICH1 1 2 3 4 5
* 1 IS SUBSTRATE
* 2 IS ENZYME
* 3 IS K2
* 4 IS KM
* 5 IS RATE
XXM1 7 6 5 XDIVV
XXM2 4 1 6 XADDV
XXM3 3 8 7 XMULTV
XXM4 2 1 8 XMULTV
.END
```

When the user-defined macro for the differential-integrator and the user-defined macro for the one-input reaction function are combined, the resulting netlist is as follows:

```
*NETLIST FOR 1-1-REACTION IN PHOSPHOLIPID-CIRCUIT
XXM1 15 6 XUNARYV
GVCIS1 15 0 4 0 1
C1 0 15 1
XXM2 3 6 5 XADDV
V8 3 0 DC 0
V2 7 0 DC 0.2V
V1 8 0 DC 0.2V
XXM3 16 11 XUNARYV
GVCIS2 16 0 0 4 1
C2 0 16 1
XXM4 10 11 9 XADDV
V18 10 0 DC 1V
XXM5 17 13 XUNARYV
GVCIS3 17 0 0 0 1
C3 0 17 1
XXM6 19 18 4 XDIVV
XXM7 7 9 18 XADDV
XXM8 8 20 19 XMULTV
XXM9 12 9 20 XMULTV
XXM10 14 13 12 XADDV
V20 14 0 DC 1V
R1 0 15 1000MEG
R2 0 16 1000MEG
R3 0 17 1000MEG
.IC V(15)=0V V(16)=0V V(17)=0V
*
* SELECTED CIRCUIT ANALYSES :
.TRAN 10M 60 0 10M
*
* MODELS/SUBCIRCUITS USED:
*
*UNARYV UNARY MINUS OF VOLTAGE
.SUBCKT XUNARYV 1 2
BX 2 0 V=-(V(1))
.ENDS XUNARYV
*
*ADDV ADD VOLTAGES
.SUBCKT XADDV 1 2 3
BX 3 0 V=V(1)+V(2)
.ENDS XADDV
*
```

```
*DIVV DIVIDE VOLTAGES
.SUBCKT XDIVV 1 2 3
BX 3 0 V=V(1)/V(2)
.ENDS XDIVV
*
*MULTV MULTIPLY VOLTAGES
.SUBCKT XMULTV 1 2 3
BX 3 0 V=V(1)*V(2)
.ENDS XMULTV
*
.END
```

8.3.2 One-Substrate, Two-Product Reaction

Figure 8.8 shows an illustrative one-substrate, two-product chemical reaction in which Phosphatidylglycerophosphate (the substrate) is transformed into orthophosphate and Phosphotidylglycerol (the two products) under control of phosphatidylglycerophosphatase (the catalyst). The catalyst in this reaction is the enzyme pyrophosphatase (EC3.1.3.27). Note that C03892 is an alternative designation herein for Phosphatidylglycerophosphate; that C00009 is an alternative designation for orthophosphate; and that C00344 is an alternative designation for Phosphotidylglycerol.

This illustrative one-substrate, two-product chemical reaction can be represented as a circuit in a manner similar to that just described in the previous section for a one-substrate, one-product reaction.

Figure 8.9 shows an electrical circuit representing a one-substrate, two-product enzymatic reaction.

This circuit is composed of

- voltage sources V23, V24, V25, V8, V2, and V1,
- voltage adders M27, M29, M30, and M8,
- sum-integrators U27, U32, U33, and U7,

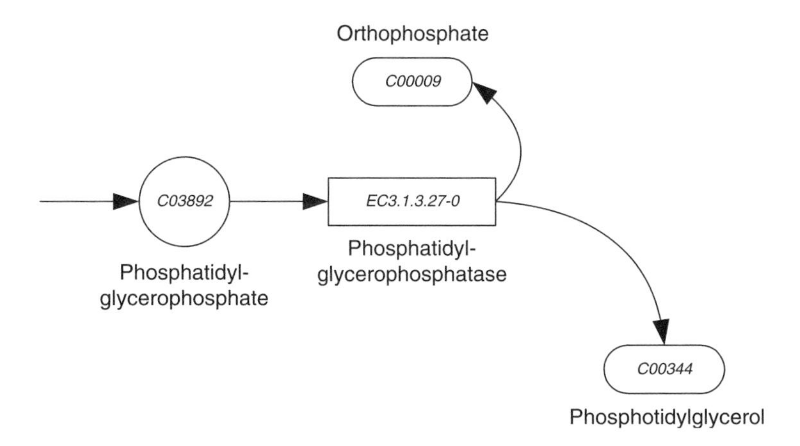

Figure 8.8 Illustrative one-substrate, two-product reaction.

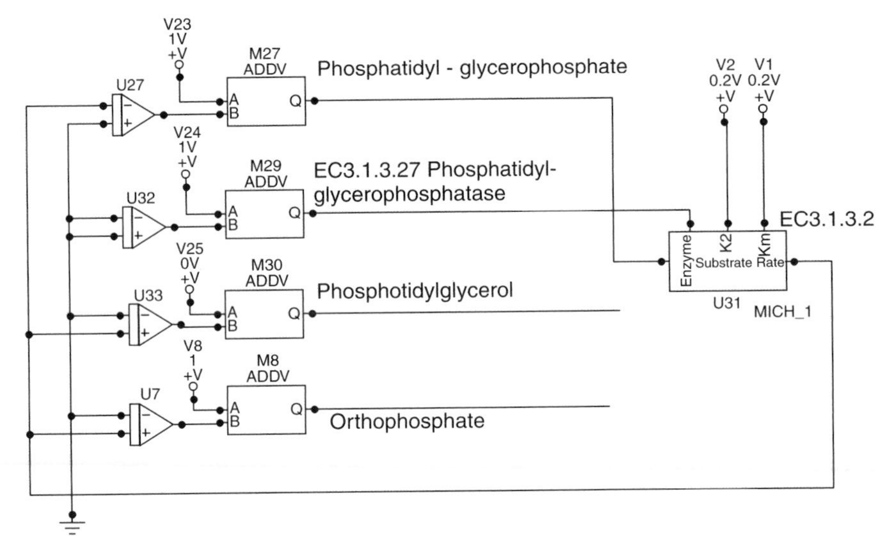

Figure 8.9 Circuit for illustrative one-substrate, two-product chemical reaction.

- a one-substrate chemical reaction function MICH_1,
- a ground point (at the bottom of the figure), and
- two output points (represented by the probe wires emanating from voltage adders M30 and M8 at the bottom middle of the figure).

Note that in this circuit

- there is one voltage adder M27 (representing the one substrate of this reaction),
- there are two voltage adders M30 and M8 (representing the two products of this reaction) and a probe wire for each of the two products of the reaction,
- the output of the reaction function MICH_1 labeled "rate" is connected to the negative lead of sum-integrator U27 (representing the one substrate) to establish that one substrate is consumed by the reaction, and
- the output of the reaction function MICH_1 labeled "rate" is connected to the positive lead of sum-integrators U33 and U7 (representing the two products) to establish that two products are produced by the reaction.

As before, there is one voltage source, sum-integrator, and voltage adder associated with the enzyme that catalyzes this reaction. Both the positive and negative leads of this sum-integrator are grounded to reflect the fact that the concentration of the enzyme remains unchanged during the reaction.

8.3.3 Two-Substrate, One-Product Reaction

Figure 8.10 shows an illustrative two-substrate, one-product chemical reaction in which fatty acid (`C00162`) and Glycerol (`C00116`) (the two substrates) are transformed into

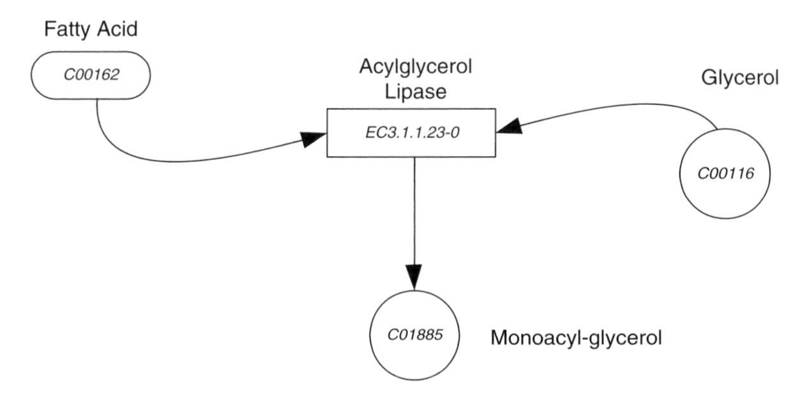

Figure 8.10 Illustrative two-substrate, one-product reaction.

Monoacyl-glycerol (C01885) (the product) under control of Acylglycerol_lipase (EC3.1.1.23) (the catalyst).

The action of an enzyme (catalyst) in a two-substrate chemical reaction can be viewed as a two-step process in which the enzyme E first binds with the substrates A and B at a rate k_1 to form ABE. The reverse reaction, in which ABE dissociates into E, A, and B occurs at a rate of k_{-1}. The formation of the product P from EAB then occurs at a rate k_2:

$$E + A + B = \underset{k_{-1}}{\overset{k_1}{\rightleftarrows}} ABE \overset{k_2}{\rightarrow} P + E.$$

The Michaelis-Menten rate law for a two-substrate chemical reaction is

$$Rate_t = \frac{[E]_0}{\frac{1}{K_0} + \frac{1}{K_A[A]_t} + \frac{1}{K_B[B]_t} + \frac{1}{K_{AB}[A]_t[B]_t}}.$$

However, when $k_{-1} \sim 0$ and $k_{-1} << k_1 << k_2$, it is often satisfactory to use a pseudo-second-order rate law such as

$$Rate_t = k_1[A][B][E].$$

This illustrative two-substrate, one-product chemical reaction can be represented as a circuit in a manner similar to that just described in the previous section.

Figure 8.11 shows an electrical circuit representing a two-substrate, one-product enzymatic reaction.

This circuit is composed of

- voltage sources V1, V6, V12, V122, V2, V3, V4, and V5,
- voltage adders M1, M6, M12, and M11,
- sum-integrators U14, U9, U2, and U4,
- a one-substrate chemical reaction function MICH_2,

- a ground point (at the bottom of the figure), and
- one output point (represented by the probe wire emanating from voltage adder M11 at the bottom middle of the figure).

This circuit makes use of

- two voltage adders M1 and M6 (representing the two substrates of this reaction),
- one voltage adder M11 (representing the one product of this reaction), and a probe wire for the product of the reaction.

The output of the reaction function MICH_1 labeled "rate" is connected to the negative lead of sum-integrators U14 and U9 (representing the two substrates) to establish that two substrates are consumed by the reaction.

The output of the reaction function MICH_1 labeled "rate" is connected to the positive lead of sum-integrator U14 (representing the one product) to establish that one product is produced by the reaction.

As before, there is one voltage source, sum-integrator, and voltage adder associated with the enzyme that catalyzes this reaction. Both the positive and negative leads of this sum-integrator (U32) are grounded to reflect the fact that the concentration of the enzyme remains unchanged during the reaction.

In practice herein, we use the pseudo-second-order rate law XSORL (defined in an earlier section) to represent the behavior of two-substrate chemical reactions.

The two-substrate Michaelis-Menten equations can be realized as an electrical circuit. Figure 8.12 shows a circuit for a two-substrate Michaelis-Menten equation MICH_1

$$Rate_t = \frac{[E]_0}{\frac{1}{K_0} + \frac{1}{K_A[A]_t} + \frac{1}{K_B[B]_t} + \frac{1}{K_{AB}[A]_t[B]_t}}.$$

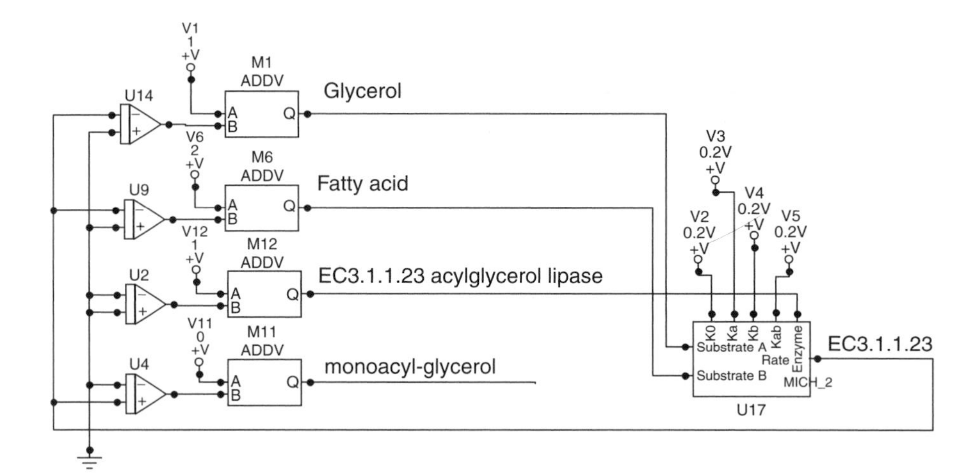

Figure 8.11 Circuit for illustrative two-substrate, one-product chemical reaction.

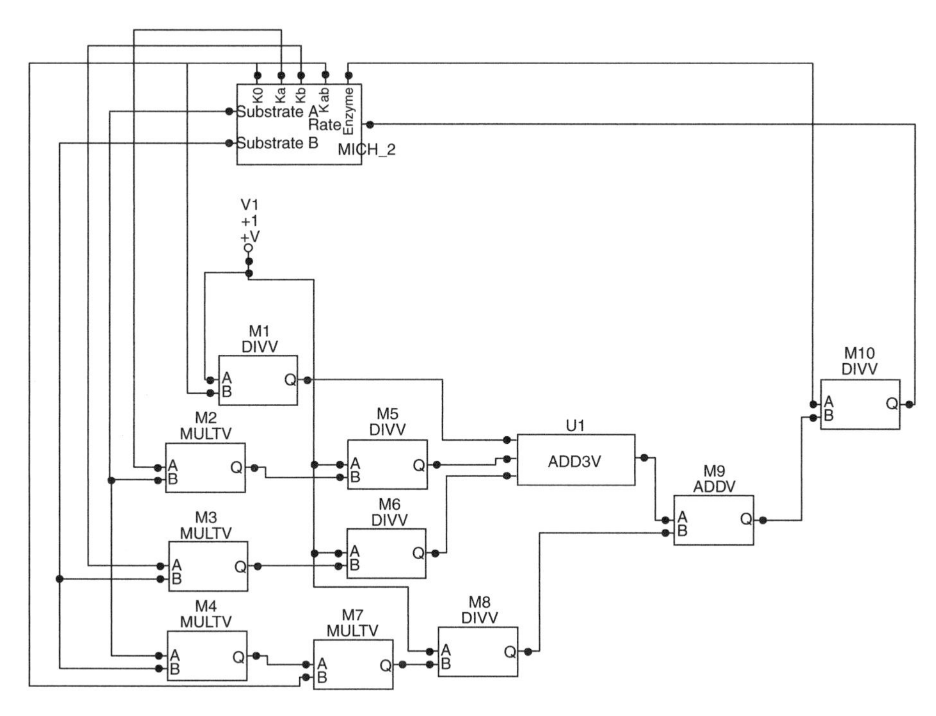

Figure 8.12 Subcircuit for the two-substrate Michaelis–Menten equation MICH_1.

This circuit includes four voltage multiplications (labeled MULTV), four voltage divisions (labeled DIVV), and one voltage addition (labeled ADDV).

The netlist in SPICE for the two-substrate Michaelis-Menten equation MICH_1 is as follows. It invokes definitions for voltage division XDIVV, voltage addition ADDV, and voltage multiplication XMULTV.

```
*NETLIST FOR MICHAELIS-MENTEN MICH_2
XXM1 4 3 2 XDIVV
XXM2 5 6 3 XADDV
XXM3 11 10 6 XDIVV
XXM4 13 12 10 XMULTV
XXM5 11 14 7 XDIVV
XXM6 11 15 8 XDIVV
XXM7 17 16 13 XMULTV
XXM8 18 16 14 XMULTV
XXM9 19 17 15 XMULTV
V1 11 0 DC +1
XXM10 11 20 9 XDIVV
XXM11 21 9 5 XADDV
XXM12 7 8 21 XADDV
* SELECTED CIRCUIT ANALYSES :
.TRAN 10M 60 0 10M UIC
```

```
* MODELS/SUBCIRCUITS USED:
*DIVV DIVIDE VOLTAGES
.SUBCKT XDIVV 1 2 3
BX 3 0 V=V(1)/V(2)
.ENDS XDIVV
*ADDV ADD VOLTAGES
.SUBCKT XADDV 1 2 3
BX 3 0 V=V(1)+V(2)
.ENDS XADDV
*MULTV MULTIPLY VOLTAGES
.SUBCKT XMULTV 1 2 3
BX 3 0 V=V(1)*V(2)
.ENDS XMULTV
.END
```

8.3.4 Two-Substrate, Two-Product Reaction

Figure 8.13 shows an illustrative two-substrate, two-product chemical reaction in which ATP (`C00002`) and Glycerol (`C00116`) (the two substrates) are transformed into ADP (`C00008`) and sn-Glycerol-3-Phosphate (`C00093`) (the two products) under control of Glycerol kinase (`EC2.7.1.30`) (the catalyst).

This illustrative two-substrate, two-product chemical reaction can be represented as a circuit in a manner similar to that just described in the previous sections.

Figure 8.14 shows the electrical circuit representing this two-substrate, two-product enzymatic reaction.

This circuit for the illustrative two-substrate, two-product chemical reaction is composed of

- voltage sources V1, V2, V7, V3, V4, V5, V8, V9, and V6,
- voltage adders M1, M2, M7, M3, and M4,

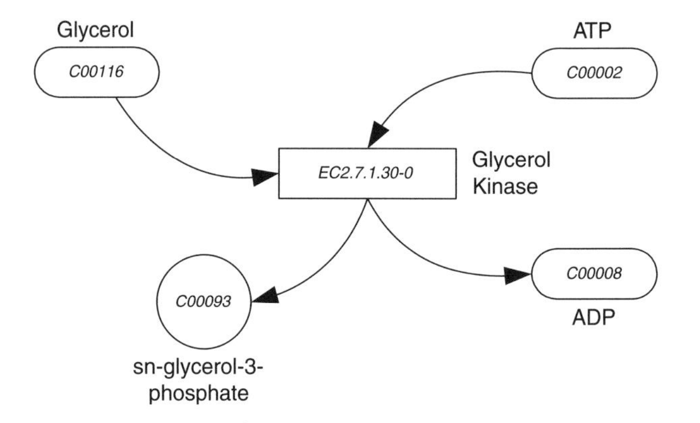

Figure 8.13 Illustrative two-substrate, two-product reaction.

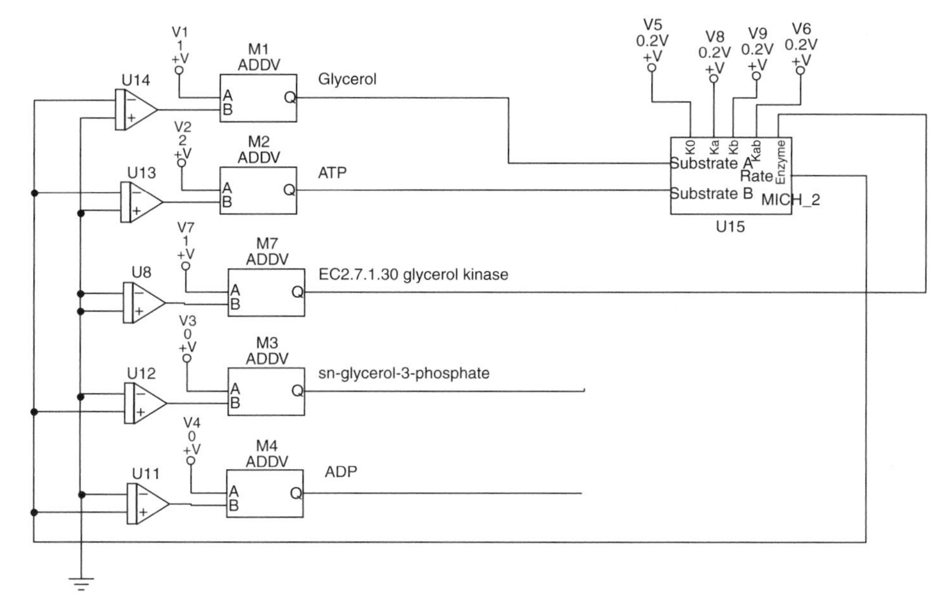

Figure 8.14 Circuit for illustrative two-substrate, two-product chemical reaction.

- sum-integrators U14, U13, U8, U12, and U21,
- a two-substrate chemical reaction function MICH_2,
- a ground point (at the bottom of the figure), and
- two output points (represented by the probe wires emanating from voltage adders M3 and M4 at the bottom middle of the figure).

This circuit uses

- two voltage adders M1 and M2 (representing the two substrates of this reaction),
- two voltage adders M3 and M4 (representing the two products of this reaction), and
- two probe wires for each of the two products of the reaction.

The output of the reaction function MICH_1 labeled "rate" is connected to the negative lead of sum-integrators U14 and U13 (representing the two substrates) to establish that two substrates are consumed by the reaction.

The output of the reaction function MICH_1 labeled "rate" is connected to the positive lead of sum-integrators U12 and U21 (representing the two products) to establish that two products are produced by the reaction.

As before, there is one voltage source, sum-integrator, and voltage adder associated with the enzyme that catalyzes this reaction. Both the positive and negative leads of this sum-integrator (U8) are grounded to reflect the fact that the concentration of the enzyme remains unchanged during the reaction.

8.4 Representation of Networks of Chemical Reactions by Computer Programs

We work with five different representations for networks of chemical reactions in this section. The first representation was presented in figures 8.1 and 8.2 in section 8.2. The remaining four ways are presented in what follows.

8.4.1 Representation as a Program Tree

This section describes one particular method for representing a network of chemical reactions as a program tree suitable for use in a run of genetic programming. Each program tree represents an interconnected network of chemical reactions involving various substances. A chemical reaction herein may consume one or two (but not more) substances and produce one or two (but not more) substances. The consumed substances may be externally supplied input substances or intermediate substances produced within the network. The chemical reactions, enzymes, and substances of a network may be completely represented by a program tree that contains

- internal nodes representing chemical reaction functions,
- internal nodes representing selector functions that select the reaction's first versus the reaction's second product (if any),
- external points (leaves) representing substances that are consumed by a reaction,
- external points (leaves) representing substances that are produced by a reaction,
- external points representing enzymes that catalyze a reaction, and
- external points representing numerical constants (reaction rates).

8.4.1.1 Repertoire of Functions in the Program Tree There are four chemical reaction functions and two selector functions.

The first argument of each chemical reaction function identifies the enzyme that catalyzes the reaction. The second argument is a numerical value that specifies the reaction's rate. There are two, three, or four additional arguments specifying the reaction's one or two substrates and the reaction's one or two products. Table 8.1 shows the number of substrate(s) and product(s) and overall arity for each of the four chemical reaction functions.

Each chemical reaction function returns a list (of length 1 or 2) composed of the reaction's one or two products. The one-argument `FIRST-PRODUCT` function returns the first of the one or two products produced by the chemical reaction function designated by its argument. The one-argument `SECOND-PRODUCT` function returns the second of the two products (or the first product, if the reaction produces only one product).

Table 8.1 Four chemical reaction functions

Function	Substrates	Products	Arity
`CR_1_1`	1	1	4
`CR_1_2`	1	2	5
`CR_2_1`	2	1	5
`CR_2_2`	2	2	6

8.4.1.2 Repertoire of Terminals Some terminals represent substances (externally supplied input substances, intermediate substances created by reactions, or output substances). Other terminals represent the enzymes that catalyze reactions. Still other terminals represent perturbable numerical values that specify the rate of the reactions.

8.4.1.3 Constrained Syntactic Structure Each program tree in the population is a composition of functions from the specified function set and terminal set. The trees are constructed in accordance with a constrained syntactic structure (section 2.3.1). The trees have one or more result-producing branches. The root of each result-producing branch must be a chemical reaction function. The enzyme that catalyzes a reaction always appears as the first argument of its chemical reaction function. A numerical value representing a reaction's rate always appears as the second argument of its chemical reaction function. The one or two substrate arguments to a chemical reaction function can be either a substance terminal or a selector function (`FIRST-PRODUCT` or `SECOND-PRODUCT`). Whenever a selector function appears as an input argument to a chemical reaction function, the effect is to create a cascade of reactions. The one or two product arguments to a chemical reaction function must be substance terminals. The argument to a selector function (`FIRST-PRODUCT` or `SECOND-PRODUCT`) is always a chemical reaction function.

8.4.1.4 Example of a Program Tree Figure 8.15 shows a program tree that corresponds to the metabolic pathway of figure 8.1. Thus, this program tree is a succinct solution to the problem of reverse engineering this particular metabolic pathway. The program tree is presented in the style of the LISP programming language (except that informative arrows have been added in certain places to aid the reader in tracing the flow of substances).

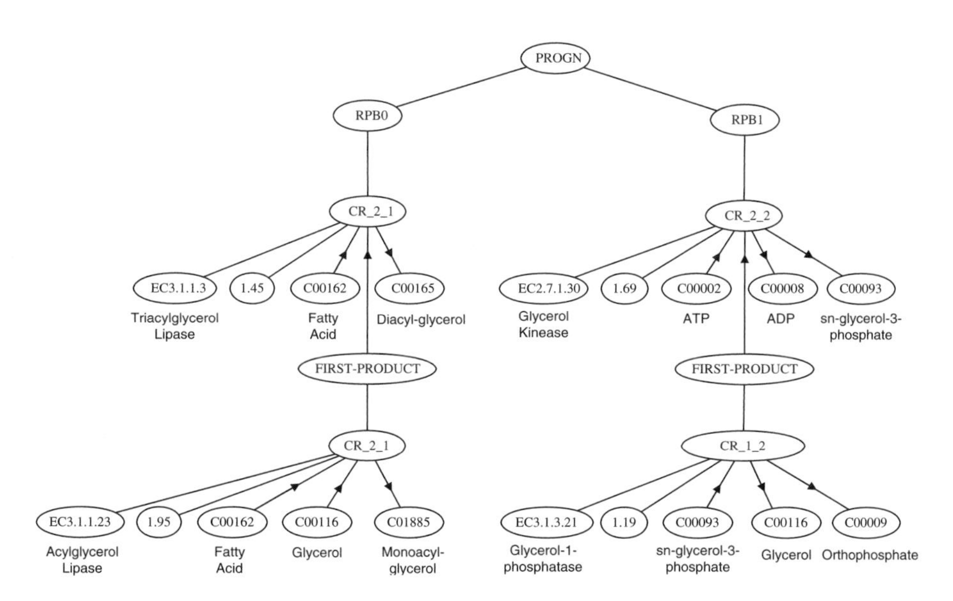

Figure 8.15 Program tree corresponding to the metabolic pathway of figure 8.1.

The program tree (figure 8.15) has two result-producing branches, RPB0 and RPB1. These two branches are connected by means of a connective PROGN function.

There is no return value for either result-producing branch or for the program tree as a whole.

As can be seen, there are four chemical reaction functions in figure 8.15. The first argument of each chemical reaction function is constrained to be an enzyme and the second argument is constrained to be a numerical rate. The remaining arguments are substances, such as externally supplied input substances, intermediate substances produced by reactions within the network, and the final output substance produced by the network. The remaining arguments are marked (as a visual aid to the reader) by an arrow. An upward arrow indicates that the substance at the tail of the arrow points to a substrate of the reaction. A downward arrow indicates that the head of the arrow points to a product of the reaction.

There is a two-substrate, one-product chemical reaction function CR_2_1 in the lower left part of figure 8.15. For this reaction, the enzyme is Acylglycerol lipase (EC3.1.1.23) (the first argument of this chemical reaction function); its rate is 1.95 (the second argument); its two substrates are fatty acid (C00162) (the third argument) and Glycerol (C00116) (the fourth argument); and its product is Monoacyl-glycerol (C01885) (the fifth argument). The fact that the third and fourth arguments of this CR_2_1 function denote substrates whereas the fifth argument denotes a product is inherent in the definition of a two-substrate, one-product chemical reaction function.

There is another two-substrate, one-product chemical reaction function CR_2_1 in the upper left part of figure 8.15 (immediately above the first chemical reaction function CR_2_1). For the higher (in the figure) reaction, the enzyme is Triacylglycerol lipase (EC3.1.1.3) (the first argument); its rate is 1.45 (the second argument); the first of its two substrates is fatty acid (C00162) (the third argument); and its product is diacyl-glycerol (C00165) (the fifth argument).

There is a FIRST-PRODUCT function between the two chemical reaction functions in the left half of figure 8.15. The FIRST-PRODUCT function selects the first of the two products of the lower CR_2_1 function. The line in the program tree from the lower chemical reaction function to the FIRST-PRODUCT function and the line between the FIRST-PRODUCT function and the higher CR_2_1 reaction means that when this tree is converted into a network of chemical reactions, the first (and, in this case, only) substance produced by the lower CR_2_1 reaction is a substrate to the higher reaction. In particular, the product of the lower reaction function (the intermediate substance Monoacyl-glycerol) is the second of the two substrates to the higher chemical reaction function (i.e., the fourth argument of the higher function). Thus, although there is no return value for any branch or for the program tree as a whole, the return value(s) of all but the top chemical reaction function of a particular branch (as well the return values of a FIRST-PRODUCT function and a SECOND-PRODUCT function) define the flow of substances in the network of chemical reactions represented by the program tree.

Notice that the fatty acid (C00162) substance terminal appears as a substrate argument to both of these chemical reaction functions (in the left half of figure 8.15 and also in the left half of figure 8.1). The repetition of a substance terminal as a substrate argument in a program tree means that when the tree is converted into a network of chemical reactions, the available concentration of this particular substrate is distributed to two reactions in the network. That is, the repetition of a substance terminal as a substrate argument in a program tree corresponds to a bifurcation point where one substance is distributed to two different reactions in the network of chemical reactions represented by the program tree.

For the one-substrate, two-product chemical reaction function CR_1_2 in the lower right part of figure 8.15, the enzyme is Glycerol-1-phosphatase (EC3.1.3.21) (the first argument); its rate is 1.19 (the second argument); its single substrate is sn-Glycerol-3-Phosphate (C00093) (the third argument); and its two products are Glycerol (C00116) (the fourth argument) and orthophosphate (C00009) (the fifth argument). The fact that the third argument of this CR_1_2 chemical reaction function denotes a substrate whereas the fourth and fifth arguments denote products is inherent in the definition of the one-substrate, two-product chemical reaction function CR_1_2.

A two-substrate, two-product chemical reaction function CR_2_2 appears above the one-substrate, two-product chemical reaction function CR_1_2 in figure 8.15. There is an intervening FIRST-PRODUCT function between CR_1_2 and CR_2_2. The FIRST-PRODUCT function selects the first of the two products of the CR_1_2 function, namely Glycerol (C00116) to be one of the substrates for the CR_2_2 chemical reaction function. If the intervening selector function was SECOND-PRODUCT (instead of FIRST-PRODUCT), then orthophosphate (C00009) (the fifth argument of the CR_1_2 function) would have been selected.

The two-substrate, two-product chemical reaction function CR_2_2 in the upper right part of figure 8.15 has six arguments. The enzyme for this reaction is Glycerol kinase (EC2.7.1.30) (the first argument); its rate is 1.69 (the second argument); its two substrates are ATP (C00002) (the third argument) and (as its fourth argument) the FIRST-PRODUCT of the two products of the lower chemical reaction function CR_1_2, namely Glycerol (C00116); and its two products are ADP (C00008) (the fifth argument) and sn-Glycerol-3-Phosphate (C00093) (the sixth argument).

Notice that Glycerol (C00116) appears as a substrate argument to both the two-substrate, one-product chemical reaction function CR_2_1 (in the lower left of figure 8.15 and in the upper left part of figure 8.1) and the two-substrate, two-product chemical reaction function CR_2_2 (in the upper right part of figure 8.15 and in the upper right part of figure 8.1). The repetition of a substance terminal as a substrate argument in a program tree means that when the tree is converted into a network of chemical reactions, the available concentration of this particular substrate is distributed to two reactions in the network. That is, the repetition of a substance terminal as a substrate argument in a program tree corresponds to a bifurcation point where one substance is distributed to two different reactions. This is the second bifurcation point in this particular network of chemical reactions.

Glycerol (C00116) has two sources in this network of chemical reactions. First, it is externally supplied (shown at the top right of figure 8.1). Second, this substance is the product of the one-substrate, two-product chemical reaction function CR_1_2 (in the middle of figure 8.1 and in the lower right of figure 8.15). When a substance in a network has two or more sources (by virtue of being externally supplied, of being a product of a reaction of a network, or a combination thereof), the substance is accumulated.

That is, the entire subtree is fully evaluated before the developmental process moves on to another subtree.

Conversely, when a subtree is evaluated using a breath-first order of evaluation in the developmental process, the evaluation of each function in the subtree is followed (in general) by the evaluation of functions (usually many) that are not part of the subtree.

The genetic algorithm and genetic programming work on the principle that there is something about the structure of a relatively fit individual that contributes to the individual's fitness. In genetic programming, the crossover operation creates new individuals by recombining subtrees of relatively fit individuals. Thus, it seems reasonable that the crossover operation would be more efficient if the parts of a program that are moved by the crossover operation (i.e., subtrees) more closely corresponded to the sequence of component-creating and topology-modifying functions.

10.1.4 New `TWO_LEAD` Function for Inserting Two-Leaded Components

In previous sections of this chapter (and in sections 11.1 and 11.2), two-leaded components, (such as resistors, capacitors, and inductors) were inserted into the developing circuit by means of separate component-creating functions (e.g., the `R`, `C`, and `L` functions). Each of these functions has a construction-continuing subtree as one of its arguments. Thus, with these functions, any crossover or mutation that changes one of these functions necessarily changes the construction-continuing subtree.

It may be advantageous for the evolutionary process to be able to change one two-leaded component into another without changing the construction-continuing subtree.

The new three-argument `TWO_LEAD` function replaces one modifiable wire (or component) with a series composition consisting of one modifiable wire, a two-leaded component, and a second modifiable wire.

The first argument of the `TWO_LEAD` function specifies the identity and numerical value (sizing) of the component. The first argument of the `TWO_LEAD` function is a subtree whose root is always one of the following new one-argument component-creating functions: `R_NEW`, `C_NEW`, or `L_NEW`. Each of these three new functions takes a value-setting subtree as its argument. The value-setting subtree may be either a single perturbable numerical value or an arithmetic-performing subtree (in particular, an arithmetic-performing subtree containing free variables).

The second and third arguments of the `TWO_LEAD` function are the construction-continuing subtrees that correspond to the two modifiable wires created by this function.

The important point is that the first argument of the `TWO_LEAD` function does not possess a construction-continuing subtree. As a result, the identity of the two-leaded component may be changed (by crossover or mutation) without changing either construction-continuing subtree. That is, a two-leaded component can be changed in a localized way.

10.1.5 New `Q` Transistor-Creating Function

Similarly, it may be advantageous for the evolutionary process to be able to change the orientation of a transistor or the transistor model without changing the construction-continuing subtrees associated with the insertion of the transistor into the developing circuit.

The new six-argument Q function inserts a transistor into a developing circuit, with both the model and orientation of the transistor specified as parameters.

The first argument of the Q function specifies the transistor model that is to be used. The set of available transistor models is specific to the problem at hand.

The second argument establishes which end (polarity) of the preexisting modifiable wire will be bifurcated (if necessary) in inserting the transistor. It can take on the values BIFURCATE_POSITIVE or BIFURCATE_NEGATIVE.

The third argument of the Q function specifies which of six possible permutations of the transistor's three leads (base, collector, and emitter) is to be used. It can take on the values B_C_E, B_E_C, C_B_E, C_E_B, E_B_C, and E_C_B. For example, C_B_E causes the first point to be connected to the transistor's collector; the second point to be connected to the transistor's base; and the third point to be connected to the transistor's emitter. As previously mentioned, the geometric coordinates define an underlying order of the three points to which the transistor's base, collector, and emitter may be connected.

Considering the second and third argument together, there are 12 possible ways of inserting a transistor.

The remaining three arguments of the Q function are the construction-continuing subtrees that correspond to the three modifiable wires created by this function.

10.2 Zobel Network with Two Free Variables

An audio power amplifier driving a loudspeaker may be modeled, at first approximation, as an inductor L1 and resistor R1 situated in series between the incoming signal and ground, as shown in figure 10.1 (Boutin 2002).

When the loudspeaker is prone to instability, it may be desirable to make the reactive load in figure 10.1 appear to be a purely resistive load to the driving source (Boutin 2002). This may be accomplished by adding additional circuitry to the LR circuit shown in figure 10.1.

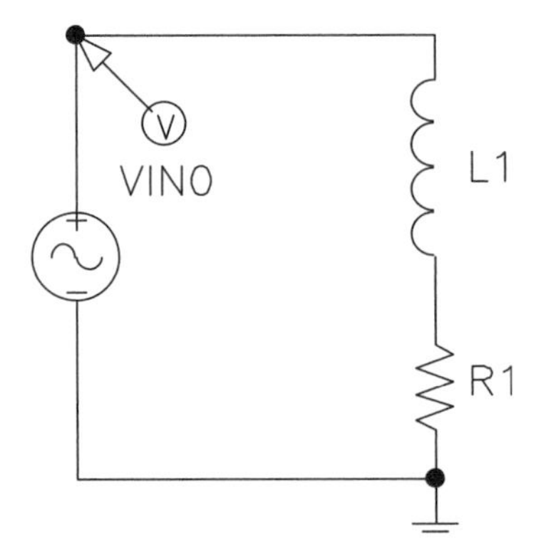

Figure 10.1 An inductor and resistor in series between the incoming signal and ground.

In this section, the problem is to find the topology of the unspecified additional circuitry as well as the sizing of each component of this additional circuitry such that the original LR circuit (figure 10.1) plus the additional circuitry together appear to the driving source to be a purely resistive load.

The desired solution to this problem must be parameterized (i.e., general) because the sizing of at least some of the additional components will necessarily depend on the particular value, L_1, of the original inductor L1 and the particular value, R_1, of the original resistor R1.

Boutin (2002) describes the mathematical solution to this problem in the form of what is called a *Zobel network* (Zobel 1926, 1928). If the original LR circuit (figure 10.1) plus the added circuitry must together have the behavior of a resistor with resistance $R_{overall}$, then the current flowing through the circuit must (according to Ohm's law) be $V_{in}/R_{overall}$, where V_{in} is the voltage of the incoming signal. If, for example, the driving source is a step function rising from 0 to V_{max} at time $t = t_{rise}$, then the current of the original LR circuit plus the added circuitry must be a step function rising from 0 to $V_{max}/R_{overall}$ at time $t = t_{rise}$.

The original LR circuit does not, of course, behave this way. Instead, its current response starts at 0 and rises exponentially (with time constant L_1/R_1) toward its final value. Thus, the to-be-discovered additional circuitry must compensate for the original circuit's exponential response (Boutin 2002).

10.2.1 Preparatory Steps for the Zobel Network Problem with Two Free Variables

10.2.1.1 Initial Circuit The initial circuit consists of an embryo and a test fixture.

We use a particularly challenging embryo (the floating embryo described in section 4.7.1.1) as the starting point of the developmental process for this problem. Specifically, as shown in figure 10.2, the floating embryo consists of a single modifiable wire Z0 that is not initially connected to the circuit's input(s) or output(s). Thus, each individual must grapple with the problem of discovering the circuit's input(s) and output(s) on its own.

Figure 10.2 shows the test fixture for the Zobel network problem. The test fixture contains the inductor L1 and resistor R1 of the original LR circuit. These two original components are in series with the input source VINO. The current probe is at IOUTO. The test fixture has two ports to which the developing circuit may connect. One of these two ports provides potential access to the input signal VINO. The other port provides potential access to ground.

10.2.1.2 Program Architecture Because there must be one result-producing branch in the program tree for each modifiable wire in the embryo and there is one modifiable wire in the embryo, the architecture of each circuit-constructing program tree has one result-producing branch. Automatically defined functions and the architecture-altering operations are not used on this problem.

10.2.1.3 Function Set A constrained syntactic structure enforces the use of one function set for the arithmetic-performing subtrees, another for the first argument of the `TWO_LEAD` function, and yet another for all other parts of the program tree.

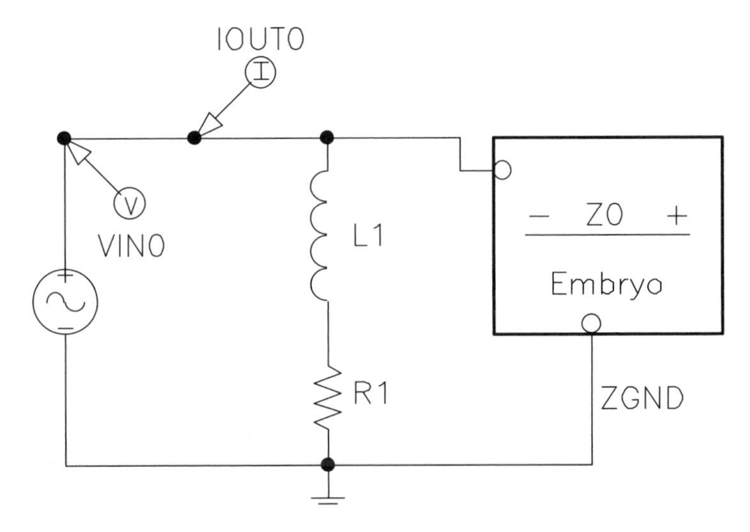

Figure 10.2 Test fixture for the Zobel network problem.

The function set, F_{aps}, for the arithmetic-performing subtrees is

$$F_{aps} = \{+, -, *, \%, \mathtt{RLOG}, \mathtt{EXP}\}.$$

Because resistors, inductors, and capacitors may be used in this problem, the function set, $F_{two\text{-}lead}$, for the first argument of the TWO_LEAD function is

$$F_{two\text{-}lead} = \{\mathtt{R\_NEW}, \mathtt{L\_NEW}, \mathtt{C\_NEW}\}.$$

The function set, F_{ccs}, for each construction-continuing subtree is

$$F_{ccs} = \{\mathtt{TWO\_LEAD}, \mathtt{SERIES}, \mathtt{PARALLEL\_NEW}, \mathtt{NODE}, \mathtt{TWO\_GROUND}, \mathtt{THREE\_GROUND}, \mathtt{INPUT\_0}\}.$$

Note that because the input source is a voltage source and the probe point measures current at the same point, there is no need for a separate output probe point in this particular test fixture (and hence no need for the OUTPUT_0 function in the function set).

10.2.1.4 Terminal Set A constrained syntactic structure enforces the use of one terminal set for the value-setting subtrees, another for the first argument of the PARALLEL_NEW function (section 10.1.2), and yet another for all other parts of the program tree.

In this problem, there are two free variables. They represent the inductance L_1 (called "L1" below) of the original inductor L1 and the resistance R_1 (called "R1") of the original resistor R1.

The component value for each component possessing a parameter is established by an arithmetic-performing subtree that may contain perturbable numerical values, arithmetic operations, and the two free variables (L1 and R1). This approach for establishing numerical parameter values is described in section 3.5.5.3.

The terminal set, T_{aps}, for the arithmetic-performing subtree for component-creating functions is

$$\mathsf{T}_{aps} = \{\Re_p, \texttt{L1}, \texttt{R1}\},$$

where $\Re_p$ denotes a perturbable numerical value.

In problems involving parameterized topologies, the nonlinear mapping (section 3.5.5) can potentially impede the evolutionary process and complicate the post-run analysis of mathematical expressions containing free variables. Thus it is not used for this problem. Note that our non-use of the nonlinear mapping on this particular problem reflects the chronology of our work and the evolution of our thinking (as opposed to any special requirement or exigency of the present problem). We believe that the nonlinear mapping is inefficient for problems involving free variables and intend to avoid it on such problems in the future.

A constrained syntactic structure specifies the terminals that may appear as the first argument of the `PARALLEL_NEW` function. The terminal set, $\mathsf{T}_{parallel}$, for the first argument of the `PARALLEL_NEW` function is

$$\mathsf{T}_{parallel} = \{\texttt{UP\_OR\_LEFT}, \texttt{DOWN\_OR\_RIGHT}\}.$$

The terminal set, T_{ccs}, for each construction-continuing subtree is

$$\mathsf{T}_{ccs} = \{\texttt{END}, \texttt{SAFE\_CUT}\}.$$

10.2.1.5 Fitness Measure The purpose of a Zobel network is to make the entire circuit (that is, the original inductor L1 and original resistor R1 in the test fixture plus any additional circuitry developed from the embryo) appear to be purely resistive to the driving source. In this problem, the inductance of inductor L1 and the resistance of resistor R1 are free variables that vary over a number of different values (fitness cases).

When there are free variables in a problem, it is necessary to ascertain the behavior and characteristics of each candidate individual for a representative sample of values of each of the free variables. Multiple combinations of values of the free variables force generalization of the to-be-evolved circuit.

The free variable for the inductance, L_1, of inductor L1 is `L1`). `L1` ranges over six values, namely 1.0, 2.5, 6.3, 15.8, 39.8, and 100.0 microhenrys.

The free variable for the resistance, R_1, of resistor R1 is `R1`. `R1` ranges over six values, namely 10, 25, 63, 158, 398, and 1,000 Ohms.

Thus, there are a total of 36 fitness cases, each representing a combination of values of the two free variables (L_1 and R_1).

The input signal (driving source) for a particular one of the 36 fitness cases is

$$1 - e^{-\frac{t}{\tau}},$$

where the time constant, τ, is L_1/R_1. This curve has an initial amplitude of 0 Volts, rises with a time constant τ, and has a potential maximum amplitude of 1 Volt.

SPICE is instructed to perform a transient (time-domain) analysis for $10L/R$ seconds. There are 100 time steps. The current at the current probe, IOUT0, is reported for 101 times.

The contribution to fitness of each of the 36 fitness cases is equalized by weighting each fitness case by $R_{overall}$. In this connection, recall that a maximum input voltage of 1 Volt corresponds to a maximum desired current of $1/R_{overall}$.

The number of hits is defined as the number of fitness cases for which the absolute error is less than or equal to 5% of the reciprocal of the value of $R_{overall}$ for that fitness case.

The fitness measure is designed to encourage two types of parsimony.

The first type of parsimony is based on the familiar idea of counting the circuit's components. For this problem, a simple count of the number components is used. However, it may be preferable to consider factors such as the component's dollar cost, power consumption, or physical size.

The second type of parsimony, called *equational parsimony*, is the total number of mathematical functions in the arithmetic-performing subtrees. This type of parsimony is often appropriate when parameterized topologies are being used.

For individuals scoring less than 36 hits, fitness is the sum, for each of the 101 time steps associated with each of the 36 fitness cases, of the product of $R_{overall}$ (the equalizing factor for the fitness case) and the weighted absolute value of the difference between the current at the current probe point, IOUT0, and the current value of $V_{in}/R_{overall}$ for the fitness case involved. The absolute value of the difference is weighted by 1.0 if the absolute error is less than or equal to 5% of the reciprocal of the value of $R_{overall}$ for that fitness case, but 10. 0 otherwise.

For individuals scoring 36 hits, fitness is one trillionth of the sum of the total number of components (including the fixed components in the test fixture) and one tenth of the total number of mathematical functions in the arithmetic-performing subtrees.

10.2.1.6 Control Parameters The population size is 500,000.

10.2.2 Results for the Zobel Network Problem with Two Free Variables

The best-of-run individual (figure 10.3) emerged in generation 15 of our one run of this problem. Genetic programming produced this circuit's overall topology as well as the two mathematical expressions (containing the free variables L_1 and R_1) for specifying the parameter values for new capacitor C2 and new resistor R2. That is, genetic programming produced a parameterized (i.e., general) solution to the problem.

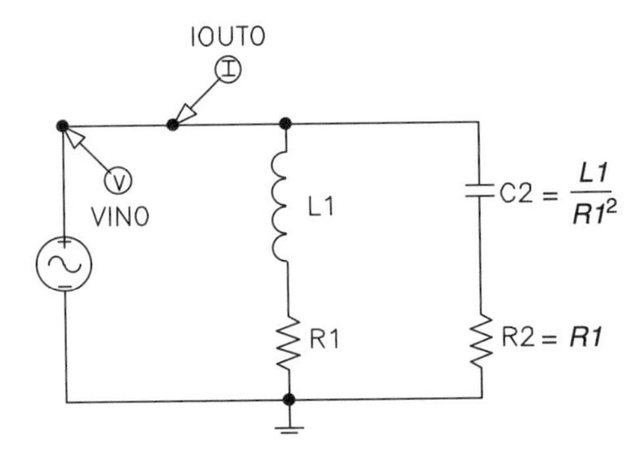

Figure 10.3 Best-of-run parameterized topology for the Zobel network problem.

The best-of-run parameterized circuit from generation 15 is a Zobel network as described by Boutin (2002). In addition to the original inductor L1 and original resistor R1, the overall circuit (figure 10.3) contains a new capacitor C2 with component value

$$L_1/R_1^2,$$

and a new resistor R2 with component value

$$R_1.$$

These values for the new components in the genetically evolved circuit are exactly those that Boutin derived by mathematical analysis.

The addition of the second shunt consisting of an appropriately valued new capacitor and an appropriately valued new resistor makes the overall load appear to be purely resistive to the driving source.

Notice that genetic programming automatically created all the following in the single run that produced the additional circuitry of figure 10.3:

- the topology of the additional circuitry, including
  - the total number of added components (two),
  - the type of each added component (i.e., one new capacitor and one new resistor),
  - all the connections between the circuit's added components, preexisting components, and accessible points of the test fixture,
- the sizing of the additional circuitry, including
  - one mathematical expression containing both of the problem's free variables (L_1 and R_1) for establishing the sizing of the added capacitor C2, and
  - another mathematical expression containing one of the problem's free variables (R_1) for establishing the sizing of the added resistor R2.

10.2.3 Routineness of the Transition from a Problem Involving a Non-Parameterized Circuit to a Problem Involving a Parameterized Circuit

There is a fundamental difference between the Zobel network problem in this section and all the problems of automatic circuit synthesis in chapter 4 and our 1999 book *Genetic Programming III* (Koza, Bennett, Andre, and Keane 1999a).

The difference is that each earlier problem calls for the discovery of a single circuit with fixed numerical values for each of the circuit's components whereas the present problem calls for the discovery of a parameterized circuit that represents the solution to an entire category of problems. Each earlier problem entailed producing a circuit (graphical structure) whose components were labeled with constant numerical component values whereas the present problem entails producing a circuit (graphical structure)whose components are labeled with mathematical expressions containing free variables.

Although this difference is fundamental, its implementation requires only two changes in the preparatory steps.

First, the problem's two free variables (`L1` and `R1`) are added to the terminal set. That is, the terminal set is changed from

$$\mathsf{T}_{\text{aps}} = \{\Re_{\text{p}}\}$$

to

$$\mathsf{T}_{\text{aps}} = \{\Re_{\text{p}}, \texttt{L1}, \texttt{R1}\}.$$

Second, fitness is measured for a representative set of combinations of values of the problem's two free variables.

These two required changes are essentially mechanical and administrative in nature. They do not draw on any knowledge of electrical engineering or any insight about how to derive mathematical expressions for the relationship between the problem's free variables and the values for the circuit's components. They are certainly not knowledge-intensive.

The other preparatory steps for the present problem are substantially the same as for other problems of circuit synthesis.

In summary, the transition from the earlier problems of circuit synthesis not involving free variables to the present problem involving free variables is nearly effortless. We consider the transition to be routine.

10.2.4 AI Ratio for the Zobel Network Problem with Two Free Variables

The problem involving the Zobel network resembles the type of problems found in many textbooks. The solution produced by genetic programming to the problem is a parameterized (i.e., general) solution. However, under the specific stringent criteria for human-competitiveness enumerated in table 1.2, these two facts (either alone or together) do not justify rating the genetically evolved result as "human-competitive." Nonetheless, it seems reasonable to say that the solution to the present problem has a moderately high amount of "A."

The preparatory steps for this problem are a straightforward rendering of the high-level statement of the problem and incorporate a small amount of "I."

Therefore, the AI ratio for the solution produced by genetic programming to the problem involving the Zobel network with two free variables is moderately high.

10.3 Third-Order Elliptic Lowpass Filter with a Free Variable for the Modular Angle

The problem in this section is to automatically create both the topology and sizing of a lowpass filter circuit with the performance characteristics of a third-order elliptic filter in which the values of the circuit's components are parameterized by mathematical expressions containing a free variable representing the filter's modular angle, Θ.

The modular angle Θ of a filter is the arcsine of the ratio of the boundaries of the passband and stopband (Williams and Taylor 1995). Table 10.1 shows the stopband boundary as a function of the modular angle Θ for a lowpass filter whose passband boundary is 1,000 Hz. For example, if the stopband boundary is 2,000 Hz, the

Table 10.1 Stopband boundary as a function of the modular angle (when the passband boundary is 1,000 Hz)

Θ	Stopband boundary
5	11,474
10	5,759
15	3,864
20	2,924
25	2,366
30	2,000
35	1,743
40	1,556
45	1,414
50	1,305
55	1,221
60	1,155

modular angle Θ is 30° because the ratio of the boundaries of the passband and stopband is $\frac{1}{2}$. As the stopband boundary approaches the 1,000-Hz passband boundary (i.e., the modular angle approaches 90°), filter design becomes increasingly difficult.

The desired filter is to have a reflection coefficient, ρ, of 20%. When the input amplitude is 1 Volt, this particular value of ρ corresponds to a passband ripple of 21 millivolts (Williams and Taylor 1995) and is close to the 30 millivolts used on other problems of filter synthesis in this book.

The solution sought in this section is analogous to what one finds in a filter cookbook. In a filter cookbook, the user first turns to the page that provides information about a particular combination of topology and reflection coefficient (e.g., the third-order elliptic filter with ρ of 20%). The user then looks at the row of the table given on that page that specifies the specific component values to be used to achieve a specific modular angle. In this section, genetic programming will create both the topology and sizing (in the form of a parameterized topology) for a third-order elliptic filter with variable modular angle. Thus, genetic programming is effectively recreating a page out of the filter cookbook.

Notice that, in this section, we are not talking about a lowpass filter that has one modular angle at some moments during the filter's operational lifetime while having a different modular angle at other times. Instead, we are talking about a circuit-constructing program tree (genome) that, at the single moment when the program tree is executed to create a fully developed circuit, yields a lowpass filter with a particular modular angle. The fully developed circuit then operates with that particular modular angle for its entire operational lifetime.

10.3.1 Preparatory Steps for the Third-Order Elliptic Lowpass Filter with a Free Variable for the Modular Angle

10.3.1.1 Initial Circuit Figure 10.4 shows a one-input, one-output initial circuit consisting of an embryo embedded in a test fixture.

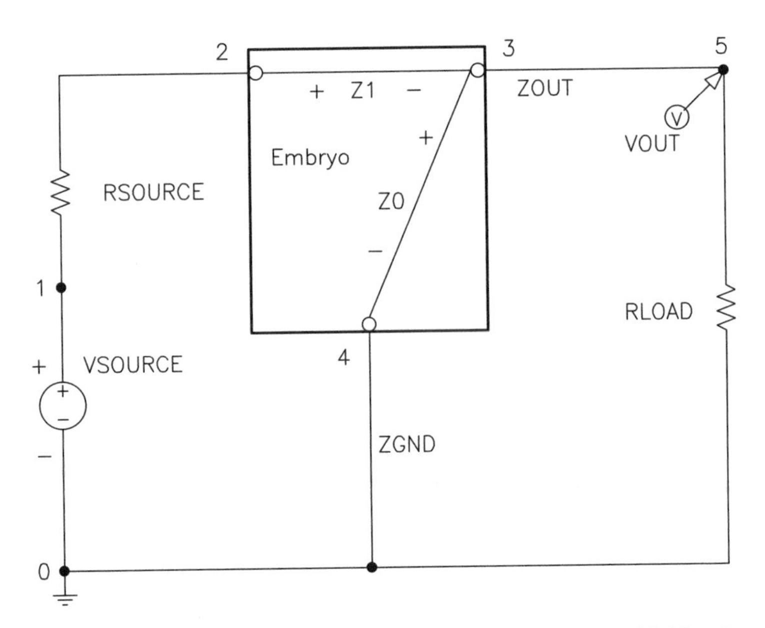

Figure 10.4 One-input, one-output initial circuit with two modifiable wires.

The embryo consists of two modifiable wires (Z0 and Z1). The test fixture has an incoming signal source VSOURCE, a 1,000-Ohm source resistor RSOURCE, a nonmodifiable wire ZOUT, a voltage probe point VOUT (the output of the overall circuit), a 1,000-Ohm load resistor RLOAD, and a nonmodifiable wire ZGND connecting to ground. The test fixture has three ports to which the developing circuit may connect. One port makes the input signal VSOURCE available to the developing circuit. A second port provides access to ground. A third port is connected to the voltage probe point VOUT.

10.3.1.2 Program Architecture Because there must be one result-producing branch in the program tree for each modifiable wire in the embryo and there are two modifiable wires in the embryo for this problem, the architecture of each circuit-constructing program tree has two result-producing branches. Automatically defined functions and the architecture-altering operations are not used.

10.3.1.3 Function Set The function set for this problem is the same as that of section 10.2.1.3, except that no resistors are used and there is no `INPUT_0` function (because of the initial circuit used for the present problem).

10.3.1.4 Terminal Set A constrained syntactic structure enforces the use of one terminal set for the value-setting subtrees, another for the first argument of the `PARALLEL_NEW` function (section 10.1.2), and yet another for all other parts of the program tree.

In this problem, there is one free variable. The free variable represents the value for *sin* Θ (called "`SIN_THETA`" below).

The component value for each component possessing a parameter is established by an arithmetic-performing subtree that may contain perturbable numerical values, arithmetic operations, and the free variable (SIN_THETA). This approach for establishing numerical parameter values is described in section 3.5.5.3.

The terminal set, T_{aps}, for the arithmetic-performing subtree for component-creating functions is

$$T_{aps} = \{\Re_p, \mathtt{SIN\_THETA}\},$$

where $\Re_p$ denotes a perturbable numerical value. The nonlinear mapping (section 3.5.5) was not used.

The terminal set, $T_{parallel}$, for the first argument of the PARALLEL_NEW function is

$$T_{parallel} = \{\mathtt{UP\_OR\_LEFT}, \mathtt{DOWN\_OR\_RIGHT}\}.$$

The terminal set, T_{ccs}, for each construction-continuing subtree is

$$T_{ccs} = \{\mathtt{END}, \mathtt{SAFE\_CUT}\}.$$

10.3.1.5 Fitness Measure The free variable Q ranges over 12 values that are equally spaced between 5° and 60°.

The passband is fixed for this problem and ends at 1,000 Hz.

The frequency at which the stopband starts is determined by the value of Θ.

Because the high-level statement of the behavior for the desired lowpass filter circuit is expressed in terms of frequencies, the output voltage VOUT is measured in the frequency domain. For each of the 12 values of Θ, SPICE is instructed to perform an AC small signal analysis (AC sweep) and report the circuit's frequency response.

The intended lowpass filter is to satisfy the special (and decidedly complex) performance requirements of a third-order elliptic filter. One of the signature characteristics of an elliptic filter is the extreme suppression of the signal in the transitional region between the boundary of the passband and the boundary of the stopband. The Ω_2 frequency of an elliptic filter is the lowest frequency at which attenuation is infinite (i.e., corresponding to the passing of 0 Volts). Note that, for this problem, we do not refer to the transitional frequencies as the "don't care" region because we do, in fact, care about what happens in the transitional region.

The range of frequencies over which the AC sweep is performed for each value of Θ is determined using the Ω_2 frequency associated with a third-order elliptic filter (with a reflection coefficient ρ of 20%) that is tuned to yield the given value of Θ. For each of the 12 values of Θ, the AC sweep is performed from four octaves below the Ω_2 frequency to three octaves above the Ω_2 frequency.

Table 10.2 shows, for each of the 12 values of Θ, the Ω_2 frequency (in Hertz), the starting point for the AC sweep (in Hertz), and the ending point for the AC sweep (in Hertz). As previously mentioned, the passband is fixed for this problem and ends at 1,000 Hz.

Each octave is divided into 20 parts. The resulting 141 frequencies end up, according to the value of Θ, in the passband, transitional region, and stopband.

Table 10.2 Ranges for the AC sweeps

Θ	Ω_2	Starting frequency	Ending frequency
5	13,242	828	105,939
10	6,637	415	53,096
15	4,442	278	35,538
20	3,351	209	26,804
25	2,700	169	21,600
30	2,270	142	18,161
35	1,967	123	15,733
40	1,742	109	13,938
45	1,571	98.2	12,568
50	1,437	89.8	11,495
55	1,330	83.1	10,642
60	1,245	77.8	9,957

The fitness measure for this problem is based on four elements measuring

- the frequency response in the passband,
- the frequency response in the transitional region,
- the frequency response in the stopband, and
- the passband ripple.

Each of these four elements is calculated separately for each of the 12 values of Θ.

The first element of the fitness measure is concerned with the part of the frequency response curve below 1,000 Hz. The detrimental contribution to fitness is the weighted sum, over all sampled frequencies below 1,000 Hz, of the absolute difference between the circuit's output and an ideal output of 1 Volt. We use a value called the *minimum acceptable passband voltage* to determine the weighting. The minimum acceptable passband voltage is obtained by multiplying the maximum passband error (i.e., the maximum deviation from 1 Volt) that is allowed for a third-order elliptic filter (with a reflection coefficient ρ of 20%) by 101% and subtracting this product from 1 Volt. If the circuit's output is greater than the minimum acceptable passband voltage, the weighting is 1.0, but 10.0 otherwise.

The second element of the fitness measure is concerned with the part of the frequency response curve between 1,000 Hz and the stopband boundary. The detrimental contribution to fitness is the weighted sum, over all sampled frequencies between 1,000 Hz and the stopband boundary, of the absolute difference between the circuit's output in decibels and the output of a third-order elliptic filter (tuned for the particular value of Θ). If the absolute difference is less than 3 dB, the weighting is 1.0, but otherwise 10.0. In order to avoid dealing with extraordinarily large negative values of attenuation, any value below -100 dB (corresponding to 10^{-5} Volts) is taken to be -100 dB.

The third element of the fitness measure is concerned with the part of the frequency response curve above the stopband boundary. The detrimental contribution to fitness is the weighted sum, over all sampled frequencies above the stopband boundary, of the absolute difference between the circuit's output in decibels and the output of a third-order elliptic filter (tuned for the particular value of Θ). If the absolute difference is less than 101% of the value for a third-order elliptic filter, the weighting is 1.0, but otherwise 10.0.

The fourth element of the fitness measure is based on the passband ripple. For each value of Θ, the minimum value of the frequency response in the passband is ascertained. The detrimental contribution is the sum, over the 12 values of Θ, of the greater of 0 and the difference between the minimum acceptable passband voltage and the minimum value of the frequency response in the passband.

In this problem, numerous mathematical expressions must be evolved to establish the parameters for components of the to-be-evolved circuit. The values of the components of the to-be-evolved circuit are presumptively complicated nonlinear functions of the free variable Θ.

Note that even if the mathematical expression being sought were, in fact, a linear function of one variable, genetic programming usually must consider more than two values of a free variable in order to discover a linear relationship. Of course, if one already knows that the relationship is linear, one can simply draw a straight line between two data points. However, genetic programming does not have such advance knowledge about the linearity of the mathematical relationships that may exist in a particular problem. Genetic programming may well create a nonlinear function to pass through the two points that are, in reality, produced by a linear relationship. Thus, a multiplicity of data points is required even if the underlying function happens to be a linear function of one variable. *A fortiori*, a multiplicity of data points is required when the underlying function is actually nonlinear (as is usually the case in non-trivial problems).

Having said that, the amount of computer time is (as an initial approximation) proportional to the number of fitness cases. Thus, there is a strong incentive to try to solve a problem with as few fitness cases as possible.

In evolving mathematical expressions containing free variables, one is always concerned about possible overfitting.

Balancing these competing considerations, we used 12 different values of Θ for this problem. We did not know in advance whether 12 different values of Θ would prove to be sufficient to enable genetic programming to unearth the complicated relationships inherent in this problem.

Recognizing that 12 fitness cases might well prove to be inadequate to discover the required curve, we doubled the weight on the lowest and highest values of Θ under consideration (i.e., 5° and 60°) in order to place additional emphasis on the two ends of this presumptively nonlinear curve.

We substantially increased the weights for the first and fourth elements of the fitness measure because these elements were in Volts (as opposed to decibels).

The fourth element of the fitness measure receives less weight than the first because it involves only the passband ripple.

Fitness is the sum of two terms. The first term is the sum, for 5° and 60°, of 200 times the first element of the fitness measure, twice the second element, twice the third element, and 100 times the fourth element. The second term is the sum, over the 10 other values of Θ, of 100 times the first element of the fitness measure, the second element, the third element, and 50 times the fourth element.

The number of hits is defined to be the number of fitness cases for which the weighting is 1.0 (as opposed to 10.0).

10.3.1.6 Control Parameters The population size is 1,000,000.

10.3.2 Results for the Lowpass Third-Order Elliptic Filter with a Free Variable for the Modular Angle

The best-of-run individual for this problem emerged in generation 293 (figure 10.5). Ignoring the fixed source and load resistors that are hard-wired into the test fixture, the genetically evolved parameterized topology has two capacitors (C21 and C66) and two inductors (L44 and L15).

The circuit of figure 10.6 is a parameterized (i.e., general) solution to the problem. The component values for the two capacitors and two inductors of the best-of-run circuit are not specified by constant numerical values, but instead, by genetically evolved mathematical expressions containing the free variable Θ.

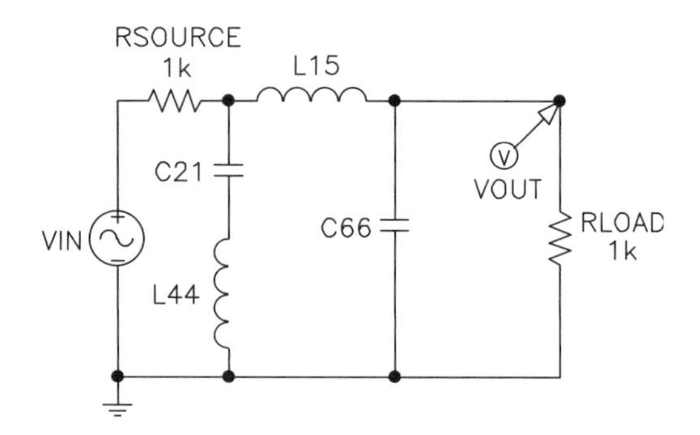

Figure 10.5 Best-of-run parameterized topology for the lowpass filter with a free variable for the modular angle.

Specifically, the sizing of capacitor C21 is given by

$$C21 = \frac{1.2059}{(2.1773 + e^{(1.2157*10\hat{}(-15) - 2*\sin(\Theta))} - \sin(\Theta) + \sin(\Theta)^2)e^{\sin(\Theta) + 14.519}}.$$

The sizing of capacitor C66 is given by

$$C66 = 0.18211*10^{-6} e^{\left(\frac{1}{5+e^{2(\sin\Theta)+2.4010+e^{INT_1}}+e^{\sin\Theta-(\sin\Theta)^2}+e^{INT_2}}\right)},$$

where

$$INT_1 = \frac{0.96061}{-4.6758 + e^{\left(\frac{2(\sin\Theta)}{\left(\frac{1.1204((\sin\Theta)-1)}{\sin\Theta}\right)+0.96376}\right)} + 2(\sin\Theta)},$$

and where

$$INT_2 = .019890 \frac{((\sin\Theta)-1)e^{0.66795(\sin\Theta)}}{\sin\Theta} - 2\sin(\Theta) \cdot$$

The sizing of inductor L15 is given by

$$L15 = 0.068877 * e^{\frac{1.2157}{1.2157 + 0.29652 * \sin(\Theta)^2}} \cdot$$

The sizing of inductor L44 is given by

$$L44 = \left| \frac{(\sin\Theta)^2}{(\sin\Theta) - 2.8032 - e^{INT3} - e^{INT4}} \right|,$$

where

$$INT_3 = \left(\frac{e^{\sin\Theta}(\sin\Theta)^2}{e^{(-0.78021(\sin\Theta)^2)} - 1.2157 - (\sin\Theta) - (\sin\Theta)^2} \right),$$

and where

$$INT_4 = \left(e^{(\sin\Theta)-(\sin\Theta)^2} + e^{\left(\frac{0.019890((\sin\Theta)-1)e^{\frac{2(\sin\Theta)}{3.3354 + \frac{1}{-2.8948 + 0.97535e}}}}{\sin\Theta} - 2(\sin\Theta) \right)} \right) \cdot$$

Notice that genetic programming automatically created all the following in the single run that produced the circuit of figure 10.6:

- the circuit's topology, including
  - the total number of components (four) in addition to those of the test fixture,
  - the type of each component (i.e., two capacitors and two inductors),
  - all the connections between the circuit's components and accessible points of the test fixture, and
- the circuit's sizing as expressed by four mathematical expressions containing the problem's free variable (Θ) for establishing the sizing of C21, C66, L44, and L15.

Averaged over the 12 values of Θ, the best-of-run circuit from generation 293 has:

- 9.17 millivolts average absolute error in the passband,
- 4.15 dB average absolute error in the stopband and transitional region, and
- 98.6% of the possible number of hits.

Note that average absolute error in the passband is based on the absolute difference between 1 Volt and the value on the evolved circuit's frequency response curve, whereas average absolute error in the stopband and "don't care" region is based on the absolute difference between the evolved circuit's frequency response and the frequency response of the textbook third-order elliptic filter.

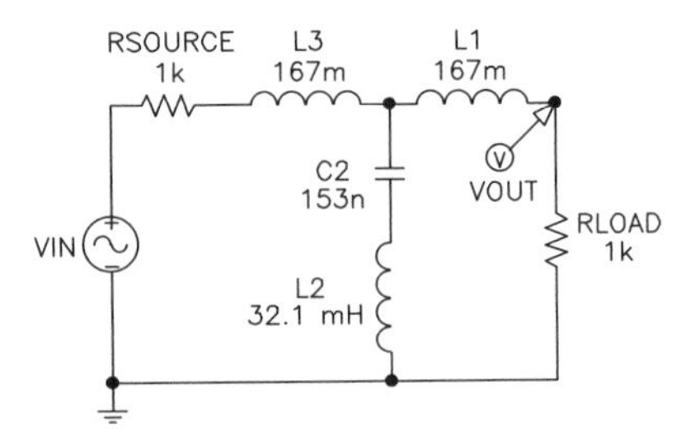

Figure 10.6 Textbook third-order elliptic filter with modular angle Θ of 30°.

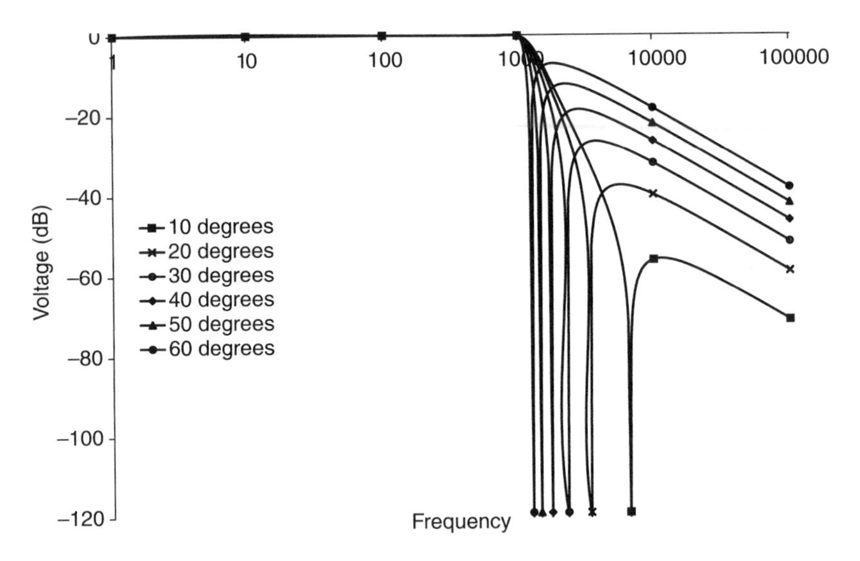

Figure 10.7 Frequency domain behavior of a textbook third-order elliptic filter for six choices of modular angle Θ between 10° and 60°.

Figure 10.6 shows the topology and sizing for a classical third-order lowpass elliptic filter for a modular angle Θ of 30° (Williams and Taylor 1995). The topology in this figure applies to any other value of Θ between 5° and 60°.

Figure 10.7 shows the behavior in the frequency domain of a textbook third-order elliptic filter for six choices of modular angle Θ between 10° and 60° (Figure 10.6). The horizontal axis shows frequencies from 1 Hz to 100,000 Hz on a logarithmic scale. The vertical axis shows gain (in decibels). As can be seen, the incoming signal is passed up to 1,000 Hz with virtually no attenuation. Note that each curve approaches a desired gain of $-\infty$dB (corresponding to the passing of 0 Volts) at the Ω_2 frequency in the transitional region above 1,000 Hz.

Figure 10.8 shows the behavior in the frequency domain of the best-of-run circuit from generation 293 for six choices of modular angle Θ between 10° and 60° (figure 10.6).

Recall that the modular angle Θ of a filter is the arcsine of the ratio of the boundaries of the passband and stopband.

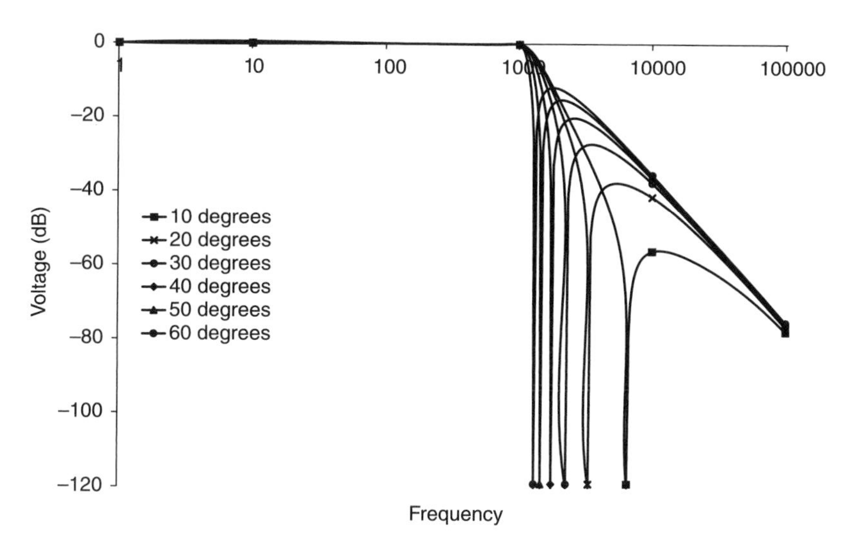

Figure 10.8 Frequency domain behavior of a best-of-run circuit for six choices of modular angle Θ between 10° and 60°.

The Θ of a third-order elliptic filter may be measured from the frequency response curve in the following way:

First, find the minimum value of the frequency response curve in the stopband and transitional region. The frequency at which this minimum value occurs is known as the Ω_2 frequency of the filter.

Second, find the maximum value of the frequency response for the region of the curve lying to the right of the Ω_2 frequency. This value is known as the *stopband attenuation* A_{min}.

Third, find the leftmost frequency in the stopband and transitional region for which the value of the frequency response curve is equal to A_{min}. This is the Ω_s frequency of the filter.

Fourth, the Θ value of a filter whose passband boundary is 1,000 Hz is given by the equation $\Theta = \sin^{-1}(1{,}000/\Omega_s)$.

Using these four steps, we can measure the Θ of the best-of-run circuit for each of the 12 in-sample fitness cases. Table 10.3 gives the actual and desired values of Θ for all 12 angles for the best-of-run filter.

The *maximum passband deviation* of a filter is defined, for each frequency in the passband, as the maximum value of the absolute difference between 1 Volt and the filter's frequency response. Table 10.4 shows the maximum passband deviation in millivolts for the best-of-run circuit for each of the 12 in-sample fitness cases.

The ρ value of a filter, the passband ripple R_{dB}, and the maximum passband deviation (*MPD*) are related by the following two equations:

$$R_{dB} = -10*\log(1-\rho^2)$$

and

$$MPD = 10^{R_{dB}/20} - 1.$$

The second equation assumes a 1-Volt source.

Table 10.3 Comparison of desired and actual Θ for the 12 in-sample fitness cases

Desired Θ	Actual Θ
5	5.06
10	9.79
15	14.7
20	19.7
25	24.7
30	29.8
35	33.7
40	38.7
45	43.6
50	47.1
55	52.3
60	55.4

Table 10.4 Maximum passband deviation for the 12 in-sample fitness cases

Desired Θ	Maximum passband deviation
5	18.5
10	17.9
15	17.7
20	17.7
25	17.7
30	17.8
35	17.9
40	18.3
45	18.7
50	19.3
55	19.9
60	21.2

A value of $\rho = 20$ percent thus corresponds to a maximum passband deviation of approximately 20.62 millivolts. As can be seen from table 10.4, the evolved filter's maximum passband deviation is within 5 percent of this value for all values of Θ.

The performance of the best-of-run circuit from generation 293 is cross-validated using 11 previously unseen values of Θ. For these 11 out-of-sample fitness cases, the best-of-run circuit has:

- 9.21 millivolts average absolute error in the passband,
- 3.47 dB average absolute error in the stopband and transitional region, and
- 97.9 percent of the possible number of hits.

Table 10.5 compares the desired and actual values of Θ for the 11 out-of-sample fitness cases.

Table 10.5 Comparison of desired and actual values of Θ for the 11 out-of-sample fitness cases

Desired Θ	Actual Θ
7	6.88
13	12.8
17	16.7
23	22.6
27	26.5
33	32.4
37	36.3
43	41.6
47	45.3
53	50.2
57	53.0

Table 10.6 Maximum passband deviation for the 11 out-of-sample fitness cases

Desired Θ	Maximum passband deviation
7	20.1
13	17.8
17	17.7
23	17.7
27	17.7
33	17.8
37	18.0
43	18.5
47	18.9
53	19.7
57	20.2

Table 10.6 shows the maximum passband deviation in millivolts for the 11 out-of-sample fitness cases.

The mathematical expressions for the sizing of C21, C66, L15, and L44 are complicated. There are several options if the user desires or requires more parsimonious expressions.

First, equational parsimony can be incorporated into the fitness measure (as we did in section 10.2.1.5 for the Zobel network problem).

Second, each separate genetically evolved equation could be converted into a set of data points. A run of symbolic regression (using with parsimony pressure in its fitness measure) can be made on the data points in order to obtain a simpler equation that passes through (or near) the same data points.

In summary, genetic programming produced a parameterized circuit that matches the behavior of a third-order elliptic filter for any value of the modular angle Θ between 5° and 60°.

When we embarked on this problem, we expected that genetic programming would rediscover the elliptic topology. However, as can be seen by comparing figures 10.5 and 10.6, the genetically evolved circuit is not an elliptic filter. In fact, the genetically evolved circuit has the topology of an *M*-derived half-section filter (patented by Zobel in 1925). However, the values of the components in the genetically evolved circuit differ by factors of more than 2-to-1 from that of the *M*-derived half-section filter. Moreover, the genetically evolved filter differs from Cauer's patented elliptic filter, Campbell's patented ladder filter (Campbell 1917), and Johnson's patented "bridged T" filter (Johnson 1926).

At the beginning of this section (and other sections in this book involving circuit-constructing program trees containing free variables), we emphasized that the solution to the problem posed in this section was not intended to be a filter that has one modular angle at some moments during the filter's operational lifetime and a different modular angle at other times. However, it should be mentioned in passing that one conceivably could use variable capacitors and variable inductors to construct a circuit having the topology shown in figure 10.5 and then control the variable capacitors and variable inductors by signals emanating from analog computational circuits that implement the mathematical functions contained in the expressions for capacitors C21 and C66 and inductors L15 and L44. In that event, one would have a circuit whose behavior would vary dynamically in response to dynamically changing values of Θ during the circuit's operational lifetime. However, such a circuit could well be impractical for several reasons.

First, implementation of the mathematical expressions for the sizing of C21, C66, L15, and L44 would require a cascade of numerous computational circuits. Each of the individual computational circuits in the cascade would have to be exceedingly precise in order to produce a precise final value.

Second, the individual computational circuits (notably the exponential circuit) would have to be accurate for values that vary over many orders of magnitude.

Third, analog computational circuits often exhibit significant transient phenomena as their inputs change. The design of analog computational circuits becomes increasingly difficult as the settling time decreases.

10.3.3 Routineness for the Lowpass Third-Order Elliptic Filter with a Free Variable for the Modular Angle

For substantially the same reasons stated in section 10.2.3, the result produced in this section is routine.

10.3.4 AI Ratio for the Lowpass Third-Order Elliptic Filter with a Free Variable for the Modular Angle

For substantially the same reasons stated in section 10.2.4, the AI ratio for the solution produced by genetic programming to the problem involving the lowpass third-order elliptic filter with a free variable is moderately high.

10.4 Passive Lowpass Filter with a Free Variable for the Passband Boundary

The previous two sections demonstrated that genetic programming can automatically discover both the topology and sizing of a circuit in which the values of the circuit's components are parameterized by mathematical expressions containing free variables. This section provides another example involving a filter.

The problem in this section is to evolve a parameterized lowpass filter circuit composed of passive components (i.e., inductors and capacitors) with 1,000-to-1 attenuation in the stopband (i.e., 60 decibels) whose passband boundary is specified by a free variable, f, such that the passband ends at frequency f and the stopband starts at frequency $2f$.

For the problem in this section, a voltage in the passband that is between 970 millivolts and 1 Volt (i.e., a passband ripple of 30 millivolts or less) and a voltage in the stopband that is between 0 Volts and 1 millivolt (i.e., a stopband ripple of 1 millivolt or less) is regarded as acceptable. Any voltage lower than 970 millivolts in the passband and any voltage above 1 millivolt in the stopband are regarded as unacceptable. The frequencies between f and $2f$ constitute the filter's "don't care" region.

Notice that, in this section, we are not talking about a lowpass filter that has one passband boundary at some moments during the filter's operational lifetime while having a different passband boundary at other times. That is, we are not talking about a filter whose passband dynamically varies, say, in response to an additional input (i.e., a control signal) that specifies the desired instantaneous passband boundary. (Note that a post-2000 patented filter with just such a dynamically varying passband appears in chapter 15). Instead, in this section, we are talking about a circuit-constructing program tree (genome) that, at the single moment when the program tree is executed to create the fully developed circuit, yields a lowpass filter with a particular passband boundary. The fully developed circuit then operates with that particular passband boundary for its entire operational lifetime.

10.4.1 Preparatory Steps for the Passive Lowpass Filter with a Free Variable for the Passband Boundary

10.4.1.1 Initial Circuit The initial circuit for this problem is the same as in section 10.3.1.1.

10.4.1.2 Program Architecture The architecture of each circuit-constructing program tree has two result-producing branches. There are no automatically defined functions in any program tree in the initial random population (generation 0). However, in later generations, the architecture-altering operations may insert (and delete) one-argument automatically defined functions. A (generous) maximum of five automatically defined functions is established for each program tree in the population.

10.4.1.3 Terminal Set A constrained syntactic structure enforces the use of one terminal set for the value-setting subtrees, another for all other parts of the result-producing branch, and yet another for all other parts of the automatically defined functions.

In this problem, there is a free variable f (called "`F`" below) representing the end of the desired filter's passband.

The numerical parameter value for each electrical component possessing a parameter is established by an arithmetic-performing subtree that may contain perturbable numerical values, arithmetic operations, and the free variable (`F`). This approach for establishing numerical parameter values is described in section 3.5.5.3.

Arithmetic-performing subtrees may appear in both result-producing branches and any automatically defined functions that may be created during the run by the architecture-altering operations.

The terminal set, T_{aps}, for the arithmetic-performing subtrees is

$$T_{aps} = \{\Re_p, \texttt{F}\},$$

where $\Re_p$ denotes a perturbable numerical value between −5.0 and +5.0.

The terminal set, T_{rpb}, for all other parts of each result-producing branch is

$$T_{rpb} = \{\texttt{END, SAFE\_CUT}\}.$$

Similarly, the terminal set, T_{adf}, for all other parts of each automatically defined function is

$$T_{adf} = \{\texttt{END, SAFE\_CUT, ARG0}\}.$$

10.4.1.4 Function Set A constrained syntactic structure enforces the use of one function set for the arithmetic-performing subtrees and another function set for all other parts of the program tree.

The function set, F_{aps}, for the arithmetic-performing subtrees is

$$F_{aps} = \{+, -, *, \%, \texttt{REXP, RLOG}\}.$$

The function set, F_{rpb}, for all other parts of each result-producing branch is

$$F_{rpb} = \{\texttt{L, C, SERIES, PARALLEL0, FLIP, NOP, PAIR\_CONNECT\_0, PAIR\_CONNECT\_1, RETAINING\_THREE\_GROUND\_0, RETAINING\_THREE\_GROUND\_1, ADF0, ADF1, ADF2, ADF3, ADF4}\}.$$

Briefly, the two three-argument THREE_GROUND functions each create a via to ground. See Koza, Bennett, Andre, and Keane 1999a for details.

The function set, F_{adf}, for all other parts of each automatically defined function is

$$F_{adf} = \{\texttt{L, C, SERIES, PARALLEL0, FLIP, NOP, PAIR\_CONNECT\_0, PAIR\_CONNECT\_1, RETAINING\_THREE\_GROUND\_0, RETAINING\_THREE\_GROUND\_1}\}.$$

10.4.1.5 Fitness Measure The free variable f ranges over nine values that are equally spaced (on a logarithmic scale) in the range between 1,000 Hz and 100,000 Hz. Specifically, the nine values of f are 1,000, 1,780, 3,160, 5,620, 10,000, 17,800, 31,600, 56,200, and 100,000 Hz.

Because the high-level statement of the behavior for the desired filter circuit is expressed in terms of frequencies, the output voltage VOUT is measured in the frequency domain.

For each of the nine values of the free variable *f*, SPICE is instructed to perform an AC small signal analysis and report the circuit's behavior over five decades with each decade being divided into 20 parts (using a logarithmic scale). Thus, there are a total of 101 sampled frequencies associated with each value of *f*. The starting frequency for each AC sweep is *f*/1,000 and the ending frequency is 100*f*. The desired lowpass filter has a passband ending at f and a stopband beginning at $2f$. For example, if f is 3,160 Hz (the third of the nine fitness cases), then the AC small signal analysis is performed between 3.16 Hz (three decades below f) and 316,000 Hz (two decades above f) and the desired lowpass filter has a passband ending at 3,160 Hz and a stopband beginning at 6,320 Hz. In the eventual solution to this problem, there should be a sharp drop-off from 1 to 0 Volts in the transitional ("don't care") region between f and $2f$.

The ideal voltage in the passband of the desired lowpass filter is 1 Volt and the ideal voltage in the desired stopband is 0 Volts. A voltage in the desired passband that is between 970 millivolts and 1 Volt (i.e., a passband ripple of 30 millivolts or less) is regarded as acceptable. A voltage in the desired stopband that is between 0 Volts and 1 millivolt (i.e., a stopband ripple of 1 millivolt or less) is regarded as acceptable. Any voltage lower than 970 millivolts in the desired passband and any voltage above 1 millivolt in the desired stopband are unacceptable.

Fitness is the sum, over all nine values of the free variable f and over all 101 frequencies associated with each of the nine values of f of the absolute weighted deviation between the actual value of the voltage that is produced by the circuit at the probe point VOUT and the target value for voltage (0 or 1 volt). A smaller value of fitness is better. Specifically,

$$F(t) = \sum_{k=1}^{9} \sum_{i=0}^{100} (W(d(f_i), f_i) d(f_i)),$$

where fi is the frequency of fitness case i; $d(x)$ is the absolute value of the difference between the target and observed values at frequency x; and $W(y, x)$ is the weighting for difference y at frequency x.

The fitness measure is designed not to penalize ideal voltage values, to penalize slightly every acceptable voltage deviation, and to penalize heavily every unacceptable voltage deviation. Specifically, for each of the points in the intended passband, if the voltage is between 970 millivolts and 1 Volt, the absolute value of the deviation from 1 Volt is weighted by a factor of 1.0. However, if the voltage is less than 970 millivolts, the absolute value of the deviation from 1 Volt is weighted by a factor of 10.0. The acceptable and unacceptable deviations for each of the points in the intended stopband are similarly weighted (by 1.0 or 10.0) based on the amount of deviation from the ideal voltage of 0 Volts and the acceptable deviation of 1 millivolt. For each of the "don't care" points between f and $2f$, the deviation is deemed to be zero.

The number of hits is defined as the number of fitness cases (0 to 909) for which the voltage is acceptable or ideal or that lie in the "don't care" band.

When the aim is to automatically create a solution to a category of problems in the form of a mathematical expression containing free variable(s), the possibility of overfitting is especially salient. Overfitting occurs when an evolved solution performs well on the fitness cases that are incorporated in the fitness measure (i.e., the training phase), but then performs poorly on previously unseen fitness cases.

In this problem, numerous mathematical expressions must be evolved to establish the parameters for components of the to-be-evolved circuit. The values of the components of the to-be-evolved circuit are presumptively nonlinear functions of the free variable, f.

Concerned about possible overfitting, we used nine different values of the free variable, *f*, all in the range between 1,000 Hz and 100,000 Hz. The choice of nine points for this problem was governed primarily by considerations of computer time. Ideally, we would have used even more points. We did not know in advance whether nine data points would prove to be sufficient to enable genetic programming to unearth the complicated relationships inherent in this problem.

10.4.1.6 Control Parameters The population size is 10,000,000.

10.4.2 Results for the Passive Lowpass Filter with a Free Variable for the Passband Boundary

The fitness of the best individual from generation 0 is 957.1.

The first individual to achieve 100 percent compliance with the problem's requirements by scoring 909 hits (out of 909) emerged in generation 78 (figure 10.9). This best-of-run parameterized circuit has a fitness of 0.18450. The circuit-constructing program tree has two result-producing branches (with 146 and 295 points, respectively) and five automatically defined functions (with 4, 3, 5, 3, and 1 point, respectively). Both result-producing branches refer to `ADF0` once. `ADF0` hierarchically refers to `ADF3` and they together evaluate to $1-f$. The other three automatically defined functions are not referenced.

Ignoring the fixed 1,000-Ohm source and fixed 1,000-Ohm load resistors that are hard-wired into the test fixture, the best-of-run parameterized circuit from generation 78 (figure 10.9) has four inductors and five capacitors. Notice that the values of these nine components are not specified by constant numerical values. Instead, each of the nine component values is specified by a mathematical expression containing the free variable *f*. Genetic programming produced this circuit's topology as well as the

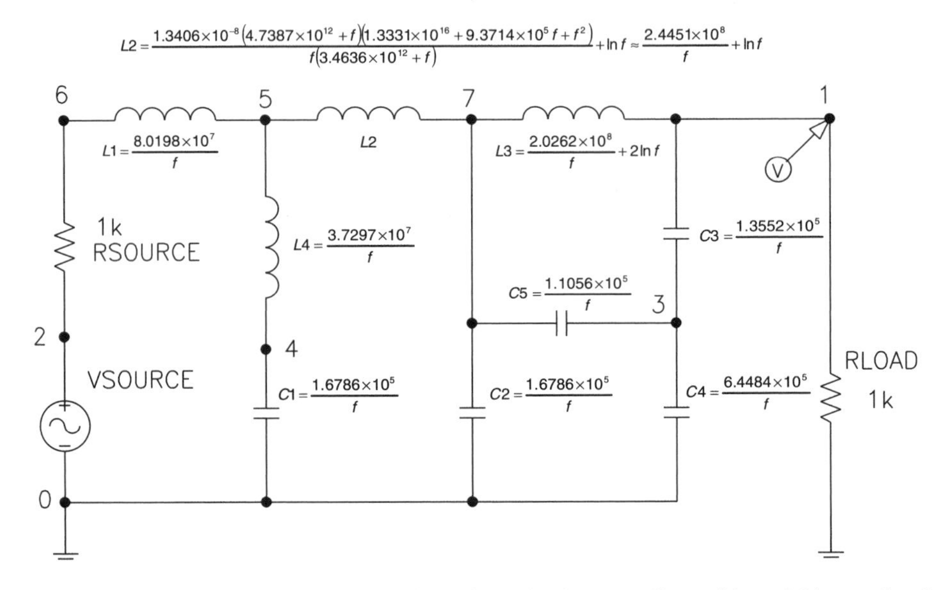

Figure 10.9 Best-of-run parameterized topology for lowpass filter with variable passband boundary.

nine mathematical expressions (each containing the free variable f) for specifying the values of the circuit's four inductors and five capacitors. That is, genetic programming produced a parameterized (i.e., general) solution to this problem.

Specifically, the mathematical expressions for the component values for the parameterized circuit's nine components (L1, L2, L3, L4, C1, C2, C3, C4, and C5) are shown below. Inductances are in microhenrys and capacitances are in nanofarads.

$$L1 = \frac{8.0198 \times 10^7}{f},$$

$$L2 = \frac{1.3406 \times 10^{-8}(4.7387 \times 10^{12} + f)(1.3331 \times 10^{16} + 9.3714 \times 10^5 f + f^2)}{f(3.4636 \times 10^{12} + f)} + \ln f \approx \frac{2.4451 \times 10^8}{f} + \ln f,$$

$$L3 = \frac{2.0262 \times 10^8}{f} + 2\ln f,$$

$$L4 = \frac{3.7297 \times 10^7}{f},$$

$$C1 = \frac{1.6786 \times 10^5}{f},$$

$$C2 = \frac{1.6786 \times 10^5}{f},$$

$$C3 = \frac{1.3552 \times 10^5}{f},$$

$$C4 = \frac{6.4484 \times 10^5}{f},$$

and

$$C5 = \frac{1.1056 \times 10^5}{f}.$$

Notice that genetic programming automatically created all the following in the single run that produced the best-of-run individual of figure 10.9:

- the circuit's topology, including
  - the total number of components (nine) in addition to those in test fixture,
  - the type of each component (i.e., four inductors and five capacitors),
  - all the connections between the circuit's components and accessible points of the test fixture, and
- the circuit's sizing as expressed by nine mathematical expressions containing the problem's free variable (f) for establishing the sizing of the circuit's components (L1, L2, L3, L4, C1, C2, C3, C4, and C5).

As can be seen, all component values (except for *L2* and *L3*) are inversely proportional to f. In the case of *L2* and *L3*, the component values are approximately inversely proportional to f for f less than 1,000,000 Hz. This is exactly what you would

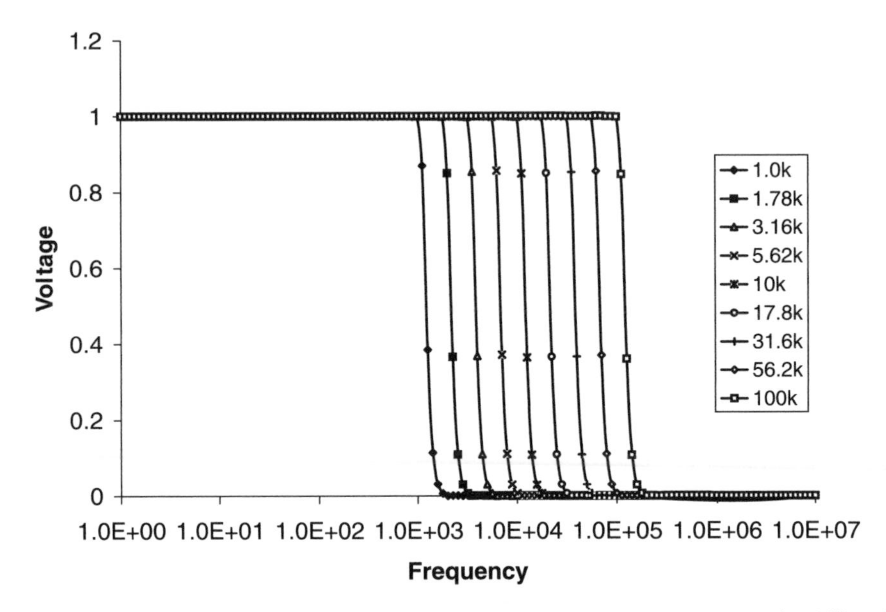

Figure 10.10 Frequency domain behavior of the genetically evolved parameterized filter for nine values of frequency *f*.

expect as a means to make a filter scalable over a wide range of frequencies when the impedance (i.e., the load and source resistance here) is fixed (Van Valkenburg 1982). Thus, the mathematical expressions for each of the component values in the evolved circuit have a reasonable interpretation in electrical engineering terms.

Figure 10.10 shows the behavior, in the frequency domain, of the genetically evolved parameterized filter from generation 78 for each of the nine values of the free variable *f*. The horizontal axis represents the frequency of the incoming signal and ranges logarithmically over the seven decades of frequency between 1 Hz and 10 MHz. The 101 equally spaced, filled circles (many of which overlap in the figure) along each of the nine curves represent the 101 frequencies. The vertical axis represents the peak voltage of the circuit's output signal and ranges linearly between 0 and +1.2 Volts. The amplitudes of the voltages produced by the genetically evolved parameterized filter for frequencies below *f* are all between 970 millivolts and 1 Volt. The amplitudes of the voltages produced by the genetically evolved parameterized filter for frequencies above 2*f* are all between 0 millivolts and 1 millivolt. Thus, the evolved parameterized filter is 100 percent compliant for all sampled frequencies for all nine values of the free variable *f*.

The question arises as to how well the genetically evolved parameterized filter from generation 78 generalizes to previously unseen values of the frequency *f*. The evolutionary process is driven by fitness as measured using nine particular in-sample values of frequency *f*. The possibility exists that the circuit that is evolved with a small number of values of the frequency *f* (i.e., the *in-sample* or *training* cases) will be overly specialized to those particular values and will prove to be inapplicable to other unseen values of *f*. This concern can be addressed by cross-validating the results on

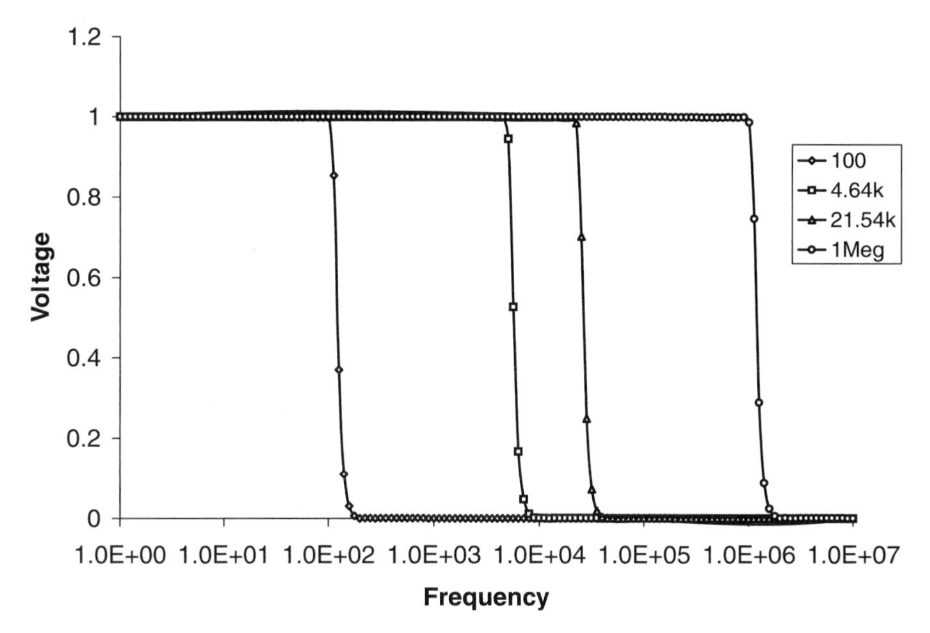

Figure 10.11 Frequency domain behavior of the genetically evolved parameterized filter for four out-of-sample values of frequency *f*.

additional unseen values of *f*. Figure 10.11 shows the behavior, in the frequency domain, of the genetically evolved parameterized filter from generation 78 for four out-of-sample values of frequency *f*. Two of the values of *f* are from inside the range between 1,000 Hz and 100,000 Hz. They are 4,640 Hz (the antilog of 3 2/3) and 21,540 Hz (the antilog of 4 1/3). Two others are from outside the range, namely 100 Hz and 1,000,000 Hz (each a full decade outside the original range of frequencies). As can be seen, the genetically evolved parameterized filter from generation 78 is 100 percent compliant for these four out-of-sample values of frequency *f*.

In summary, this section demonstrates that genetic programming can automatically create the design for both the topology and component values for an analog electrical circuit in which the value of each component in the evolved circuit is specified by a mathematical expression containing a free variable. That is, genetic programming evolved a circuit that is parameterized (i.e., general) in the sense that it represents the solution to all instances of a problem (instead of just the solution to a single instance of the problem). The mathematical expressions for each of the component values in the evolved circuit have a reasonable interpretation in electrical engineering terms. The evolved circuit was cross-validated on unseen values of the free variable.

For additional details, see Koza, Keane, Yu, and Mydlowec 2000.

10.4.3 Routineness for the Passive Lowpass Filter with a Free Variable for the Passband Boundary

For substantially the same reasons stated in section 10.2.3, the result produced in this section is routine.

10.4.4 AI Ratio for the Passive Lowpass Filter with a Free Variable for the Passband Boundary

For substantially the same reasons stated in section 10.2.4, the AI ratio for the solution produced by genetic programming to the problem involving the lowpass third-order elliptic filter with a free variable is moderately high.

10.5 Active Lowpass Filter with a Free Variable for the Passband Boundary

The problem in this section is to evolve a one-input parameterized lowpass filter composed of transistors, capacitors, and resistors whose passband boundary is specified by a free variable.

The specifications for the lowpass filter desired in this section originate from U.S. patent 6,225,859—one of the six post-2000 inventions discussed in detail in chapter 15. The patented circuit (Irvine and Kolb 2001) in chapter 15 is (unlike the circuit being sought in this section) a dynamically tunable integrated active filter. The patented circuit has the behavior of the RLC circuit shown in figure 10.12 (hereafter called the "target circuit"). The target circuit consists of a series composition of a capacitor C11, inductor L11, and two resistors (R11 and R12).

The values of the four components (R11, R12, L11, and C11) of the target circuit (figure 10.12) are functions of the passband boundary frequency, *f*. The values of the four components are specified in terms of two intermediate variables, R_{source} and I_c, as shown in the following equations:

$$C_{11} = 100 \text{ nF},$$

$$R_{source} = 50{,}000\ \Omega,$$

$$R_{12} = \frac{R_{source}}{100},$$

$$L_{11} = \frac{1}{(2\pi f)^2 * C_{11}},$$

$$I_c = \frac{C_{11} * R_{source} * (26 \text{ mV})}{L_{11}},$$

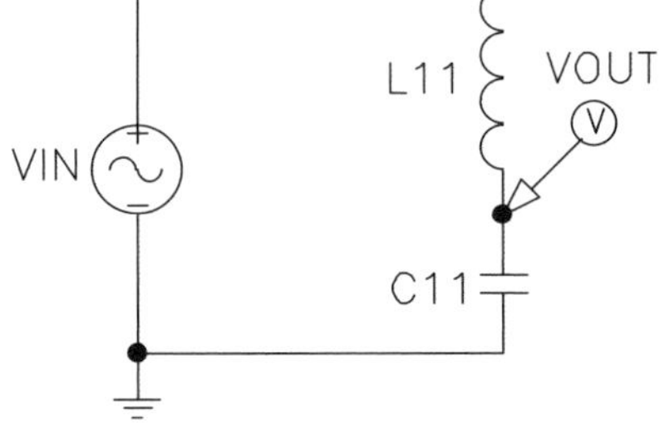

Figure 10.12 Target circuit for U.S. patent 6,225,859.

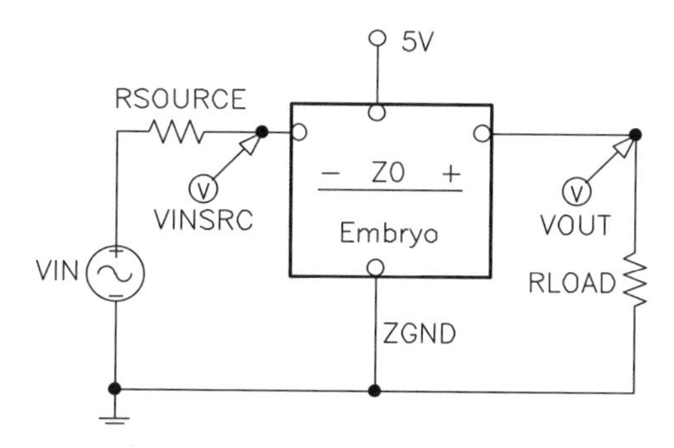

Figure 10.13 One-input, one-output initial circuit with one floating modifiable wire.

and

$$R_{11} = \frac{26\,\mathrm{mV}}{I_c}.$$

Note that the companion problem in chapter 15 involves a two-input tunable integrated active filter circuit in which the passband boundary varies dynamically in response to the control signal, I_c, that varies during the circuit's operational lifetime. However, in this section, we are not talking about a filter that has one passband boundary at some moments during the filter's operational lifetime while having a different passband boundary at other times. Instead, we are talking about a circuit-constructing program tree (genome) that, at the single moment when the program tree is executed to create a fully developed circuit, yields a lowpass filter with a particular passband boundary. The fully developed circuit then operates with that particular passband boundary for its entire operational lifetime.

10.5.1 Preparatory Steps for the Active Lowpass Filter with a Free Variable for the Passband Boundary

10.5.1.1 Initial Circuit The initial circuit (figure 10.13) has a floating embryo (section 4.7.1.1). The floating embryo consists of a single modifiable wire that is not initially connected to the circuit's input(s) or output(s). Thus, each individual circuit must master the problem of discovering the circuit's input(s) and output(s) on its own. The initial circuit also has an incoming signal source VIN, a 1-micro-Ohm source resistor RSOURCE, a +5-Volt power source, a voltage probe point VOUT, and a 1,000-Ohm load resistor RLOAD.

10.5.1.2 Program Architecture Because there must be one result-producing branch in the program tree for each modifiable wire in the embryo and there is one modifiable wire in the embryo, the architecture of each circuit-constructing program tree has one result-producing branch. Automatically defined functions and the architecture-altering operations are not used.

10.5.1.3 Function Set A constrained syntactic structure enforces the use of one function set for the arithmetic-performing subtrees, another for the first argument of the `TWO_LEAD` function, and yet another for all other parts of the program tree.

The function set, F_{aps}, for the arithmetic-performing subtrees is

$$F_{aps} = \{+, -, *, \%\}.$$

Because resistors, inductors, and capacitors may be used in this problem, the function set, $F_{two\text{-}lead}$, for the first argument of the TWO_LEAD function is

$$F_{two\text{-}lead} = \{\text{R\_NEW, C\_NEW}\}.$$

The function set, F_{ccs}, for each construction-continuing subtree is

$$F_{ccs} = \{\text{Q, TWO\_LEAD, SERIES, PARALLEL\_NEW, NODE, TWO\_GROUND, THREE\_GROUND, TWO\_POS5V, INPUT\_0, OUTPUT\_0}\}.$$

Note that the power supply appears because this problem involves transistors.

10.5.1.4 Terminal Set In this problem, the free variable is *log f* (called "LOG_F" below), where *f* is the passband boundary.

There are six types of terminals for this problem. A constrained syntactic structure specifies the terminals that may appear in arithmetic-performing subtrees; the first argument of the PARALLEL_NEW function; the first, second, and third arguments of the transistor-creating Q function; and each construction-continuing subtree.

The component value for each component possessing a parameter is established by an arithmetic-performing subtree that may contain perturbable numerical values, arithmetic operations, and the free variable. This approach for establishing numerical parameter values is described in section 3.5.5.3.

The terminal set, T_{aps}, for the arithmetic-performing subtrees of component-creating functions is

$$T_{aps} = \{\Re_p, \text{LOG\_F}\},$$

where $\Re_p$ denotes a perturbable numerical value and LOG_F is the common logarithm of the passband boundary *f*.

The terminal set, $T_{parallel}$, for the first argument of the PARALLEL_NEW function is

$$T_{parallel} = \{\text{UP\_OR\_LEFT, DOWN\_OR\_RIGHT}\}.$$

The first argument of the Q function specifies the transistor model used. We use the commercially popular 2N3904 (*npn*) and 2N3906 (*pnp*) transistors for this problem. That is, the terminal set, T_{model}, is

$$T_{model} = \{\text{2N3904, 2N3906}\}.$$

The second argument of the transistor-creating Q function establishes which end (polarity) of the preexisting modifiable wire will be bifurcated (if necessary) in inserting the transistor. That is, the terminal set, $T_{bifurcate}$, is

$$T_{bifurcate} = \{\text{BIFURCATE\_POSITIVE, BIFURCATE\_NEGATIVE}\}.$$

The third argument of the Q function specifies which of six possible permutations of the transistor's three leads (base, collector, and emitter) is to be used. That is, the terminal set, $T_{permutation}$, is

$$T_{permutation} = \{\text{B\_C\_E, B\_E\_C, C\_B\_E, C\_E\_B, E\_B\_C, E\_C\_B}\}.$$

The terminal set, T_{ccs}, for each construction-continuing subtree is

$$T_{ccs} = \{\texttt{END, SAFE\_CUT}\}.$$

10.5.1.5 Fitness Measure The free variable f ranges over the following nine values: 441, 588, 784, 1,046, 1,395, 1,861, 2,482, 3,310, and 4,414 Hz. Ic acts as a control signal in the circuit described in U.S. patent 6,225,859 (Irvine and Kolb 2001). The starting value of 441 Hz corresponds to a value of 100 microamperes of current for Ic. The ending value of 4,414 Hz corresponds to a value of 10 milliamperes of current for Ic. This is a reasonable range of values over which the circuit described in U.S. patent 6,225,859 operates. (Chapter 15 describes another run related to this same patent).

Because the high-level statement of the behavior for the desired filter circuit is expressed in terms of frequencies, the output voltage is measured in the frequency domain.

For each of the nine values of the free variable *f*, SPICE is instructed to perform an AC small signal analysis and report the circuit's behavior from *f*/100 to 10*f* with each of these three decades being divided into 20 parts (using a logarithmic scale), so there are a total of 61 sampled frequencies associated with each value of *f*. Note that the 41st point is *f*.

Fitness is the sum, over the 61 frequencies for each of the nine values of the free variable *f*, of the absolute weighted deviation between the voltage (in decibels) that is produced by the circuit at the probe point VOUT and the voltage (in decibels) produced by the target circuit in figure 10.12. For the 31st through the 40th (of the 61) points, the weighting is 100.0 if the voltage (in decibels) that is produced by the individual circuit at the probe point VOUT is within 3 dB of the voltage (in decibels) produced by the target circuit, but 10.0 otherwise. For all other points, the weighting is 10.0 if the voltage (in decibels) that is produced by the individual circuit at the probe point VOUT is within 3 dB of the voltage (in decibels) produced by the target circuit, but 1.0 otherwise.

The number of hits is defined as the number of points for which the voltage (in decibels) that is produced by the individual circuit at the probe point VOUT is within 3 dB of the voltage (in decibels) produced by the target circuit.

10.5.1.6 Control Parameters The population size is 5,000,000.

10.5.2 Results for the Active Lowpass Filter with a Free Variable for the Passband Boundary

The best-of-run individual (figure 10.14) emerged at generation 100.

The best-of-run parameterized circuit (figure 10.14) has 10 transistors. It has six capacitors and two resistors whose values are specified by the expressions below involving the free variable, *f*. The units for capacitors are nanofarads. The units for resistors are kilo-Ohms. *NLM* is the nonlinear mapping described in section 3.5.5.

$$R_1 = f$$

$$R_2 = NLM\ (2 \log f)$$

$$C_1 = NLM\ ((\log f)^2 + \log f)$$

$$C_2 = NLM\ ((\log f)^2 + \log f)$$

$$C_3 = NLM\ (\log f - (\log f)^2 + 9.6707)$$
$$C_4 = NLM\ (2 \log f)$$
$$C_5 = NLM\ (5(\log f)^2 - 14.509)$$
$$C_6 = NLM\ (3.6900 + (\log f)^2 - \log f)$$

Notice that genetic programming automatically created all the following in the single run that produced the parameterized circuit of figure 10.14:

- the circuit's topology, including
  - the total number of components (18) in addition to those of the test fixture,
  - the type of each component (i.e., 10 transistors, six capacitors, and two resistors),
  - all the connections between the circuit's components, accessible points of the test fixture, and the power source, and
- the circuit's sizing as expressed by eight mathematical expressions containing the problem's free variable (*f*) for establishing the sizing of the circuit's components (R1, R2, C1, C2, C3, C4, C5, and C6).

That is, genetic programming produced a parameterized (i.e., general) solution to the problem.

Figure 10.15 shows the frequency response of the best-of-run circuit for the nine in-sample values of frequency. Averaged over the nine in-sample frequencies, the best-of-run circuit has 0.316 dB average absolute error.

Figure 10.16 shows the frequency response of the target circuit (depicted in figure 10.12) for the nine in-sample values of frequency.

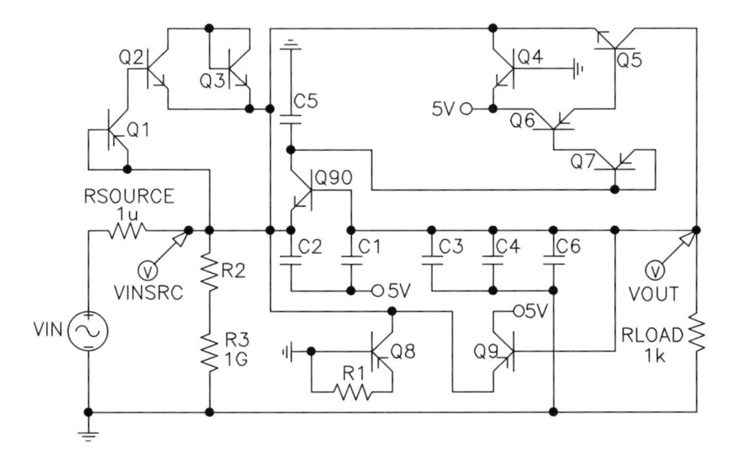

Figure 10.14 Best-of-run parameterized topology for the active lowpass filter with a free variable for the passband boundary.

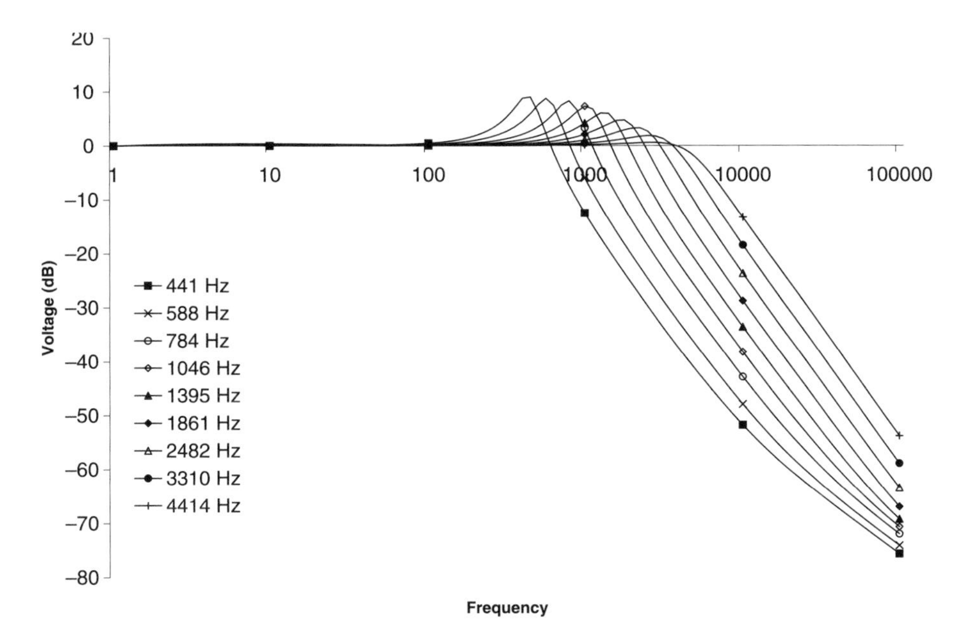

Figure 10.15 In-sample frequency response for the best-of-run circuit for the active lowpass filter with a free variable for the passband boundary.

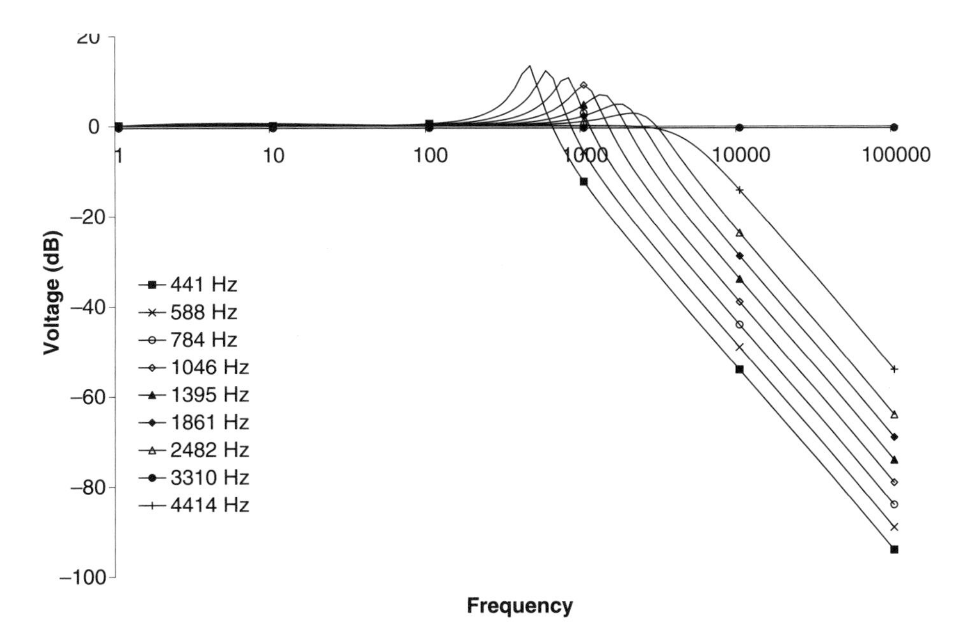

Figure 10.16 In-sample frequency response for the target circuit of figure 10.12.

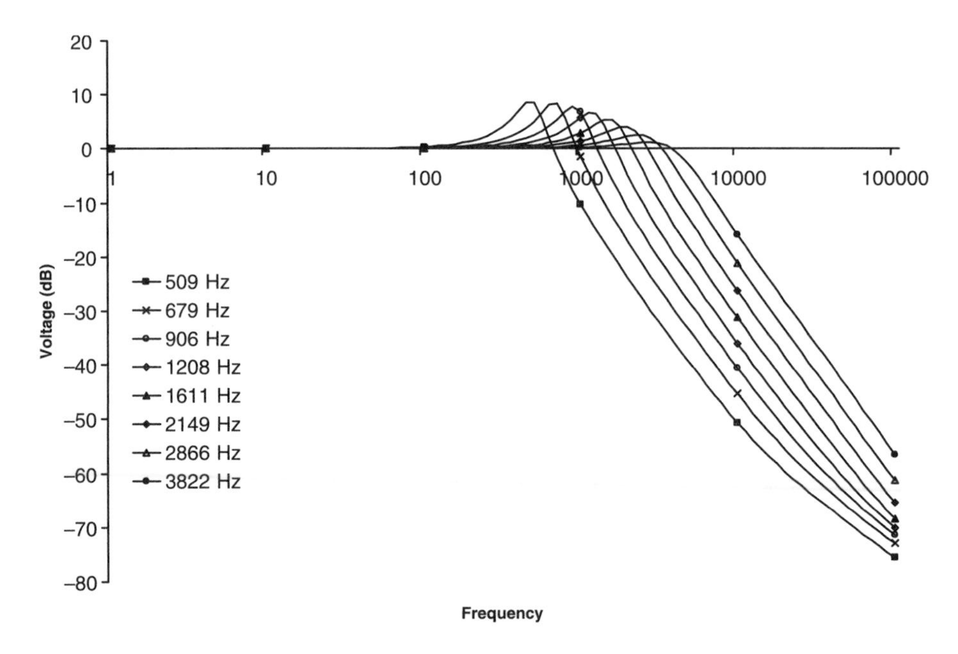

Figure 10.17 Out-of-sample frequency response for the best-of-run circuit for the active lowpass filter with a free variable for the passband boundary.

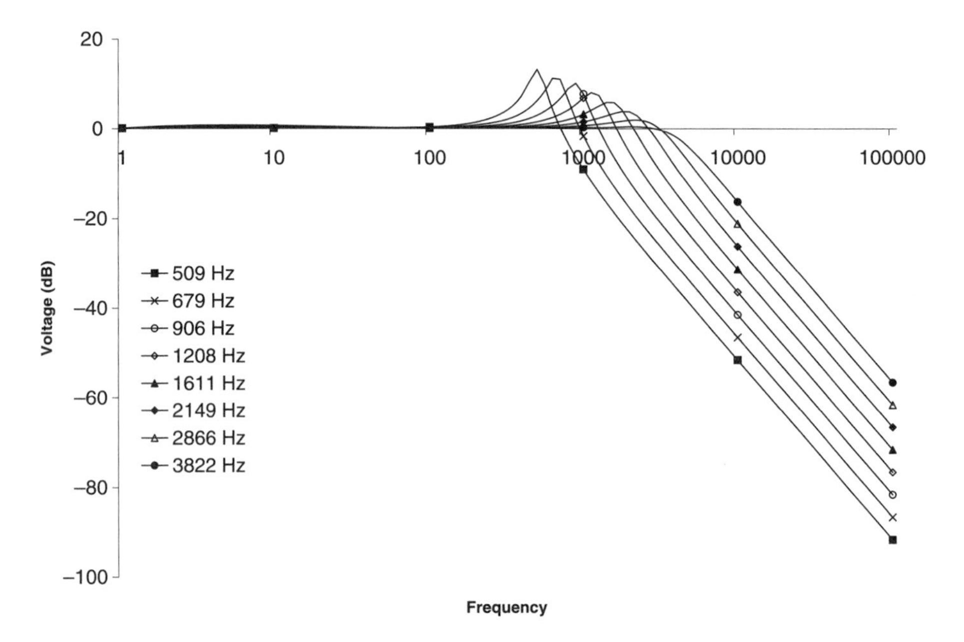

Figure 10.18 Out-of-sample frequency response for target circuit for the active lowpass filter with a free variable for the passband boundary.

Eight out-of-sample frequencies were used to cross-validate the best-of-run circuit. These frequencies (509, 679, 906, 1,208, 1,611, 2,149, 2,866, and 3,822 Hz) are spaced halfway (on a logarithmic scale) between each consecutive pair of the original nine in-sample frequencies.

Figure 10.17 shows the frequency response for the best-of-run circuit from generation 100 for the eight out-of-sample frequencies. Averaged over the eight out-of-sample frequencies, the best-of-run circuit has 0.42 dB average absolute error.

Figure 10.18 shows the frequency response for the target circuit (figure 10.12) for the eight out-of-sample frequencies.

10.5.3 Routineness for the Active Lowpass Filter with a Free Variable for the Passband Boundary

For substantially the same reasons stated in section 10.2.3, the result produced in this section is routine.

10.5.4 AI Ratio for the Active Lowpass Filter with a Free Variable for the Passband Boundary

For substantially the same reasons stated in section 10.2.4, the AI ratio for the solution produced by genetic programming to the problem involving the active lowpass filter with a free variable is moderately high.

11

Automatic Synthesis of Parameterized Topologies with Conditional Developmental Operators for Circuits

Most computer programs contain conditional operators. Conditional operators enable a single program to execute alternative sequences of steps based on different circumstances. The particular sequence that is executed is usually dictated (directly or indirectly) by the program's inputs (i.e., instantiations of values for the program's free variables).

When conditional operators appear in computer programs that have free variables as inputs and that produce a complex structure during a developmental process, a single program may yield topologically and functionally different fully developed structures. In this connection, the reader is referred to early work by Spector and Stoffel (1996a, 1996b) on ontogenetic programming.

This chapter demonstrates that circuit-constructing programs that are evolved using genetic programming may contain conditional developmental operators that enable a single genetically evolved program to be expressed as different circuits. Specifically, genetic programming can be used to create circuit-constructing programs that contain conditional developmental operators that create topologically and functionally different fully developed circuits depending on the program's inputs (free variables) at the moment when the program is executed. Thus, conditional operators enable genetic programming to create a different graph (topological structure) in response to different instantiations of values for the free variables.

This chapter demonstrates the automatic synthesis of parameterized topologies with conditional developmental operators by means of illustrative problems in which the circuit-constructing program trees yield

- either a lowpass or highpass filter (section 11.1),
- either a lowpass filter with a variable passband boundary or a highpass filter with a variable passband boundary (section 11.2),
- either a quadratic or cubic computational circuit (section 11.3), and
- either a 40 dB or 60 dB amplifier (section 11.4).

Notice that, in this chapter, we are not talking about a circuit that acts as a lowpass filter, quadratic computational circuit, or 40 dB amplifier at some moments during its operational lifetime (say, in response to a varying incoming control signal), while acting as a highpass filter, cubic computational circuit, or 60 dB amplifier, respectively, at other times. Instead, we are talking about a circuit-constructing program tree (genome) that, at the single moment that the program is executed to create the fully developed circuit, yields a circuit that operates throughout its entire lifetime as either a lowpass filter or a highpass filter, either a quadratic or cubic computational circuit, or either a 40 dB amplifier or a 60 dB amplifier. Using a biological analogy, we are not talking about a single cell that acts like a liver cell at certain moments during its lifetime and acts like an eye cell at other times. Instead, we are talking about the development of either a liver cell or an eye cell such that the fully developed cell operates as either a liver cell or an eye cell throughout its entire lifetime.

11.1 Lowpass/Highpass Filter Circuit

This section describes the automatic creation of a circuit-constructing program tree with free variables and conditional developmental operators that yields two functionally different circuits (a lowpass filter and highpass filter) depending on the particular values of two free variables.

Specifically, the inputs to the program tree are the boundary of the passband (`F1`) and the boundary of the stopband (`F2`). In this section, there are only two possible combinations of values for the two free variables.

- If the boundary of the passband (`F1`) is 10,000 Hz and the boundary of the stopband (`F2`) is 20,000 Hz, a lowpass filter is desired.
- If the boundary of the passband (`F1`) is 20,000 Hz and the boundary of the stopband (`F2`) is 10,000 Hz, a highpass filter is desired.

The three-argument `IFGTZ_DEVELOPMENTAL` operator enables different parts of the LISP S-expression (genome) to be expressed (executed) depending on the outcome of a numerical test. Specifically, if the first (numerical) argument of the `IFGTZ_DEVELOPMENTAL` operator is greater than zero, the operator executes its second (developmental) argument (but not its third argument); otherwise, this operator executes its third (developmental) argument (but not its second argument).

In this section, both the desired lowpass filter and the desired highpass filter are to have a stopband attenuation of 60 decibels (1,000-to-1).

11.1.1 Preparatory Steps for the Lowpass/Highpass Filter

11.1.1.1 Initial Circuit The initial circuit for this problem has one input and one output and is shown in figure 25.6 of *Genetic Programming III* (Koza, Bennett, Andre, and Keane 1999a).

The embryo consists of two modifiable wires.

The test fixture has an incoming signal source VSOURCE, a nonmodifiable 1,000-Ohm source resistor, a voltage probe point VOUT, a nonmodifiable 1,000-Ohm load resistor, and other nonmodifiable wires. The test fixture has three ports. One port makes the input signal VSOURCE available to the developing circuit. A second port provides access to ground. A third port is connected to the voltage probe point VOUT.

11.1.1.2 Program Architecture The program architecture is the same as in section 10.4.

11.1.1.3 Terminal Set A constrained syntactic structure enforces the use of one terminal set for the value-setting subtrees, another for all other parts of the result-producing branch, and yet another for all other parts of the automatically defined functions.

In this problem, there are two free variables (F1 and F2) representing the boundary of the passband and stopband, respectively.

The numerical parameter value for each electrical component possessing a parameter is established by an arithmetic-performing subtree that may contain perturbable numerical values, arithmetic operations, and the two free variables (F1 and F2). This approach for establishing numerical parameter values is described in section 3.5.5.3.

Arithmetic-performing subtrees may appear in both result-producing branches and any automatically defined functions that may be created during the run by the architecture-altering operations.

The terminal set, T_{aps}, for the arithmetic-performing subtrees is

$$T_{aps} = \{\Re, \texttt{F1}, \texttt{F2}\},$$

where $\Re$ denotes perturbable numerical values between -5.0 and $+5.0$.

The terminal set, T_{rpb}, for all other parts of each result-producing branch is

$$T_{rpb} = \{\texttt{END}, \texttt{SAFE\_CUT}\}.$$

Similarly, the terminal set, T_{adf}, for all other parts of each automatically defined function is

$$T_{adf} = \{\texttt{END}, \texttt{SAFE\_CUT}\}.$$

11.1.1.4 Function Set A constrained syntactic structure enforces the use of one function set for the arithmetic-performing subtrees and another function set for all other parts of the program tree.

The function set, F_{aps}, for the arithmetic-performing subtrees is

$$F_{aps} = \{+, -, *, \%, \texttt{REXP}, \texttt{RLOG}\}.$$

The function set, F_{ccs}, for each construction-continuing subtree is

$$F_{ccs} = \{\texttt{IFGTZ\_DEVELOPMENTAL}, \texttt{L}, \texttt{C}, \texttt{SERIES}, \texttt{PARALLEL0}, \texttt{FLIP}, \texttt{NOP}, \texttt{PAIR\_CONNECT\_0}, \texttt{PAIR\_CONNECT\_1}, \texttt{THREE\_GROUND\_0}, \texttt{THREE\_GROUND\_1}, \texttt{ADF0}, \texttt{ADF1}, \texttt{ADF2}, \texttt{ADF3}, \texttt{ADF4}\}.$$

11.1.1.5 Fitness Measure Because the high-level statement of the behavior of the desired circuit is expressed in terms of frequencies, the output voltage VOUT is measured in the frequency domain.

The two free variables, F1 and F2, may assume the values (10,000, 20,000) or (20,000, 10,000). For each of these two combinations, SPICE is instructed to perform an AC small signal analysis and report the circuit's behavior over five decades of frequency with each decade being divided into 20 parts (using a logarithmic scale). Thus, there are a total of 101 sampled frequencies for each such combination of values of `F1` and `F2`. The starting frequency for each AC sweep is 1 Hz and the ending frequency is 100,000 Hz. The desired lowpass filter has a passband ending at 10,000 Hz and a stopband beginning at 20,000 Hz. There should be a sharp drop-off from 1 to 0 Volts in its transitional ("don't care") region between 10,000 Hz and 20,000 Hz. The desired highpass filter has a stopband ending at 10,000 Hz and a passband beginning at 20,000 Hz. There should be a sharp rise from 0 to 1 Volt in the transitional ("don't care") region between 10,000 Hz and 20,000 Hz.

The ideal voltage in the desired passband is 1 Volt and the ideal voltage in the desired stopband is 0 Volts. A voltage in the desired passband that is between 970 millivolts and 1 Volt (i.e., a passband ripple of 30 millivolts or less) is regarded as acceptable. A voltage in the desired stopband that is between 0 Volts and 1 millivolt (i.e., a stopband ripple of 1 millivolt or less) is regarded as acceptable. Any voltage lower than 970 millivolts in the desired passband and any voltage above 1 millivolt in the desired stopband are regarded as unacceptable.

Fitness is the sum, over the two combinations of values of the free variables and over all 101 frequencies associated with each combination, of the absolute weighted deviation between the actual value of the voltage that is produced by the circuit at the probe point VOUT and the target value for voltage (0 or 1 Volt). Specifically,

$$F(t) = \sum_{k=1}^{2} \sum_{i=0}^{100} (W(d(f_i), f_i) d(f_i)),$$

where f_j is the frequency of fitness case i; $d(x)$ is the absolute value of the difference between the target and observed values at frequency x; and $W(y,x)$ is the weighting for difference y at frequency x. A smaller value of fitness is better.

The fitness measure is designed to slightly penalize every acceptable voltage deviation and heavily penalize every unacceptable deviation. Specifically, for each of the points in the intended passband, if the voltage is between 970 millivolts and 1 Volt, the absolute value of the deviation from the ideal voltage of 1 Volt is weighted by a factor of 1.0. However, if the voltage is less than 970 millivolts, the absolute value of the deviation from 1 Volt is weighted by a factor of 10.0. The acceptable and unacceptable deviations for each of the points in the intended stopband are similarly weighted (by 1.0 or 10.0) based on the amount of deviation from the ideal voltage of 0 Volts and the acceptable deviation of 1 millivolt. The "don't care" points are ignored.

The number of hits is defined as the number (0 to 202) of fitness cases for which the voltage is acceptable or ideal or that lie in the "don't care" band.

11.1.1.6 Control Parameters The population size is 10,000,000.

11.1.2 Results for the Lowpass/Highpass Filter

Before proceeding, let us be clear about the nature of the results that we are expecting in this section. The outcome of the run of genetic programming is a designated best-of-run circuit-constructing program tree. The inputs to this best-of-run program tree (or, in fact, any other program tree in the population) will consist of the problem's free variables (`F1` and `F2`). Depending on the values of `F1` and `F2`, one of two topologies will be constructed when the circuit-constructing program tree is executed. One topology will be applicable when a lowpass filter is desired (i.e., when the boundary of the passband, F1, is 10,000 Hz and the boundary of the stopband, F2, is 20,000 Hz). The other topology will be applicable when a highpass filter is desired. In addition, the run of genetic programming will yield the sizing of all components in both topologies. Some components will be sized in terms of the problem's free variables whereas the sizing of other components will be constant-valued.

The best individual in generation 0 has a fitness of 214.9 and scores 102 hits (out of 202).

The first glimmerings of the ultimate solution to the problem can be seen as early as generation 1 in one of the result-producing branches of the following pace-setting individual:

```
(FLIP
 (IFGTZ_DEVELOPMENTAL
  (* F1 (- F1 F2))
  (IFGTZ_DEVELOPMENTAL
   (* F2 F2)
   (L F2 END)
   (FLIP END))
  (C (DIVIDE_NUMERIC F2 F2) (PARALLEL0 END END END END))))
```

The first argument of the first IFGTZ_DEVELOPMENTAL operator in this S-expression performs the subtraction F1−F2 and then multiplies the difference by the always-positive quantity F1. If F1−F2 is non-positive (i.e., a lowpass filter is desired), then the component-creating C function is unconditionally executed (with the arithmetic-performing subtree returning a value of 1.0), thereby inserting a shunt capacitor. If the result is positive (i.e., a highpass filter is desired), the second IFGTZ_DEVELOPMENTAL operator is executed. Because the first argument of the second IFGTZ_DEVELOPMENTAL operator squares F2, the FLIP is never executed and the component-creating L function is unconditionally executed (with the arithmetic-performing subtree returning a value of F2), thereby inserting a shunt inductor. Thus, even at this early stage of the run, the evolutionary process has discovered that it is advantageous to insert a capacitor into the developing circuit as a shunt to ground if a lowpass filter is desired, but to instead insert an inductor if a highpass filter is desired. The other branch of the circuit-constructing program tree unconditionally inserts an inductor into the developing circuit in series with the incoming signal VSOURCE. If F1<F2, the result is a poor lowpass filter composed of a single series inductor and a single shunt capacitor. If F1>F2, the result is an extremely poor highpass filter composed of a single series inductor and a single shunt inductor.

The best-of-run individual emerged in generation 47. The circuit developed from this best-of-run individual has a near-zero fitness of 0.0519 and scores 202 hits (i.e., it is 100% compliant with the problem's requirements). The result-producing branches of this best-of-run individual have 230 and 196 points, respectively. There are also two small automatically defined functions that are not referenced.

The best-of-run individual from generation 47 is a parameterized circuit-constructing program that yields different topologies based on the particular values of two free variables, F1 and F2. Moreover, the sizing of the components in the resulting topologically different circuits is not specified entirely by constant numerical values. Instead, many of the component values are parameterized by mathematical expressions containing the free variables.

In presenting the best-of-run circuit, there are two cases to consider, namely

- F1 is 10,000 Hertz and F2 is 20,000 Hz (calling for a lowpass filter), or
- F1 is 20,000 Hz and F2 is 10,000 Hz (calling for a highpass filter).

Figure 11.1 shows the best-of-run circuit from generation 47 that develops when the two inputs to the 426-point program call for a lowpass filter—that is, where the combination of values for the two free variables (F1 and F2) is 10,000 and 20,000, respectively. Note that the values of the components in this figure (and other similar figures in this chapter) are established by mathematical expressions involving the problem's free variables (F1 and F2 here). The circuits in this chapter are presented after instantiating the free variables with their appropriate values.

Figure 11.2 shows the behavior in the frequency domain of the best-of-run circuit from generation 47. The horizontal axis represents the frequency of the incoming

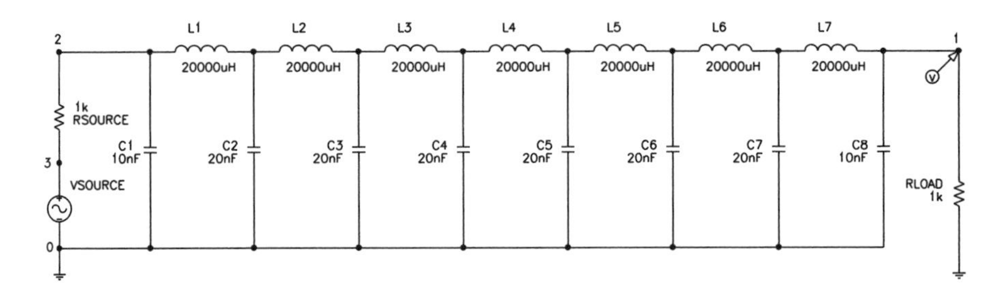

Figure 11.1 Best-of-run circuit when inputs call for a lowpass filter.

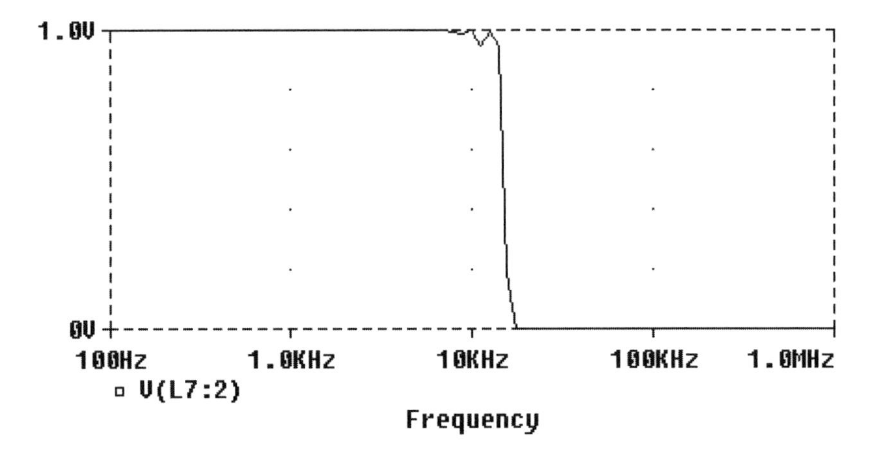

Figure 11.2 Behavior of the best-of-run circuit when inputs call for a lowpass filter.

signal and ranges logarithmically over five decades of frequency between 1 Hz and 100,000 Hz. The vertical axis represents the peak voltage of the circuit's output signal at each particular frequency and ranges linearly between 0 and +1.0 Volts. As can be seen, the circuit is a 100%-compliant lowpass filter in that all voltages are near 1 Volt (i.e., between 970 millivolts and 1 Volt) for frequencies up to 10,000 Hz and all voltages are near 0.0 Volts (i.e., between 0 millivolts and 1 millivolt) for frequencies above 20,000 Hz.

This evolved circuit (figure 11.1) is electrically reasonable. As can be seen, this circuit is the classical ladder design for a lowpass filter (i.e., inductors in series at the top of the figure and capacitors as shunts to ground). This circuit is a cascade of seven identical symmetric *π-sections* (Johnson 1950; Williams and Taylor 1995; Koza, Bennett, Andre, and Keane 1999a). These sections are called *constant K* ladder sections. Each π-section has an equal series inductor L (with inductance of 20,000 microhenrys) forming the horizontal bar along the top of the π. Each π-section also has two equal capacitors ($C/2 =$ 10 nanofarads each) in parallel as the shunts forming the vertical legs of each π. (Two 10-nanofarad capacitors in parallel are equivalent to one 20-nanofarad capacitor). Such π-sections are characterized by a characteristic impedance (resistance) and a nominal cutoff frequency. The section's characteristic impedance should approximately match the circuit's fixed load (1,000 Ohms here). The section's nominal cutoff frequency should lie in the transition region between the end of the passband and the beginning of the stopband. The formula for the characteristic resistance, R, of each π-section is $R = \sqrt{L/C}$. In this formula, C is 20 nanofarads because the half-shunt value, $C/2$, is 10 nanofarads. This formula yields a value of 1,000 Ohms (exactly the value of the load resistor here). The formula for the nominal cutoff frequency, f_c, of each π-section of a lowpass filter is $f_c = 1/(\pi\sqrt{LC})$. This formula yields a nominal cutoff frequency, f_c, of 15,915 Hz (approximately the midpoint of the transition region here). In other words, genetic programming produced both a reasonable topology as well as component values that are close to those that a human would come up with using classical filter design rules.

Figure 11.3 shows the best-of-run circuit from generation 47 that develops when the inputs to the circuit-constructing program tree call for a highpass filter—that is, when the combination of values for the two free variables (`F1` and `F2`) is 10,000 and 20,000, respectively.

Figure 11.4 shows the behavior in the frequency domain of the best-of-run circuit from generation 47. As can be seen, the circuit is a 100%-compliant highpass filter in that all voltages are near 0.0 Volts (i.e., between 0 millivolts and 1 millivolt) for frequencies up to 10,000 Hz and all voltages are near 1 Volt (i.e., between 970 millivolts and 1 Volt) for frequencies above 20,000 Hz.

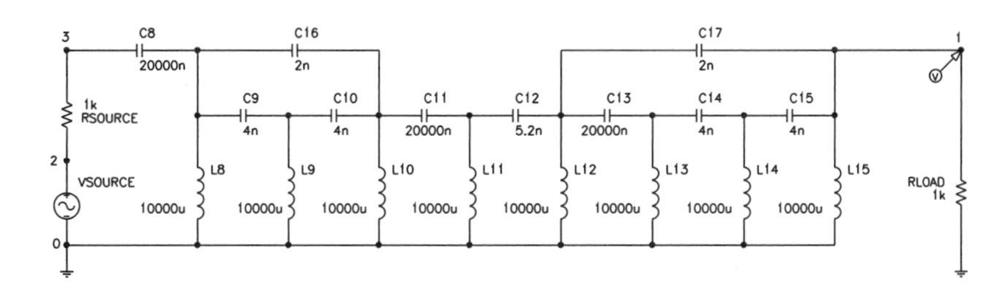

Figure 11.3 Best-of-run circuit when inputs call for a highpass filter.

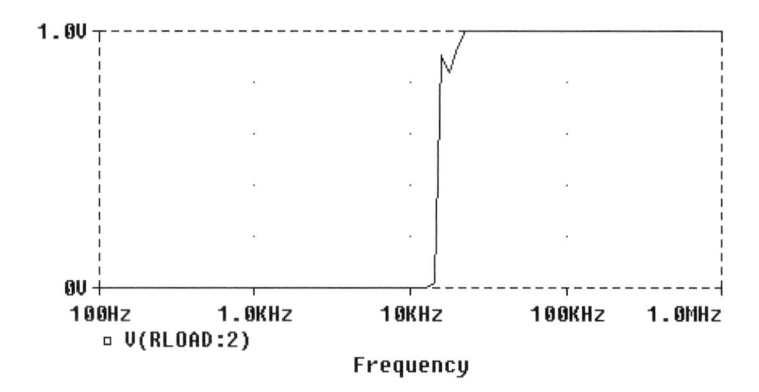

Figure 11.4 Behavior of the best-of-run circuit when inputs call for a highpass filter.

In summary, this section demonstrated that genetic programming can automatically create a two-input computer program that correctly produces either a 100%-compliant lowpass filter or a 100%-compliant highpass filter depending on the values of the inputs.

11.1.3 Routineness of the Transition from a Parameterized Topology Problem without Conditional Developmental Operators to a Problem with Conditional Developmental Operators

The difference between the present lowpass/highpass filter problem and the problems in chapters 9 and 10 is that the present problem uses a conditional developmental operator. This difference is implemented by adding the conditional developmental operator to the function set of the present problem. That is, the transition required to get genetic programming to successfully handle the present problem is virtually effortless and therefore, the transition is routine.

11.1.4 AI Ratio for the Lowpass/Highpass Filter Problem

The solution produced by genetic programming to the problem involving the design of a lowpass/highpass filter with conditional developmental operators is a general solution to a category of problems. This solution has a moderately high amount of "A." Except for the relatively mechanical and minor matter of using a conditional developmental operator, the preparatory steps for the present problem are substantially the same as the preparatory steps for the problems in chapters 9 and 10. The solution produced by genetic programming incorporates a small amount of "I." As a result, the AI ratio for the solution produced by genetic programming to the problem involving a lowpass filter with a conditional developmental operator is moderately high.

11.2 Lowpass/Highpass Filter with Variable Passband Boundary

This section combines some of the features of the problems of section 11.1 (involving a program that can create either a lowpass or highpass filter) and section 10.4 (involving a lowpass filter with variable passband boundary). This section describes the automatic creation of a circuit-constructing program with free variables and conditional developmental operators that yields either a lowpass filter with a particular passband boundary or a highpass filter with a particular passband boundary.

In this section, two functionally and topologically different circuits (lowpass versus highpass filters) are created depending on the particular values of two free variables. The boundaries of the passband and the stopband may vary over the range between 1,000 Hz and 100,000 Hz. The program's inputs are the boundary of the passband (`F1`) and the boundary of the stopband (`F2`). If `F1<F2`, then a lowpass filter is desired, whereas if `F2>F1`, then a highpass filter is desired. Both desired filters are to have stopband attenuation of 60 decibels (1,000-to-1).

11.2.1 Preparatory Steps for the Lowpass/Highpass Filter with Variable Passband Boundary

The preparatory steps for this problem are the same as in section 11.1, except as noted below.

11.2.1.1 Fitness Measure When there are free variables in a problem, it is necessary to ascertain the behavior and characteristics of each candidate individual for a representative sample of values for each of the free variables. Multiple combinations of values of the free variables force generalization of the to-be-evolved circuit.

The boundary of the passband for a lowpass filter (`F1`) may range over nine equally spaced values (on a logarithmic scale) between 1,000 Hz and 100,000 Hz, namely 1,000, 1,780, 3,160, 5,620, 10,000, 17,800, 31,600, 56,200, and 100,000 Hz and the corresponding boundary of the stopband of the lowpass filter (`F2`) will then be 2,000, 3,560, 6,320, 11,240, 20,000, 35,600, 63,200, 112,400, and 200,000 Hz, respectively.

The boundary of the stopband for a highpass filter (`F2`) may range over the values of 1,000, 1,780, 3,160, 5,620, 10,000, 17,800, 31,600, 56,200, and 100,000 Hz and the corresponding boundary of the passband of the highpass filter (`F1`) will then be 2,000, 3,560, 6,320, 11,240, 20,000, 35,600, 63,200, 112,400, and 200,000 Hz, respectively.

For each of the 18 combinations of values of the two free variables, `F1` and `F2`, SPICE is instructed to perform an AC small signal analysis and report the circuit's behavior over five decades with each decade being divided into 20 parts (using a logarithmic scale). Thus, there are a total of 101 sampled frequencies for each of the 18 combinations of values. The starting frequency for each AC sweep is 1,000 Hz and the ending frequency is 100,000,000 Hz.

The desired lowpass filter has a passband ending at `F1` and a stopband beginning at 2`F1`. There should be a sharp drop-off from 1 to 0 Volts in its transitional ("don't care") region between `F1` and 2`F1`. The desired highpass filter has a stopband ending at `F2` and a passband beginning at 2`F2`. There should be a sharp rise from 0 to 1 Volt in its transitional ("don't care") region between `F2` and 2`F2`.

The ideal voltage in the desired passband is 1 Volt and the ideal voltage in the desired stopband is 0 Volts. A voltage in the desired passband that is between 970 millivolts and 1 Volt (i.e., a passband ripple of 30 millivolts or less) is regarded as acceptable. A voltage in the desired stopband that is between 0 Volts and 1 millivolt (i.e., a stopband ripple of 1 millivolt or less) is regarded as acceptable. Any voltage lower than 970 millivolts in the desired passband and any voltage above 1 millivolt in the desired stopband are unacceptable.

Fitness is the sum, over all 18 combinations of values of the free variables, F1 and F2, and over all 101 frequencies associated with each combination, of the absolute weighted deviation between the actual value of the voltage that is produced by the circuit at the probe point VOUT and the target value for voltage (0 or 1 Volt). Specifically,

$$F(t) = \sum_{k=1}^{2} \sum_{j=1}^{9} \sum_{i=0}^{100} (W(d(f_i), f_i) d(f_i)),$$

where *fi* is the frequency of fitness case *i*; *d(x)* is the absolute value of the difference between the target and observed values at frequency *x*; and *W*(*y*,*x*) is the weighting for difference *y* at frequency *x*. A smaller value of fitness is better.

The fitness measure is designed not to penalize ideal voltage values, to penalize slightly every acceptable voltage deviation, and to penalize heavily every unacceptable voltage deviation. Specifically, for each of the points in the intended passband, if the voltage is between 970 millivolts and 1 Volt, the absolute value of the deviation from 1 Volt is weighted by a factor of 1.0. However, if the voltage is less than 970 millivolts, the absolute value of the deviation from 1 Volt is weighted by a factor of 10.0. The acceptable and unacceptable deviations for each of the points in the intended stopband are similarly weighted (by 1.0 or 10.0) based on the amount of deviation from the ideal voltage of 0 Volts and the acceptable deviation of 1 millivolt. For each "don't care" point, the deviation is deemed to be zero.

The number of hits is defined as the number of fitness cases (0 to 1,818) for which the voltage is acceptable or ideal or that lie in the "don't care" band.

11.2.2 Results for the Lowpass/Highpass Filter with a Variable Passband Boundary

The best individual in generation 0 has a fitness of 3,332.3 and scores 803 hits (out of 1,818).

The first of the two result-producing branches from a pace-setting individual from generation 1 (with a fitness of 3,186 and 792 hits) contains the first glimmerings of this problem's ultimate solution.

```
(IFGTZ_DEVELOPMENTAL
 (*(% F1 F1)
   (- F2 F1))
 (C (RLOG F1)
    (PARALLEL0
      (PARALLEL0 END END END END)
      END
      END
      END))
 (IFGTZ_DEVELOPMENTAL
   (RLOG F2)
   (L F2 END)
   (L F1 END)
 )
)
```

This branch examines the difference between F2 and F1 (after superfluously multiplying the difference by the quotient of two identical quantities). If F1<F2 (i.e., a lowpass filter is desired), a capacitor is inserted as a shunt. If F2<F1 (i.e., a highpass filter is desired), an inductor is inserted. This approach is a reasonable ingredient of the ultimate solution.

The best-of-run individual emerged in generation 93. The result-producing branches of this best-of-run individual have 284 and 282 points (i.e., functions and terminals), respectively. There are three automatically defined functions with 1, 92, and 25 points, respectively. The first result-producing branch calls all three automatically defined functions. The second result-producing branch calls ADF0 and ADF2. ADF0 hierarchically calls ADF2.

The circuit that develops from the best-of-run individual from generation 93 has a near-zero fitness of 0.63568, and scores 1,818 hits (i.e., it is 100% compliant with the problem's requirements).

Figure 11.5 shows the circuit that develops when the inputs to the best-of-run circuit-constructing program tree from generation 93 call for a highpass filter (i.e., F1>F2). Ignoring the fixed source and load resistors that are hard-wired into the test fixture, the resulting circuit has six capacitors and six inductors. The values of the 12 components in the resulting circuit are not specified by constant numerical values, but instead, by mathematical expressions containing one or more of the problem's free variables. Genetic programming produced this circuit's overall topology as well as the 12 mathematical expressions for specifying the component values for the circuit's six inductors and six capacitors. As it happens, all the genetically evolved mathematical expressions involve F2 in this figure. That is, genetic programming produced a parameterized (i.e., general) circuit.

As can be seen in figure 11.5, when the inputs to the program tree call for a highpass filter, the best-of-run circuit-constructing program tree from generation 93 develops into the classical ladder design for a highpass filter (i.e., capacitors in series at the top of the figure and inductors as shunts to ground). In addition, the rungs of the ladder have approximately identical sections. This ladder approximates a classical "constant *K*" ladder whose inductors have approximately a value of *L* of 56.2/F2 Henrys

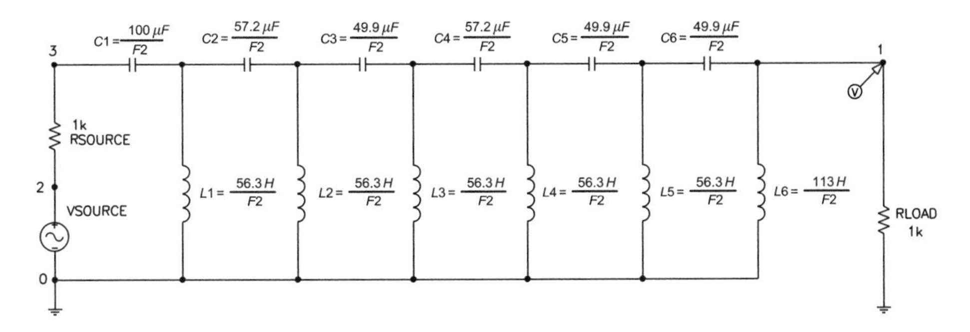

Figure 11.5 Best-of-run circuit when inputs call for a highpass filter (i.e., F1>F2).

and whose capacitors have approximately a value of C of 51/F2 microfarads. The design formulas for "constant K" highpass filters are

$$R = \sqrt{L/C}$$

and

$$f_c = \frac{1}{4\pi\sqrt{LC}}.$$

Inserting these values of L and C into these formulae yields a value for R of 1,050 Ohms (approximately the value of the load resistor) and a value for f_c of 1.5F2 (the midpoint of the transition region). In other words, genetic programming produced both a reasonable topology as well as component values that are close to those that a human would come up with using classical rules of filter design.

Figure 11.6 shows the behavior in the frequency domain of the best-of-run circuit from generation 93 that develops when the inputs to the program tree call for a highpass filter for the nine frequencies. The horizontal axis of the figure represents the frequency of the incoming signal and ranges logarithmically over the four decades of frequency between 100 Hz and 1,000,000 Hz. The vertical axis represents the peak voltage of the circuit's output signal for each frequency and ranges linearly between 0 and +1.2 Volts. As can be seen, the circuit is a 100%-compliant highpass filter for all nine frequencies. In particular, for frequencies below F2, the amplitude of the voltages are all near 0.0 Volts (i.e., between 0 millivolts and 1 millivolt), as required by the design specifications of the desired highpass filter. Also note that, for frequencies above F1, the amplitude of the voltages are all near 1 Volt (i.e., between 970 millivolts and 1 Volt).

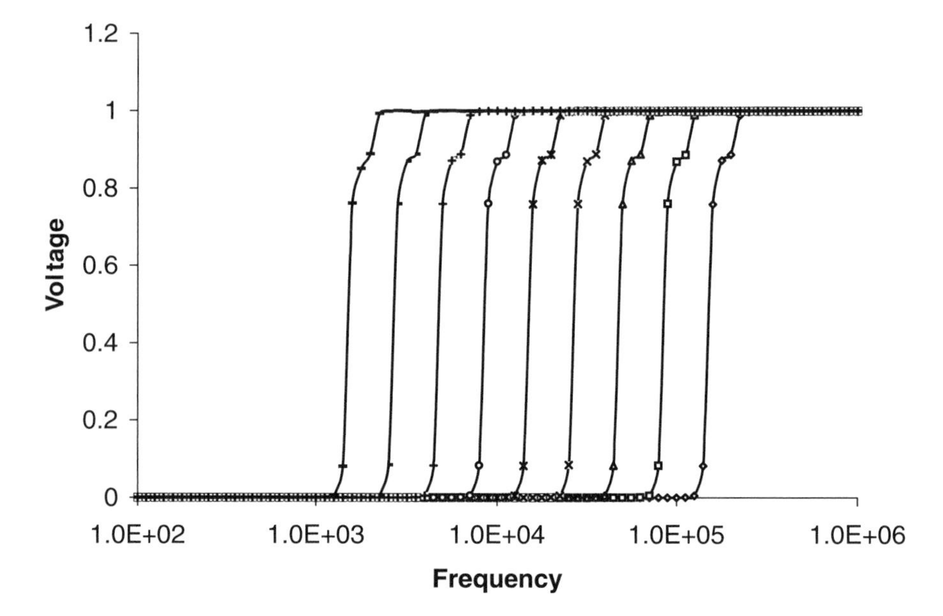

Figure 11.6 Frequency domain behavior of the best-of-run circuit when inputs call for a highpass filter.

Table 11.1 Component values for the best-of-run circuit when inputs call for a highpass filter

Start of transition band (Hertz)	C1 (nF)	C2 and C4 (nF)	C3, C5, and C6 C6 (nF)	L1, L2, L3, L4, and L5 (μH)	L6 (μH)
1,000	100.0	57.2	49.9	56,300	100,000
1,778	56.2	32.1	28.1	31,700	63,300
3,162	31.6	18.1	15.8	17,800	35,600
5,623	17.8	10.2	8.88	10,000	20,000
10,000	10.0	5.72	4.99	5,630	11,300
17,782	5.62	3.21	2.81	3,170	6,330
31,622	3.16	1.81	1.58	1,780	3,560
56,234	1.78	1.02	0.888	1,000	2,000
100,000	1.0	0.572	0.499	563	1,130

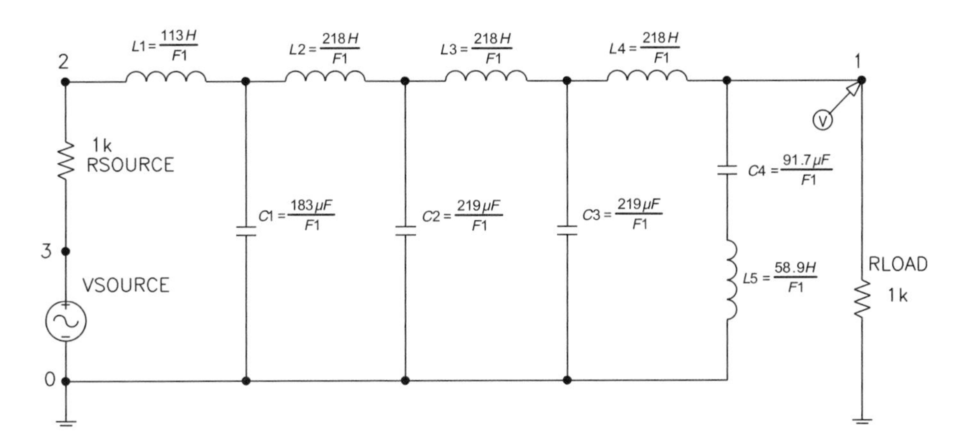

Figure 11.7 Best-of-run circuit when inputs call for a lowpass filter.

Table 11.1 shows the values for the circuit's 12 components (six inductors and six capacitors) when the program's inputs call for a highpass filter. Frequencies are in Hertz; inductances are in microhenrys; and capacitances are in nanofarads. Several of the components have identical values, so the table shows only five distinct values for the highpass filter.

Figure 11.7 shows the circuit that develops when the inputs to the best-of-run circuit-constructing program tree from generation 93 call for a lowpass filter. Notice that, except for the fixed 1,000-Ohm source and 1,000-Ohm load resistors that are hard-wired into the test fixture, the component sizing is not specified by constant numerical values. Instead, each component value is specified by a mathematical expression containing one or more of the problem's free variables. Genetic programming produced this circuit's overall topology as well as the nine mathematical expressions for specifying the component values for the circuit's five inductors and four capacitors. As it happens, all the genetically evolved mathematical expressions involve `F1` in this figure. That is, genetic programming produced a parameterized (i.e., general) circuit.

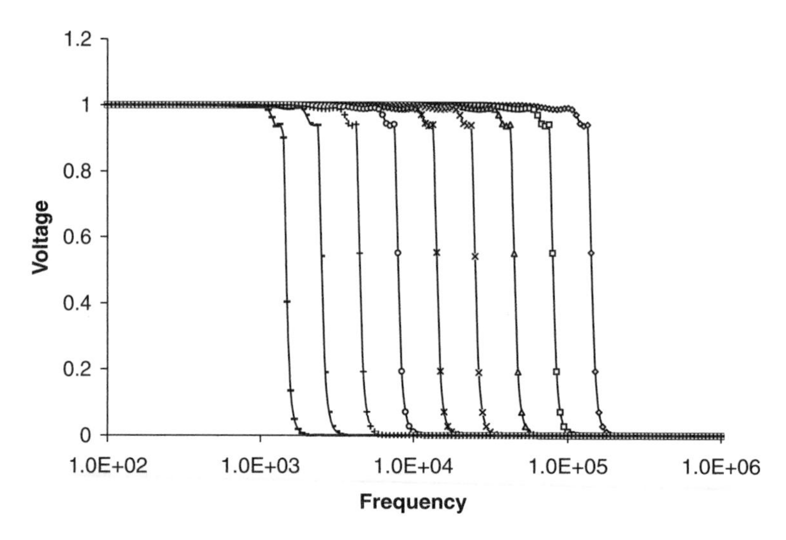

Figure 11.8 Frequency domain behavior of the best-of-run circuit when inputs call for a lowpass filter.

Table 11.2 Component values for the best-of-run circuit when inputs call for a lowpass filter

Start of transition band (Hertz)	L1 (μH)	L2, L3, and L4 (μH)	L5 (μH)	C1 (nF)	C2 and C3 (nF)	C4 (nF)
1,000	100,000	200,000	58,900	183	219	91.7
1,778	63,400	123,000	33,100	103	123	51.6
3,162	35,600	69,000	18,600	58	69.2	29.0
5,623	20,000	38,800	10,500	32.6	38.9	16.3
10,000	11,300	21,800	5,890	18.3	21.9	9.17
17,782	6,340	12,300	3,310	10.3	12.3	5.16
31,622	3,560	6,900	1,860	5.8	6.92	2.90
56,234	2,000	3,880	1,050	3.26	3.89	1.63
100,000	1,130	2,180	589	1.83	2.19	0.917

The circuit in figure 11.7 is (approximately) another "constant *K*" lowpass ladder filter with $L = 220/\texttt{F1}$ henrys and $C = 201/\texttt{F1}$ microfarads. Using the formulae mentioned earlier, the calculated value of R is 1,046 Ohms and the calculated value of f_c is 1.51F1.

Figure 11.8 shows the behavior in the frequency domain of the circuit in figure 11.7 for the nine frequencies used as the start of the transition band. As can be seen, the circuit is a 100%-compliant lowpass filter for each of the nine frequencies.

Table 11.2 shows the values for the circuit's nine components (five inductors and four capacitors) when the program's inputs call for a lowpass filter. Several of the components have identical values, so the table shows only six distinct values for the highpass filter.

Figure 11.9 shows component values in table 11.1 (for the highpass filter) for the five components with distinct values. Both axes use a logarithmic scale.

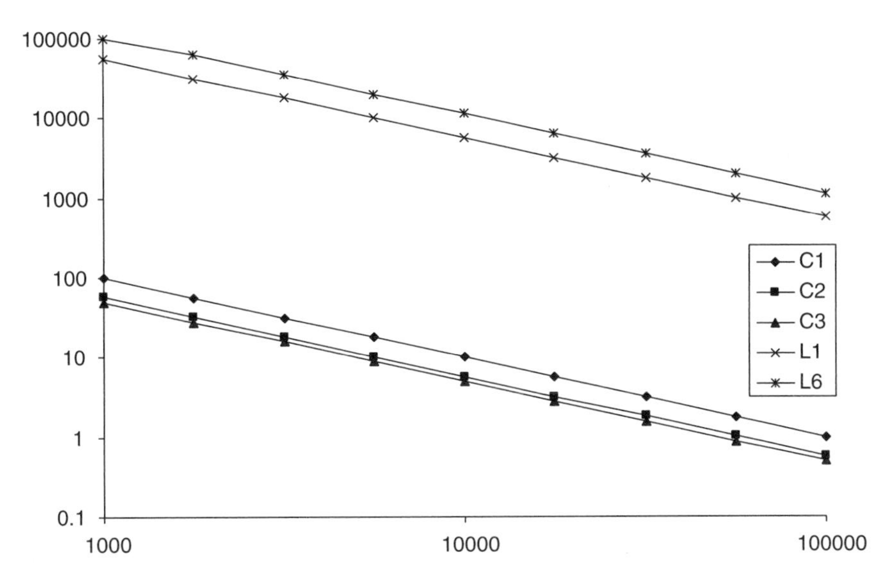

Figure 11.9 Comparison of component values for the highpass filter.

As can be seen from figure 11.9, the component values for all five distinct components for all nine frequencies lie along straight lines. The following formulae (all of which have the frequency F2 in the denominator) represent the relationships between the component values and frequencies for the highpass filter:

$$C1 = \frac{100{,}000 \text{ nF}}{F2} = \frac{100 \text{ µF}}{F2},$$

$$C2{,}C4 = \frac{57{,}200 \text{ nF}}{F2} = \frac{57.2 \text{ µF}}{F2},$$

$$C3{,}C5{,}C6 = \frac{49{,}900 \text{ nF}}{F2} = \frac{49.9 \text{ µF}}{F2},$$

$$L1, L2, L3, L4, L5 = \frac{56{,}300{,}000 \text{ µH}}{F2} = \frac{56.3 \text{ H}}{F2},$$

$$L6 = \frac{113{,}000{,}000 \text{ µH}}{F2} = \frac{113 \text{ H}}{F2}.$$

This inverse proportionality is exactly what you would expect as a means to make a filter scalable over a wide range of frequencies when the impedance (i.e., the load and source resistance here) is fixed (Van Valkenburg 1982).

Note that the above equations yield a value of 113,000 μH for L6 for F2= 1,000 Hz, whereas table 11.1 shows a value of 100,000 μH. This slight discrepancy for L6 is an artifact of our limiting of component values in the final circuit to a prespecified range.

Six plots for the parameterized lowpass filter (not shown here, but similar to those of figure 11.9) reveal that the component values for the six distinct components for all nine frequencies also lie along straight lines. The following formulae (all of which have the frequency F1 in the denominator) represent the relationships between the component values and frequencies in the parameterized lowpass filter:

$$L1 = \frac{113{,}000{,}000\ \mu\text{H}}{F1} = \frac{113\ \text{H}}{F1},$$

$$L2, L3, L4 = \frac{218{,}000{,}000\ \mu\text{H}}{F1} = \frac{218\ \text{H}}{F1},$$

$$L5 = \frac{58{,}900{,}000\ \mu\text{H}}{F1} = \frac{58.9\ \text{H}}{F1},$$

$$C1 = \frac{183{,}000\ \text{nF}}{F1} = \frac{183\ \mu\text{F}}{F1},$$

$$C2, C3 = \frac{219{,}000\ \text{nF}}{F1} = \frac{219\ \mu\text{F}}{F1},$$

$$C4 = \frac{91{,}700\ \text{nF}}{F1} = \frac{91.7\ \mu\text{F}}{F1}.$$

The question arises as to how well the genetically evolved parameterized filter from generation 93 generalizes to previously unseen values of F1 and F2. The evolutionary process is driven by fitness as measured by a particular set of 18 combinations of in-sample values of F1 and F2. The possibility exists that the circuit that is evolved with a small number of in-sample values of these variables will be overly specialized to those particular values and will prove to be inapplicable to other unseen values of those variables.

Figure 11.10 shows the behavior, in the frequency domain, of the genetically evolved parameterized filter from generation 93 when the inputs call for a highpass filter for four out-of-sample values of the frequencies F1 and F2. Two of the values of

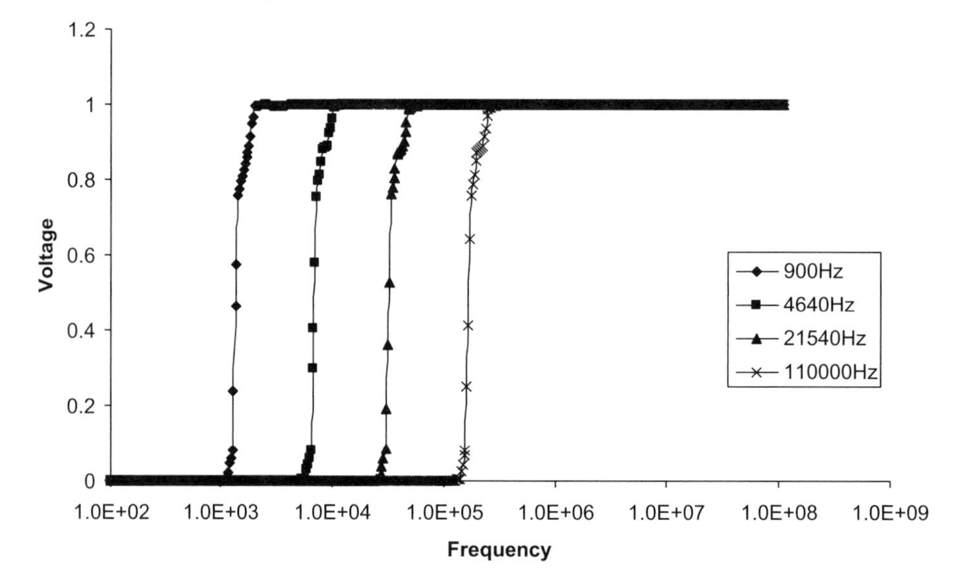

Figure 11.10 Frequency domain behavior of the best-of-run circuit for four out-of-sample values of frequency when inputs call for a highpass filter.

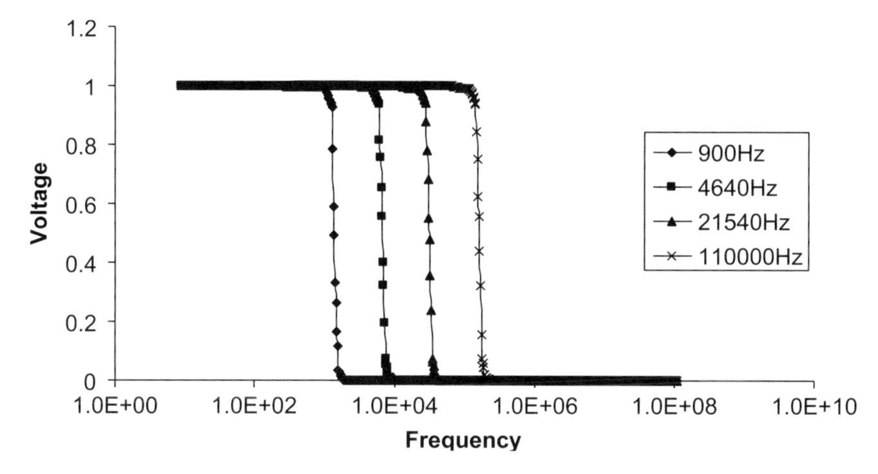

Figure 11.11 Frequency domain behavior of the best-of-run circuit for four out-of-sample values of frequency when inputs call for a lowpass filter.

`F1` are from inside the range between 1,000 Hz and 100,000 Hz. They are 4,640 Hz (the antilog of 3 2/3) and 21,540 Hz (the antilog of 4 1/3). Two others are from outside the range, namely 900 Hz and 110,000 Hz. This best-of-run circuit from generation 93 is 100% compliant for these four out-of-sample frequencies.

Figure 11.11 shows the behavior, in the frequency domain, of the genetically evolved parameterized filter from generation 93 when inputs call for a lowpass filter for the same four out-of-sample values of frequency `F1` and `F2`. Again, the genetically evolved parameterized filter is 100% compliant for these four out-of-sample frequencies.

For additional information, see Koza, Yu, Keane, and Mydlowec 2000b.

This section demonstrated that genetic programming could automatically create a single two-input circuit-constructing program that contains both conditional developmental operators and inputs (free variables) and that produces either a lowpass or highpass filter with specified boundaries for its passband and stopband. The single evolved computer program represents the solution to multiple instances of the problem. The ability of a single evolved program to represent the solution to multiple instances of a problem is made possible by the fact that the evolutionary search was conducted in the space of computer programs containing conditional developmental operators and free variables.

11.2.3 Routineness for the Lowpass/Highpass Filter with a Variable Passband Boundary

For substantially the same reasons stated in section 11.1.3, the transition required to obtain the result produced in this section is routine.

11.2.4 AI Ratio for the Lowpass/Highpass Filter with a Variable Passband Boundary

For substantially the same reasons stated in section 11.1.4, the AI ratio for the solution produced by genetic programming to the problem involving the lowpass third-order elliptic filter with a free variable is moderately high.

11.3 Quadratic/Cubic Computational Circuit

This section describes the automatic creation of a circuit-constructing program with free variables and conditional developmental operators that yields either a quadratic or cubic computational circuit.

Specifically, the free variable TARGET is the input to the program tree. This free variable may assume the value of either 2 (for quadratic) or 3 (for cubic).

11.3.1 Preparatory Steps for the Quadratic/Cubic Computational Circuit

11.3.1.1 Initial Circuit This problem uses a one-input, one-output initial circuit (figure 11.12) with one modifiable wire. The one-input, one-output initial circuit has an incoming signal source VIN, a 1-micro-Ohm source resistor RSOURCE, a −15V power source, a +15-Volt power source, a voltage probe point VOUT, and a 1,000-Ohm load resistor RLOAD.

11.3.1.2 Program Architecture Because there must be one result-producing branch in the program tree for each modifiable wire in the embryo and there is one modifiable wire in the embryo, the architecture of each circuit-constructing program tree has one result-producing branch. Automatically defined functions and the architecture-altering operations are not used.

11.3.1.3 Function Set A constrained syntactic structure enforces the use of one function set for the arithmetic-performing subtrees associated with the IFGTZ_DEVELOPMENTAL function, another for the first argument of the TWO_LEAD function, and yet another for all other parts of the program tree.

The function set, $F_{\text{aps-ifgtz}}$, for the arithmetic-performing subtrees is

$$F_{\text{aps-ifgtz}} = \{+, -, *, \%, \text{RLOG}, \text{EXP}\}.$$

The function set, $F_{\text{two-lead}}$, for the first argument of the TWO_LEAD function is

$$F_{\text{two-lead}} = \{\text{R\_NEW}, \text{C\_NEW}\}.$$

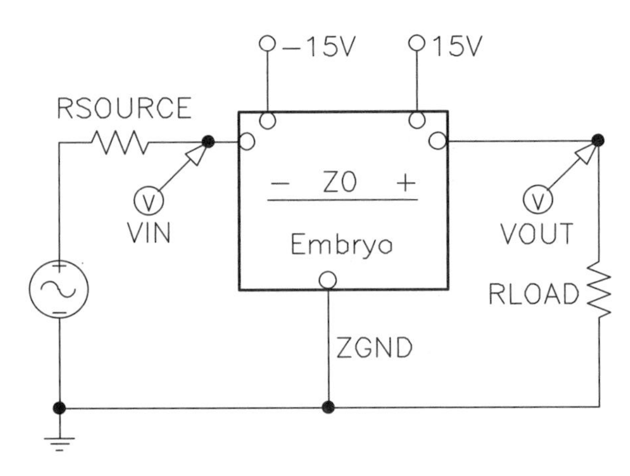

Figure 11.12 One-input, one-output initial circuit with one floating modifiable wire.

The function set, F_{ccs}, for each construction-continuing subtree is

F_{ccs} = {IFGTZ_DEVELOPMENTAL, Q, TWO_LEAD, SERIES, PARALLEL_NEW, TWO_GROUND, THREE_GROUND, NODE, INPUT_0, OUTPUT_0, TWO_POS15V, TWO_NEG15V}.

The two-argument TWO_POS15V ("positive reference voltage source") function enables any part of a circuit to be connected to the constant +15.0-Volt DC power source (e.g., a battery). The TWO_POS1V, TWO_POS2V, TWO_POS5V, and TWO_NEG15V functions operate in a similar way.

11.3.1.4 Terminal Set In this problem, the free variable is TARGET. This free variable may assume the value of either 2 (for quadratic) or 3 (for cubic).

A constrained syntactic structure enforces the use of one terminal set for the arithmetic-performing subtrees associated with the IFGTZ_DEVELOPMENTAL function, another for the value-setting subtree for the component-creating functions, and yet another for all other parts of the program tree.

The terminal set, $T_{vss\text{-}ifgtz}$, for the value-setting subtrees associated with the IFGTZ_DEVELOPMENTAL function is

$$T_{vss\text{-}ifgtz} = \{\Re_p, \text{TARGET}\},$$

where $\Re_p$ denotes a perturbable numerical value.

The terminal set, $T_{vss\text{-}components}$, for the value-setting subtrees for component-creating functions is

$$T_{vss\text{-}components} = \{\Re_p\},$$

where $\Re_p$ denotes a perturbable numerical value.

The remainder of the specification for the terminal set for this problem is the same as that found in section 10.5.1.4.

11.3.1.5 Fitness Measure The purpose of the quadratic/cubic computational circuit is to produce an output voltage (probed at VOUT) that is equal to the square or cube of the input voltage, depending on whether the value of the free variable terminal (TARGET) is 2 or 3, respectively.

Fitness is measured using two sets (one for squaring, one for cubing) of four time-domain fitness cases involving input signals of various shapes and time scales. The four fitness cases for each desired function (i.e., squaring or cubing) are the same except that the target voltage is either the square or cube of the input voltage, respectively.

The first fitness case is a ramp that rises from 0 Volts to 1 Volt over a period of 1 millisecond.

The second fitness case is a single full cycle of a sine wave ranging between 0 Volts and +1 Volt over a period of 1 millisecond.

The third fitness case is a ramp that falls from 1 Volt to 0 Volts over a period of 10 milliseconds.

The fourth fitness case is a constant input of 0.5 Volts over a period of 10 milliseconds.

There are 100 time steps associated with each fitness case.

Fitness is the sum, over the four fitness cases and 101 time values for each fitness case, of the absolute value of the weighted difference between the square or cube (as appropriate) of the input voltage and circuit's actual output. The weight is 1.0 if the absolute value of the difference is less than 1 millivolt or within 1% of the correct value, but 10.0 otherwise.

11.3.1.6 Control Parameters The population size is 1,000,000.

11.3.2 Results for the Quadratic/Cubic Computational Circuit

All occurrences of the `IFGTZ_DEVELOPMENTAL` operator for which the conditional part always evaluates to true or false have been removed from the LISP S-expressions presented in this section.

The LISP S-expression for the best-of-generation individual for generation 0 codes for a circuit that consists of a single 776-Ohm resistor located between the circuit's input and output. It contains no non-degenerate occurrences of the `IFGTZ_DEVELOPMENTAL` operator.

```
(TWO-LEAD (R_NEW -0.1157337) (SERIES (SERIES END SAFE-CUT
END) (SERIES SAFE-CUT END END) (INPUT_0 END END)) (SERIES
(OUTPUT_0 SAFE-CUT SAFE-CUT) (INPUT_0 END SAFE-CUT)
(SERIES SAFE-CUT END SAFE-CUT)))
```

Prior to generation 32, there is no instance of a functioning genetic switch in any pace-setting individual. In this initial part of the run, individuals typically achieve modest levels of fitness by producing a single output (independent of the free variable) that is either near the quadratic or cubic curve or intermediate between the two.

The first effective instance of the use of a genetic switch in a pace-setting individual appears in a pace-setting individual from generation 32. In the S-expression below (and throughout this section), the operative genetic switch is underlined; the code specific to the squaring circuit is in bold; and the code specific to the cubing circuit is in italics.

```
(TWO-LEAD (R_NEW 0.4040401) (SERIES (SERIES (TWO_POS15V
END SAFE-CUT) SAFE-CUT (TWO_NEG15V END SAFE-CUT)) SAFE-CUT
(SERIES SAFE-CUT SAFE-CUT SAFE-CUT)) (INPUT_0 (OUTPUT_0
(Q_NEW Q2N3906 BIFURCATE_POSITIVE E_C_B END SAFE-CUT SAFE-
CUT) END) (INPUT_0 SAFE-CUT (IFGTZ_DEVELOPMENTAL (* (log
(+ (% (log (log TARGET)) TARGET) (+ -0.9049953 TARGET)))
(exp (% TARGET TARGET))) (SERIES END (TWO-LEAD (R_NEW
0.4973120) END (OUTPUT_0 (NODE END END) (SERIES END END
SAFE-CUT))) SAFE-CUT) (TWO-LEAD (R_NEW 0.4040401) END
(OUTPUT_0 (NODE END END) (SERIES END END SAFE-CUT))))))))
```

When a squaring circuit is called for, the S-expression for the pace-setting individual from generation 32 produces a circuit that has one transistor and two resistors and that makes one connection to the positive power supply, one connection to the negative power supply, two connections to the input, and two connections to the output.

When a cubing circuit is called for, the S-expression for the pace-setting individual from generation 32 produces a circuit with the same topology as for the squaring circuit. However, the difference between the code in italics and the code in bold in the above S-expression causes one of the resistors in the squaring circuit to be 2,540 Ohms, but 3,140 Ohms in the cubing circuit.

The best-of-run circuit emerged in generation 241. When the genetic switch calls for a squaring circuit, this best-of-run individual yields a circuit with an average absolute error of 4.68 millivolts. When the genetic switch calls for a cubing circuit, this best-of-run individual yields a circuit with an average absolute error of 5.27 millivolts.

Figure 11.13 shows the best-of-run circuit from generation 241 when the genetic switch calls for a squaring circuit.

Figure 11.14 shows the best-of-run circuit from generation 241 when the genetic switch calls for a cubing circuit.

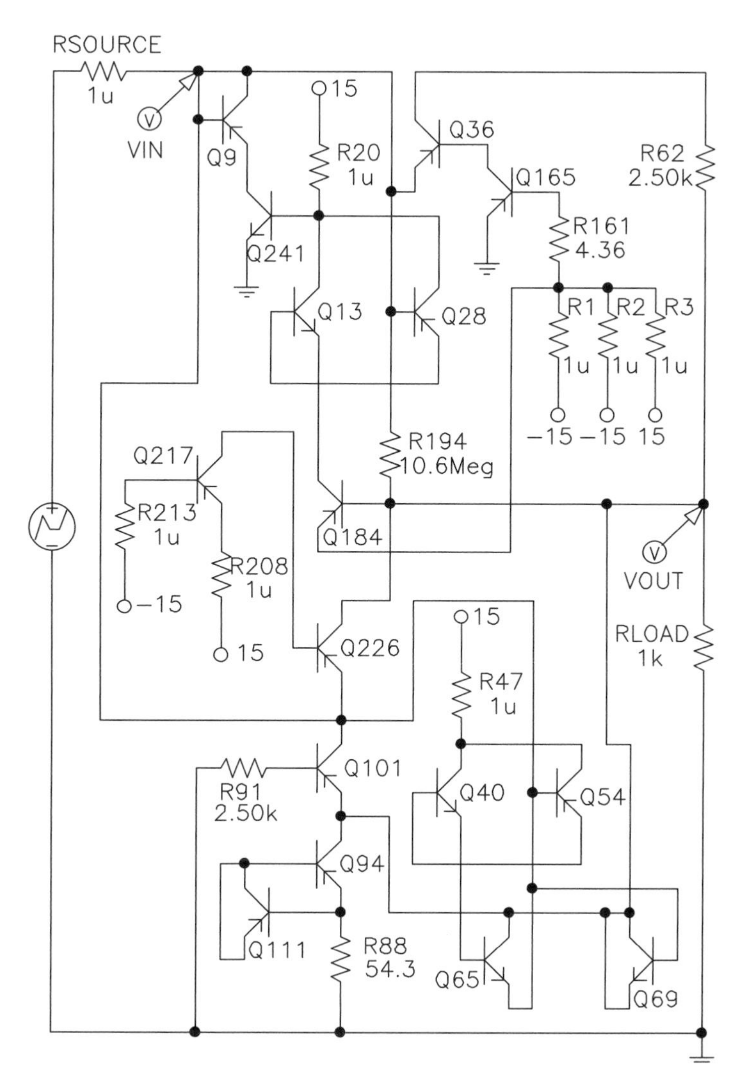

Figure 11.13 Best-of-run circuit when the genetic switch calls for a squaring circuit.

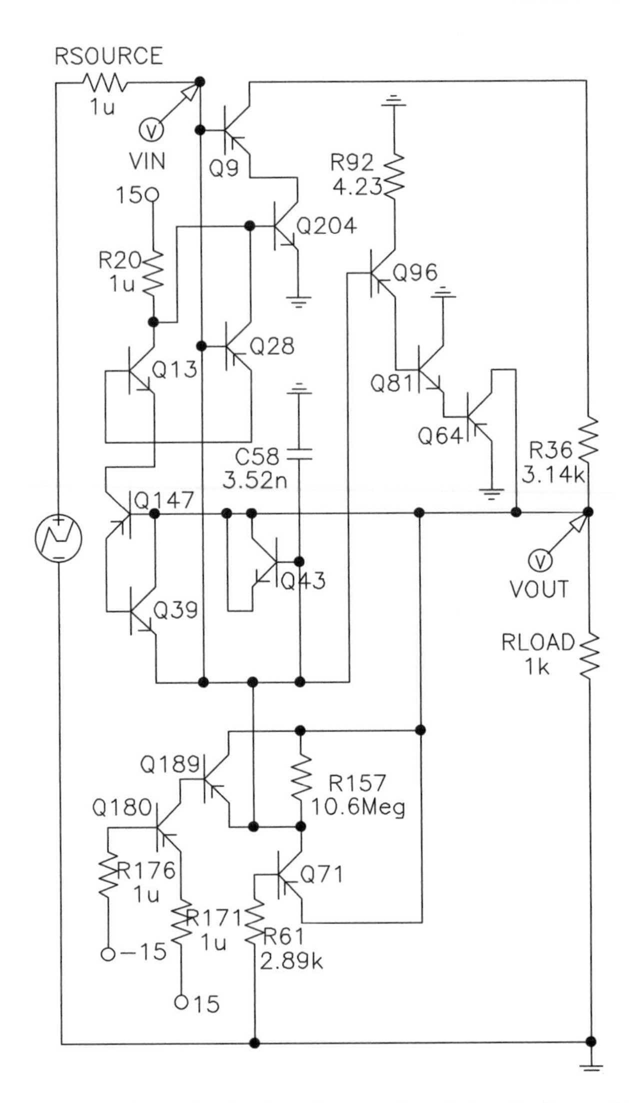

Figure 11.14 Best-of-run circuit when the genetic switch calls for a cubing circuit.

The LISP S-expression for the best of run individual from generation 241 appears below.

```
(Q_NEW Q2N3906 BIFURCATE_NEGATIVE E_C_B (Q_NEW Q2N3904
BIFURCATE_POSITIVE C_B_E END (INPUT_0 (TWO_POS15V END
(NODE END SAFE-CUT)) (Q_NEW Q2N3906 BIFURCATE_NEGATIVE
B_C_E SAFE-CUT END END)) (NODE (IFGTZ_DEVELOPMENTAL (+
-4.410578 (+ -0.8668906 (+ TARGET TARGET))) (TWO-LEAD
(R_NEW 0.4973120) END (Q_NEW Q2N3904 BIFURCATE_NEGATIVE
E_B_C (Q_NEW Q2N3904 BIFURCATE_NEGATIVE B_E_C (OUTPUT_0
END END) (INPUT_0 SAFE-CUT END) (OUTPUT_0 END END))
```

SAFE-CUT (TWO-LEAD (C_NEW 0.5471030) (TWO-LEAD (R_NEW 0.4610370) (Q_NEW Q2N3906 BIFURCATE_NEGATIVE C_E_B (INPUT_0 (Q_NEW Q2N3906 BIFURCATE_NEGATIVE C_E_B SAFE-CUT END SAFE-CUT) (OUTPUT_0 END SAFE-CUT)) (Q_NEW Q2N3904 BIFURCATE_POSITIVE B_C_E END (INPUT_0 (TWO_POS15V END (TWO-LEAD (R_NEW −2.373404) END SAFE-CUT)) (Q_NEW Q2N3906 BIFURCATE_NEGATIVE B_C_E SAFE-CUT END END)) (SERIES END END SAFE-CUT)) (PARALLEL_NEW DOWN_OR_RIGHT SAFE-CUT (Q_NEW Q2N3904 BIFURCATE_POSITIVE C_B_E END (TWO_NEG15V (TWO_NEG15V SAFE-CUT END) (SERIES END END END)) END) (THREE_GROUND END END END) (SERIES END END END))) (TWO_GROUND SAFE-CUT (TWO_POS15V END (NODE END SAFE-CUT)))) (INPUT_0 SAFE-CUT SAFE-CUT)))) **(Q_NEW Q2N3906 BIFURCATE_NEGATIVE C_B_E (Q_NEW Q2N3904 BIFURCATE_POSITIVE C_B_E END (INPUT_0 (TWO_POS15V END (NODE END END)) (Q_NEW Q2N3906 BIFURCATE_NEGATIVE B_C_E SAFE-CUT END END)) (NODE (TWO-LEAD (R_NEW 0.3975176) END (Q_NEW Q2N3904 BIFURCATE_NEGATIVE E_B_C (Q_NEW Q2N3904 BIFURCATE_NEGATIVE B_E_C (OUTPUT_0 END END) (INPUT_0 SAFE-CUT SAFE-CUT) (OUTPUT_0 END END)) SAFE-CUT (NODE (TWO-LEAD (R_NEW −1.265278) (TWO-LEAD (R_NEW 0.3975176) (Q_NEW Q2N3906 BIFURCATE_NEGATIVE C_E_B (INPUT_0 (Q_NEW Q2N3906 BIFURCATE_NEGATIVE C_E_B SAFE-CUT END SAFE-CUT) (OUTPUT_0 END SAFE-CUT)) (Q_NEW Q2N3906 BIFURCATE_NEGATIVE C_E_B SAFE-CUT END SAFE-CUT) (NODE SAFE-CUT SAFE-CUT)) (TWO_GROUND (THREE_GROUND END SAFE-CUT SAFE-CUT) (THREE_GROUND SAFE-CUT SAFE-CUT END))) (NODE SAFE-CUT SAFE-CUT)) (INPUT_0 SAFE-CUT SAFE-CUT)))) END)) (TWO_NEG15V (TWO_NEG15V (TWO_NEG15V SAFE-CUT END) (TWO_POS15V END (TWO-LEAD (R_NEW −2.360147) END SAFE-CUT))) (Q_NEW Q2N3906 BIFURCATE_POSITIVE E_B_C SAFE-CUT SAFE-CUT (THREE_GROUND END END END))) (SERIES SAFE-CUT (INPUT_0 END SAFE-CUT) END)))** (OUTPUT_0 (Q_NEW Q2N3906 BIFURCATE_POSITIVE B_C_E SAFE-CUT END SAFE-CUT) SAFE-CUT))) (INPUT_0 (TWO-LEAD (R_NEW 4.023496) (NODE (NODE (SERIES END END SAFE-CUT) (SERIES (TWO_POS15V END SAFE-CUT) END (TWO_NEG15V END (Q_NEW Q2N3906 BIFURCATE_NEGATIVE E_C_B SAFE-CUT SAFE-CUT END)))) (INPUT_0 (Q_NEW Q2N3906 BIFURCATE_NEGATIVE E_C_B END END SAFE-CUT) (OUTPUT_0 SAFE-CUT SAFE-CUT))) SAFE-CUT) END) (THREE_GROUND SAFE-CUT END (Q_NEW Q2N3904 BIFURCATE_NEGATIVE C_B_E END END END))).

As can be seen from figures 11.13 and 11.14, this S-expression codes for two distinctly different circuits depending on the value of the free variable TARGET. These two different circuits deliver the appropriate quadratic output and the appropriate cubic output, respectively.

11.4 A 40/60 dB Amplifier

This section describes the automatic creation of a circuit-constructing program with free variables and conditional developmental operators that yields either a 40 dB or 60 dB amplifier.

Specifically, the free variable TARGET is the input to the program tree. This free variable may assume the value of either 40 or 60.

11.4.1 Preparatory Steps for the 40/60 dB Amplifier

The preparatory steps are the same as for quadratic/cubic computational circuit in the 11.3.1, except as noted below.

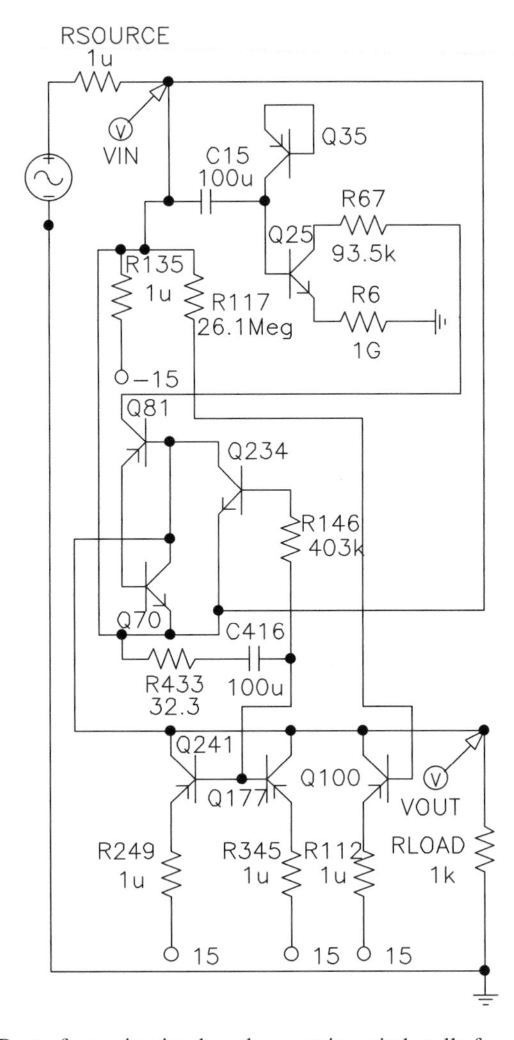

Figure 11.15 Best-of-run circuit when the genetic switch calls for a 40 dB amplifier.

11.4.1.1 Initial Circuit The 40–60 dB amplifier problem uses a one-input, one-output initial circuit with one modifiable wire. The initial circuit has an incoming signal source, a 1-micro-Ohm source resistor RSOURCE, a voltage probe point VOUT, and a 1,000-Ohm load resistor RLOAD. Figure 11.12 shows the topology of this initial circuit.

11.4.1.2 Terminal Set The free variable `TARGET` may assume the value of either 40 or 60.

11.4.1.3 Fitness Measure The purpose of the 40–60 dB amplifier is to produce an output voltage (probed at VOUT) with an amplification factor of 100 (40 dB) or 1,000

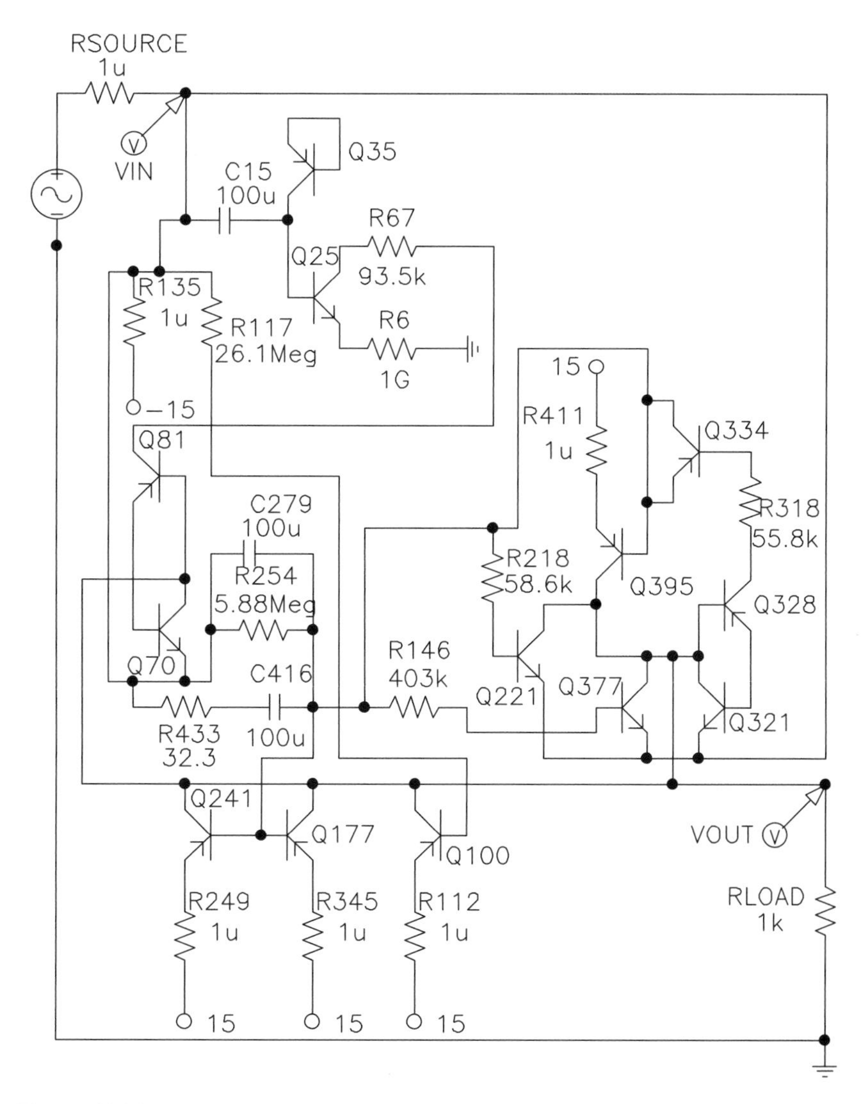

Figure 11.16 Best-of-run circuit when the genetic switch calls for a 60 dB amplifier.

(60 dB) depending on whether the value of the free variable terminal (`TARGET`) is 40 or 60, respectively.

Fitness is measured using two time-domain fitness cases, one for each desired amplification factor.

The first fitness case (associated with a desired amplification of 40 dB) is a 1,000 Hz sine wave ranging between −10 millivolts and +10 millivolts over a period of 1 millisecond.

The second fitness case (associated with a desired amplification of 60 dB) is a 1,000 Hz sine wave ranging between −10 millivolts and +10 millivolts over a period of 1 millisecond.

When a 40 dB amplifier is desired, the designed output is a sine wave between −1 and +1 Volt that is 180 degrees out of phase with respect to the input voltage. When a 60 dB amplifier is desired, the desired output is a sine wave between −10 and +10 Volts that is 180 degrees out of phase with respect to the input voltage.

There are 100 time steps associated with each fitness case.

Fitness is the sum, over the two fitness cases and 101 time values for each fitness case, of the absolute value of the weighted difference between the desired output and circuit's actual output. The weight is 1.0 if the absolute value of the difference is within 5% of the correct value, but 10.0 otherwise.

11.4.1.4 Control Parameters The population size is 2,000,000.

11.4.2 Results for 40/60 dB Amplifier

The best-of-run individual emerged at generation 332.

When the genetic switch calls for a 40 dB amplifier, this best-of-run individual yields a circuit with an amplification of 40.1 dB. This circuit (figure 11.15) produces an inverted sine wave whose minimum value is 967 millivolts and whose maximum value is 1.06 Volts.

When the genetic switch calls for a 60 dB amplifier, this best-of-run individual yields a circuit with an amplification of 59.3 dB. This circuit (figure 11.16) produces an inverted sine wave whose minimum value is −7.27 Volts and whose maximum value is 11.1 Volts.

12

Automatic Synthesis of Improved Tuning Rules for PID Controllers

The PID controller was patented in 1939 by Albert Callender and Allan Stevenson of Imperial Chemical Limited of Northwich, England (Callender and Stevenson 1939). The PID controller was an enormous improvement over previous manual and automatic methods for control.

In 1942, Ziegler and Nichols published a paper entitled "Optimum Settings for Automatic Controllers" in which they developed a set of mathematical rules for automatically selecting the parameter values associated with the proportional, integrative, and derivative blocks of a PID controller (Ziegler and Nichols 1942). The Ziegler-Nichols PID tuning rules do not require an analytic model of the plant. Instead, they are based on several parameters that provide a simple characterization of the to-be-controlled plant. These parameters have the practical advantage of being measurable for plants in the real world by means of relatively straightforward testing in the field. The Ziegler-Nichols rules have been in widespread use for tuning PID controllers since World War II.

The quality of PID tuning rules is of considerable practical importance because a small percentage improvement in the operation of a plant can translate into large economic savings or other (e.g., environmental) benefits.

Thus, the question arises as to whether it is possible to improve upon the Ziegler-Nichols tuning rules.

Åström and Hägglund answered that question in the affirmative in their important 1995 book *PID Controllers: Theory, Design, and Tuning*. In that book, Åström and Hägglund identified four families of plants "that are representative for the dynamics of typical industrial processes."

The first of the four families of plants in Åström and Hägglund 1995 consists of plants represented by transfer functions of the form

$$G(s) = \frac{e^{-s}}{(1 + sT)^2}, \qquad [A]$$

where $T = 0.1, \ldots, 10$.

The second of the four industrially representative families of plants consists of the n-lag plants represented by transfer functions of the form

$$G(s) = \frac{1}{(1+s)^n}, \qquad [B]$$

where $n = 3$, 4, and 8.

The third family consists of plants represented by transfer functions of the form

$$G(s) = \frac{1}{(1+s)(1+\alpha s)(1+\alpha^2 s)(1+\alpha^3 s)}, \qquad [C]$$

where $\alpha = 0.2$, 0.5, and 0.7.

The fourth family consists of plants represented by transfer functions of the form

$$G(s) = \frac{1-\alpha s}{(s+1)^3}, \qquad [D]$$

where $\alpha = 0.1$, 0.2, 0.5, 1.0, and 2.0.

In their 1995 book, Åström and Hägglund developed rules for automatically tuning PID controllers for the 16 plants from these four industrially representative families of plants.

The PID tuning rules developed by Åström and Hägglund (like those of Ziegler and Nichols) are based on several parameters representing important overall characteristics of the plant that can be obtained by straightforward testing in the field.

In one version of their method, Åström and Hägglund characterize a plant by two frequency-domain parameters. The first is the ultimate gain, K_u (the minimum value of the gain that must be introduced into the feedback path to cause the system to oscillate). The second is the ultimate period, T_u (the period of this lowest frequency oscillation).

In another version of their method, Åström and Hägglund characterize a plant by two time-domain parameters, namely the time constant, T_r, and the dead time, L (the period before the plant output begins to respond significantly to a new reference signal). Åström and Hägglund describe a procedure for estimating these two time-domain parameters from the plant's response to a simple step input.

Åström and Hägglund set out to develop new tuning rules to yield improved performance with respect to the multiple (often conflicting) issues associated with most practical control systems including setpoint response, disturbance rejection, sensor noise attenuation, and robustness in the face of plant model changes as expressed by the stability margin. In their 1995 book, Åström and Hägglund approached the challenge of improving on the 1942 Ziegler-Nichols tuning rules with a shrewd combination of mathematical analysis, domain-specific knowledge, rough-and-ready approximations, creative flair, and intuition sharpened over years of practical experience. They started by identifying a large number of analytic plant models that are representative of the plants encountered in industry. Next, they characterized each of the plants in their test bed in terms of a small number of parameters that are easily measured in the field. Then they applied the known analytic design technique of dominant pole design to the analytic plant models and recorded the resulting parameters for PID controllers produced by this technique. They decided that functions of the form

$$f(x) = a_0 * e^{a_1 x + a_2 x^2},$$

where $x = 1/K_u$, were the appropriate form for the tuning rules. Finally, they fit approximating functions of the chosen form to the PID controller parameters produced by the dominant pole design technique.

The tuning rules developed by Åström and Hägglund in their 1995 book outperform the 1942 Ziegler-Nichols tuning rules on all 16 industrially representative plants used by Åström and Hägglund. As Åström and Hägglund observe,

> "[Our] new methods give substantial improvements in control performance while retaining much of the simplicity of the Ziegler-Nichols rules."

Figure 12.1 shows the topology of the PID controller used by Åström and Hägglund. The inputs to this controller consist of the reference signal 200 and the plant output 202 (the feedback). The square 260 containing the expression $1/s$ represents integration whereas the square 290 containing the expression s represents differentiation. The rectangles 210, 230, 250, and 280 each represent gain blocks. The square 270 is a gain block with an amplification factor of -1. The circle 232 with positive signs on all three of its inputs is a three-argument adder. The circles (220 and 240) with a positive sign on one input and a negative sign on the other input represent subtractors. The controller's output 234 is the three-argument sum 232 of the result produced by gain block 230, integrator 260, and differentiator 290.

The controller of figure 12.1, like many controllers, employs setpoint weighting in the proportional (P) block. Specifically, reference signal 200 is weighted by gain 210 prior to being fed into subtractor 220. The fed-back plant output 202 is then subtracted (at 220) from the weighted reference signal coming out of gain block 210.

Setpoint weighting is sometimes also used in the derivative (D) portion of some PID controllers. However, in the Åström-Hägglund tuning rules, the weighting applied to the reference signal in the derivative block is zero. The weighting of zero is manifested in figure 12.1 by the presence of a gain block 270 with a gain of -1. That is, the input to the gain block 280 associated with the derivative block 290 is merely the negated plant output 202 (instead of the difference between a weighted reference signal and the plant output).

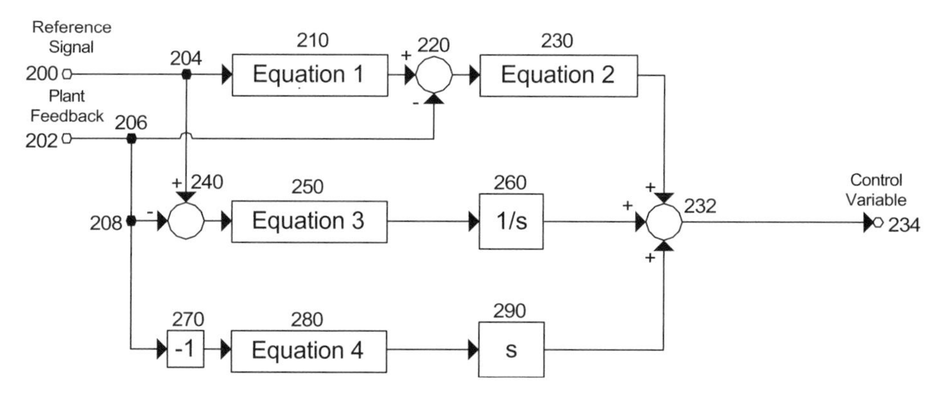

Figure 12.1 The topology of a PID controller with nonzero setpoint weighting of the reference signal in the proportional block but no setpoint weighting for the derivative block.

Setpoint weighting is generally not used for the input to the gain block associated with a controller's integrative (I) block.

As mentioned earlier, the PID tuning rules developed by Åström and Hägglund in their 1995 book can be succinctly expressed by four equations of the form

$$f(x) = a_0 * e^{a_1 x + a_2 x^2},$$

where $x = 1/K_u$

Equation 1 implements setpoint weighting 210 of the reference signal 200 (the setpoint) for the proportional (P) part of figure 12.1. Equation 1 specifies that the setpoint weighting, b, is given by

$$b = 0.25 * e^{((0.56/K_u) + (-0.12/K_u^2))}. \quad [1]$$

Equation 2 in figure 12.1 specifies the gain, K_p, for the proportional (P) block 230 of the controller.

$$K_{\mathrm{p}} = 0.72 * K_u * e^{((-1.6/K_u) + (1.2/K_u^2))}. \quad [2]$$

Equation 3 in figure 12.1 specifies the gain 250, K_i, that is associated with the input to the integrative (I) block 260.

$$K_i = \frac{0.72 * K_u * e^{((-1.6/K_u) + (1.2/K_u^2))}}{0.59 * T_u * e^{((-1.3/K_u) + (0.38/K_u^2))}}. \quad [3]$$

Equation 4 in figure 12.1 specifies the gain 280, K_d, that is associated with the input to the derivative (D) block 290.

$$K_{\mathrm{d}} = 0.108 * K_u * T_u * e^{((-1.6/K_u) + (1.2/K_u^2))} * e^{((-1.4/K_u) + (0.56/K_u^2))}. \quad [4]$$

The transfer function for the Åström-Hägglund PID controller is

$$A = K_p(b*R - P) + \frac{K_{\mathrm{i}}}{s}(R - P) + K_d * s * (-P),$$

where A is the output of the Åström-Hägglund controller, R is the reference signal, P is the plant output, and b, K_p, K_i, and K_d are given by the four equations above. Note that only $-P$ (as opposed to $R - P$) appears in the derivative term of this transfer function because Åström and Hägglund apply a setpoint weighting of zero to the reference signal R in the derivative block.

In this chapter, genetic programming is used to discover PID tuning rules that improve upon the tuning rules developed by Åström and Hägglund in their 1995 book. The topology of the to-be-evolved controller is not open-ended in this chapter, but instead, is prespecified to be the PID topology. In the next chapter (chapter 13), the topology will be open-ended and genetic programming will be used to synthesize both topology and tuning of improved controllers. As it turns out, all these improved controllers will be non-PID controllers.

The fitness measure used in this chapter considers plants from the same four families used by Åström and Hägglund. The resulting parameterized tuning rules are parameterized (i.e., general) in the sense that they incorporate free variables and provide a solution to an entire category of problems (i.e., the control of all the plants in all the families).

12.1 Test Bed of Plants

Before proceeding, we present a test bed of plants that are used to compare the performance of a PID controller tuned using the Åström-Hägglund tuning rules with that of a PID controller tuned using the genetically evolved tuning rules. This test bed is used in this chapter and in chapter 13.

Table 12.1 shows the characteristics of 33 plants that are employed, during the evolutionary process, in connection with evaluating the fitness of controllers in this chapter and chapter 13. All the plants in this table are members of Åström and Hägglund's four families of plants. Column 1 of this table identifies the plant family (A, B, C, or D). Column 2 shows the particular parameter value for the plant within its family. For

Table 12.1 Characteristics of 33 plants used in various runs

Family	Parameter value	K_u	T_u	L	T_r	Runs in which the plant is used
A	$T=0.1$	1.07	2.37	1.00	0.103	A, P, 1, 2, 3
A	$T=0.3$	1.40	3.07	1.01	0.299	A, P, 1, 2, 3
A	$T=1$	2.74	4.85	1.00	1.00	A, P, 1, 2, 3
A	$T=3$	6.80	7.87	1.02	2.99	A, P, 1, 2, 3
A	$T=4.5$	9.67	9.60	1.00	4.50	P
A	$T=6$	12.7	11.1	1.00	6.00	P, 2, 3
A	$T=7.5$	15.6	12.3	1.00	7.50	P
A	$T=9$	18.7	13.4	1.01	9.00	P
A	$T=10$	20.8	14.2	0.916	10.1	A, P, 1, 2, 3
B	$n=3$	8.08	3.62	0.517	1.24	A, P, 1, 2, 3
B	$n=4$	4.04	6.27	1.13	1.44	A, P, 1, 2, 3
B	$n=5$	2.95	8.62	1.79	1.61	P, 2, 3
B	$n=6$	2.39	10.9	2.45	1.78	P, 1, 2, 3
B	$n=7$	2.09	13.0	3.17	1.92	P, 2, 3
B	$n=8$	1.89	15.2	3.88	2.06	A, P, 1, 2, 3
B	$n=11$	1.57	21.6	6.19	2.41	1
C	$\alpha=0.1$	113	0.198	−0.244	0.674	1
C	$\alpha=0.2$	30.8	0.562	−0.137	0.691	A, P, 1, 2, 3
C	$\alpha=0.215$	26.6	0.626	−0.155	0.713	P
C	$\alpha=0.23$	23.6	0.693	−0.116	0.705	P
C	$\alpha=0.26$	19.0	0.833	−0.099	0.722	P
C	$\alpha=0.3$	15.0	1.04	−0.024	0.720	P, 2, 3
C	$\alpha=0.4$	9.62	1.59	0.111	0.759	P, 2, 3
C	$\alpha=0.5$	6.85	2.23	0.267	0.804	A, P, 1, 2, 3
C	$\alpha=0.6$	5.41	2.92	0.431	0.872	P, 2, 3
C	$\alpha=0.7$	4.68	3.67	0.604	0.962	A, P, 1, 2, 3
C	$\alpha=0.9$	4.18	5.31	3.44	0.685	1
D	$\alpha=0.1$	6.21	4.06	0.644	1.22	A, P, 1, 2, 3
D	$\alpha=0.2$	5.03	4.44	0.739	1.23	A, P, 1, 2, 3
D	$\alpha=0.5$	3.23	5.35	1.15	1.17	A, P, 1, 2, 3
D	$\alpha=0.7$	2.59	5.81	1.38	1.16	P, 2, 3
D	$\alpha=1$	2.02	6.29	1.85	1.07	A, P, 1, 2, 3
D	$\alpha=2$	1.15	7.46	3.46	0.765	A, P, 1, 2, 3

example, family B is the family of n-lag plants and the first entry in the table for family B is the three-lag plant (i.e., $n = 3$). Columns 3, 4, 5, and 6 show each plant's ultimate gain, K_u; ultimate period, T_u; dead time, L; and time constant, T_r; respectively. Column 7 identifies the 16 plants (with an "A") belonging to the test bed used by Åström and Hägglund in their 1995 book; the 30 plants (with a "P") that are used to evolve the improved PID tuning rules in this chapter; and the plants that are used to evolve the three controllers in chapter 13 (indicated by "1," "2," or "3," respectively). There are 20 plants marked "1," 24 plants marked "2," and 24 plants marked "3."

The possibility exists that a controller that is created using an automated process that relies on certain plants (the *in-sample* or *training* cases) will be overly specialized to those particular plants and that the automatically created controller will prove not to operate correctly on other plants. This concern can be addressed by cross-validating the controller on additional unseen plants.

Table 12.2 shows the characteristics of 18 additional plants that are used to cross-validate the performance of the genetically evolved PID tuning rules in this chapter and the three genetically evolved controllers in chapter 13. All 18 additional plants are members of Åström and Hägglund's families A, C, and D. Columns 3, 4, 5, and 6 of the table describe each plant in terms of its ultimate gain, K_u; ultimate period, T_u; dead time, L; and time constant, T_r; respectively. Because the plant parameter for family B (the n-lag plants) is an integer and all integer values within the range of values used by Åström and Hägglund already appear in table 12.1, there are no additional plants within family B in this table.

Table 12.3 shows the performance of a PID controller tuned using the Åström-Hägglund tuning rules on the 33 plants of table 12.1 in terms of the fitness measure

Table 12.2 Characteristics of 18 additional plants used for cross-validation

Family	Parameter value	K_u	T_u	L	T_r
A	0.15	1.13	2.57	0.993	0.153
A	0.5	1.74	3.65	0.982	0.509
A	0.9	2.51	4.60	1.011	0.894
A	2.5	5.69	7.25	0.999	2.50
A	4.0	8.68	9.07	1.002	4.00
A	9.0	18.7	13.4	1.005	9.00
C	0.25	20.3	0.786	−0.099	0.713
C	0.34	12.0	1.25	0.005	0.744
C	0.43	8.35	1.77	0.144	0.775
C	0.52	6.40	2.36	0.287	0.821
C	0.61	5.30	3.00	0.439	0.884
C	0.69	4.72	3.60	0.563	0.965
D	0.15	5.52	4.26	0.680	1.23
D	0.3	4.21	4.77	0.846	1.23
D	0.6	2.86	5.54	1.24	1.18
D	0.85	2.25	6.03	1.62	1.12
D	1.2	1.74	6.57	2.16	1.02
D	1.8	1.25	7.25	3.15	0.824

Table 12.3 Performance of the Åström–Hägglund PID controller on the 33 plants of table 12.1

Plant	Plant parameter value	ITAE 1	ITAE 2	ITAE 3	ITAE 4	ITAE 5	ITAE 6	Stability margin	Sensor noise
A	0.1	0.740	0.740	0.749	0.743	0.269	0.478	0.980	10.1
A	0.3	0.701	0.701	0.702	0.701	0.266	0.434	0.669	1.75
A	1	0.510	0.510	0.511	0.510	0.237	0.292	0.660	1.11
A	3	0.331	0.331	0.331	0.331	0.211	0.122	0.594	1.49
A	4.5	0.282	0.282	0.282	0.282	0.201	0.082	0.410	1.34
A	6	0.259	0.259	0.259	0.259	0.198	0.062	0.279	1.25
A	7.5	0.244	0.244	0.244	0.244	0.195	0.049	0.158	1.16
A	9	0.236	0.236	0.236	0.236	0.195	0.041	0.070	1.13
A	10	0.232	0.232	0.232	0.232	0.195	0.037	0.016	1.11
B	3	0.334	0.334	0.335	0.334	0.225	0.108	0.368	0
B	4	0.456	0.456	0.456	0.456	0.238	0.215	0.707	0
B	5	0.527	0.526	0.526	0.527	0.243	0.281	0.690	0
B	6	0.570	0.571	0.569	0.570	0.245	0.321	0.646	0
B	7	0.600	0.599	0.597	0.599	0.248	0.348	0.642	0
B	8	0.623	0.623	0.621	0.623	0.251	0.369	0.643	0
B	11	0.667	0.667	0.663	0.665	0.257	0.405	0.641	0
C	0.1	0.187	0.187	0.187	0.187	0.181	0.006	0	0.659
C	0.2	0.193	0.194	0.193	0.193	0.171	0.022	0	0.024
C	0.215	0.196	0.196	0.196	0.196	0.171	0.025	0.039	0
C	0.23	0.201	0.201	0.200	0.201	0.173	0.028	0.081	0
C	0.26	0.211	0.210	0.210	0.211	0.175	0.035	0.156	0
C	0.3	0.225	0.225	0.224	0.225	0.178	0.046	0.241	0
C	0.4	0.272	0.272	0.272	0.272	0.191	0.079	0.421	0
C	0.5	0.325	0.325	0.325	0.325	0.207	0.117	0.542	0
C	0.6	0.374	0.373	0.374	0.374	0.219	0.152	0.594	0
C	0.7	0.414	0.413	0.412	0.414	0.231	0.181	0.675	0
C	0.9	0.452	0.452	0.451	0.452	0.238	0.213	0.711	0
D	0.1	0.187	0.187	0.187	0.187	0.181	0.006	0.585	0
D	0.2	0.193	0.193	0.194	0.193	0.171	0.022	0.699	0
D	0.5	0.225	0.225	0.225	0.225	0.178	0.046	0.728	0
D	0.7	0.272	0.272	0.272	0.272	0.191	0.079	0.746	0
D	1	0.325	0.325	0.325	0.325	0.207	0.116	0.839	0
D	2	0.374	0.373	0.374	0.374	0.219	0.152	4.18	0

used in this chapter (and in chapter 13). The fitness measure is the same as in section 9.2. The fitness measure seeks to optimize the integral of the time-weighted absolute error for a step input and disturbance rejection while simultaneously imposing constraints on maximum sensitivity and sensor noise attenuation. The fitness of each controller is measured by means of eight separate invocations of the SPICE simulator for each plant under consideration. Columns 3 through 10 correspond to the eight elements of the fitness measure used for each plant. Columns 3 through 8 show the six elements of the fitness measure that are measured by means of the integral of the time-weighted absolute error. Column 9 shows the stability margin penalty. Column 10 shows the sensor noise penalty.

Table 12.4 Performance of the Åström–Hägglund PID controller on the 18 additional plants of table 12.2

Plant	Plant parameter value	ITAE 1	ITAE 2	ITAE 3	ITAE 4	ITAE 5	ITAE 6	Stability margin	Sensor noise
A	0.15	0.740	0.740	0.746	0.743	0.269	0.478	0.860	4.72
A	0.5	0.701	0.701	0.702	0.701	0.266	0.434	0.621	0.89
A	0.9	0.510	0.510	0.511	0.51	0.237	0.292	0.638	1.24
A	2.5	0.331	0.331	0.331	0.331	0.211	0.122	0.657	1.05
A	4.0	0.259	0.259	0.259	0.259	0.198	0.062	0.459	1.44
A	9.0	0.232	0.232	0.232	0.232	0.195	0.037	0.070	1.13
C	0.25	0.335	0.334	0.333	0.335	0.225	0.108	0.128	0
C	0.34	0.457	0.457	0.455	0.457	0.239	0.215	0.297	0
C	0.43	0.527	0.526	0.526	0.527	0.243	0.282	0.471	0
C	0.52	0.570	0.571	0.569	0.570	0.245	0.321	0.546	0
C	0.61	0.600	0.599	0.597	0.599	0.248	0.348	0.600	0
C	0.69	0.623	0.623	0.621	0.623	0.251	0.369	0.665	0
D	0.15	0.187	0.187	0.187	0.187	0.181	0.006	0.585	0
D	0.3	0.193	0.193	0.194	0.193	0.171	0.022	0.699	0
D	0.6	0.225	0.225	0.225	0.225	0.178	0.046	0.728	0
D	0.85	0.272	0.272	0.272	0.272	0.191	0.079	0.746	0
D	1.2	0.325	0.325	0.325	0.325	0.207	0.116	0.839	0
D	1.8	0.374	0.373	0.374	0.374	0.219	0.152	4.18	0

Similarly, table 12.4 shows the performance of a PID controller tuned using the Åström-Hägglund tuning rules on the 18 additional plants shown in table 12.2 in terms of the fitness measure used in this chapter (and in chapter 13).

12.2 Preparatory Steps for Improved PID Tuning Rules

In this chapter, we are seeking improved PID tuning rules. That is, the controller's topology is not subject to evolution, but instead, is prespecified to be the PID topology of figure 12.1. Specifically, in this chapter, we are seeking four mathematical expressions (for K_p, K_i, K_d, and b). Each of the expressions may contain free variables representing the plant's ultimate gain, K_u, and the plant's ultimate period, T_u. Note that the problem in this chapter differs from the problems in chapters 9 and 13 in that there is no search for, or evolution of, the topology. The problem in this chapter is simply a search for four surfaces in three-dimensional space.

Initial experimentation quickly confirmed the fact that the Åström and Hägglund tuning rules are highly effective. Starting from scratch, genetic programming readily evolved three-dimensional surfaces that closely match the Åström-Hägglund surfaces and that outperform the Åström-Hägglund tuning rules *on average*. However, these genetically evolved surfaces did not satisfy our goal, namely outperforming the Åström-Hägglund tuning rules for every plant in the test bed.

For that reason, we decided to approach the problem of discovering improved tuning rules by building on the known and highly effective Åström-Hägglund results.

Specifically, in this chapter, we use genetic programming to evolve four mathematical expressions (containing the free variables K_u and T_u) which, when added to the corresponding mathematical expressions developed by Åström and Hägglund, yield improved performance on every plant in the test bed. Although we anticipated that the resulting three-dimensional surfaces would be similar to the Åström-Hägglund surfaces (i.e., the magnitudes of the added numbers would be relatively small), this additive approach can, in fact, yield any surface (or, more precisely, any surface that can be represented by the generously large number of points that a single program tree may contain). We similarly build on the Åström-Hägglund results in the next chapter (chapter 13).

It should be noted that this technique for incorporating a known outstanding result can be used for problems from many different fields. For additional discussion on solving problems by progressively building on previously known solutions, see Streeter, Keane, and Koza 2002a.

12.2.1 Program Architecture

The architecture of each program tree in the population has four result-producing branches (one associated with each of the problem's four independent variables, namely K_p, K_i, K_d, and b).

Note that although there are free variables in this problem, this problem is not included among the 11 problems in this book listed in section 1.3 in connection with parameterized topologies because the present problem does not entail discovery of the solution's topology.

12.2.2 Terminal Set

In this problem, there are two free variables. They are the plant's ultimate gain K_u (called "KU" below) and the plant's ultimate period T_u (called "TU").

The terminal set, T, for each of the four result-producing branches is

$$\mathsf{T} = \{\Re, \texttt{KU}, \texttt{TU}\},$$

where $\Re$ denotes a perturbable numerical value between -5.0 and $+5.0$.

12.2.3 Function Set

The function set, F, for each of the four result-producing branches contains functions for performing arithmetic, exponential, and logarithmic operations.

$$\mathsf{F} = \{+, -, *, \%, \texttt{REXP}, \texttt{RLOG}, \texttt{POW}\}.$$

The two-argument POW function returns the value of its first argument to the power of the value of its second argument. The POW function has the advantage (as compared to the EXP function) of permitting the free variables to appear in the base (as well as the exponent) of an exponential expression.

12.2.4 Fitness Measure

The fitness measure is the same as in section 9.2. Fitness is measured by means of eight separate invocations of the SPICE simulator for each plant under consideration.

When the aim is to automatically create a solution to a category of problems in the form of mathematical expressions containing free variable(s), the possibility of overfitting is especially salient. Overfitting occurs when an evolved solution performs well on the fitness cases that are incorporated in the fitness measure (i.e., in the training phase), but then performs poorly on previously unseen fitness cases.

In this problem, four mathematical expressions (each incorporating two free variables) must be evolved. The to-be-evolved mathematical expressions are presumptively nonlinear and complicated functions of the two free variables.

In applying genetic programming to this program, we were especially concerned about overfitting because there are only 16 plants in the test bed used by Åström and Hägglund. Moreover, there are only three plants in two of the four families. As discussed in section 10.3.1.5, genetic programming must usually consider a multiplicity of data points in order to discover even a linear relationship. *A fortiori*, a multiplicity of data points is required when the underlying function is actually nonlinear (as is usually the case in non-trivial problems).

Concerned about possible overfitting, we used the 30 plants marked "P" in column 7 of table 12.1 for this problem. Thus, the fitness measure entails 240 separate invocations of the SPICE simulator. The choice of 30 plants for this first run was governed primarily by considerations of computer time. Ideally, we would have used even more plants. We did not know in advance whether 30 plants would prove to be sufficient to enable genetic programming to unearth the complicated relationships inherent in this problem.

12.2.5 Control Parameters

The population size is 100,000.

12.3 Results for Improved PID Tuning Rules

In our one run of this problem, the best-of-run individual emerged in generation 76.

Because the population size is 100,000, the fitness of approximately 7,700,000 individuals was evaluated during the run. The average time to evaluate the fitness of an individual (i.e., to perform the 240 SPICE simulations) was 51 seconds. It took 106.8 hours (4.4 days) to produce the best-of-run individual from generation 76.

The improved PID tuning rules are obtained by adding the genetically evolved adjustments below to the values of K_p, K_i, and K_d, and b developed by Åström and Hägglund in their 1995 book.

The quantity, $K_{p\text{-}adj}$, that is to be added to K_p for the proportional part of the controller is

$$K_{p\text{-}adj} = -.0012340{*}T_u - 6.1173{*}10^{-6}.$$

The quantity, $K_{i\text{-}adj}$, that is to be added to K_i for the integrative part is

$$K_{i\text{-}adj} = -\ .065825{*}\frac{K_u}{T_u}.$$

The quantity, $K_{d\text{-}adj}$, that is to be added to K_d for the derivative part is

$$K_{d\text{-}adj} = -0.0026640\left(e^{T_u}\right)^{\log\left(1.6342^{\log K_u}\right)}.$$

The quantity, b_{adj}, that is to be added to b for the setpoint weighting of the reference signal in the proportional block is

$$b_{adj} = \frac{K_u}{e^{K_u}}.$$

In other words, the final values ($K_{p\text{-}final}$, $K_{i\text{-}final}$, $K_{d\text{-}final}$, and b_{final}) for the genetically evolved PID tuning rules are as follows:

$$K_{p\text{-}final} = 0.72{*}K_u{*}e^{((-1.6/K_u)+(1.2/K_u^2))} - .0012340{*}T_u - 6.1173{*}10^{-6},$$

$$K_{i\text{-}final} = \frac{0.72{*}K_u{*}e^{((-1.6/K_u)+(1.2/K_u^2))}}{0.59{*}T_u{*}e^{((-1.3/K_u)+(0.38/K_u^2))}} - .068525{*}\frac{K_u}{T_u},$$

$$K_{d\text{-}final} = 0.108{*}K_u{*}T_u{*}e^{((-1.6/K_u)+(1.2/K_u^2){*}e^{(-1.4/K_u)+(0.56/K_u^2))}} - 0.0026640(e^{T_u})^{\log(1.6342^{\log K_u})},$$

and

$$b_{final} = 0.25{*}e^{((0.56/K_u)+(-0.12/K_u^2))} + \frac{K_u}{e^{K_u}}.$$

The authors refer to these genetically evolved PID tuning rules as the Keane-Koza-Streeter (KKS) PID tuning rules.

The automatically designed controller is parameterized (i.e., general) in the sense that it contains free variables (K_u and T_u) and thereby provides a solution to an entire category of problems (i.e., the control of all the plants in all the families).

Averaged over the 30 plants used in this run (i.e., those marked "P" in column 7 of table 12.1), the best-of-run tuning rules from generation 76 have

- 91.6% of the setpoint ITAE of the Åström-Hägglund tuning rules,
- 96.2% of the disturbance rejection ITAE of the Åström-Hägglund tuning rules,
- 99.5% of the reciprocal of minimum attenuation of the Åström-Hägglund tuning rules, and
- 98.6% of the maximum sensitivity, M_s, of the Åström-Hägglund tuning rules.

Averaged over the 18 additional plants of table 12.2, the best-of-run tuning rules from generation 76 have

- 89.7% of the setpoint ITAE of the Åström-Hägglund tuning rules,
- 95.6% of the disturbance rejection ITAE of the Åström-Hägglund tuning rules,
- 99.5% of the reciprocal of minimum attenuation of the Åström-Hägglund tuning rules, and
- 98.5% of the maximum sensitivity, M_s, of the Åström-Hägglund tuning rules.

As can be seen, the results obtained for the 18 previously unseen additional plants are similar to those for the results for the plants used by the evolutionary process.

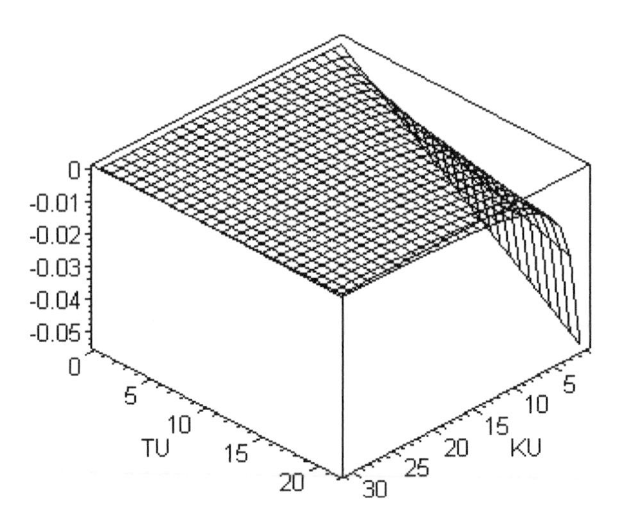

Figure 12.2 The genetically evolved adjustment, $K_{p\text{-}adj}$, for the proportional (P) part of the PID controller.

We now provide additional comparative details.

We first consider the overall pattern of values for the four genetically evolved adjusting quantities ($K_{p\text{-}adj}$, $K_{i\text{-}adj}$, $K_{d\text{-}adj}$, and b_{adj}).

The genetically evolved adjusting quantity, $K_{p\text{-}adj}$, for the proportional (P) part of the PID controller is the smallest and most localized adjustment to the corresponding value for a PID controller tuned according to the Åström-Hägglund tuning rules. Figure 12.2 shows, as a function of K_u and T_u, the genetically evolved adjusting quantity, $K_{p\text{-}adj}$, for the proportional part of the PID controller divided by the value of the gain, K_p, for the proportional block of a PID controller tuned according to the Åström-Hägglund tuning rules. The ranges of values for K_u and T_u in this figure (and the next three figures) are those encountered for the 33 plants of table 12.1 and the 18 additional plants in table 12.2. As can be seen from the figure, the genetically evolved adjustment is near zero for most values of K_u and T_u (i.e., the genetically evolved solution is about the same as the Åström-Hägglund solution). The genetically evolved adjustment becomes as large as -5% only in the combination of circumstances when K_u is at the low end of its range and T_u is at the high end of its range.

The genetically evolved adjusting quantity, $K_{i\text{-}adj}$, for the integrative (I) part of the PID controller is more significant than that for $K_{p\text{-}adj}$. Figure 12.3 shows, as a function of K_u and T_u, the genetically evolved adjusting quantity, $K_{i\text{-}adj}$, for the integrative part of the PID controller divided by the value of the gain, K_i, for the integrative block of a PID controller tuned according to the Åström-Hägglund tuning rules. As can be seen from the figure, the genetically evolved adjustments range between -4% and -5.5% and depend only on K_u. Specifically, the genetically evolved adjustment is about -5.5% for most values of K_u, but drops to as little as -4% when K_u is at the low end of its range.

The genetically evolved adjusting quantity, $K_{d\text{-}adj}$, for the derivative (D) part of the PID controller is also more significant than that for $K_{p\text{-}adj}$. Figure 12.4 shows, as a

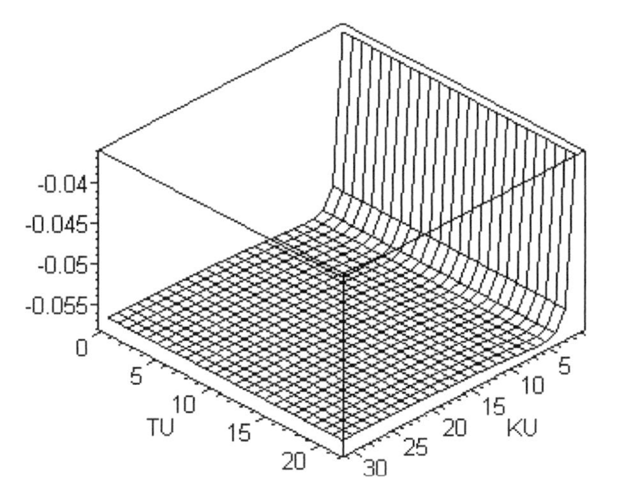

Figure 12.3 The genetically evolved adjustment, $K_{i\text{-}adj}$, for the integrative (I) part of the PID controller.

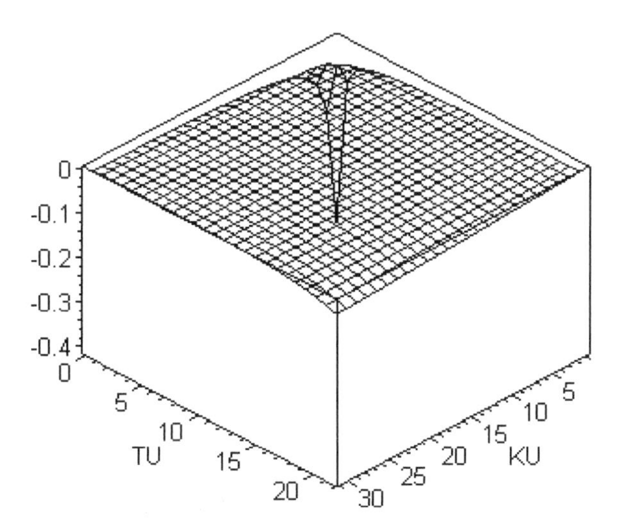

Figure 12.4 The genetically evolved adjustment, $K_{d\text{-}adj}$, for the derivative (D) part of the PID controller.

function of K_u and T_u, the genetically evolved adjusting quantity, $K_{d\text{-}adj}$, for the derivative part of the PID controller divided by the value of the gain, K_d, for the derivative block of a PID controller tuned according to the Åström-Hägglund tuning rules. As can be seen from the figure, the genetically evolved adjustment is near zero for most values of K_u and T_u (i.e., the genetically evolved solution is about the same as the Åström-Hägglund solution). However, the genetically evolved adjustment becomes as large as -40% for the combination of circumstances when both K_u and T_u are at the low ends of their ranges.

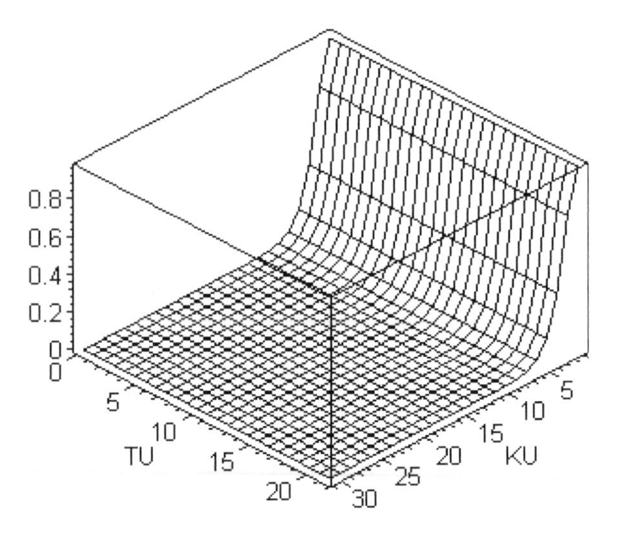

Figure 12.5 The genetically evolved adjustment, b_{adj}, for the setpoint weighting of the reference signal of the controller's proportional block.

Finally, the genetically evolved adjusting quantity, b_{adj}, for the setpoint weighting applied to the reference signal of the controller's proportional block is a substantial quantity. Figure 12.5 shows, as a function of K_u and T_u, the genetically evolved adjusting quantity, b_{adj}, for the setpoint weighting for the PID controller divided by the value of the gain, b, for the setpoint weighting of a PID controller tuned according to the Åström-Hägglund tuning rules. As can be seen from the figure, the genetically evolved adjustments range between 0% and about +80% and depend only on K_u. Specifically, the genetically evolved adjustment is near zero for most values of K_u, but rises sharply when K_u is at the low end of its range.

Table 12.5 shows the performance of the best-of-run PID tuning rules from generation 76 for the 30 plants used in this run (i.e., those marked "P" in column 7 of table 12.1).

Table 12.6 compares the performance of the best-of-run PID tuning rules from generation 76 as a percentage of the value for the Åström and Hägglund controller for the 30 plants used in this run (i.e., those marked "P" in column 7 of table 12.1). An "OK" appears in the table for the cases where the penalties associated with both tuning rules are zero. As can be seen, all percentages are either below 100% (indicating improvement) or "OK."

Table 12.7 shows the performance of the best-of-run PID tuning rules from generation 76 on the 18 additional plants from table 12.2.

Table 12.8 compares the performance of the best-of-run PID tuning rules from generation 76 as a percentage of the value for the Åström and Hägglund controller for the 18 additional plants from table 12.2. An "OK" appears in the table for the cases where the penalties associated with both tuning rules are zero. As can be seen, all percentages are either below 100% (indicating improvement) or "OK."

Averaged over the 16 plants used by Åström and Hägglund in their 1995 book (i.e., the 16 plants marked "A" in column 7 of table 12.1), the best-of-run tuning rules

Table 12.5 Performance (for 30 plants) of the best-of-run PID tuning rules

Plant	Plant parameter value	ITAE 1	ITAE 2	ITAE 3	ITAE 4	ITAE 5	ITAE 6	Stability margin	Sensor noise
A	0.1	0.698	0.698	0.705	0.701	0.234	0.469	0.934	9.79
A	0.3	0.674	0.674	0.675	0.674	0.240	0.433	0.623	1.73
A	1	0.460	0.460	0.461	0.460	0.197	0.281	0.624	1.10
A	3	0.310	0.310	0.310	0.310	0.195	0.117	0.559	1.49
A	4.5	0.266	0.266	0.266	0.266	0.188	0.079	0.393	1.34
A	6	0.246	0.246	0.246	0.246	0.187	0.059	0.260	1.25
A	7.5	0.233	0.233	0.233	0.233	0.186	0.048	0.139	1.16
A	9	0.225	0.225	0.225	0.225	0.186	0.040	0.056	1.13
A	10	0.222	0.222	0.222	0.222	0.186	0.036	0.009	1.11
B	3	0.320	0.319	0.320	0.320	0.215	0.104	0.339	0
B	4	0.433	0.432	0.434	0.433	0.222	0.208	0.650	0
B	5	0.498	0.498	0.499	0.498	0.222	0.273	0.640	0
B	6	0.538	0.537	0.536	0.538	0.221	0.314	0.564	0
B	7	0.564	0.564	0.561	0.564	0.218	0.342	0.536	0
B	8	0.585	0.585	0.582	0.584	0.218	0.363	0.530	0
C	0.2	0.188	0.188	0.188	0.188	0.166	0.021	0	0.023
C	0.215	0.190	0.191	0.191	0.190	0.166	0.024	0.020	0
C	0.23	0.195	0.194	0.194	0.195	0.167	0.027	0.062	0
C	0.26	0.204	0.203	0.203	0.204	0.168	0.034	0.140	0
C	0.3	0.216	0.217	0.217	0.216	0.171	0.045	0.227	0
C	0.4	0.261	0.260	0.260	0.261	0.183	0.077	0.389	0
C	0.5	0.310	0.311	0.312	0.310	0.196	0.113	0.518	0
C	0.6	0.355	0.356	0.354	0.355	0.206	0.147	0.541	0
C	0.7	0.394	0.393	0.393	0.394	0.217	0.175	0.616	0
D	0.1	0.352	0.353	0.351	0.352	0.216	0.135	0.467	0
D	0.2	0.386	0.387	0.386	0.386	0.218	0.166	0.577	0
D	0.5	0.495	0.495	0.496	0.495	0.231	0.262	0.660	0
D	0.7	0.581	0.581	0.583	0.581	0.248	0.330	0.664	0
D	1	0.720	0.719	0.721	0.720	0.281	0.436	0.659	0
D	2	1.48	1.48	1.48	1.48	0.533	0.944	5.18	0

from generation 76 have

- 90.5% of the setpoint ITAE of the Åström-Hägglund tuning rules,
- 96% of the disturbance rejection ITAE of the Åström-Hägglund tuning rules,
- 99.3% of the reciprocal of minimum attenuation of the Åström-Hägglund tuning rules, and
- 98.5% of the maximum sensitivity, M_s, of the Åström-Hägglund tuning rules.

It is true that the four genetically evolved mathematical expressions are more complicated than the four Åström-Hägglund expressions of the form

$$f(x) = a_0 * e^{a_1 x + a_2 x^2}.$$

However, most present-day controllers are implemented by means of programmable microprocessors. Thus, making use of more complicated mathematical tuning rules

Table 12.6 Percentage comparison (for 30 plants) of the best-of-run PID tuning rules and the Åström–Hägglund tuning rules

Plant	Plant parameter value	ITAE 1	ITAE 2	ITAE 3	ITAE 4	ITAE 5	ITAE 6	Stability margin	Sensor noise
A	0.1	94.4	94.4	94.1	94.2	86.9	98.2	95.4	96.7
A	0.3	96.1	96.1	96.1	96.1	90.3	99.7	93.1	98.9
A	1	90.2	90.2	90.2	90.2	83.2	96.3	94.6	99.4
A	3	93.8	93.8	93.8	93.8	92.7	95.8	94.1	99.8
A	4.5	94.2	94.2	94.2	94.2	93.6	96	95.9	99.8
A	6	94.8	94.8	94.8	94.8	94.5	95.5	93.2	99.8
A	7.5	95.4	95.4	95.4	95.4	95.3	96.3	88	99.7
A	9	95.5	95.5	95.5	95.5	95.4	96.6	81.2	99.7
A	10	95.6	95.6	95.6	95.6	95.4	96.6	60.2	99.6
B	3	95.7	95.5	95.7	95.7	95.6	96	92.2	OK
B	4	94.9	94.8	95.1	94.9	93.4	96.8	91.9	OK
B	5	94.4	94.6	94.8	94.4	91.5	97.2	92.9	OK
B	6	94.2	94	94.1	94.2	90.1	97.9	87.2	OK
B	7	94.1	94.2	94	94.1	87.9	98.3	83.6	OK
B	8	93.8	93.8	93.8	93.7	87	98.5	82.5	OK
C	0.2	97.6	97.1	97.6	97.6	97.6	97.4	OK	93.5
C	0.21	97.1	97.3	97.3	97.1	97	97.7	52.1	OK
C	0.23	96.8	96.8	97.1	96.8	96.6	97.3	77.6	OK
C	0.26	96.4	96.8	96.8	96.4	96.3	97.4	89.5	OK
C	0.3	96.2	96.2	96.6	96.2	95.9	97.4	94.3	OK
C	0.4	96.1	95.8	95.8	96.1	95.8	96.9	92.4	OK
C	0.5	95.4	95.7	95.9	95.4	94.7	96.5	95.6	OK
C	0.6	95	95.3	94.8	95	94	96.4	91.1	OK
C	0.7	95.1	95	95.3	95.1	93.9	96.7	91.3	OK
D	0.1	94.9	95.1	94.9	94.9	94.3	95.7	92.4	OK
D	0.2	94.2	94.2	93.9	94.2	93.1	95.8	90	OK
D	0.5	91.8	91.7	91.9	91.8	88.6	94.6	93.7	OK
D	0.7	91	90.9	91.4	91	86.4	94.7	88.8	OK
D	1	90.1	90.1	90.1	90.1	84.6	94.4	87.4	OK
D	2	91.4	91.4	91.2	91.5	88.4	93.2	73.3	OK

merely involves a change in software. The relatively small cost of writing a few lines of additional code is insignificant in comparison to the economic benefits of more efficient control.

In summary, the genetically evolved tuning rules are an improvement over the PID tuning rules developed by Åström and Hägglund in their 1995 book.

12.4 Human-Competitiveness of the Results for the Improved PID Tuning Rules

The Ziegler-Nichols tuning rules (Ziegler and Nichols 1942) for PID controllers were a significant development in the field of control engineering. These rules have been in widespread use since they were invented.

Table 12.7 Performance (on the 18 additional plants) of the best-of-run PID tuning rules

Plant	Plant parameter value	ITAE 1	ITAE 2	ITAE 3	ITAE 4	ITAE 5	ITAE 6	Stability margin	Sensor noise
A	0.15	0.687	0.687	0.694	0.691	0.234	0.453	0.821	4.40
A	0.5	0.546	0.546	0.546	0.546	0.198	0.387	0.564	0.876
A	0.9	0.477	0.477	0.477	0.477	0.198	0.300	0.607	1.23
A	2.5	0.334	0.334	0.334	0.334	0.198	0.139	0.627	1.05
A	4.0	0.277	0.277	0.277	0.277	0.190	0.089	0.443	1.43
A	9.0	0.225	0.225	0.225	0.225	0.186	0.040	0.056	1.13
C	0.25	0.200	0.200	0.200	0.200	0.167	0.032	0.111	0
C	0.34	0.233	0.234	0.233	0.233	0.175	0.057	0.286	0
C	0.43	0.276	0.276	0.275	0.276	0.188	0.087	0.441	0
C	0.52	0.320	0.319	0.321	0.320	0.199	0.120	0.524	0
C	0.61	0.359	0.360	0.358	0.359	0.207	0.150	0.538	0
C	0.69	0.390	0.389	0.389	0.390	0.216	0.173	0.605	0
D	0.15	0.369	0.370	0.368	0.369	0.217	0.150	0.517	0
D	0.3	0.420	0.420	0.421	0.420	0.220	0.198	0.643	0
D	0.6	0.537	0.537	0.538	0.537	0.239	0.295	0.643	0
D	0.85	0.648	0.648	0.650	0.648	0.263	0.382	0.669	0
D	1.2	0.833	0.833	0.834	0.833	0.312	0.517	0.726	0
D	1.8	1.26	1.26	1.259	1.262	0.453	0.805	2.489	0

Table 12.8 Percentage comparison (on the 18 additional plants) of the best-of-run PID tuning rules and the Åström–Hägglund tuning rules

Plant	Plant parameter value	ITAE 1	ITAE 2	ITAE 3	ITAE 4	ITAE 5	ITAE 6	Stability margin	Sensor noise
A	0.15	94.7	94.7	94.3	94.5	88	98.6	95.5	93.2
A	0.5	90.7	90.7	90.7	90.7	79.7	98.8	91	98.5
A	0.9	90.1	90.1	90.1	90.1	82.3	96.8	95.1	99.5
A	2.5	93.2	93.2	93.2	93.2	91.6	95.4	95.4	99.8
A	4.0	94	94	94	94	93.2	96.2	96.5	99.8
A	9.0	95.5	95.5	95.5	95.5	95.4	96.6	81.2	99.7
C	0.25	96.6	96.9	96.9	96.6	96.5	97.7	86.7	OK
C	0.34	96.4	96.6	96.5	96.4	96.1	96.9	96.2	OK
C	0.43	96	95.9	95.7	96	95.7	96.6	93.7	OK
C	0.52	95.8	95.3	95.7	95.8	95.3	96.3	95.9	OK
C	0.61	95.1	95.3	94.9	95.1	94.2	96.4	89.6	OK
C	0.69	95.2	95.1	95.4	95.2	94	96.7	91.1	OK
D	0.15	94.7	94.9	94.3	94.7	94.1	95.7	88.4	OK
D	0.3	93.2	93.3	93.4	93.2	91.2	95.7	91.9	OK
D	0.6	91.2	91.3	91.5	91.2	87.4	94.6	88.3	OK
D	0.85	90.2	90	90.4	90.2	85.3	94.4	89.8	OK
D	1.2	90.4	90.4	90.4	90.4	84.4	94.2	86.6	OK
D	1.8	91.1	91	90.8	91.1	87	93.5	59.6	OK

The 1995 Åström-Hägglund tuning rules were another significant development. They outperform the 1942 Ziegler-Nichols tuning rules on the industrially representative plants used by Åström and Hägglund. Åström and Hägglund developed their improved tuning rules by applying mathematical analysis, shrewdly chosen approximations, and considerable creative flair.

The genetically evolved PID tuning rules are an improvement over the 1995 Åström-Hägglund tuning rules.

Referring to the eight criteria in table 1.2 for establishing that an automatically created result is competitive with a human-produced result, the creation by genetic programming of improved tuning rules for PID controllers satisfies the following five of the eight criteria:

(B) The result is equal to or better than a result that was accepted as a new scientific result at the time when it was published in a peer-reviewed scientific journal.
(D) The result is publishable in its own right as a new scientific result—*independent* of the fact that the result was mechanically created.
(E) The result is equal to or better than the most recent human-created solution to a long-standing problem for which there has been a succession of increasingly better human-created solutions.
(F) The result is equal to or better than a result that was considered an achievement in its field at the time it was first discovered.
(G) The result solves a problem of indisputable difficulty in its field.

A patent application (Keane, Koza, and Streeter 2002a) was filed on July 12, 2002, for the best-of-run PID tuning rules from generation 76 (as well as for the three non-PID controllers described in chapter 13). The applicants believe that the genetically evolved PID tuning rules are patentable because they satisfy the statutory requirements of being "new," "useful," "improved," and "unobvious" to someone "having ordinary skill in the art."

U.S. law suggests that inventions created by automated means are patentable by saying:

> "Patentability shall not be negatived by the manner in which the invention was made." (35 *United States Code* 103a)

If (as expected) a patent is granted, it will (we believe) be the first patent granted for an invention created by genetic programming.

We believe that the creation by genetic programming of improved tuning rules for PID controllers satisfies the following additional criterion from table 1.2:

(A) The result was patented as an invention in the past, is an improvement over a patented invention, or would qualify today as a patentable new invention.

The creation by genetic programming of the improved PID tuning rules satisfies Arthur Samuel's criterion (1983) for artificial intelligence and machine learning:

> "[T]he aim [is]... to get machines to exhibit behavior, which if done by humans, would be assumed to involve the use of intelligence."

12.5 Routineness of the Transition from Problems Involving Parameterized Topologies for Controllers to a Problem Involving PID Tuning Rules

Genetic programming previously created parameterized topologies for a controller for a single plant (section 9.1) and two families of plants (section 9.2).

What is the nature of the transition from this previous work to a problem requiring the automatic synthesis of tuning rules for controllers for four families of plants?

The fitness measure for the present problem is the same as in section 9.2. The choices of functions and terminals are similar to the choices for elementary problems of symbolic regression. Therefore very little effort was required to make the transition to the present problem. Therefore, we consider this transition to be routine.

12.6 AI Ratio for the Improved PID Tuning Rules

As mentioned in section 12.4, the result produced by genetic programming on the present problem is considered to be human-competitive for six reasons. The fact that PID tuning rules have been the target of intense ongoing work in the field of control, that there was a 63 year gap between the 1942 Ziegler-Nichols tuning rules and the 1995 Åström-Hägglund tuning rules, and that an additional seven years has elapsed between 1995 and the genetically evolved PID tuning rules reported in this chapter indicate that the results produced by the artificial system (genetic programming) represent a high amount of "A" in the numerator of the AI ratio.

The preparatory steps in this chapter are simply not noteworthy. The human-supplied "I" amounts to little more than a request to produce four mathematical expressions involving ordinary arithmetic and mathematical functions. The same preparatory steps are applicable to a many similar problems from a variety of fields and similar to the preparatory steps for elementary problems of symbolic regression. The fitness measure is straightforward (and is the same as in section 9.2).

However, because the results in this chapter bootstrapped on an important domain-specific result from the field of control (the Åström-Hägglund tuning rules), the results in this chapter benefited from more "I" than would have been the case if the evolutionary process had started from scratch. Thus, the solution produced by genetic programming in this chapter has a non-trivial amount of "I" in the denominator of the AI ratio. Hence, the AI ratio for the problem in this chapter is moderately high.

13

Automatic Synthesis of Parameterized Topologies for Improved Controllers

This chapter presents three genetically evolved parameterized topologies for controllers. Unlike the work in chapter 12 (where the topology was prespecified to be that of a PID controller), the topology of the to-be-evolved controller in this chapter is open-ended. That is, genetic programming creates the topology dynamically during the run. Accordingly, genetic programming is free to create either a PID or non-PID topology. As it turns out, all three controllers evolved in this chapter employ non-PID topologies.

The three controllers evolved in this chapter are parameterized (i.e., general) in the sense that they incorporate free variables and provide a solution to an entire category of problems (i.e., the control of all the plants in all the families).

The fitness measure used in this chapter is based on plants from the same four families used by Åström and Hägglund.

The first of the three genetically evolved controllers in this chapter has 66.4% of the setpoint ITAE of the Åström-Hägglund controller and 85.7% of the disturbance rejection ITAE of the Åström-Hägglund controller.

In the early years of the field of control, controllers were typically composed of mechanical, pneumatic, physical, or other non-programmable devices. Although the parameter values of the blocks in these controllers were usually variable, the topologies of these devices were typically not easily modifiable. Because of this inflexibility, there was a strong tendency to consider only PID controllers. In contrast, most present-day controllers are implemented by means of programmable microprocessors. Thus, more complex topologies (i.e., non-PID topologies) can be readily implemented today merely by changing software.

13.1 Preparatory Steps for Improved General-Purpose Controllers

The preparatory steps are, with 10 differences, the same as in section 9.2. The differences (except the fifth one) represent either minor adjustments in technique or changes dictated by the problem's difficulty (e.g., population size, concerns about overfitting,).

Four of the 10 differences (namely the third, fourth, sixth, and ninth) relate to a single change (i.e., the handling of takeoff points). All the minor adjustments in technique occurred to us after our work on the control problems in chapter 3. Our use of these different techniques reflects the chronology of our work and the evolution of our thinking (as opposed to any special requirement or exigency of the present problem). We intend to use these new techniques for our future work on problems of automatic synthesis of controllers.

13.1.1 Function Set

The first difference is that the POW function (section 12.2.3) is added to the function set for the arithmetic-performing subtrees.

The second difference is that the ULIMIT function is deleted from the function set.

The third difference relates to the handling of takeoff points. In chapter 3, takeoff points were implemented using automatically defined functions. This approach was attractive to us in our early work on the automatic synthesis of controllers because the architecture-altering operations provide a convenient way to create automatically defined functions. However, because takeoff points are so common in controllers and the architecture-altering operations are invoked at a relatively low frequency, we came to believe that this approach is inefficient. Thus, a new one-argument function was defined. The one-argument TAKEOFF function acts as an identity function in that it returns the value of its argument. At the same time, the TAKEOFF function stores the value of its argument so as to make this value potentially available to other points in the block diagram. If a subsequent *takeoff point reference terminal* (defined below) appears in the program tree, the effect is to connect points in the controller's block diagram.

13.1.2 Terminal Set

The fourth difference (which goes hand in hand with the new approach to takeoff points mentioned above in section 13.1.1) is that eight new takeoff point reference terminals (LEFT_1, ..., LEFT_4 and RIGHT_1, ..., RIGHT_4) are added to the terminal set. Whenever one of the four terminals (LEFT_1, ..., LEFT_4) is encountered, it returns the value stored by the first, second, third, or fourth TAKEOFF function (if any), respectively, occurring earlier (i.e., to the left) in the overall program tree. For this purpose, the earlier TAKEOFF function is determined on the basis of the usual depth-first order of evaluation (from left to right) used in the LISP programming language. If there is no such stored value, the terminal defaults to the value of the root of the entire program tree (i.e., the controller's output). Similarly, the value returned by RIGHT_1, ..., RIGHT_4 is based on occurrences of the TAKEOFF function that occur later (i.e., to the right) in the overall program tree.

The fifth (and most substantial) difference enables us to bootstrap on the 1995 Åström-Hägglund tuning rules. Specifically, a terminal (called AH) is inserted into the terminal set representing the time-domain output of a PID controller that is tuned for the plant under consideration according to tuning rules developed by Åström and Hägglund in their 1995 book. That is, the terminal causes a connection to be made to the output port of a PID controller that was tuned using the Åström-Hägglund rules for the particular instantiations (for the plant under consideration) of the free variables for ultimate gain, K_u; ultimate period, T_u; dead time, L; and time constant, T_r. Thus, the terminal set, T, is (except for the arithmetic-performing subtrees)

```
T = {AH, REFERENCE_SIGNAL, CONTROLLER_OUTPUT,
  PLANT_OUTPUT, LEFT_1,…, LEFT_4, RIGHT_1,…,
  RIGHT_4}.
```

It should be noted that this technique for incorporating a known outstanding result can be used for problems from many different fields.

13.1.3 Program Architecture

The sixth difference (which goes hand in hand with the new approach to takeoff points mentioned in section 13.1.1) is that automatically defined functions (and hence architecture-altering operations) are not used. Thus, each program tree in the population simply has one result-producing branch.

13.1.4 Fitness Measure

The seventh difference is that we increase the number of plants considered by the fitness measure. In this problem, numerous mathematical expressions (each incorporating up to four free variables) must be evolved to establish the parameters for signal-processing functions of the to-be-evolved controller. We were especially concerned about overfitting because there are only three plants in two of the four families in the test bed of 16 plants used by Åström and Hägglund. Because each additional plant adds eight SPICE simulations to the evaluation of fitness for each individual in the population, we necessarily counterbalanced our concern about overfitting with practical considerations of computer time. Consequently, we used the 20 plants marked "1" in column 7 of table 12.1 for the first run in this chapter. Thus, the fitness measure for the first run entails 160 separate invocations of the SPICE simulator. Of course, we did not know in advance whether 20 plants would prove to be sufficient to produce an improved controller. Also, if an improved controller can be evolved at all, we did not know in advance the number of generations or amount of time required to accomplish this. After the first run succeeded in producing an improved controller, we decided to further increase the number of plants. We decided to use the 24 plants marked "2" and "3" in column 7 of table 12.1 for the second and third runs (i.e., a total of 192 separate invocations of the SPICE simulator are used to evaluate the fitness of each individual in the population).

The eighth difference applies only to the second and third runs. After the first run succeeded, we noticed that the evolved controller has a rather large number of signal-processing blocks. Consequently, we added a 193rd element (measuring parsimony) to the fitness measure for the second and third runs. The parsimony penalty places more weight on the major signal-processing functions (e.g., gain, integrator, differentiator, lead, lag) than on the ordinary arithmetic functions (e.g., addition, subtraction). The parsimony penalty is the sum, taken over each signal-processing block in the controller, of 0.1 for each addition or subtraction block and 1.0 for all other blocks. Note that this parsimony penalty counts signal-processing blocks and does not consider the number of TAKEOFF functions or the functions in the arithmetic-performing subtrees of the signal-processing blocks.

The number of hits is defined as the number of fitness cases for which the individual controller outperforms a PID controller tuned using the rules developed by Åström and Hägglund in their 1995 book.

For an individual that scores less than 100% of the possible number of hits (160 for the first run and 192 for the second and third runs), fitness is the sum of the elements of the fitness measure associated with the SPICE simulations plus one trillionth of the parsimony penalty. Thus, the parsimony penalty has virtually no effect on individuals that do not outperform Åström and Hägglund on all SPICE simulations.

For an individual that scores 100% of the possible number of hits, the fitness is one trillionth of the parsimony penalty (and nothing else). That is, among individuals that outperform the Åström-Hägglund controller, the more parsimonious individual will always be deemed to be better (i.e., have a smaller value of fitness). Note that this arrangement means that, once 192 hits is achieved, there is no additional selective pressure in the direction of increasing the amount of improvement over the Åström-Hägglund controller.

In the hope of achieving a parsimonious controller, our intent was to permit the second and third runs to continue for one or two weeks after the time when the run produced its first fully compliant individual.

13.1.5 Control Parameters

The ninth difference (which goes hand in hand with the new approach to takeoff points mentioned in section 13.1.1) is that necessary minor adjustments are made in the percentages of genetic operations to reflect the absence of the architecture-altering operations and automatically defined functions. The percentages of genetic operations are given in appendix B.

The 10th difference is the population size. For difficult problems, we typically choose the population size such that genetic programming can execute a reasonably large number of generations within the amount of computer time we are willing to devote to the problem. Advance testing indicated that the time required to measure the fitness of an individual would exceed 2 minutes per individual. Our intent was to spend several weeks on our 1,000-Pentium Beowulf-style parallel computer on each run of this manifestly difficult problem. We chose a small population size of 100 for each node on our 1,000-Pentium Beowulf-style parallel computer (for a total population size of 100,000). Given this choice, it was anticipated that each generation would take a little over three hours and that there would therefore be about eight generations per day and about 56 generations per week. Note that there is considerable uncertainly surrounding all advance estimates of the length of time of runs of genetic programming because the complexity of the individuals that are actually generated during the run significantly affects simulation time.

13.2 Results for Improved General-Purpose Controllers

We made three runs of this problem.

13.2.1 Results for First Run for Improved General-Purpose Controllers

The best-of-run individual from the first run of this problem emerged in generation 88.

The fitness measure for this first run entails 160 separate invocations of the SPICE simulator (i.e., eight simulations for each of 20 plants). The average time to evaluate the fitness of an individual was 2 minutes and 10 seconds. Because the population size is 100,000, the fitness of about 8,900,000 individuals was evaluated during the run. It took 320 hours (13.3 days) to produce the best-of-run individual. As in all the problems in this book, most of the computer time was consumed in evaluating the fitness of individuals in the population—not on genetic operations.

Figure 13.1 shows the best-of-run parameterized controller from generation 88 of the first run. This first parameterized controller is composed of lags (with transfer functions of the form $1/(1 + \tau s)$) as well as gain, differentiator, integrator, adder, and subtractor blocks. Genetic programming produced this controller's overall topology consisting of four adders, nine subtractors, eight gain blocks parameterized by a constant, five gain blocks parameterized by non-constant mathematical expressions containing free variables, three lag blocks parameterized by non-constant mathematical expressions containing free variables, and one Åström-Hägglund sub-controller.

The output of the improved parameterized controller of figure 13.1 is control variable 390. There are three time-domain inputs to this controller, namely the reference signal 300, the plant output 304, and controller output 390 (i.e., internal feedback of the output of this controller into itself).

The four free variables available to each arithmetic-performing subtree of each signal-processing function possessing a parameter value are the plant's ultimate gain, K_u; ultimate period, T_u; dead time, L; and time constant, T_r.

Note that the plant output indirectly enters the best-of-run parameterized controller from generation 88 through block 306. Block 206 is a PID controller that is tuned with the tuning rules developed by Åström and Hägglund in their 1995 book for the plant under consideration.

The first parameterized controller has eight gain blocks (326, 327, 332, 334, 336, 382, 384, and 385) that are parameterized by a constant numerical amplification factor. In particular, gain blocks 326, 332, 382, and 384 each have a gain of 3. Gain blocks 327, 334, 336, and 385 each have a gain of 2.

The first parameterized controller has four two-argument adders (324, 352, 388, and 387) and seven two-argument subtractors (308, 312, 338, 346, 366, 376, and 386). It also has two three-argument subtractors (328 and 335) in which one signal is subtracted from the sum of the two other signals.

In addition, the first parameterized controller has five gain blocks (310, 330, 340, 360, and 370) that are parameterized by genetically evolved non-constant mathematical expressions containing free variables.

Gain block 310 in figure 13.1 is parameterized by equation 11:

$$10^{e^{\log|\log|\log(e^{K_u * L})/L||}}. \quad [11]$$

Gain block 330 in figure 13.1 is parameterized by equation 13:

$$10^{e^{\log|\log|K_u * L||}}. \quad [13]$$

Gain block 340 in figure 13.1 is parameterized by equation 14:

$$e^{\log|K_u/L|}. \quad [14]$$

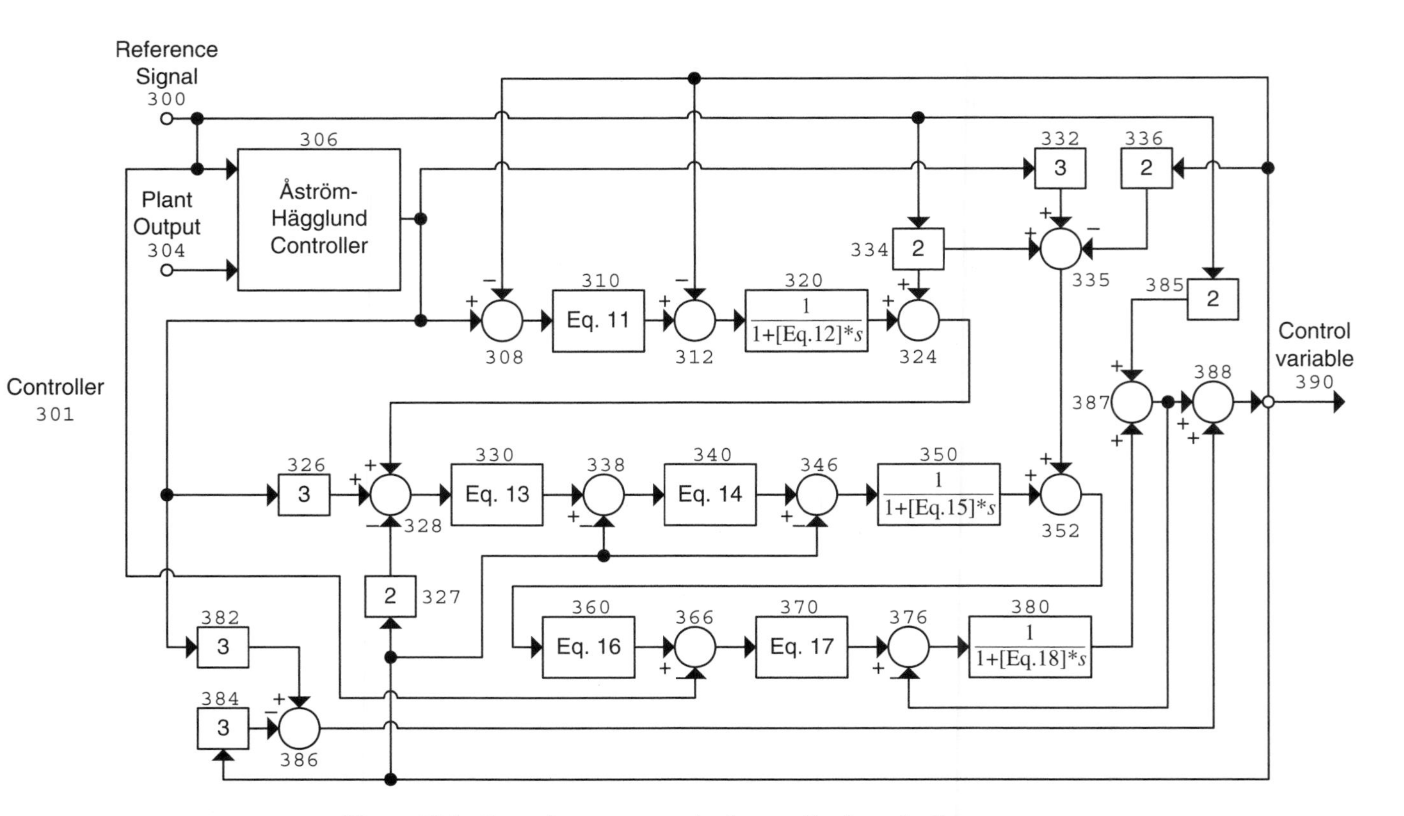

Figure 13.1 Best-of-run parameterized controller from the first run.

Gain block 360 in figure 13.1 is parameterized by equation 16:

$$10^{e^{\log|\log|K_u * L||}}. \qquad [16]$$

Gain block 370 in figure 13.1 is parameterized by equation 17:

$$e^{\log(K_u)}. \qquad [17]$$

The first parameterized controller also has three lag blocks (320, 350, and 380) that are parameterized by genetically evolved non-constant mathematical expressions containing free variables.

Equation 12 for lag block 320, equation 15 for lag block 350, and equation 18 for lag block 380 are the same:

$$T_r.$$

The first parameterized controller has internal feedback in several places. For one thing, its own final output 390 is fed back to seven other blocks (308, 312, 327, 336, 338, 346, and 384) within this controller.

In addition, there is internal feedback involving subtractor 376, lag block 380, and adder 387 in which the signal that is fed back is not the controller's own final output. That is, this internal feedback loop lies entirely inside the controller.

In addition, the first parameterized controller contains Åström-Hägglund controller 306 (which contains additional adder, subtractor, integration, differentiation, and gain blocks).

Note that figure 13.1 (and other similar figures in this chapter) is simplified by replacing multiple blocks adding the same signal with a single gain block (with an appropriate amplification factor) and by deleting pairs of blocks that cancel each other (e.g., the same signal is first added and then subtracted).

Notice that genetic programming automatically created all the following in the single run that produced the controller of figure 13.1:

- the controller's topology, including
  - the total number of processing blocks (30),
  - the type of each block,
  - all the connections (directed lines) that exist between the controller's two external input points (i.e., the reference signal and the plant output), the controller's 30 blocks, and the controller's external output point,
- the controller's tuning, including
  - three mathematical expressions containing free variables (for the three lag blocks),
  - five mathematical expressions containing free variables (for five of the 13 gain blocks),
  - eight constant mathematical expressions (for eight of the 13 gain blocks).

As will be seen, the three parameterized controllers in this chapter have several features in common.

First, each controller includes an Åström-Hägglund block as well as a second sub-controller that receives the output from the Åström-Hägglund controller.

Second, the difference between the output of the Åström-Hägglund controller and output of the second sub-controller is computed at least once.

Third, each second sub-controller makes use of internal feedback in the sense that it feeds its own output back into itself.

Fourth, each second sub-controller includes at least one lag block or at least one lead block (in addition to the usual gain, integrative, derivative, additive, and subtractive blocks).

The best-of-run parameterized controller from generation 88 of the first run can be described in terms of its transfer function. The transfer function is given by the equation:

$$U = \frac{A(3 + 3G_{LAG}^2 E_{13}E_{14}E_{16}E_{17} + G_{LAG}^3 E_{11}E_{13}E_{14}E_{16}E_{17} + 3G_{LAG}(1 + E_{16}E_{17})) + (2 + 2G_{LAG}^2 E_{13}E_{14}E_{16}E_{17} + G_{LAG}(-1 + 2E_{16})E_{17})R}{4 + G_{LAG}^3(1 + E_{11})E_{13}E_{14}E_{16}E_{17} + G_{LAG}^2(1 + E_{14} + 2E_{13}E_{14})E_{16}E_{17} + 2G_{LAG}(2 + E_{16}E_{17})},$$

where U is the controller output, A is the output of the Åström-Hägglund controller, R is the reference signal, E_{11} through E_{17} correspond to equations 11 through 17, respectively, and G_{LAG} is the transfer function for the identical lag blocks, namely

$$G_{LAG} = \frac{1}{1 + T_r{*}s}.$$

Averaged over the 20 plants used in this run (i.e., those marked "1" in column 7 of table 12.1), the first parameterized controller has

- 66.4% of the setpoint ITAE of the Åström-Hägglund controller,
- 85.7% of the disturbance rejection ITAE of the Åström-Hägglund controller,
- 94.6% of the reciprocal of minimum attenuation of the Åström-Hägglund controller, and
- 92.9% of the maximum sensitivity, M_s, of the Åström-Hägglund controller.

Averaged over the 18 additional plants of table 12.2, the first parameterized controller has

- 64.1% of the setpoint ITAE of the Åström-Hägglund controller,
- 84.9% of the disturbance rejection ITAE of the Åström-Hägglund controller,
- 95.8% of the reciprocal of minimum attenuation of the Åström-Hägglund controller, and
- 93.5% of the maximum sensitivity, M_s, of the Åström-Hägglund controller.

As can be seen, the results obtained for the plants used in this run are very similar to those for the 18 previously unseen additional plants.

Thus, the best-of-run parameterized controller from generation 88 is an improvement over the PID controller developed by Åström and Hägglund in their 1995 book.

We now provide additional comparative details.

Figure 13.2 compares the cumulative ITAE for the best-of-run parameterized controller from generation 88 of the first run (solid line) with that for the Åström and Hägglund controller (dotted line) for the three-lag plant in response to a reference signal that steps up from 0 Volts to 1 Volt. As can be seen, the genetically evolved controller virtually eliminates the error at about 6 seconds whereas the Åström and Hägglund controller continues to accumulate error.

Table 13.1 shows the performance of the best-of-run parameterized controller from generation 88 of the first run on the 20 plants used in this run (i.e., those marked "1" in column 7 of table 12.1).

Table 13.2 compares the performance of the best-of-run parameterized controller from generation 88 of the first run as a percentage of the value for the Åström and Hägglund controller for the 20 plants used in this run (i.e., those marked "1" in column 7 of table 12.1). An "OK" appears in this table (and other similar tables later in this chapter) for the cases where the penalties associated with both controllers are 0. As can be seen in the table, all percentages are either below 100% (indicating improvement) or "OK."

Table 13.3 shows the performance of the best-of-run parameterized controller from generation 88 of the first run on the 18 additional plants from table 12.2.

Table 13.4 compares the performance of the best-of-run parameterized controller from generation 88 of the first run as a percentage of the value for the Åström and Hägglund controller for the 18 additional plants from table 12.2. As can be seen, there is one value over 100% (100.3) for one plant from family A concerning sensitivity and another value over 100% (235.9) for one plant from category C concerning stability.

For additional information, see Keane, Koza, and Streeter 2002b.

When a human engineer conceives of the design for a controller, circuit, antenna, or other complex structure, the engineer usually has sufficient mathematical and intuitive understanding of the design to be confident about its behavior over a wide range of situations. In contrast, when an automated method generates a design (particularly a novel design), confidence must be established in other ways.

The amount of time and attention that is devoted to any particular individual in the population is necessarily limited because of the millions of individuals that are

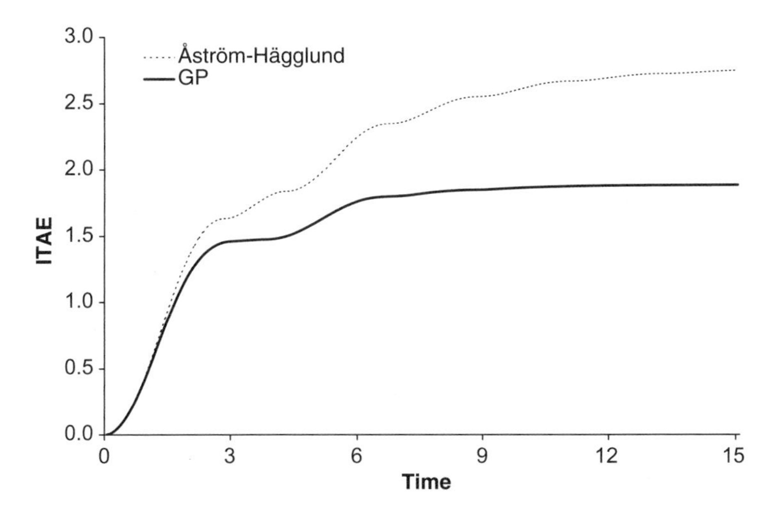

Figure 13.2 Comparison of cumulative ITAE for the best-of-run parameterized controller from the first run and the Åström–Hägglund controller for the three-lag plant.

Table 13.1 Performance (for 20 plants) of the best-of-run controller from the first run

Plant	Parameter value	ITAE 1	ITAE 2	ITAE 3	ITAE 4	ITAE 5	ITAE 6	Stability margin	Sensor noise
A	0.1	0.661	0.663	0.664	0.665	0.216	0.445	0.581	8.53
A	0.3	0.606	0.606	0.604	0.605	0.171	0.434	0.472	1.54
A	1	0.385	0.385	0.385	0.365	0.138	0.246	0.590	0.85
A	3	0.222	0.222	0.221	0.222	0.131	0.095	0.426	1.09
A	10	0.165	0.166	0.165	0.165	0.137	0.028	0	0.974
B	3	0.225	0.225	0.225	0.225	0.145	0.079	0	0
B	4	0.320	0.320	0.319	0.319	0.151	0.168	0.336	0
B	6	0.449	0.449	0.448	0.449	0.172	0.274	0.510	0
B	7	0.485	0.485	0.482	0.484	0.173	0.308	0.515	0
B	8	0.512	0.512	0.509	0.511	0.175	0.333	0.511	0
C	0.1	0.141	0.142	0.142	0.141	0.135	0.005	0	0.203
C	0.2	0.144	0.144	0.144	0.144	0.126	0.018	0	0
C	0.5	0.221	0.221	0.221	0.221	0.130	0.090	0.177	0
C	0.7	0.284	0.284	0.284	0.284	0.143	0.140	0.288	0
C	0.9	0.319	0.318	0.318	0.318	0.151	0.166	0.336	0
D	0.1	0.248	0.249	0.249	0.248	0.142	0.105	0.071	0
D	0.2	0.275	0.274	0.274	0.275	0.143	0.131	0.210	0
D	0.5	0.364	0.364	0.364	0.364	0.148	0.215	0.452	0
D	1	0.579	0.578	0.578	0.579	0.196	0.383	0.534	0
D	2	1.32	1.32	1.32	1.32	0.441	0.874	4.343	0

Table 13.2 Percentage comparison (for 20 plants) of the best-of-run parameterized controller from the first run and the Åström–Hägglund controller

Plant	Parameter value	ITAE 1	ITAE 2	ITAE 3	ITAE 4	ITAE 5	ITAE 6	Stability margin	Sensor noise
A	0.1	86.8	87	86.1	87	77.8	90.8	59.4	84.2
A	0.3	86.3	86.3	86	86.3	64.1	99.9	70.5	88.3
A	1	78.9	78.9	78.8	74.7	60.6	88.2	89.4	76.9
A	3	67.8	67.8	67.6	67.8	62.4	79.3	71.7	73.3
A	10	73.1	73.3	73.1	73	72	78.3	0	87.6
B	3	67.7	67.8	67.9	67.7	64.8	74.3	0	OK
B	4	69.7	69.7	69.9	69.7	62.8	78	47.5	OK
B	6	78.4	78.5	78.5	78.4	70	85.4	78.9	OK
B	7	80.1	80.2	80.2	80.2	69	88	80.2	OK
B	8	83.3	83.3	83.2	83.2	70.9	91.7	79.5	OK
C	0.1	77.4	78.2	78.5	77.4	77	90.4	OK	30.9
C	0.2	76	76	75.8	76	74.7	86.2	OK	0
C	0.5	70.5	70.4	70.3	70.5	64.7	80.7	32.8	OK
C	0.7	68.3	68.3	68.6	68.3	61.6	77.2	42.7	OK
C	0.9	69.2	69.1	69	69.1	61.9	77.1	47.3	OK
D	0.1	67.1	67.3	67.3	66.8	62.2	74.7	14.2	OK
D	0.2	67.4	67.3	67.1	67.4	61.3	76	32.8	OK
D	0.5	70.3	70.4	70.3	70.4	59.2	80.7	64.2	OK
D	1	73.9	73.6	73.7	73.9	60.4	84.4	70.9	OK
D	2	83.3	83.2	83.4	83.1	74.9	88.2	61.5	OK

Table 13.3 Performance (on the 18 additional plants) of the best-of-run controller from the first run

Plant	Parameter value	ITAE 1	ITAE 2	ITAE 3	ITAE 4	ITAE 5	ITAE 6	Stability margin	Sensor noise
A	0.15	0.653	0.653	0.658	0.657	0.207	0.447	0.545	1.94
A	0.5	0.533	0.533	0.533	0.533	0.155	0.377	0.449	0.635
A	0.9	0.406	0.406	0.382	0.384	0.140	0.266	0.581	1.135
A	2.5	0.237	0.237	0.237	0.238	0.130	0.113	0.484	0.793
A	4.0	0.203	0.203	0.202	0.202	0.134	0.071	0.322	1.44
A	9.0	0.168	0.168	0.168	0.167	0.136	0.031	0	0.991
C	0.25	0.150	0.150	0.150	0.150	0.123	0.027	0.302	0
C	0.34	0.168	0.168	0.168	0.168	0.122	0.045	0.011	0
C	0.43	0.203	0.203	0.203	0.203	0.129	0.073	0.174	0
C	0.52	0.228	0.228	0.228	0.227	0.131	0.096	0.187	0
C	0.61	0.258	0.258	0.258	0.258	0.137	0.119	0.226	0
C	0.69	0.281	0.281	0.281	0.281	0.142	0.138	0.278	0
D	0.15	0.261	0.262	0.262	0.261	0.143	0.118	0.147	0
D	0.3	0.302	0.302	0.302	0.302	0.143	0.158	0.316	0
D	0.6	0.399	0.399	0.399	0.400	0.154	0.244	0.490	0
D	0.85	0.502	0.502	0.502	0.503	0.181	0.321	0.549	0
D	1.2	0.694	0.695	0.694	0.693	0.227	0.465	0.593	0
D	1.8	1.12	1.12	1.12	1.12	0.367	0.748	1.38	0

Table 13.4 Percentage comparison (on the 18 additional plants) of the best-of-run controller from the first run and the Åström–Hägglund controller

Plant	Parameter value	ITAE 1	ITAE 2	ITAE 3	ITAE 4	ITAE 5	ITAE 6	Stability margin	Sensor noise
A	0.15	90	90	89.4	89.9	77.8	97.2	63.5	41.2
A	0.5	88.6	88.6	88.5	88.6	62.5	96.4	72.4	71.4
A	0.9	76.8	76.8	72.2	72.6	58.2	85.7	91	91.6
A	2.5	66.3	66.3	66.2	66.4	60.2	77.8	73.6	75.6
A	4.0	68.9	68.9	68.8	68.8	65.8	77.2	70.2	100.3
A	9.0	71.2	71.3	71.1	71	70	76.5	0	87.8
C	0.25	72.9	72.8	72.8	72.9	71.3	82.1	235.9	OK
C	0.34	69.5	69.2	69.4	69.5	67	77.8	3.9	OK
C	0.43	70.8	70.8	70.8	70.5	66	81.2	37.1	OK
C	0.52	68.1	68.1	67.9	68.1	62.6	77.1	34.3	OK
C	0.61	68.4	68.5	68.3	68.4	62.4	77	37.7	OK
C	0.69	68.6	68.6	69	68.6	61.9	77.3	41.9	OK
D	0.15	67.1	67.1	67.2	67.1	61.9	74.8	25.2	OK
D	0.3	67	67	67	67	59.2	76.4	45.2	OK
D	0.6	67.9	67.8	67.9	67.9	56.4	78.3	67.4	OK
D	0.85	70	69.8	69.8	70.1	58.6	79.3	73.7	OK
D	1.2	75.4	75.3	75.2	75.3	61.3	84.7	70.7	OK
D	1.8	80.6	80.6	80.6	80.8	70.5	86.9	33.1	OK

processed during a run of genetic programming. However, once a single individual emerges as the best-of-run individual from a run, it becomes practical (and advisable) to analyze and test that individual more extensively. Post-run testing may involve even more cross-validation (i.e., additional plants, additional values for the height of the reference signal and disturbance signals, additional values for noise, different types of added noise signals). The additional post-run testing may involve types of analysis that were not part of the original fitness measure. The genetically evolved individual may also be analyzed mathematically. Of course, there will be considerable motivation and justification to do such post-run testing and analysis if the automatically created design outperforms previously known designs.

13.2.2 Results for Second Run for Improved General-Purpose Controllers

The best-of-run individual from the second run of this problem emerged in generation 38.

The fitness measure for the second run entails 192 separate invocations of the SPICE simulator (i.e., eight simulations for each of 24 plants). The average time to evaluate the fitness of an individual is about 2 minutes and 36 seconds. Note that the time is longer than that of the first run because there are more plants. It took 397 hours (16.5 days) to produce the best-of-run individual.

The fitness measure for this second run contains a 193rd element that considers the controller's parsimony. The parsimony element of the fitness measure is dominant only after a controller outperforms the Åström-Hägglund PID controller for all 192 SPICE simulations. Unfortunately, within a few hours after the second run evolved its first individual scoring 192 hits, a citywide power failure (lasting longer than the 15 minutes covered by our uninterruptable power supply) prematurely terminated this 16-day run. Thus, although the second controller outperforms the Åström-Hägglund controller (by a small margin), it is not particularly parsimonious. Nonetheless, this second controller is noteworthy because it is another topologically different controller that outperforms the Åström-Hägglund PID controller and because it helps establish the common features of the three controllers in this chapter (section 13.2.1).

Figure 13.3 shows the best-of-run parameterized controller from generation 38 of the second run. This second controller is composed of lags as well as gain, differentiator, integrator, adder, and subtractor blocks. Genetic programming produced this controller's overall topology consisting of four adders, six subtractors, three gain blocks parameterized by a constant, eight lag blocks parameterized by non-constant mathematical expressions containing free variables, and one Åström-Hägglund sub-controller.

The output of the second controller is control variable 490. There are three time-domain inputs to this controller, namely the reference signal 400, the plant output 404, and controller output 490 (i.e., internal feedback of the output of the controller into itself).

Note that the plant output indirectly enters the best-of-run parameterized controller from generation 38 through block 406. Block 406 is a PID controller that is tuned with the Åström-Hägglund tuning rules for the plant under consideration.

The second controller has three gain blocks (412, 416, and 442) that are parameterized by a constant numerical amplification factor. In particular, gain block 412 has a gain of 3. Gain block 416 has a gain of 15. Gain block 442 has a gain of 2.

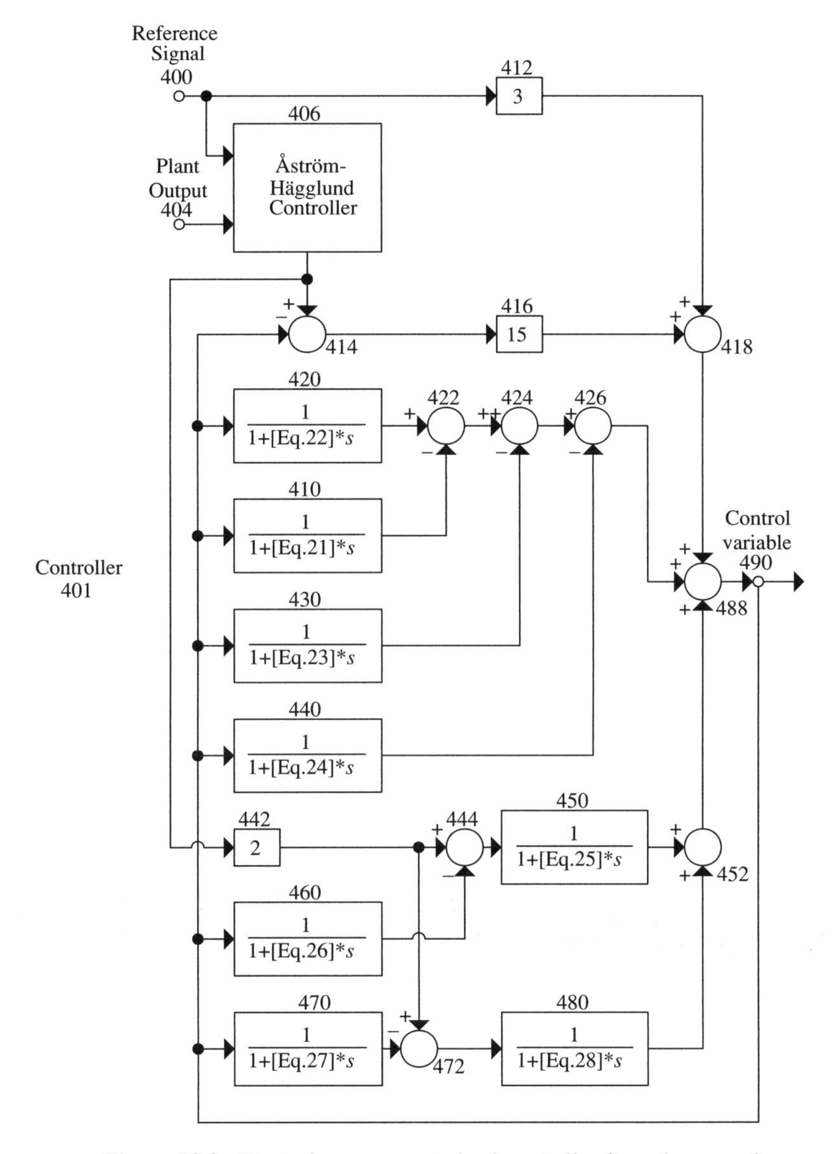

Figure 13.3 Best-of-run parameterized controller from the second run.

The second controller has one three-argument adder 488, two two-argument adders (418 and 452), and six two-argument subtractors (414, 422, 424, 426, 444, and 472).

The second controller also has eight lag blocks (410, 420, 430, 440, 450, 460, 470, and 480) that are parameterized by genetically evolved non-constant mathematical expressions containing free variables.

Lag block 410 in figure 13.3 is parameterized by equation 21:

$$\log|2T_r + K_u^L|. \quad [21]$$

Lag block 420 in figure 13.3 is parameterized by equation 22:

$$\left|\log\left|T_r + K_u^{\left(\frac{0.68631}{\frac{1}{K_u^{INT_1}} - T_r^L - T_u}\right)}\right|\right|, \quad [22]$$

where

$$INT_1 = \frac{\frac{0.13031}{\frac{T_u}{K_u^{\log\left|\log\left|T_r + K_u^{(K_u^L)}\right|\right|}} - (K_u^L)^L - T_u}}{\frac{1}{\frac{0.69897}{T_r + K_u^{T_r + K_u^L}}} - T_r^L - T_u}.$$

Lag block 430 in figure 13.3 is parameterized by equation 23:

$$\log|2T_r + K_u^{\log|T_r + K_u^L|}|. \quad [23]$$

Lag block 440 in figure 13.3 is parameterized by equation 24:

$$\log|T_r + K_u^L|. \quad [24]$$

Lag block 450 in figure 13.3 is parameterized by equation 25:

$$|\log|T_r + \log(T_r + 1.2784)||. \quad [25]$$

Lag block 460 in figure 13.3 is parameterized by equation 26:

$$\log|T_r + (T_r + (x)^{\log|\log|K_u^L||})^L|, \quad [26]$$

where

$$x = T_r + K_u^{\log|\log|T_r + K_u^L||}.$$

Lag block 470 in figure 13.3 is parameterized by equation 27:

$$\log|T_r + (T_r + K_u^L)^L|. \quad [27]$$

Lag block 480 in figure 13.3 is parameterized by equation 28:

$$\log|2T_r + K_u^L|. \quad [28]$$

The second controller has internal feedback of its own final output 490 back into itself in seven places. Specifically, controller output 490 is fed back as input to six lag blocks (410, 420, 430, 440, 460, and 470). In addition, controller output 490 is subtracted, by subtractor 414, from the output of PID controller 406 tuned using the Åström-Hägglund tuning rules. The difference becomes the input to gain block 416.

In addition, the second controller contains Åström-Hägglund controller 406 (which contains additional adder, subtractor, integration, differentiation, and gain blocks).

The best-of-run parameterized controller from generation 38 from the second run can be described in terms of its transfer function, namely

$$U = -\frac{-3R + A(-15 - 2/(1 + E_{25}{*}s) - 2/(1 + E_{28}{*}s))}{16 + \frac{1}{1 + E_{21}{*}s} - \frac{1}{1 + E_{22}{*}s} + \frac{1}{1 + E_{23}{*}s} + \frac{1}{1 + E_{24}{*}s} + \frac{1}{(1 + E_{25}{*}s)(1 + E_{26}{*}s)} + \frac{1}{(1 + E_{27}{*}s)(1 + E_{28}{*}s)}},$$

where U is the controller output, A is the output of the Åström-Hägglund controller, R is the reference signal, and E_{21} through E_{28} correspond to equations 21 through 28, respectively.

Averaged over the 24 plants used in this run (i.e., those marked "2" in column 7 of table 12.1), the second controller has

- 85.5% of the setpoint ITAE of the Åström-Hägglund controller,
- 91.8% of the disturbance rejection ITAE of the Åström-Hägglund controller,
- 98.9% of the reciprocal of minimum attenuation of the Åström-Hägglund controller, and
- 97.5% of the maximum sensitivity, M_s, of the Åström-Hägglund controller.

Averaged over the 18 additional plants of table 12.2, the second controller has

- 84% of the setpoint ITAE of the Åström-Hägglund controller,
- 90.6% of the disturbance rejection ITAE of the Åström-Hägglund controller,
- 98.9% of the reciprocal of minimum attenuation of the Åström-Hägglund controller, and
- 97.5% of the maximum sensitivity, M_s, of the Åström-Hägglund controller.

As can be seen, the results obtained for the plants used in this run are very similar to those for the 18 previously unseen additional plants.

Thus, the best-of-run parameterized controller from generation 38 from the second run is an improvement over the PID controller developed by Åström and Hägglund in their 1995 book.

We now provide additional comparative details.

Table 13.5 shows the performance of the best-of-run parameterized controller from generation 38 of the second run on the 24 plants used in this run (i.e., those marked "2" in column 7 of table 12.1).

Table 13.6 presents the performance of the best-of-run parameterized controller from generation 38 of the second run as a percentage of the value for the Åström and Hägglund controller for the 24 plants used in this run (i.e., those marked "2" in column 7 of table 12.1). As can be seen in the table, all percentages are either below 100% (indicating improvement) or "OK."

Table 13.7 shows the performance of the best-of-run parameterized controller from generation 38 of the second run on the 18 additional plants from table 12.2.

Table 13.8 compares the performance of the best-of-run parameterized controller from generation 38 of the second run as a percentage of the value for the Åström and Hägglund controller for the 18 additional plants from table 12.2. As can be seen, there

Table 13.5 Performance (for 24 plants) of the best-of-run controller from the second run

Plant	Parameter value	ITAE 1	ITAE 2	ITAE 3	ITAE 4	ITAE 5	ITAE 6	Stability margin	Sensor noise
A	0.1	0.698	0.698	0.700	0.692	0.231	0.464	0.737	9.76
A	0.3	0.658	0.658	0.660	0.659	0.230	0.428	0.571	1.71
A	1	0.446	0.446	0.447	0.446	0.203	0.260	0.603	1.05
A	3	0.291	0.291	0.291	0.291	0.182	0.111	0.546	1.44
A	6	0.229	0.229	0.229	0.229	0.174	0.057	0.247	1.19
A	10	0.216	0.216	0.216	0.216	0.181	0.036	0	1.06
B	3	0.288	0.287	0.287	0.288	0.193	0.093	0.321	0
B	4	0.407	0.406	0.406	0.407	0.209	0.195	0.578	0
B	5	0.478	0.477	0.479	0.478	0.216	0.261	0.634	0
B	6	0.526	0.525	0.524	0.526	0.218	0.305	0.603	0
B	7	0.562	0.562	0.560	0.562	0.221	0.338	0.573	0
B	8	0.581	0.580	0.579	0.580	0.221	0.357	0.577	0
C	0.2	0.170	0.170	0.170	0.170	0.150	0.019	0	0
C	0.3	0.190	0.190	0.190	0.190	0.149	0.040	0.163	0
C	0.4	0.232	0.232	0.232	0.232	0.159	0.072	0.266	0
C	0.5	0.260	0.261	0.261	0.260	0.163	0.095	0.432	0
C	0.6	0.317	0.317	0.317	0.317	0.183	0.132	0.526	0
C	0.7	0.361	0.360	0.360	0.361	0.197	0.162	0.53	0
D	0.1	0.319	0.320	0.319	0.319	0.196	0.123	0.459	0
D	0.2	0.354	0.354	0.354	0.354	0.199	0.153	0.531	0
D	0.5	0.450	0.450	0.451	0.450	0.210	0.238	0.634	0
D	0.7	0.530	0.530	0.531	0.530	0.225	0.303	0.625	0
D	1	0.672	0.672	0.672	0.672	0.262	0.409	0.662	0
D	2	1.31	1.31	1.31	1.32	0.460	0.852	4.628	0

is one value over 100% (102.2) for one plant from category A concerning sensor noise.

After the power failure that unexpectedly interrupted this run, we started up a new run (called a *Noah run*) whose initial population was pre-populated with a mixture of best-of-node individuals existing at (or just before) the time of the power failure. Because this Noah run did not yield any significant additional progress, we made a third run.

13.2.3 Results for Third Run for Improved General-Purpose Controllers

The best-of-run controller emerged in generation 199.

The fitness measure for the third run (like that of the second run) entails 192 separate invocations of the SPICE simulator (i.e., eight simulations for each of 24 plants). The average time to evaluate the fitness of an individual was about 2 minutes and 5 seconds. It took 692 hours (28.8 days) to produce the best-of-run individual from generation 199. This is the longest run of genetic programming that we have ever made.

In addition to the 192 elements of the fitness measure that require a SPICE simulation, the fitness measure for this third run contains a 193rd element that considers the controller's parsimony. Unlike the interrupted second run, this 193rd element came into play during this third run. The first best-of-generation controller with 192 hits

Table 13.6 Percentage comparison (for 24 plants) of the best-of-run controller from the second run and the Åström–Hägglund controller

Plant	Parameter value	ITAE 1	ITAE 2	ITAE 3	ITAE 4	ITAE 5	ITAE 6	Stability margin	Sensor noise
A	0.1	91.6	91.6	90.7	90.5	83.5	94.6	75.2	96.4
A	0.3	93.8	93.8	93.9	93.9	86.3	98.5	85.2	97.9
A	1	91.4	91.4	91.4	91.4	89.3	93.2	91.5	94.7
A	3	88.9	88.9	88.9	88.9	86.9	92.4	92	96.4
A	6	91.9	91.9	91.9	91.9	90.9	94.8	88.5	95.3
A	10	95.2	95.2	95.3	95.3	94.7	98.2	0	95
B	3	86.4	86.2	86.5	86.4	86.2	87.2	87.4	OK
B	4	88.7	88.6	88.8	88.7	86.9	90.7	81.7	OK
B	5	90.8	90.9	91.1	90.8	88.4	93.5	92	OK
B	6	91.9	91.7	91.9	92	88.6	95.1	93.4	OK
B	7	92.8	93	93.1	93.1	88.3	96.3	89.3	OK
B	8	94.5	94.4	94.5	94.4	89.3	98.1	89.8	OK
C	0.2	89.5	89.5	89.4	89.5	89.1	92.8	OK	0
C	0.3	88.3	88.3	88.4	88.4	87.6	91.3	67.7	OK
C	0.4	89.5	89.5	89.6	89.5	86.7	97.1	63.3	OK
C	0.5	82.8	82.8	82.8	82.8	81.4	85.3	79.8	OK
C	0.6	85.5	85.6	85.4	85.5	84.1	87.8	88.7	OK
C	0.7	86.6	86.6	86.9	86.6	84.7	89	78.5	OK
D	0.1	86.2	86.4	86.2	86.2	85.3	87.5	90.9	OK
D	0.2	86.9	86.9	86.6	86.9	85.4	88.8	82.9	OK
D	0.5	86.8	86.9	87.1	86.8	83.5	89.6	90	OK
D	0.7	87.3	87.2	87.4	87.3	82.4	91.1	83.5	OK
D	1	85.8	85.6	85.6	85.8	80.3	90.3	87.8	OK
D	2	83	82.9	82.9	83.1	78.2	85.9	65.5	OK

Table 13.7 Performance (for the 18 additional plants) of the best-of-run controller of the second run

Plant	Parameter value	ITAE 1	ITAE 2	ITAE 3	ITAE 4	ITAE 5	ITAE 6	Stability margin	Sensor noise
A	0.15	0.653	0.653	0.661	0.658	0.208	0.445	0.658	4.23
A	0.5	0.566	0.566	0.566	0.566	0.226	0.379	0.522	0.823
A	0.9	0.478	0.478	0.478	0.478	0.209	0.289	0.592	1.27
A	2.5	0.318	0.318	0.318	0.318	0.187	0.134	0.618	0.96
A	4.0	0.265	0.265	0.265	0.265	0.181	0.086	0.455	1.38
A	9.0	0.221	0.221	0.221	0.221	0.181	0.04	0.048	1.08
C	0.25	0.183	0.183	0.183	0.183	0.152	0.03	0.062	0
C	0.34	0.213	0.213	0.213	0.213	0.159	0.053	0.242	0
C	0.43	0.241	0.240	0.240	0.241	0.162	0.077	0.400	0
C	0.52	0.279	0.279	0.28	0.279	0.172	0.106	0.462	0
C	0.61	0.323	0.324	0.323	0.323	0.185	0.136	0.525	0
C	0.69	0.356	0.355	0.355	0.356	0.195	0.159	0.527	0
D	0.15	0.338	0.339	0.338	0.338	0.198	0.139	0.505	0
D	0.3	0.392	0.392	0.392	0.392	0.205	0.185	0.568	0
D	0.6	0.512	0.512	0.513	0.511	0.227	0.282	0.638	0
D	0.85	0.618	0.618	0.618	0.619	0.251	0.365	0.647	0
D	1.2	0.788	0.788	0.787	0.788	0.294	0.491	0.703	0
D	1.8	1.16	1.16	1.16	1.16	0.406	0.748	2.02	0

Table 13.8 Percentage comparison (for the 18 additional plants) of the best-of-run controller from the second run and the Åström–Hägglund controller

Plant	Parameter value	ITAE 1	ITAE 2	ITAE 3	ITAE 4	ITAE 5	ITAE 6	Stability margin	Sensor noise
A	0.15	90	90	89.9	90	78.4	96.8	76.5	89.7
A	0.5	94.1	94.1	94	94.1	91	96.7	84.1	92.5
A	0.9	90.3	90.3	90.3	90.3	87.1	93.3	92.7	102.2
A	2.5	88.8	88.8	88.8	88.8	86.5	92.2	94	91.9
A	4.0	90.2	90.2	90.2	90.2	88.7	93.8	99	96.5
A	9.0	93.6	93.6	93.6	93.6	93	96.5	69	96
C	0.25	88.4	88.4	88.4	88.5	88.1	91.5	48.4	OK
C	0.34	88.1	87.8	88.1	88.2	87.3	91.3	81.5	OK
C	0.43	83.7	83.6	83.4	83.7	82.7	85.7	85	OK
C	0.52	83.7	83.3	83.4	83.7	82.4	85.7	84.6	OK
C	0.61	85.7	85.8	85.6	85.7	84.3	87.8	87.5	OK
C	0.69	86.8	86.7	87.1	86.8	84.8	89.2	79.3	OK
D	0.15	86.8	86.9	86.6	86.8	85.7	88.4	86.4	OK
D	0.3	87	87	87.1	87	84.8	89.6	81.2	OK
D	0.6	87	86.9	87.2	86.9	83.1	90.3	87.6	OK
D	0.85	86.1	85.9	86	86.2	81.2	90.3	86.8	OK
D	1.2	85.5	85.5	85.3	85.5	79.4	89.6	83.8	OK
D	1.8	83.6	83.6	83.6	83.7	78.1	86.9	48.4	OK

emerged in generation 105 after 422 hours (about 61% through the full 28.8-day run). The remaining 39% of the run was concerned with parsimony. Figure 13.4 shows the progressive improvement in parsimony during the portion of the third run that occurred after the creation of the first best-of-generation controller with 192 hits. The horizontal axis is the time (in hours) starting at 422 hours (17.6 days). The vertical axis is the parsimony element of the fitness measure. Each point represents a new best-so-far individual encountered during the run. The parsimony element (weighted block count) of the fitness measure for the best-of-generation individual from generation 105 is 22.6. The parsimony element of the fitness measure of the eventual best-of-run individual from generation 199 is 13.1.

Figure 13.5 shows the best-of-run parameterized controller from generation 199 of the third run. Genetic programming produced this controller's overall topology consisting of three adders, three subtractors, four gain blocks parameterized by a constant, two gains blocks parameterized by non-constant mathematical expressions containing free variables, two lead blocks parameterized by non-constant mathematical expressions containing free variables, and one Åström-Hägglund sub-controller.

The third controller is composed of leads (with transfer functions of the form $1 + \tau s$) as well as gain, differentiator, integrator, adder, and subtractor blocks.

The output of the third controller is control variable 790. There are three time-domain inputs to this controller, namely the reference signal 700, the plant output 704, and controller output 790 (i.e., internal feedback of the output of this controller into itself).

Note that the plant output indirectly enters the best-of-run parameterized controller from generation 199 through PID block 706. Block 706 is a PID controller tuned with the Åström-Hägglund tuning rules for the plant under consideration.

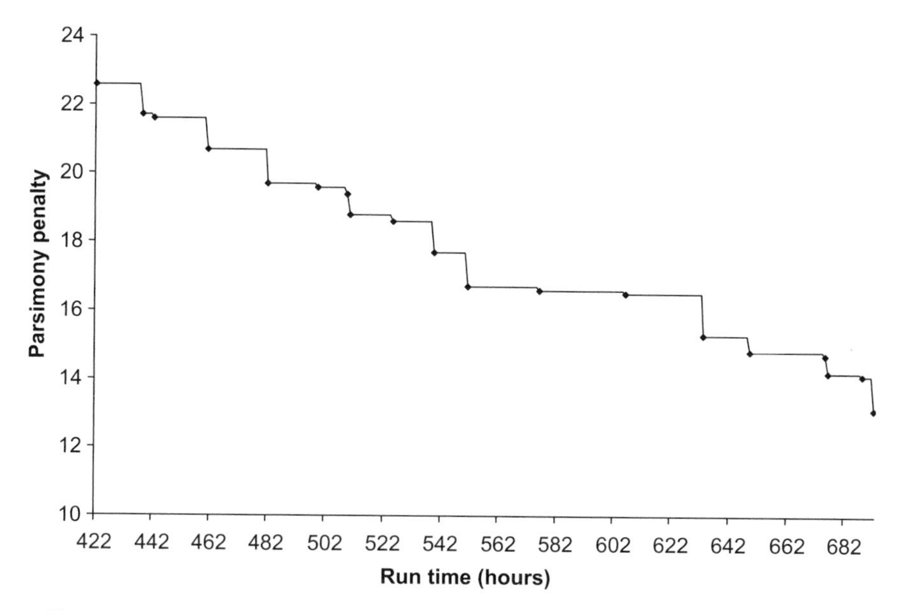

Figure 13.4 Progressive improvement in parsimony during the last part of the third run.

The third controller has four gain blocks (710, 720, 770, and 780) that are parameterized with a constant numerical amplification factor. In particular, gain block 710 has a gain of 3. Each of gain blocks 720 and 770 has a gain of 2. Gain block 780 has a gain of 10.

The third controller has three three-argument adders (738, 748, and 788) and three two-argument subtractors (734, 736, and 778).

The third controller also has two gain blocks (730 and 760) whose gain is expressed as an equation involving the four free variables that describe a particular plant (i.e., the plant's ultimate gain, K_u; ultimate period, T_u; dead time, L; and time constant, T_r). Specifically, gain block 730 has a gain (equation 31) of

$$\left|\log|T_r - T_u + \log\left|\frac{\log(|L|^L)}{T_u + 1}\right||\right|, \qquad [31]$$

whereas gain block 760 has a gain (equation 34) of

$$|\log|T_r + 1||. \qquad [34]$$

The third controller also has two lead blocks (i.e., blocks with transfer functions of the form $1 + \tau s$) that are parameterized by genetically evolved mathematical expressions. They are blocks 740 and 750.

Lead block 740 in figure 13.5 is parameterized by equation 32:

$$NLM(\log|L| - (\text{abs}(L)^L)^2 T_u^3 (T_u + 1) T_r e^L - 2T_u e^L), \qquad [32]$$

where *NLM* is the nonlinear mapping described in section 3.5.5.

Note that this nonlinear mapping was not mentioned in the discussion of the first or second run of this problem because the equations evolved in the first and second

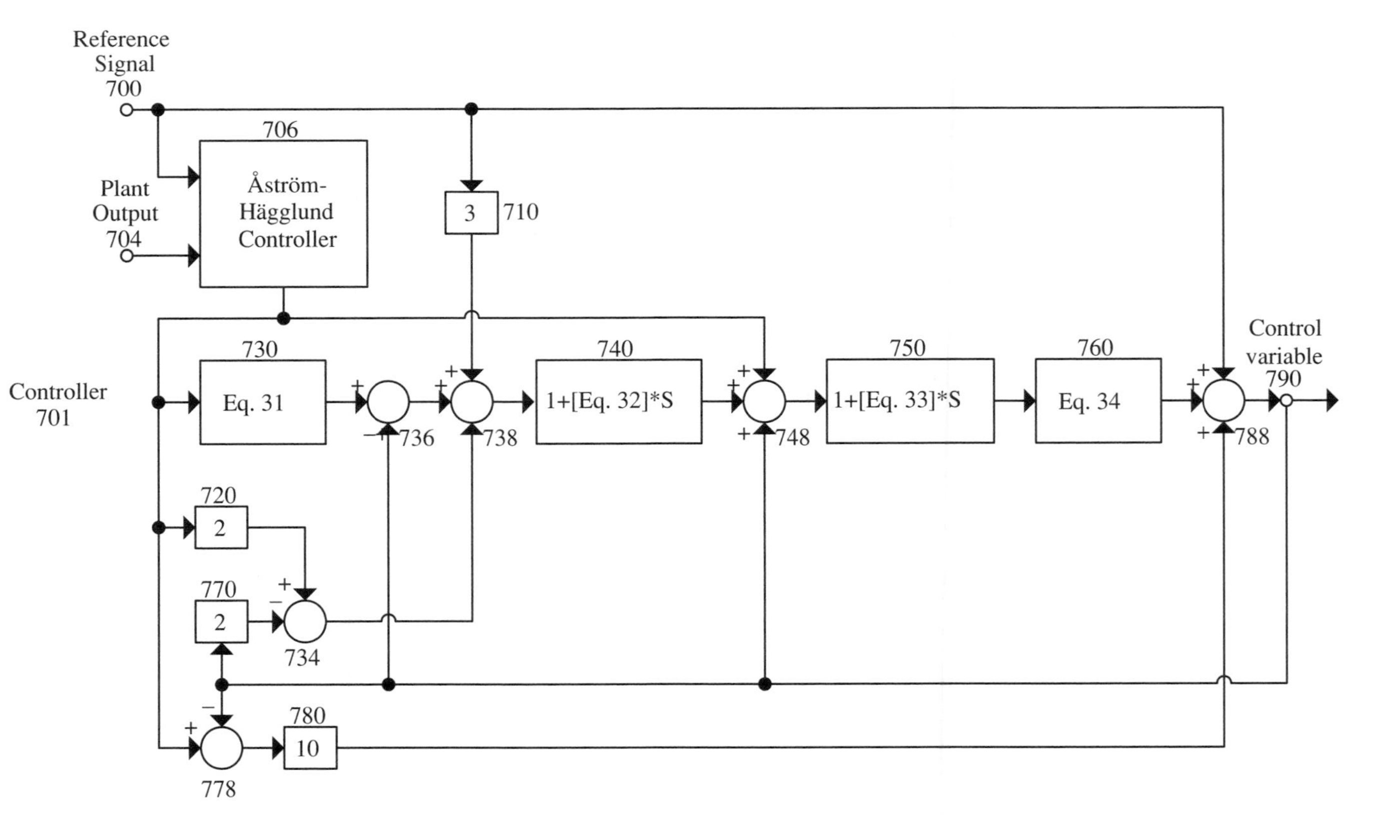

Figure 13.5 Best-of-run parameterized controller from the third run.

runs all had the property that the nonlinear mapping could be removed in simplifying the evolved equation. This simplification is possible when the argument x to the *NLM* is always in the range $-5<x<5$ (so the *NLM* has the form 10^x) and when x has the form $x=\log(y)$, where y is some mathematical expression. The value of the expression is then $NLM(x)=10^x=10^{\log(y)}=y$. However, for some of the equations evolved in this third run, this simplification is not possible.

Lead block 750 in figure 13.5 is parameterized by equation 33:

$$NLM(\log|L| - 2T_u e^L(2K_u(\log|K_u e^L|-\log|L|)T_u + K_u e^L)). \quad [33]$$

The third controller has internal feedback of its own output (control variable 790) back into itself. Specifically, controller output 790 is subtracted from the output of PID controller 706 tuned using the Åström-Hägglund tuning rules. The difference (amplified by a factor of 10 by gain block 780) becomes one of the three signals that are added together (by adder 788) to create controller output 790. Similarly, controller output 790 (amplified by a factor of 2 by gain block 770) is subtracted from the output of the Åström-Hägglund controller 706 (amplified by a factor of 2 by gain block 720) by subtractor 734.

In addition, the third controller contains Åström-Hägglund controller 706 (which contains additional adder, subtractor, integration, differentiation, and gain blocks).

The best-of-run parameterized controller from generation 199 of the third run can be described in terms of its transfer function:

$$U=\frac{R(1+3E_{34}(1+E_{32}{*}s)(1+E_{33}{*}s))+A(10+E_{34}(3+E_{31}+2E_{32}{*}s+E_{31}E_{32}{*}s)(1+E_{33}{*}s))}{11+E_{34}(2+3E_{32}{*}s)(1+E_{33}{*}s)},$$

where R is the reference signal; A is the output of the Åström-Hägglund controller; U is the controller output; P is the plant output (which is not used explicitly); and E_{31}, E_{32}, E_{33}, and E_{34} refer to equations 31, 32, 33, and 34 respectively.

Averaged over the 24 plants used in this run (i.e., those marked "3" in column 7 of table 12.1), the third controller has

- 81.8% of the setpoint ITAE of the Åström-Hägglund controller,
- 93.8% of the disturbance rejection ITAE of the Åström-Hägglund controller,
- 98.8% of the reciprocal of minimum attenuation of the Åström-Hägglund controller, and
- 93.4% of the maximum sensitivity, M_s, of the Åström-Hägglund controller.

Averaged over the 18 additional plants of table 12.2, the third controller has

- 81.8% of the setpoint ITAE of the Åström-Hägglund controller,
- 94.2% of the disturbance rejection ITAE of the Åström-Hägglund controller,
- 99.7% of the reciprocal of minimum attenuation of the Åström-Hägglund controller, and
- 92.5% of the maximum sensitivity, M_s, of the Åström-Hägglund controller.

As can be seen, the results obtained for the plants used in this run are very similar to those for the 18 previously unseen additional plants.

Thus, the best-of-run parameterized controller from generation 199 of the third run is an improvement over the PID controller developed by Åström and Hägglund in their 1995 book.

Table 13.9 shows the performance of the best-of-run parameterized controller from generation 199 of the third run on the 24 plants used in this run (i.e., those marked "3" in column 7 of table 12.1).

Table 13.10 presents the performance of the best-of-run parameterized controller from generation 199 of the third run as a percentage of the value for the Åström and Hägglund controller for the 24 plants used in this run (i.e., those marked "3" in column 7 of table 12.1). As can be seen in the table, all percentages are either below 100% (indicating improvement) or "OK."

Table 13.11 shows the performance of the best-of-run parameterized controller from generation 199 of the third run on the 18 additional plants from table 12.2.

Table 13.12 compares the performance of the best-of-run parameterized controller from generation 199 of the third run as a percentage of the value for the Åström and Hägglund controller for the 18 additional plants from table 12.2. As can be seen in the table, all percentages are either below 100% (indicating improvement) or "OK."

The authors refer to all three of the genetically evolved non-PID controllers in this chapter as the Keane-Koza-Streeter (KKS) controllers.

Table 13.9 Performance (for 24 plants) of the best-of-run controller from the third run

Plant	Parameter value	ITAE 1	ITAE 2	ITAE 3	ITAE 4	ITAE 5	ITAE 6	Stability margin	Sensor noise
A	0.1	0.694	0.700	0.705	0.695	0.235	0.460	0.726	9.77
A	0.3	0.678	0.679	0.679	0.678	0.245	0.430	0.480	1.70
A	1	0.434	0.434	0.434	0.434	0.182	0.266	0.489	1.14
A	3	0.275	0.275	0.275	0.275	0.170	0.108	0.457	1.46
A	6	0.212	0.212	0.212	0.212	0.163	0.050	0.122	1.26
A	10	0.188	0.187	0.187	0.188	0.158	0.030	0	1.08
B	3	0.270	0.272	0.271	0.270	0.175	0.094	0.144	0
B	4	0.391	0.391	0.391	0.391	0.191	0.197	0.493	0
B	5	0.461	0.461	0.461	0.461	0.197	0.263	0.616	0
B	6	0.510	0.510	0.509	0.510	0.200	0.306	0.569	0
B	7	0.544	0.544	0.542	0.544	0.205	0.337	0.572	0
B	8	0.557	0.558	0.555	0.556	0.202	0.352	0.568	0
C	0.2	0.158	0.158	0.158	0.158	0.138	0.020	0	0
C	0.3	0.187	0.187	0.187	0.187	0.144	0.042	0	0
C	0.4	0.225	0.225	0.225	0.225	0.154	0.070	0	0
C	0.5	0.274	0.273	0.274	0.274	0.168	0.105	0.157	0
C	0.6	0.322	0.321	0.322	0.322	0.180	0.140	0.263	0
C	0.7	0.349	0.350	0.350	0.349	0.181	0.166	0.413	0
D	0.1	0.307	0.309	0.308	0.307	0.181	0.125	0.299	0
D	0.2	0.345	0.346	0.345	0.345	0.187	0.158	0.379	0
D	0.5	0.452	0.451	0.451	0.452	0.201	0.248	0.511	0
D	0.7	0.540	0.540	0.540	0.540	0.219	0.317	0.503	0
D	1	0.699	0.699	0.698	0.699	0.262	0.433	0.587	0
D	2	1.37	1.37	1.37	1.38	0.463	0.897	3.892	0

Table 13.10 Percentage comparison (for 24 plants) of the best-of-run controller from the third run and the Åström–Hägglund controller

Plant	Parameter value	ITAE 1	ITAE 2	ITAE 3	ITAE 4	ITAE 5	ITAE 6	Stability margin	Sensor noise
A	0.1	91.1	91.9	91.3	90.9	84.9	93.8	74.1	96.5
A	0.3	96.6	96.8	96.7	96.6	92.1	99	71.8	97.3
A	1	88.9	88.9	88.9	88.9	80.3	95.1	74.2	96.5
A	3	84.2	84.2	84.2	84.2	81.1	90	76.9	98.2
A	6	85.2	85.2	85.2	85.2	85.5	83.4	43.7	99.6
A	10	82.9	82.6	82.6	82.9	83.1	82.7	0	96.8
B	3	81.1	81.8	81.5	81.1	78	88.4	39.3	OK
B	4	85.2	85.3	85.5	85.2	79.6	91.4	69.8	OK
B	5	87.6	87.8	87.7	87.6	81	94.2	89.3	OK
B	6	89.2	89.1	89.2	89.2	81.5	95.3	88	OK
B	7	90	90.1	90.1	90	81.8	96	89.1	OK
B	8	90.6	90.7	90.6	90.6	81.6	96.8	88.4	OK
C	0.2	83.3	83.2	82.9	83.3	81.9	94.1	OK	0
C	0.3	87	87.1	87.1	87	84.6	96.3	0	OK
C	0.4	87	86.8	86.8	87	84.1	94.3	0	OK
C	0.5	87.2	86.7	87	87.2	83.6	94.1	29	OK
C	0.6	86.8	86.6	86.9	86.8	82.7	92.9	44.3	OK
C	0.7	83.9	84.1	84.6	83.9	78	91.1	61.1	OK
D	0.1	83	83.3	83.2	83	78.9	89.3	59.2	OK
D	0.2	84.8	84.7	84.4	84.8	80	91.5	59.1	OK
D	0.5	87.2	87.1	87	87.2	80.3	93.3	72.5	OK
D	0.7	88.8	88.7	88.9	88.8	80.5	95.3	67.2	OK
D	1	89.2	89.1	88.9	89.2	80.5	95.5	77.8	OK
D	2	86.7	86.6	86.6	86.8	78.8	90.4	55.1	OK

This third run demonstrates the principle that parsimony can be incorporated into the fitness measure of this problem. Parsimony can be important in controller design because many real-time controllers embedded in manufactured products are implemented by inexpensive microprocessors with very limited processing capability (e.g., eight-bit chips). Such inexpensive microprocessors can sometimes only accommodate controllers whose transfer function can be represented by a low-order polynomial. This limitation is especially pertinent at high sampling rates.

On the other hand, a significant percentage of present-day controllers are implemented by means of microprocessors that are fast enough to accommodate any controller that is ever likely to be evolved by genetic programming. When this is the case, the parsimony of a controller is of little importance because the only cost associated with a non-parsimonious controller is the relatively small cost of writing a few lines of additional code. This cost will be generally inconsequential in comparison to the economic benefits of more efficiently controlling and operating the plant. When parsimony is not important, the first genetically evolved controller in this chapter is, by far, the best of the three controllers presented in this chapter because it has only 66.4% of the setpoint ITAE of the Åström-Hägglund controller and only 85.7% of the disturbance rejection ITAE of the Åström-Hägglund controller.

Table 13.11 Performance (on the 18 additional plants) of the best-of-run controller from the third run

Plant	Plant parameter value	ITAE 1	ITAE 2	ITAE 3	ITAE 4	ITAE 5	ITAE 6	Stability margin	Sensor noise
A	0.15	0.684	0.680	0.687	0.687	0.237	0.450	0.599	4.14
A	0.5	0.568	0.566	0.566	0.568	0.220	0.387	0.413	0.806
A	0.9	0.472	0.472	0.472	0.472	0.192	0.301	0.469	1.22
A	2.5	0.302	0.302	0.303	0.302	0.174	0.130	0.443	0.993
A	4.0	0.249	0.249	0.249	0.249	0.170	0.080	0.294	1.44
A	9.0	0.192	0.192	0.192	0.192	0.159	0.034	0	1.16
C	0.25	0.178	0.178	0.178	0.178	0.146	0.032	0	0.093
C	0.34	0.207	0.207	0.207	0.207	0.151	0.055	0	0
C	0.43	0.251	0.250	0.251	0.251	0.164	0.085	0.021	0
C	0.52	0.291	0.290	0.291	0.291	0.174	0.115	0.178	0
C	0.61	0.326	0.325	0.326	0.326	0.181	0.144	0.271	0
C	0.69	0.344	0.345	0.345	0.344	0.180	0.163	0.411	0
D	0.15	0.327	0.328	0.327	0.327	0.183	0.143	0.337	0
D	0.3	0.386	0.384	0.385	0.386	0.194	0.191	0.427	0
D	0.6	0.515	0.515	0.515	0.515	0.219	0.295	0.515	0
D	0.85	0.639	0.638	0.638	0.639	0.250	0.386	0.561	0
D	1.2	0.825	0.825	0.826	0.825	0.297	0.524	0.646	0
D	1.8	1.22	1.22	1.22	1.22	0.414	0.796	1.55	0

Table 13.12 Percentage comparison (for the 18 additional plants) of the best-of-run controller from the third run and the Åström–Hägglund controller

Plant	Plant parameter value	ITAE 1	ITAE 2	ITAE 3	ITAE 4	ITAE 5	ITAE 6	Stability margin	Sensor noise
A	0.15	94.3	93.8	93.4	94	89.2	97.8	69.7	87.8
A	0.5	94.4	94	94	94.4	88.4	98.8	66.5	90.6
A	0.9	89.2	89.2	89.2	89.2	80	96.9	73.5	98.3
A	2.5	84.5	84.5	84.5	84.5	80.8	89.5	67.4	94.7
A	4.0	84.6	84.6	84.6	84.6	83.6	86.9	64	100.4
A	9.0	81.4	81.4	81.4	81.4	81.5	81.9	0	102.3
C	0.25	86.1	86.2	85.9	86.1	84.3	96.4	0	OK
C	0.34	85.6	85.5	85.8	85.6	82.9	94.2	0	OK
C	0.43	87.2	86.8	87.3	87.2	83.9	94.8	4.5	OK
C	0.52	87.1	86.6	86.9	87.1	83.4	93.1	32.7	OK
C	0.61	86.5	86.1	86.4	86.5	82.1	92.9	45.2	OK
C	0.69	84	84.1	84.7	84	78.1	91.6	61.9	OK
D	0.15	83.9	84.2	83.7	83.9	79.3	90.9	57.6	OK
D	0.3	85.6	85.3	85.4	85.6	80.1	92.2	61	OK
D	0.6	87.6	87.5	87.6	87.6	80	94.3	70.8	OK
D	0.85	89	88.7	88.7	89	80.8	95.4	75.3	OK
D	1.2	89.6	89.5	89.6	89.6	80.2	95.6	77	OK
D	1.8	87.9	87.9	87.7	87.9	79.6	92.4	37.2	OK

13.3 Human-Competitiveness of the Results for the Improved General-Purpose Controllers

The three genetically evolved improved non-PID controllers described in this chapter outperform a PID controller tuned using the rules developed by Åström and Hägglund in their 1995 book. As previously mentioned, the tuning rules for the Åström and Hägglund PID controller, in turn, outperform the 1942 Ziegler-Nichols tuning rules on the 16 industrially representative plants used by Åström and Hägglund.

The development of the Ziegler-Nichols tuning rules (Ziegler and Nichols 1942) for PID controllers was a significant development in the field of control engineering. These rules have been in widespread use since they were invented.

The development by Åström and Hägglund of the tuning rules described in their 1995 book *PID Controllers: Theory, Design, and Tuning* was another significant development. Åström and Hägglund developed their improved rules by applying mathematical analysis, shrewdly chosen approximations, and considerable creative flair.

Referring to the eight criteria in table 1.2 for establishing that an automatically created result is competitive with a human-produced result, the creation by genetic programming of improved tuning rules for PID controllers satisfies the following five of the eight criteria:

- **(B)** The result is equal to or better than a result that was accepted as a new scientific result at the time when it was published in a peer-reviewed scientific journal.
- **(D)** The result is publishable in its own right as a new scientific result—independent of the fact that the result was mechanically created.
- **(E)** The result is equal to or better than the most recent human-created solution to a long-standing problem for which there has been a succession of increasingly better human-created solutions.
- **(F)** The result is equal to or better than a result that was considered an achievement in its field at the time it was first discovered.
- **(G)** The result solves a problem of indisputable difficulty in its field.

A patent application (Keane, Koza, and Streeter 2002a) was filed on July 12, 2002, for the three genetically evolved non-PID controllers described in this chapter (as well as the PID tuning rules described in chapter 12). The applicants believe that the three genetically evolved controllers are patentable because they satisfy the statutory requirements of being "useful," "new," "improved," and "unobvious" to someone "having ordinary skill in the art." If (as expected) a patent is granted, it will (we believe) be the first patent granted for an invention created by genetic programming.

We believe that the creation by genetic programming of improved tuning rules for PID controllers satisfies the following additional criterion from table 1.2:

(A) The result was patented as an invention in the past, is an improvement over a patented invention, or would qualify today as a patentable new invention.

The creation by genetic programming of the three improved non-PID controllers satisfies Arthur Samuel's criterion (1983) for artificial intelligence and machine learning:

> "[T]he aim [is]... to get machines to exhibit behavior, which if done by humans, would be assumed to involve the use of intelligence."

13.4 Routineness for the Improved General-Purpose Controllers

The implementation of these runs of genetic programming for the problem of synthesizing improved parameterized controllers was routine. The best known human-created solution was made available to genetic programming merely by adding one terminal to the terminal set during the preparatory steps. One could bootstrap from any similar known solution to any problem in any field in substantially the same way. Because the preparatory steps in this chapter are very similar to the preparatory steps for all the other examples of automatic controller synthesis in this book, very little effort was required to make the transition from earlier problems to the problem of synthesizing improved non-PID parameterized controllers in this chapter.

13.5 AI Ratio for the Improved General-Purpose Controllers

As mentioned in the section 13.3, the result produced by genetic programming on this particular problem is considered to be human-competitive for five reasons. Thus, the results produced by the artificial system (genetic programming) in this chapter put a high amount of "A" in the numerator of the AI ratio.

The preparatory steps for this problem are similar to the preparatory steps for the other problems of controller synthesis in chapter 3. However, because the results in this chapter bootstrapped on an important domain-specific r esult from the field of control (the Åström and Hägglund tuning rules for PID controllers), the results in this chapter benefited from more "I" than would have been the case if the evolutionary process had started from scratch. Thus, the solution produced by genetic programming in this chapter has a non-trivial amount of "I" in the denominator of the AI ratio. Hence, the AI ratio for this problem is moderately high.

14

Reinvention of Negative Feedback

Section 1.2 recounts the history of Harold S. Black's many years of work at AT&T on the problem of reducing distortion in amplifiers.

As Black recounted on the 50th anniversary of his 1927 solution to the problem (Black 1977):

> "Then came the morning of Tuesday, August 2, 1927, when the concept of the negative feedback amplifier came to me in a flash while I was crossing the Hudson River on the Lackawanna Ferry, on my way to work".

Prior to Black's invention of negative feedback in 1927, almost all the work on amplifiers in the then-new field of electronics was based on the "nearly universal idiom" of positive feedback (Lee 1998). In fact, even after the 1927 invention of negative feedback (Black 1977):

> "...more than nine years would elapse before the patent was issued...One reason for the delay was that *the concept was so contrary to established beliefs*". (Emphasis added.)

We believe that one reason why it took an inordinate amount of time for negative feedback to gain acceptance was that human thinking often becomes channeled along the well-traveled paths of "established beliefs". One of the virtues of genetic programming is that it approaches a problem in an open-ended way that is not encumbered by previous human thinking. Genetic programming is not aware, much less concerned, about whether a solution is "contrary to established beliefs". For this reason, genetic programming often unearths solutions that might have never occurred to human scientists and engineers who are steeped in the thinking of the day.

So, now let's see how the problem of reducing distortion in amplifiers can be solved by means of genetic programming. As will be seen, Black's solution flows almost effortlessly from a genetic search based on a high-level statement of Black's problem.

14.1 Genetic Programming Takes a Ride on the Lackawanna Ferry

14.1.1 Fitness Measure

In Black's own account of the history of the invention of negative feedback (quoted at length in section 1.2), Black mentioned that he was greatly "impressed" by how the famous scientist and engineer Steinmetz got down to "the fundamentals". The preparatory steps of genetic programming are effective in forcing the human user to focus on "the fundamentals". In particular, the construction of a problem's fitness measure requires the human user to identify exactly what is wanted.

Focusing on "the fundamentals" immediately leads to a three-element fitness measure based on the degree to which a candidate circuit

- amplifies the incoming signal,
- minimizes distortion, and
- contains a small number of expensive components.

As to the first element of the fitness measure, if the specified amount of amplification is, say, 10 dB, then the desired output would be an inverted sine wave whose amplitude is 3.16 times that of the input. If the average absolute difference between a candidate circuit's output and the desired output is reasonably small (say, less than 0.1 Volt), the circuit is deemed to deliver a satisfactory amount of amplification and this first element of the fitness measure is 0. Otherwise, the first element is equal to the average absolute difference between the desired output and the actual output.

The second element of the fitness measure is based on total harmonic distortion (*THD*):

$$THD = \frac{\sqrt{\sum_{i=2}^{N} A_i^2}}{A_1},$$

where A_1 is the magnitude of the first harmonic (i.e., the fundamental frequency) and A_i is the magnitude of the nth harmonic (Vladimirescu 1994). For the audio signals of interest over a telephone, it would be reasonable to choose 1,000 Hz as the fundamental frequency and to consider $N = 9$ harmonics. If the total harmonic distortion is less than, say, -45 dB, a circuit is deemed to be satisfactory in terms of reducing distortion and the second element of the fitness measure is 0. Otherwise, the second element is equal to

$$10*(1 + |THD - (-45)|).$$

The third element of the fitness measure assigns a cost of 1.0 for each vacuum tube (transistor) and 0.01 for each resistor and capacitor.

Fitness is the sum of the first element (amplification), 10^{-6} times the second element (distortion), and 10^{-18} times the third element (parsimony). As usual, the smaller the total value of fitness, the better.

Note that this fitness measure was created by focusing the problem's high-level requirements—amplification, distortion, and parsimony. The fitness measure is concerned with "what needs to be done"—not with "how to do it". In particular, notice that the fitness measure does not mandate a feed-forward approach, feedback approach, or any other approach. And, if feedback is used at all, the fitness measure is agnostic as to whether the feedback is Armstrong's positive type of feedback (Armstrong 1914) or Black's negative type of feedback.

14.1.2 Initial Circuit, Function Set, Terminal Set, and Control Parameters

Although part numbers for vacuum tubes are mentioned in Black's patents, accurate models for the vacuum tubes used by Black between 1921 and 1927 were not readily available to us. However, present-day transistors operate in a manner similar to vacuum tubes. Because field-effect transistors (FETs) are voltage-controlled, their behavior closely more resembles the behavior of vacuum tubes (which are voltage-controlled) than do, say, bipolar junction transistors (which are current-controlled). Therefore, we used a model `IRFZ44` FET transistor in lieu of vacuum tubes.

It is reasonable (in light of the performance levels discussed in Black's patents) to use an initial circuit with a voltage source delivering a sine wave with a 4-Volt amplitude, a 50-Ohm source resistor, a 100-Ohm load resistor, and a 60-Volt power supply. It would also be reasonable to employ the floating embryo shown in figure 4.30 in the initial circuit and the function set common to all six problems involving the post-2000 patented analog circuits employed in section 15.3.3. The population size is 1,000,000.

14.2 Results for the Problem of Reducing Amplifier Distortion

The best circuit from among the 1,000,000 individuals in generation 0 delivers amplification of −2.91 dB. That is, the best-of-generation circuit from generation 0 acts as an attenuator rather than an amplifier.

The first best-of-generation circuit that acts as an amplifier appeared in generation 9. This individual acts as a 5.37 dB amplifier. However, it has a total harmonic distortion of −5.65 dB.

The first circuit in the run satisfying the amplification criterion and having a total harmonic distortion of less than −45 dB appeared in generation 46. This circuit has a total harmonic distortion of −54.2 dB. This circuit's average absolute error (measuring amplification) is 0.098 Volts.

This best-of-generation circuit from generation 46 (figure 14.1) consists of three field-effect transistors and two resistors (ignoring the source resistor RSRC and the load resistor RLOAD in the test fixture). In viewing the figures in this section, note that a FET's source corresponds to a vacuum tube's cathode, the FET's drain corresponds to the tube's anode, and the FET's gate corresponds to the tube's grid. Additional information on field-effect transistors may be found in chapter 51 of *Genetic Programming III* (Koza, Bennett, Andre, and Keane 1999a). In this circuit, Q3 is biased off so its removal does not affect the circuit's behavior. However, Q1 and Q2 are overlaid in parallel and together act as a single transistor, $Q_{equivalent}$, with twice the transconductance and interelectrode capacitance of Q1. The removal of Q2 decreases the amplification slightly (to 9.64 dB), changes the bias, and adversely affects total harmonic distortion (changing it to −51.9 dB).

The 174-Ohm resistor R2 in figure 14.1 is the mechanism for providing negative feedback. Because $Q_{equivalent}$ reverses the phase of the incoming signal, the feedback

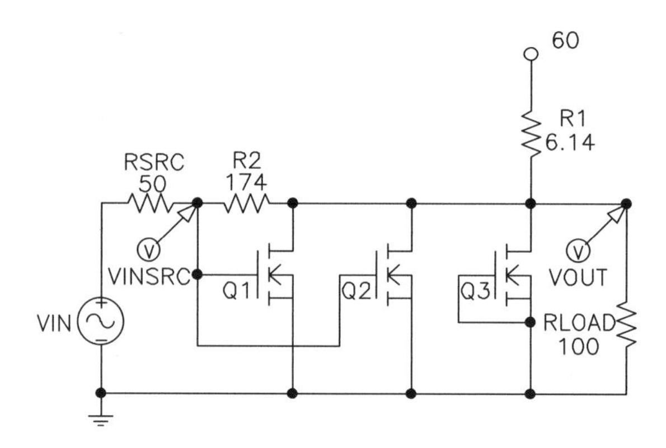

Figure 14.1 Best-of-generation circuit from generation 46 for the problem of reducing amplifier distortion.

Table 14.1 Distortion for the best-of-generation circuit from generation 46 for the problem of reducing amplifier distortion

Harmonic	Amplitude	Amplitude with respect to fundamental frequency
1,000 (fundamental frequency)	9.96	0
2,000	−44.3	−54.3
3,000	−61.5	−71.5
4,000	−68.7	−78.7
5,000	−71.0	−81.0
6,000	−78.8	−88.8
7,000	−85.6	−95.6
8,000	−81.6	−91.6
9,000	−81.4	−91.4

is negative—that is, the signal from the drain (corresponding to a vacuum tube's plate) of $Q_{equivalent}$ is subtracted from the incoming signal VINSRC.

Table 14.1 shows, in column 2, the amplitude (in decibels) of the fundamental frequency (1,000 Hz) and various harmonics for the best-of-generation circuit from generation 46 (figure 14.1). Column 3 shows the amplitude (in decibels) with respect to the fundamental frequency.

The best-of-run circuit (figure 14.2) emerged in generation 48. The average absolute error (measuring amplification) of this best-of-run circuit from generation 48 is 0.065 Volts (about two-thirds of that of the best-of-generation circuit from generation 46). The best-of-run circuit has amplification of 10.06 dB and total harmonic distortion of −51.2 dB. This circuit is more parsimonious than the best-of-generation circuit from generation 46 in that it has only one transistor and two resistors (ignoring the source resistor RSRC and the load resistor RLOAD in the test fixture). As can be seen, the 182-Ohm resistor R2 is the mechanism for providing the negative feedback.

Table 14.2 shows, in column 2, the amplitude (in decibels) of the fundamental frequency (1,000 Hz) and various harmonics for the best-of-run circuit from generation

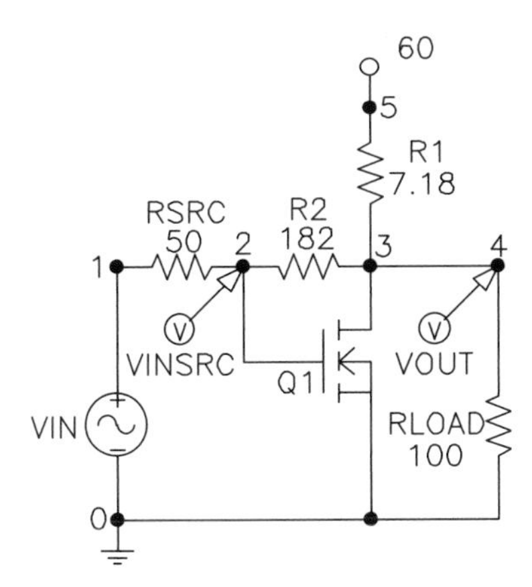

Figure 14.2 Parsimonious best-of-run circuit from generation 48 for the problem of reducing amplifier distortion.

Table 14.2 Distortion of the best-of-run circuit from generation 48 for the problem of reducing amplifier distortion

Harmonic	Amplitude	Amplitude with respect to fundamental frequency
1,000 (fundamental frequency)	10.06	0
2,000	−41.3	−51.4
3,000	−54.7	−64.8
4,000	−73.2	−83.3
5,000	−66.9	−77.0
6,000	−68.8	−18.9
7,000	−73.2	−83.3
8,000	−69.9	−80.0
9,000	−73.3	−83.4

48 (figure 14.2). Column 3 shows the amplitude (in decibels) with respect to the fundamental frequency.

Black received U.S. patents 2,003,282 (Black 1935), 2,102,670 (Black 1937a), and 2,102,671 (Black 1937b) relating to his work on the problem of reducing distortion in amplifiers (as well as U.S. patent 1,686,792 (Black 1928) for the earlier impractical solution described in section 1.2). The overall goal of all these efforts is stated in the description of U.S. patent 2,102,670 (Black 1937a),

> "It is common experience that increase of the power output of vacuum tubes or electric space discharge devices tends to increase distortion of signaling or other waves transmitted by the devices, and tends to lower the gain of the circuits of the devices. . . .
>
> "Therefore, a major problem in devising vacuum tube systems, as for example vacuum tube amplifier systems, is the securing of high output of power without

attendant disadvantages, as for example without increase of first cost or decrease of operating efficiency of the systems, and especially in the case of vacuum tube amplifiers and repeaters, without sacrifice of quality of signal reproduction".

The best-of-run circuit from generation 48 (figure 14.2) infringes claims 1 and 3 of U.S. patent 2,102,671 (Black 1937b). Claim 1 covers,

"In a wave translating device or system having amplifying properties, an input portion and an output portion, means to apply fundamental waves to said input portion, said system carrying fundamental components in said output portion, and having means producing other wave components in said output portion, and means controlling the relative magnitudes of said components in said output portion comprising means to feed waves from said output portion to said input portion to decrease the gain of the system".

The field-effect transistor Q1 is the "wave translating device or system". The incoming voltage signal source is the "means to apply fundamental waves to said input portion". Resistor R2 is the "means to feed waves from said output portion to said input portion to decrease the gain of the system". The negative feedback "decrease[s] the gain of the system".

Claim 3 of U.S. patent 2,102,671 (Black 1937b) covers,

"In a wave translating system operating to amplify applied fundamental waves, and to produce distortion components as a function of non-linearity in the system, means to increase the ratio of the amplified fundamental wave component to distortion components comprising means to utilize a portion of the waves translated by said system to reduce the gain of the system below the gain with zero feedback in the system of the waves translated by the system".

14.3 Human-Competitiveness of the Result for the Problem of Reducing Amplifier Distortion

Referring to the eight criteria in table 1.2 for establishing that an automatically created result is competitive with a human-produced result, the rediscovery by genetic programming of negative feedback for reducing amplifier distortion satisfies the following four of the eight criteria:

(A) The result was patented as an invention in the past, is an improvement over a patented invention, or would qualify today as a patentable new invention.

(E) The result is equal to or better than the most recent human-created solution to a long-standing problem for which there has been a succession of increasingly better human-created solutions.

(F) The result is equal to or better than a result that was considered an achievement in its field at the time it was first discovered.

(G) The result solves a problem of indisputable difficulty in its field.

The rediscovery by genetic programming of negative feedback for reducing amplifier distortion came over seven decades after Black took his now-famous ride on the Lackawanna Ferry. Nonetheless, the fact that the original human-conceived solution to the problem (eventually) satisfied the Patent Office's criteria for patent-worthiness means that the genetically evolved duplicate would also have satisfied the Patent Office's criteria for patent-worthiness (if only it had arrived before Black took his ferryboat ride).

The fact that genetic programming rediscovered a solution that was unobvious "to a person having ordinary skill in the art" establishes that the genetically evolved result satisfies Arthur Samuel's criterion (1983) for artificial intelligence and machine learning:

> "[T]he aim [is]...to get machines to exhibit behavior, which if done by humans, would be assumed to involve the use of intelligence".

14.4 Routineness for the Problem of Reducing Amplifier Distortion

The preparatory steps for the problem of evolving a circuit that reduces amplifier distortion are substantially the same as the preparatory steps for other problems of circuit synthesis in chapter 4 and elsewhere in this book, except, of course, for the fitness measure. Thus, aside from the usual effort required to construct a fitness measure appropriate to the problem at hand, relatively little effort was required to make the transition from other problems of circuit synthesis in this book (chapters 4, 10, and 11) to the problem of reducing amplifier distortion that was so vexatious in the early part of the 20th century. That is, the transition was routine.

14.5 AI Ratio for the Problem of Reducing Amplifier Distortion

As previously mentioned, the result produced by genetic programming on this problem is human-competitive for four reasons. Thus, it is fair to say that the solution produced by genetic programming to this problem has a high amount of "A". The preparatory steps for the problem of evolving a circuit reducing amplifier distortion reflect *de minimus* information about electrical circuits. Thus, the solution produced by genetic programming for the present problem incorporates only a small amount of "I". The high amount of "A" represented by the result in conjunction with the small amount of "I" provided by the human user means that the AI ratio for the solution produced by genetic programming is high.

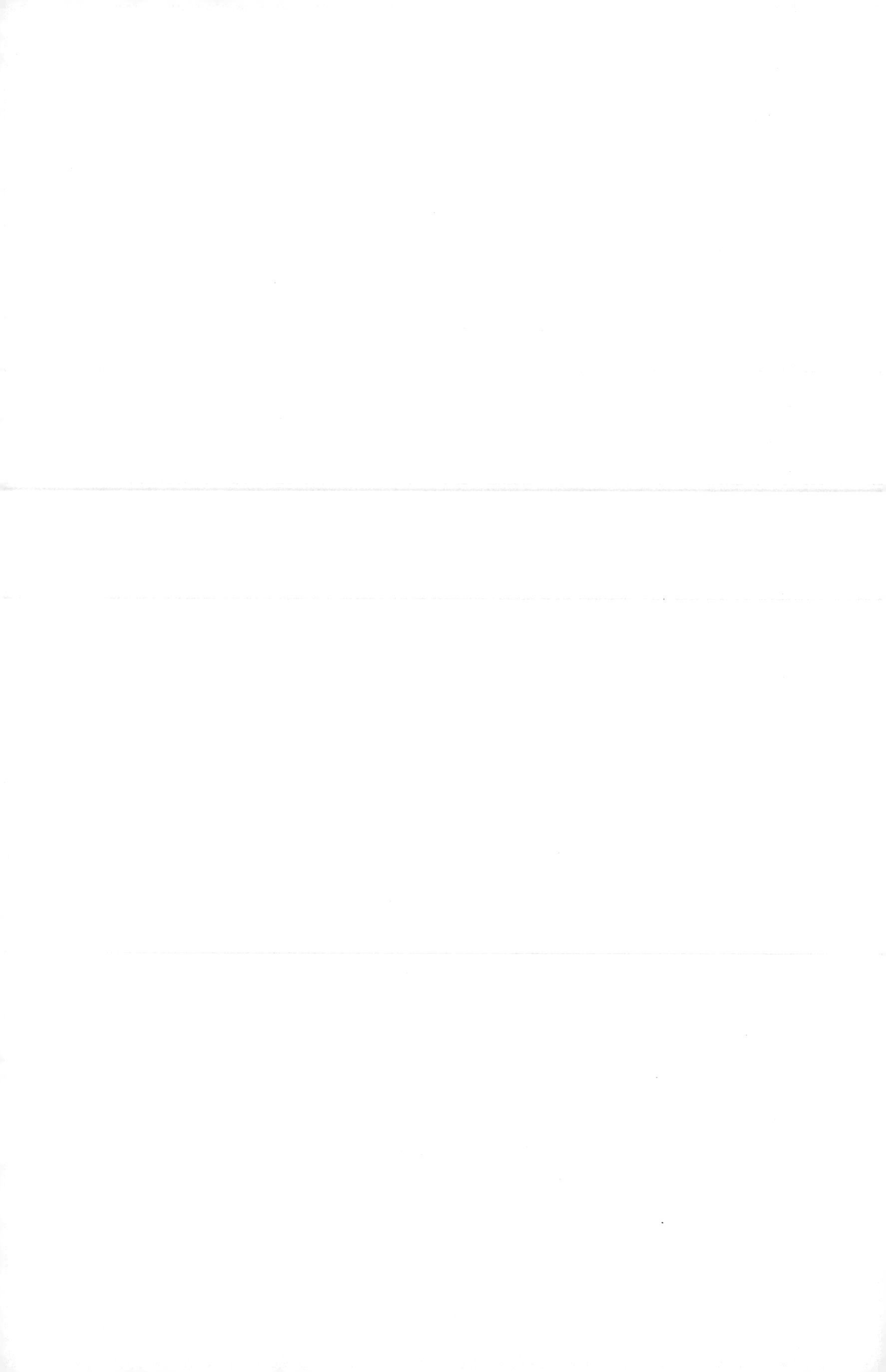

15

Automated Reinvention of Six Post-2000 Patented Circuits

Filing for a patent entails the expenditure of a considerable amount of time and money. Therefore, a patent is generally sought only if an individual, business, or institution believes that the invention is potentially useful in the real world. Patents are only issued if an arms-length examiner is convinced that the proposed invention satisfies the statutory tests of being "new," "useful," "improved," and "unobvious." Recently issued patents represent current research and development efforts by the engineering and scientific communities.

Sections 4.3 and 4.4 and chapters 5 and 14 of this book and parts of *Genetic Programming III* (Koza, Bennett, Andre, and Keane 1999a) demonstrate that genetic programming can automatically create analog electrical circuits that duplicate the functionality of, or infringe, patents that were issued during the first two thirds of the 20th century. This chapter reports that genetic programming is capable of creating inventions that duplicate the functionality of, or infringe, six patents for analog electrical circuits that were issued after January 1, 2000.

When genetic programming creates a design whose detailed structure matches that described in the claims of an unexpired patent, it has created an invention that, if manufactured, would infringe the patent. When genetic programming creates a design that duplicates the functionality of a previously patented invention (but not the detailed structure specified in the patent's claims), it has either rediscovered prior art (patented or not) or created a potentially patentable new invention. Such a design may provide a way to avoid infringing an existing patent (assuming that the differences between it and the patented invention are not close enough to be covered by the doctrine of equivalence).

The six post-2000 patented inventions are shown in table 15.1.

The six circuits in table 15.1 were selected after browsing the numerous patents that were issued for analog circuits during the year and a half following January 1, 2000. The group was selected so as to include different types of circuits, including at least one mixed analog-digital circuit, at least one multi-input circuit, at least one multi-output circuit, and at least one circuit whose behavior is measured by each of the major types of analyses that are commonly used in evaluating circuits (i.e., time-domain

Table 15.1 Six post-2000 patented inventions

Invention	Date	Inventor	Place	Patent
Low-voltage balun circuit	2001	Sang Gug Lee	Information and Communications University	6,265,908
Mixed analog-digital variable capacitor circuit	2000	Turgut Sefket Aytur	Lucent Technologies Inc.	6,013,958
Voltage-current conversion circuit	2000	Akira Ikeuchi and Naoshi Tokuda	Mitsumi Electric Co., Ltd.	6,166,529
Low-voltage high-current circuit for testing a voltage source	2001	Timothy Daun-Lindberg and Michael Miller	International Business Machines Corporation	6,211,726
Low-voltage cubic function generator	2000	Stefano Cipriani and Anthony A. Takeshian	Conexant Systems, Inc.	6,160,427
Tunable integrated active filter	2001	Robert Irvine and Bernd Kolb	Infineon Technologies AG	6,225,859

analysis, frequency-domain analysis, DC operating point analysis, and Fourier analysis). Additionally, each circuit of the group satisfies the following criteria:

- The circuit is of some importance and general interest.
- The specific technological problem addressed by the patented invention can be succinctly explained on a stand-alone basis (i.e., without recourse to a lengthy history of previous patents and research).
- The patent document contains a clear quantitative statement of the circuit's intended behavior and characteristics.
- The patent document contains sufficient information to enable us to build and simulate the circuit using simulation software.

The last requirement was necessary to enable us to compare the behavior of the genetically evolved circuit to that of the patented circuit. This requirement caused us to pass over a number of otherwise interesting patents because some patents left us uncertain as to the exact range of operation, component values, and component types. Even for the patents that we eventually selected, we had to risk making assumptions concerning certain details of the circuit's operation.

Although multiple runs are often required to solve a problem when working with a probabilistic algorithm, genetic programming produced a satisfactory solution to all six problems in this chapter on the very first run. Also, we made additional runs (in some cases with added requirements) for four of the problems. All the additional runs were also successful. As will be discussed in section 15.5, the fact that all these runs were successful suggests that we have not yet reached the limits of what is possible in the realm of automated circuit synthesis by means of genetic programming (using currently available computing resources).

15.1 The Six Circuits

15.1.1 Low-Voltage Balun Circuit

U.S. patent 6,265,908, covering a low voltage balun circuit, was issued to Sang Gug Lee of the Information and Communications University in South Korea.

The purpose of a balun (balance/unbalance) circuit is to produce two outputs from a single input, each output having half the amplitude of the input, one output being in phase with the input and the other being 180 degrees out of phase with the input, with both outputs having the same DC offset.

Commercially useful balun circuits in battery-powered devices (e.g., cellular telephones) must minimize power consumption in order to maximize battery life. Contemporary balun circuits typically operate at DC voltages of 3 Volts or less.

The balun circuit for which Lee received a patent in 2001 is noteworthy in that it operates using a power supply of only 1 Volt.

15.1.2 Mixed Analog-Digital Variable Capacitor

U.S. patent 6,013,958 covers a mixed analog-digital circuit whose behavior is equivalent to that of a capacitor whose capacitance is dynamically controlled by the value stored in a digital register. The inventor is Turgut Sefket Aytur of Lucent Technologies Inc. of Murray Hill, New Jersey.

Variable capacitors are useful (Aytur 2000) because

> "...the nature of the process of making integrated circuits causes the capacitance of the individual unit capacitors to be consistent, improving the linearity of the capacitance. Capacitors according to embodiments of the invention are useful in circuits that need to be accurately trimmed once (e.g., filters) and in circuits in which variable capacitance is required, such as voltage controlled oscillators."

Variable capacitance devices that possess both accuracy and a wide tuning range have been primarily available in the past only as discrete components.

15.1.3 Voltage-Current Conversion Circuit

The purpose of the voltage-current conversion circuit patented by Ikeuchi and Tokuda of Mitsumi Electric Co., Ltd. (U.S. patent 6,166,529) is to take two time-varying voltages as inputs and to produce as output a stable current whose magnitude is proportional to the difference between the two voltages.

In their patent, Ikeuchi and Tokuda (2000) note that there is a limit on the values of the input voltages that can be correctly handled by previously known voltage-current conversion circuits:

> "A requirement of the conventional circuit described above is that the input voltages V_{in1}, V_{in2} be lower than the power source voltage V_{cc}. If this requirement is not met, the circuit does not operate properly. As a result, the size of the input voltages V_{in1}, V_{in2} is limited and hence the range of applications of the circuit is limited as well.

"Accordingly, it is a general object of the present invention to provide an improved and useful voltage-current conversion circuit in which the disadvantages described above are eliminated.

"A more specific object of the present invention is to provide an improved and useful voltage-current conversion circuit not restricted by the requirement that the input voltages be lower than the power source voltage and hence capable of an expanded range of applications."

15.1.4 Low-Voltage High-Current Transistor Circuit

U.S. patent 6,211,726 covers a circuit designed to sink a time-varying amount of current through a power supply in response to a control signal. Such a circuit is useful for testing that a power supply is capable of delivering specified amounts of current while maintaining a constant voltage.

Daun-Lindberg and Miller of IBM state:

"In the design and production of power supply modules for computer systems and the like, it is necessary to have load circuits which simulate the operation of the computer itself in the way the voltage and current levels presented by the load vary with time and other conditions. For example, the load presented to the power supply by the computer system may switch between high-current and low-current when the computer goes into a power-down or sleep mode as power management mechanisms go into effect. Or, current spikes may occur when starting up equipment such as a hard drive, or at boot-up when large capacitive loads are being charged. Likewise, power-up self test creates a widely variable load as all of the peripheral equipment is exercised.

"Previous load circuits used for test purposes in the manufacture of computer systems or the like have not been able to operate at low voltages such as the supply voltage levels being specified for contemporary microprocessor chips and memory chips. For example, supply voltage levels of 3.3 V have been used for some time, and levels of 1.7 V or 1.5 V are coming into common usage. At these low voltage levels, and at high currents of sometimes hundreds of amps, the available equipment is not able to accurately simulate varying load currents, nor represent the loading exhibited by a computer system."

To satisfy the high-current requirements described in this patent, Daun-Lindberg and Miller of IBM employed a number of field-effect transistors (FETs) arranged in a parallel structure, each of which sinks a fraction of the desired current.

15.1.5 Cubic Function Generator

U.S. patent 6,160,427 covers a low-voltage cubic function generator having "a voltage drop across only two active devices." The inventors were Stefano Cipriani and Anthony A. Takeshian of Conexant Systems, Inc.

The purpose of a cubic function generator (an analog computational circuit) is to produce an output that is equal to the cube of its input signal.

Cipriani and Takeshian (2000) state the high-level requirements of their cubic computational circuit in their patent:

> "In many systems today, especially in the digital environment, function generators are required to perform at high speeds. For example, many circuits operate at frequencies in the gigahertz range. In addition to the speed requirements, many digital systems are low voltage, requiring a function generator capable of operating at approximately 2 Volts. If the function generator only needs to produce a quadratic function, the voltage criteria are easily met. However, existing circuits do not meet these specifications when a cubic function is required.
>
> "What is needed is a compact, cubic function generator capable of operating at high frequencies and low voltages. Specifically, a circuit is required that generates a cubic function while operating at approximately 2 Volts and at frequencies up to and including the gigahertz range."...
>
> "The present invention generates a cubic transfer function while maintaining a voltage drop across only two active devices. This allows the present invention to operate with low voltage applications, specifically applications requiring a voltage drop of approximately 2 Volts."

The design of analog computational circuits is exceedingly difficult even for seemingly mundane mathematical functions. Success usually relies on the inspired discovery and exploitation of some aspect of the underlying device physics of the components (e.g., transistors) that yields the particular desired mathematical function (Gilbert 1968, 1979; Sheingold 1976; Babanezhad and Temes 1986). Because of this, the implementation of each different mathematical function (e.g., a squaring function versus a cubing function, a squaring function versus a square root function) typically requires an entirely different (clever) insight. In other words, there is nothing "routine" about the transition from designing one analog computational circuit to the next when humans design analog computational circuits.

In general, analog computational circuits are especially useful when the mathematical function must be performed more rapidly than is possible with digital circuitry (e.g., for real-time signal-processing at extremely high frequencies). Analog computational circuits are also especially useful when the need for a mathematical function in an otherwise entirely analog circuit does not warrant the cost (in terms of both circuitry and processing time) of the three-step process of converting the analog signal into a digital signal (using an analog-to-digital converter), performing the mathematical function in the digital domain (requiring a general-purpose digital processor consisting of perhaps millions of transistors as well as appropriate software), and then converting the result to the analog domain (using a digital-to-analog converter).

In discussing a logarithmic computational circuit, Robert Pease (1999) explained the reasons why analog computational circuits are so useful in practice:

> "In the digital, DSP-driven-algorithm world of today's design, you may think that determining the logarithm of a signal value without using numerical calculation is a throwback to an earlier era. But, doing it digitally is not a viable option for many reasons, and you have to use a single all-analog function block.

"Why not do it digitally? First, it would take considerable floating-point MIPS to calculate the logarithm of signals with bandwidth of even 100 MHz, and calculations on gigahertz signals would be impossible. Second, you'd need a very fast A/D converter ahead of the DSP to convert the analog signal. Finally, even if these factors were not obstacles, you would also need resolution of 12 bits to reach 72 dB of range under ideal conditions. You would need 16 bits for 96 dB of range, which means using a costly, power-hungry converter. And the design still wouldn't work! Just because the converter has the resolution doesn't mean that your system would actually realize that resolution, because the actual resolution of the A/D conversion for millivolt signals is different from what it is for volt-level signals. You'd need to put a variable-gain stage ahead of the converter, so, your design would be back to where it started in terms of signal-path gain, range, and resolution."

Of course, genetic programming is given no knowledge concerning the underlying device physics of the transistors to aid it in solving this problem (or any other problem of circuit synthesis). Genetic programming simply has access to the fitness of the entire circuit containing numerous transistors and other components (e.g., capacitors, resistors). In contrast to the approach used by knowledgeable, experienced, inventive, and creative human designers, genetic programming simply searches, over a series of generations, for ever-better circuits under the guidance of a high-level statement of the specifications of the desired circuit.

15.1.6 Tunable Integrated Active Filter

U.S. patent 6,225,859 covering a tunable integrated active filter was issued to Robert Irvine and Bernd Kolb of Infineon Technologies AG of Germany.

As the inventors state:

"Multipolar single-stage filters are difficult to implement in integrated bipolar high-frequency circuits. One essential goal is a steep edge dropoff in the amplitude response of the filter toward high frequencies, but without having to have recourse to multistage filter arrays.

"In the past, the art has either dispensed with explicit filter action entirely, or else used unipolar filters in the form of RC circuits. ...

"It is accordingly an object of the invention to provide an integrated low-pass filter, which overcomes the above-mentioned disadvantages of the heretofore-known devices and methods of this general type and that has a high selectivity in the depletion range, and an optimized characteristic curve with minimal damping in the conducting range, and that can be integrated in an integrated circuit."

15.2 Uniformity of Treatment of the Six Problems

All six problems in this chapter employ the same embryo, program architecture, function set, terminal set, minor control parameters, termination criteria, and parallel computing machinery.

The only significant difference in the human-supplied preparatory steps for the six problems is that each problem has an appropriate problem-specific fitness measure and associated test fixture. The fitness measure and test fixture together provide the means to measure the behavior of a candidate circuit. After this measurement process is complete, the circuit's fitness is communicated to the genetic programming system in the form of a single numerical value.

There are three minor departures from strict uniformity.

The first departure arises because the number of inputs and outputs varies from problem to problem. The circuit and test fixture must necessarily possess the appropriate number of input and output ports.

The second departure arises because the simulation time required to evaluate the fitness of an individual varies considerably among the six problems. In order to ensure the reasonableness of the total elapsed time for each run, we used a smaller population size on some problems. This departure from strict uniformity reflects practical considerations of managing available computational resources—not any foreknowledge that one particular problem would be facilitated by a particular choice of population size.

The third departure arises because the very essence of some inventions requires unique treatment.

For example, balun circuits have been in widespread use for decades. The specific aspect that makes Lee's balun circuit noteworthy (and patentable) in the year 2001 is that it operates at the low voltage and power levels required by today's wireless communication needs. Thus, we use a 1-Volt power source for this particular problem. Our departure from strict uniformity does not make it easier for genetic programming to solve this particular problem. Quite the contrary. This change makes the problem harder. In particular, the use of a low voltage source makes the problem as hard as the problem solved by Lee and forces genetic programming to address the essential characteristic of this particular post-2000 invention.

Similarly, low voltage is the essential issue in the patent for the cubic function generator. In fact, Cipriani and Takeshian (2000) discuss in detail a previously known (prior art) cubic function generator that operates at 3 Volts and state:

> "A prior art circuit 300 used to provide a cubic function to compensate for these power amplifiers is shown in FIG. 3....
>
> "In operation, the circuit 300 generates the cubic part of the transfer function... However, the circuit 300 only functions at voltages at or above approximately 3 Volts.... The 3-Volt limit makes the circuit unusable for many applications. For example, many low-voltage power amplifiers have a maximum voltage requirement of 2.7 Volts."

Thus, we use a 2-Volt power source for the cubic function generator problem because Cipriani and Takeshian specifically claim that their circuit operates at 2 Volts.

The same issue arises in connection with transistor models. We mainly use the commercially standard `2N3904` (*npn*) and `2N3906` (*pnp*) transistors for problems of circuit synthesis in this book. However, as detailed in the section 15.3.4 concerning the terminal set, different transistor models are required for certain problems. For example, we use a MOSFET transistor for the high-current load problem because Daun-Lindberg and Miller (2001) specifically state that they use this type of transistor. We chose a MOSFET transistor with a value of r_{dson} (measuring the transistor's drain source resistance when it is fully turned on) so that no one transistor could handle the problem's demands. Our departure from strict uniformity and our use of this particular transistor model forces genetic programming to create a solution that is in the spirit of the patent.

15.3 Preparatory Steps for the Six Post-2000 Patented Circuits

15.3.1 Initial Circuit

The initial circuit consists of an embryo and a test fixture.

The test fixture provides the means to enable the fitness measure to evaluate the circuit's behavior and characteristics. For each problem, the test fixture consists of certain fixed components that are connected to the input port(s) and the output port(s).

A "floating" embryo (section 4.7.1.1) is used as the starting point of the developmental process for all six problems in this chapter. The floating embryo consists of a single modifiable wire that is not initially connected to the circuit's input(s) or output(s). Thus, each individual circuit must master the problem of discovering the circuit's input(s) and output(s) on its own.

Note that several of the initial circuits in this chapter are unnecessarily complex in that they contain, for our convenience, current-to-voltage or voltage-to-current converters. These converters enable us to conveniently use certain simulation software to make measurements on the scales specified by the patent documents. They are not necessary parts of the setup of the problems.

15.3.1.1 Test Fixture for the Low Voltage Balun Circuit Figure 15.1 shows the test fixture for the low voltage balun circuit. The test fixture has five ports to which the developing circuit may connect. One of the ports makes the input voltage source available to the developing circuit. One port connects to the in-phase output VOUT0 and another port connects to the out-of-phase output VOUT1. Another port makes ground available to the developing circuit. Because the essential feature of the balun circuit for which Lee received a patent in 2001 is its ability to operate at a low voltage, the remaining port makes a 1-Volt power source available to the developing circuit.

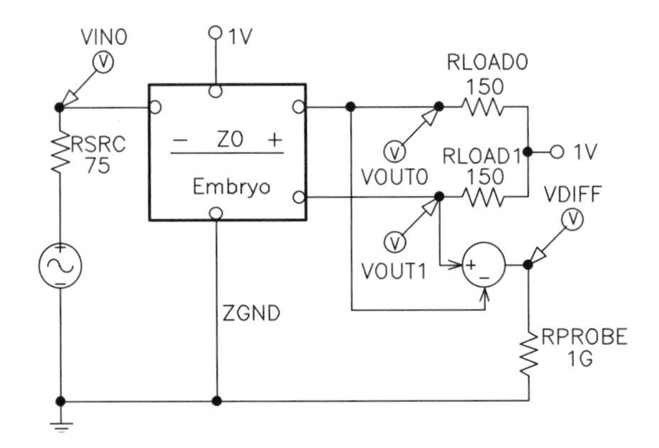

Figure 15.1 Test fixture for the low voltage balun circuit.

The input source in figure 15.1 is in series with a 75-Ohm resistor (the characteristic impedance of a common *unbalanced* coaxial video source). VIN0 is the port associated with the input source.

The circuit is probed at three probe points (the in-phase output VOUT0, the out-of-phase output VOUT1, and VDIFF). The ports labeled VOUT0 and VOUT1 are each in series with a 150-Ohm load resistor and a 1-Volt power source. The voltage at the third probe point, VDIFF, is equal to the difference between VOUT1 and VOUT0. There is a 1-giga-Ohm resistor RPROBE between VDIFF and ground. Note that 300 Ohms (twice 150) is the characteristic impedance of the common balanced twin-lead video cable.

15.3.1.2 Test Fixture for the Mixed Analog-Digital Variable Capacitor Circuit This circuit has four inputs. VREG0, VREG1, and VREG2 are the digital values in the three-bit digital register whereas VIN0 is an analog signal.

Figure 15.2 shows the test fixture for the mixed analog-digital variable capacitor circuit. The test fixture has six ports to which the developing circuit may connect. Three of the ports make the values (VREG0, VREG1, and VREG2) in a three-bit digital register available to the embryo (and, later, the fully developed circuit). Another port provides access to the analog input signal VIN0 that is used to test the circuit's behavior. Yet another port provides a means of connection to a current probe point IOUT0. The remaining port provides access to ground.

In figure 15.2, three input sources (VREG0, VREG1, and VREG2) are associated with the three-bit digital register. Each of these input sources is in series with a 1-micro-Ohm resistor. The input sources take on the value of either +15 Volts or −15 Volts

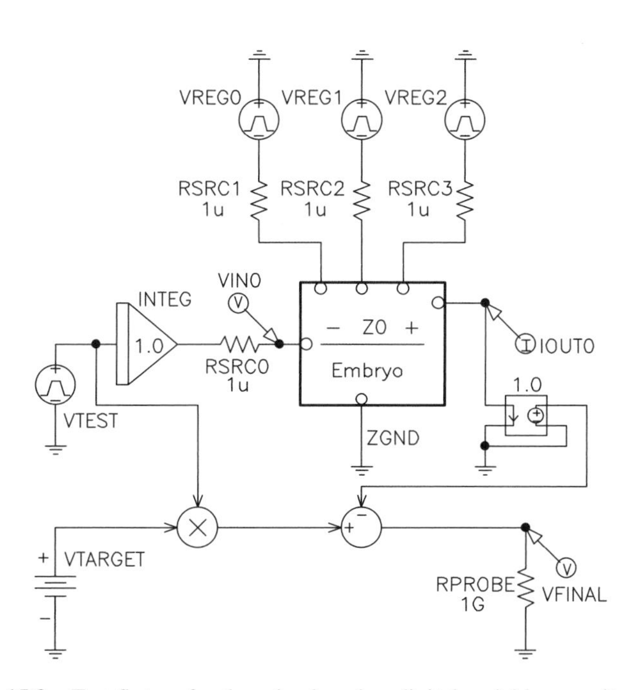

Figure 15.2 Test fixture for the mixed analog-digital variable capacitor circuit.

(representing the bits 1 and 0, respectively). There are no explicit power sources in the test fixture for this problem.

A capacitor is a two-leaded electrical component whose voltage, V, and current, I, are related by

$$I = C\frac{dV}{dt}.$$

The test fixture applies a test voltage to the evolved circuit and compares the current that flows through the evolved circuit to the current that would flow through an ideal capacitor whose capacitance is equal to 10 nanofarads times the value (0 to 7) specified by the three-bit digital register. Specifically, the input signal VTEST (representing the quantity dV/dt in the above equation for I) is connected in series to an integrator INTEG (whose output represents the voltage, V, in the equation) and a 1-micro-Ohm resistor. The output of the integrator INTEG is probed at VIN0 in the test fixture. The circuit is probed at current probe point IOUT0 (representing the current, I, in the equation).

The output port is fed into a current-to-voltage converter with a conversion factor of 1 Volt per Ampere.

The output of VTEST is multiplied (by multiplier X) by a DC voltage VTARGET that is always set to be equal to 10 nanofarads times the value (0 to 7) in the three-bit digital register. In other words, VTEST corresponds to dV/dt in the above equation and VTARGET corresponds to C. Their product, $C(dV/dt)$, is equal to I, the desired output current. Error is computed in the circuit by subtracting the actual output current from this ideal output current. Note that because the desired output current is computed as a voltage in the circuit, it is necessary to pass the current output IOUT0 through a current-to-voltage converter with a conversion factor of 1 Ampere per Volt before performing the subtraction. The error is probed at VFINAL.

15.3.1.3 Test Fixture for High-Current Load Circuit Figure 15.3 shows the test fixture for the high-current load circuit. The test fixture has six ports to which the

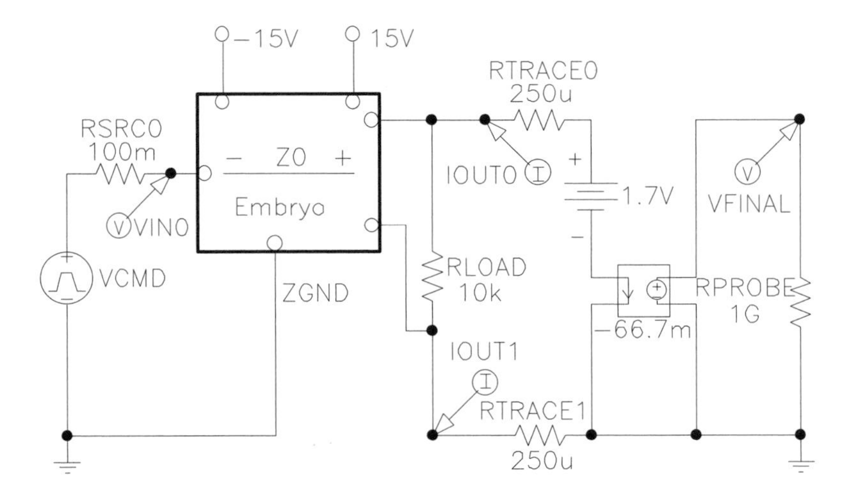

Figure 15.3 Test fixture for the high-current load circuit.

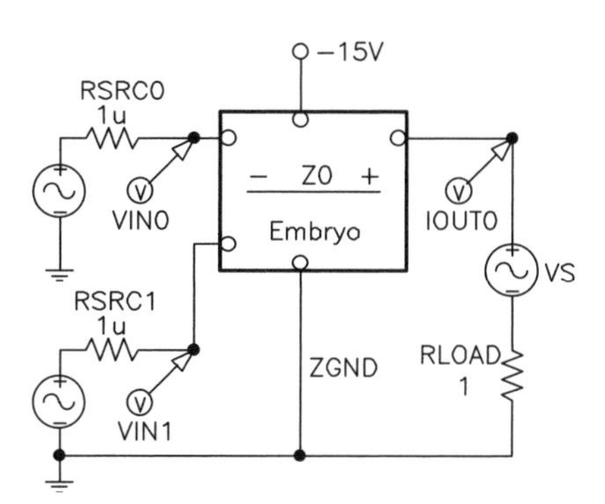

Figure 15.4 Test fixture for the voltage-current conversion circuit.

developing circuit may connect. One of the ports makes the input voltage available to the developing circuit. The input voltage is a control signal indicating how much current is to be sunk through the power supply that is being tested. Specifically, an input voltage V_{input} indicates that a current of V_{input} (15 Amperes/Volt) should be sunk through the power supply. Two ports make the −15-Volt and +15-Volt power sources available to the developing circuit. Another port makes ground available to the developing circuit.

The input voltage signal VCMD is fed through a 100-milliohm source resistor which is in turn connected to input port VIN0 in figure 15.3.

In figure 15.3, the output current probe point IOUT0 is in series with a 250-micro-Ohm trace resistor RTRACE0 which is connected to the positive end of the 1.7-Volt power supply that is to be tested. The probe point IOUT0 is connected through a 10,000-Ohm load resistor RLOAD to another 250-micro-Ohm resistor RTRACE1, which is in turn connected to ground. There is also a second current probe point IOUT1 between RLOAD and RTRACE1.

The negative end of the 1.7-Volt power supply being tested is connected to the high end of a current-to-voltage converter with a conversion factor of −0.0667 Volts per Ampere (in other words, 1 Volt per −15 Amperes). The output of this current-to-voltage converter is labeled VFINAL. The 1-giga-Ohm resistor RPROBE prevents the creation of a dangling wire at VFINAL. Large resistors such as this are commonly inserted into circuits to enable SPICE to simulate them.

15.3.1.4 Test Fixture for the Voltage-Current Conversion Circuit Figure 15.4 shows the test fixture for the voltage-current conversion circuit. The test fixture has five ports to which the developing circuit may connect. Two of the ports make the time-domain input sources available to the developing circuit. One provides a connection to the output point. The remaining two ports make the −15-Volt power source and ground available to the developing circuit.

Each input source in figure 15.4 is in series with a 1-micro-Ohm resistor. VIN0 is the port associated with the first input source and VIN1 is the port associated with the second input source.

The circuit is probed at current probe point IOUT0. The output port is in series with a time-varying voltage source VS, a 1-Ohm load resistor, and a ground point. The purpose of VS is to force the evolved circuit to produce a stable current at the probe point IOUT0 (i.e., a current which is independent of the value of VS).

Note that Ikeuchi and Tokuda (2000) employ a current sink (which contains a power source) in their design (figure 15.28) whereas we employ a *negative* 15-Volt power source in our test fixture. Neither the Ikeuchi-Tokuda circuit nor the genetically evolved circuit employs a positive power source.

15.3.1.5 Test Fixture for the Cubic Function Generator Figure 15.5 shows the test fixture for the cubic function generator. The test fixture has five ports to which the developing circuit may connect. One of the ports makes the input current available to the developing circuit. Two of the ports provide a means for measuring the flow of output current. The remaining two ports make the 2-Volt power source and ground available to the developing circuit.

Because we find it convenient to represent both input and output signals as voltages (rather than currents), the test fixture contains both a voltage-to-current converter at the input and a current-to-voltage converter at the output.

Specifically, in figure 15.5, the input signal VIN0 is fed into a voltage-to-current converter with a conversion factor of 100 microamperes per Volt. The high-output of the voltage-to-current converter is connected to the 2-Volt power source. The low-output is passed through a 1-micro-Ohm resistor to input port IIN0.

The output current probe point IOUT0 is in series with a 1-micro-Ohm resistor RCON0 and feeds into the high-input of a current-to-voltage converter with a conversion factor of 10.5 kilovolts per Ampere. There is also a second current probe point IOUT1 in series with a 1-micro-Ohm resistor RCON1 that feeds into the low-input of a current-to-voltage converter. The output of the current-to-voltage converter is VFINAL. There is a 1-giga-Ohm resistor RPROBE between VFINAL and ground.

15.3.1.6 Test Fixture for the Tunable Integrated Active Filter The test fixture (figure 15.6) for this problem has an incoming signal source VIN, an incoming control signal

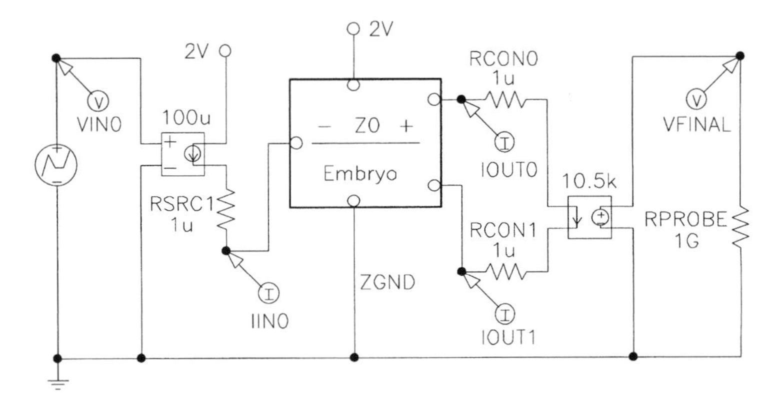

Figure 15.5 Test fixture for the cubic function generator.

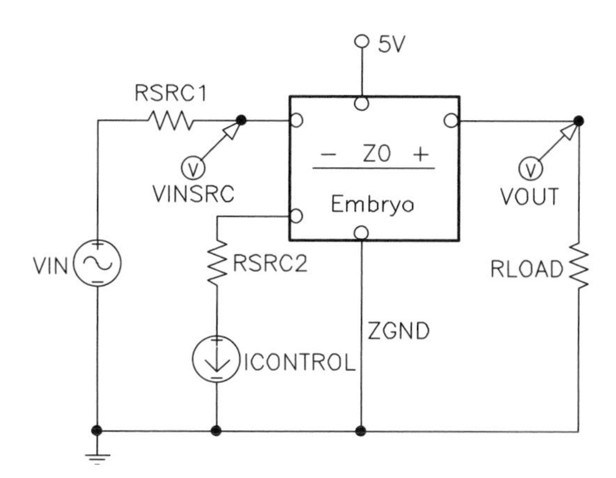

Figure 15.6 Test fixture for the tunable integrated active filter.

ICONTROL, a 50,000-Ohm source resistor RSRC1 associated with the incoming signal VIN, a 1-micro-Ohm source resistor RSRC2 associated with the control signal, a +5-Volt power source, a voltage probe point VOUT, and a 50,000-Ohm load resistor RLOAD.

15.3.2 Program Architecture

Because there must be one result-producing branch in the program tree for each modifiable wire in the embryo and there is one modifiable wire in the embryo for all six problems, the architecture of each circuit-constructing program tree for all six problems has one result-producing branch. Automatically defined functions and the architecture-altering operations are not used.

15.3.3 Function Set

A constrained syntactic structure enforces the use of one function set for the first argument of the `TWO_LEAD` function and another function set for all other parts of the program tree.

Because capacitors and resistors are the two-leaded functions used in this chapter, the function set, $F_{\text{two-lead}}$, for the first argument of the `TWO_LEAD` function is

$$F_{\text{two-lead}} = \{\texttt{R\_NEW}, \texttt{C\_NEW}\}.$$

Some functions are common to all six problems in this chapter. The common functions for each construction-continuing subtree for the six problems include the following:

$$F_{\text{common}} = \{\texttt{Q}, \texttt{TWO\_LEAD}, \texttt{SERIES}, \texttt{PARALLEL\_NEW}, \texttt{TWO\_GROUND}, \texttt{THREE\_GROUND}, \texttt{NODE}\}.$$

In addition to these common functions, table 15.2 shows the additional input, output, and power source functions required for particular problems.

The `INPUT_0`, `INPUT_1`, `INPUT_2`, `INPUT_3`, `OUTPUT_0` and `OUTPUT_1` functions are described in section 4.7.1.1.

Table 15.2 Inputs, outputs, and power sources for the six problems

Circuit	Input(s)	Output(s)	Power sources(s)
Low-voltage balun circuit	INPUT_0	OUTPUT_0, OUTPUT_1	TWO_POS1V
Mixed analog-digital variable capacitor circuit	INPUT_0, INPUT_1, INPUT_2, INPUT_3	OUTPUT_0	None
High-current load circuit	INPUT_0	OUTPUT_0, OUTPUT_1	TWO_POS15V, TWO_NEG15V
Voltage-current conversion circuit	INPUT_0, INPUT_1	OUTPUT_0	TWO_NEG15V
Low-voltage cubic function generator	INPUT_0	OUTPUT_0, OUTPUT_1	TWO_POS2V
Tunable integrated active filter	INPUT_0, INPUT_1	OUTPUT_0	TWO_POS5V

The two-argument TWO_POS15V ("positive reference voltage source") function enables any part of a circuit to be connected to the constant +15.0-Volt DC power source (e.g., a battery) as described in section 11.3.1.3. The TWO_POS5V, TWO_POS2V, TWO_POS1V, and TWO_NEG15V functions operate in a similar way.

15.3.4 Terminal Set

There are six types of terminals for the six problems in this chapter. A constrained syntactic structure specifies the terminals that may appear in arithmetic-performing subtrees; the first argument of the PARALLEL_NEW function; the first, second, and third arguments of the transistor-creating Q function; and each construction-continuing subtree.

First, the numerical parameter value(s) for each component possessing a parameter (i.e., resistors and capacitors) are established by a value-setting subtree containing a single perturbable numerical value (as described in section 3.5.5.2). The terminal set, T_{vss}, for the value-setting subtrees is

$$T_{vss} = \{\Re_p\},$$

where $\Re_p$ denotes a perturbable numerical value.

Second, a constrained syntactic structure specifies the terminals that may appear as the first argument of the PARALLEL_NEW function. The terminal set, $T_{parallel}$, for the first argument of the PARALLEL_NEW function is

$$T_{parallel} = \{\texttt{UP\_OR\_LEFT, DOWN\_OR\_RIGHT}\}.$$

Third, a constrained syntactic structure specifies the terminals that may appear as the first argument of the transistor-creating Q function. The first argument of the Q function specifies the transistor model used. We mainly use the commercially popular 2N3904 (*npn*) and 2N3906 (*pnp*) transistor models for problems of circuit synthesis in this book. That is, the terminal set, T_{model}, is usually

$$T_{model} = \{\texttt{2N3904, 2N3906}\}.$$

Table 15.3 Transistor models for various runs of the six problems

Circuit	Transistors
Low-voltage balun circuit	`2N3904, 2N3906`
Mixed analog-digital variable capacitor	`IRF511, IRF9230`
High-current load circuit—First run	`IRFZ44, MTP50P03HDL`
High-current load circuit—Second run	Same as first run
Voltage-current conversion circuit	`2N3904, 2N3906`
Low-voltage cubic function generator—First run	`2N3904, 2N3906`
Low-voltage cubic function generator—Second run	`HFA3046, HFA3128`
Tunable integrated active filter—First run	`2N3904, 2N3906`
Tunable integrated active filter—Second run	Same as first run
Tunable integrated active filter—Third run	Same as first run
Tunable integrated active filter—Fourth run	Same as first run

However, because the description in the patent of the high-current load circuit (Daun-Lindberg and Miller 2001) specifically calls for a MOSFET transistor, we used the `IRFZ44` (*nmos*) and a complementary `MTP50P03HDL` (*pmos*) transistor models for that particular problem. Similarly, for the mixed analog-digital variable capacitor problem, we used a standard `IRF511` (*nmos*) and a complementary `IRF9230` (*pmos*) transistor model. After the first run of the cubic function generator problem was successful, we tried an additional experiment involving the high-frequency transistor models `HFA3046` (*npn*) and `HFA3128` (*pnp*) on the second run of this problem.

Table 15.3 shows the transistor model used for each of the six problems in this chapter.

Fourth, the second argument of the transistor-creating `Q` function establishes which end (polarity) of the preexisting modifiable wire will be bifurcated (if necessary) in inserting the transistor. The terminal set, $T_{bifurcate}$, is

$$T_{bifurcate} = \{\texttt{BIFURCATE\_POSITIVE, BIFURCATE\_NEGATIVE}\}.$$

Fifth, the third argument of the `Q` function specifies which of six possible permutations of the transistor's three leads (base, collector, and emitter) is to be used. The terminal set, $T_{permutation}$, is

$$T_{permutation} = \{\texttt{B\_C\_E, B\_E\_C, C\_B\_E, C\_E\_B, E\_B\_C, E\_C\_B}\}.$$

Sixth, the terminal set, T_{ccs}, for each construction-continuing subtree is

$$T_{ccs} = \{\texttt{END, SAFE\_CUT}\}.$$

15.3.5 Fitness Measure

The third preparatory step prior to launching a run of genetic programming involves the construction of a fitness measure for the problem. The fitness measure is the primary mechanism for communicating the high-level statement of the problem's requirements to the genetic programming system. The fitness measure specifies what needs to be done. The fitness measure is, for all the problems in the chapter, multiobjective in the sense that it combines two or more different elements. The different elements of the fitness measure are typically in competition with one another to some degree. Each fully developed candidate circuit is evaluated over a representative sample of different elements of the different *fitness cases*.

The fitness measure is used to establish a partial order among candidate individuals (i.e., that one individual is better than another). This partial order is used to probabilistically select individuals to participate in the various genetic operations (i.e., crossover, reproduction, mutation, and the architecture-altering operations) during the run of genetic programming. The fitness measure accomplishes this by assigning a single numeric value that reflects the extent to which the individual circuit satisfies the high-level requirements of the problem.

15.3.5.1 Fitness Measure for Low Voltage Balun Circuit The circuit described in U.S. patent 6,265,908 entitled "Low Voltage Balun Circuit" by Sang Gug Lee is a balun circuit that operates using a power supply of only 1 Volt.

The balun (balance/unbalance) circuit in this problem is to produce two outputs from a single input, each output having half the amplitude of the input, one output being in phase with the input with the other being 180 degrees out of phase with the input, with both outputs having the same DC offset. One of the two outputs is probed at the in-phase output point VOUT0 and the other is probed at the out-of-phase output point VOUT1 (of figure 15.1).

Fitness is measured using seven fitness cases, five of which are designed to ensure the correct magnitude and phase at the circuit's two outputs. These five fitness cases involve frequency-sweep analyses. Additionally, there is a fitness case that uses a DC operating point analysis to ensure that the circuit's output has the correct DC offset. Finally, there is a fitness case that uses a Fourier analysis that is designed to penalize harmonic distortion.

All five frequency-sweep analyses range over two decades of frequency from 1 MHz to 100 MHz. Each decade is divided equally on a logarithmic scale into 20 parts (for a total of 41 frequencies). All five frequency-sweep analyses employ an input signal whose amplitude is 10 millivolts.

Two of the five frequency-sweep analyses probe the in-phase output (one for amplitude and one for phase); two probe the out-of-phase output (again, one for amplitude and one for phase); and one probes the phase of the difference between the out-of-phase and in-phase outputs.

The first fitness case for this problem probes the amplitude of the in-phase output. The desired amplitude of the in-phase output is 5 millivolts (i.e., one half of the amplitude of the input signal). This corresponds to -46 dB (assuming a 1 Volt reference) and corresponds to -6 dB (using the 10-millivolt input signal as a reference).

The second fitness case is identical to the first except that it probes the amplitude of the out-of-phase output.

The third fitness case probes the phase of the in-phase output. The desired phase of the in-phase output is 0 degrees.

The fourth fitness case probes the phase of the out-of-phase output. The desired phase of the out-of-phase output is 180 degrees.

The fifth fitness case probes the phase of the difference between the out-of-phase and the in-phase outputs (probed at VDIFF). The desired phase of the difference is 180 degrees.

The sixth fitness case is a DC operating point analysis that determines the DC offset associated with the difference between the out-of-phase and in-phase outputs. The desired value for the DC operating point is 0 Volts.

The seventh fitness case involves a Fourier analysis designed to penalize total harmonic distortion (defined in chapter 14 and Vladimirescu 1994). The Fourier analysis measures the 2nd through 9th order harmonics of a 10-MHz test frequency. The duration for the Fourier analysis is 200 nanoseconds corresponding to two cycles of the 10-MHz test frequency. The number of time steps is 100. For the Fourier analysis, the point probed was VDIFF (the difference between the out-of-phase and in-phase outputs). A total harmonic distortion of -30 dB is considered acceptable for this problem.

Fitness is the sum of a detrimental contribution to fitness associated with each of these seven fitness cases.

For the first five fitness cases (i.e., those involving frequency-sweep analyses), the detrimental contribution to fitness consists of the sum, over the 41 frequencies between 1 MHz to 100 MHz, of the weighted absolute difference between the desired value (defined in the description of each fitness case) and the value associated with that frequency for that fitness case.

For the two fitness cases that relate to amplitude and that are measured in decibels, the weight is 1.0 if the absolute value of the difference is less than 3 dB and the weight is 10.0 otherwise.

For the three fitness cases that relate to phase and that are measured in degrees, the weight is 1.0 if the absolute value of the difference is less than 10 degrees and the weight is 10.0 otherwise.

For the sixth fitness case (involving the DC operating point analysis), the detrimental contribution to fitness is equal to 1,000 times the difference between the probed value and the desired value of 0 Volts.

For the seventh fitness case (involving the Fourier analysis designed to penalize harmonic distortion), the detrimental contribution to fitness is

- $10(1 + |THD - (-30)|)$ if the total harmonic distortion is ≥ -30 dB, or
- $1/(1 + |THD - (-30)|)$ if the distortion ratio is < -30 dB.

Notice that an ideal value of total harmonic distortion of 0 (i.e., $-\infty$ dB) corresponds to a detrimental contribution to fitness of 0; that a value of total harmonic distortion of exactly -30 dB corresponds to a detrimental contribution to fitness of 10; and that if the total harmonic is acceptable (i.e., less than -30 dB), then the detrimental contribution to fitness is at most 1.

15.3.5.2 Fitness Measure for Mixed Analog-Digital Variable Capacitor U.S. patent 6,013,958 by Turgut Sefket Aytur of Lucent Technologies Inc. covers a mixed analog-digital variable capacitor.

The purpose of the mixed analog-digital variable capacitor circuit is to provide behavior (probed at IOUTO) that is equivalent to that of a capacitor whose capacitance is dynamically controlled by the value stored in a digital register.

There are 16 fitness cases for this problem. Each fitness case employs a combination of one of two time-domain input signals in conjunction with one of the eight possible values (000 to 111) in the three-bit digital register.

The two time-domain input signals vary by shape and time scale.

A capacitor is a two-leaded electrical component whose voltage, V, and current, I, are related by

$$I = C\frac{dV}{dt}.$$

The input signal for each fitness case is based on the time-domain value of the voltage source VTEST (corresponding to dV/dt in the above equation). The voltage that is actually made available to the circuit is the integral of VTEST (i.e., V in the above equation for I). The integral of VTEST is probed at VIN0 in figure 15.2. Additionally, note that although VTEST is implemented in our test fixture as a voltage source, its value actually represents a derivative. Thus, the value of VTEST is expressed in "Volts per second" rather than "Volts." Note that at time $t = 0$, the output of the integrator is zero.

Figure 15.7 shows the integral (i.e., V) of the first input signal. The horizontal axis is in microseconds and the vertical axis is in Volts. The signal is divided into 100 time steps (over 10 microseconds).

Figure 15.8 shows the curve for the first input signal (i.e., dV/dt). The horizontal axis is in microseconds and the vertical axis is in Volts per second.

Figure 15.9 shows the integral (i.e., V) of the second input signal. The horizontal axis is in microseconds and the vertical axis is in Volts.

Figure 15.10 shows the curve for the second input signal (i.e., dV/dt). The horizontal axis is in microseconds and the vertical axis is in Volts per second.

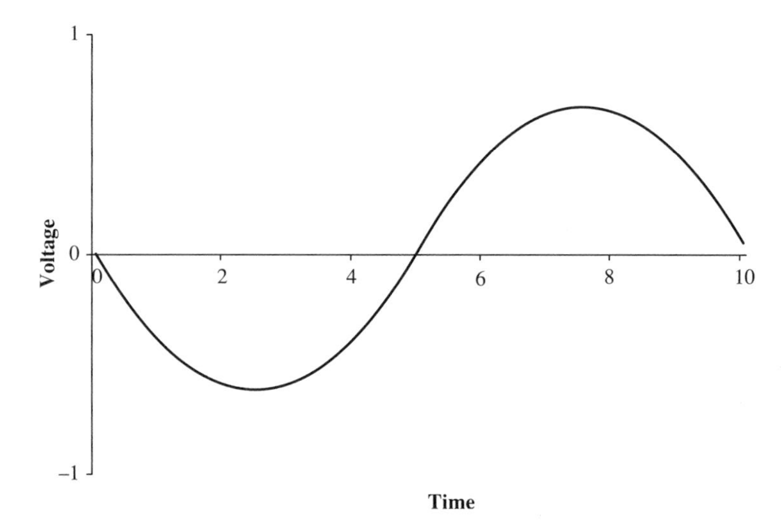

Figure 15.7 Integrated input signal for fitness case 1 for the mixed analog-digital variable capacitor problem.

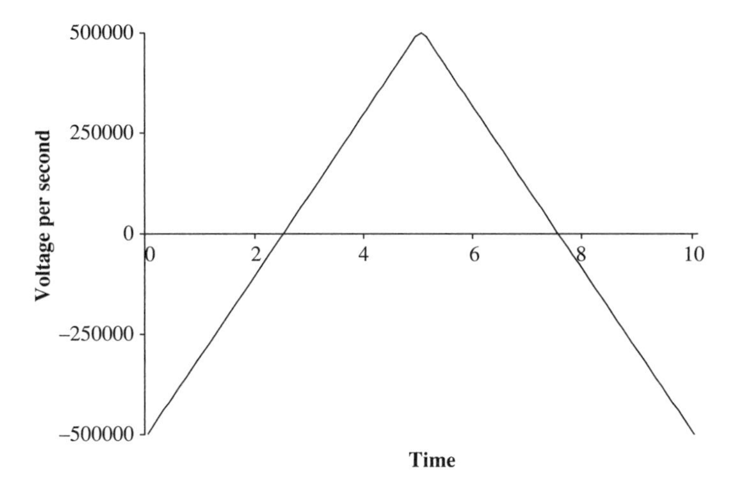

Figure 15.8 Input signal for fitness case 1 for the mixed analog-digital variable capacitor problem.

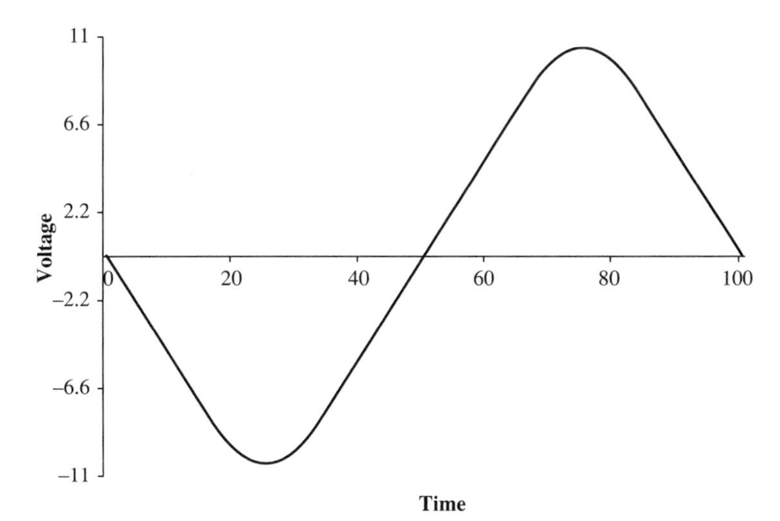

Figure 15.9 Integrated input signal for fitness case 2 for the mixed analog-digital variable capacitor problem.

One time-domain simulation is performed for each of the eight values of the 3-bit digital register in conjunction with each of two possible time-domain input signals.

The eight values of the 3-bit digital register specify a desired capacitance that is between 0 and 70 nanofarads. The output current desired at the probe point is the product of the desired capacitance and the input voltage (which represents d*V*/d*t*). For example, if the 3-bit digital register is 001 (making the desired capacitance 10 nanofarads), then the output voltage desired at the probe point IOUT would range between −5 and +5 milliamperes (for both time-domain input signals).

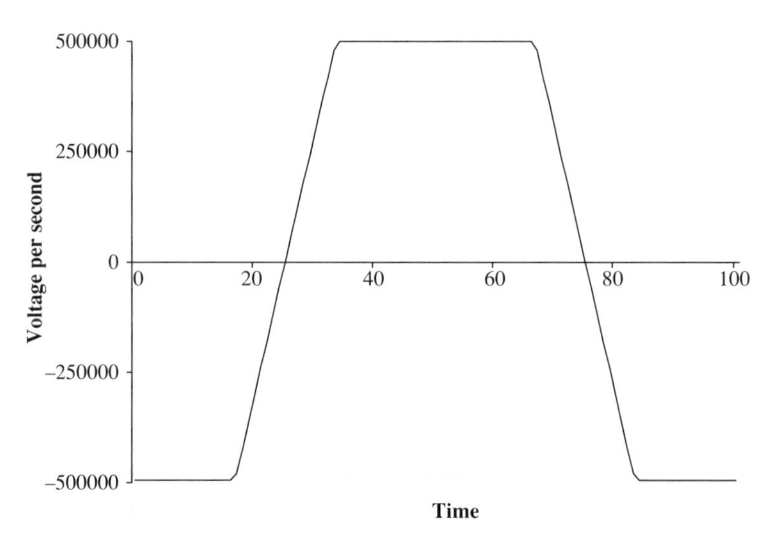

Figure 15.10 Input signal for fitness case 2 for the mixed analog-digital variable capacitor problem.

Fitness is the sum, over 101 time values for each of the 16 fitness cases, of the absolute value of the weighted difference between ideal output current (i.e., that which would flow through a capacitor whose capacitance was equal to 10 nanofarads times the value in the three-bit digital register) and the circuit's actual output current. The weight is 1.0 if the absolute value of the difference is less than 0.25 milliamperes and the weight is 10.0 otherwise.

15.3.5.3 Fitness Measure for High-Current Load Circuit U.S. patent 6,211,726 covers a circuit designed to sink a time-varying amount of current in response to an input signal. Toward this end, Daun-Lindberg and Miller of IBM employed a number of FET transistors arranged in a parallel structure, each of which sinks a small amount of the desired current.

The amount of current is probed at VFINAL (the output of a current-to-voltage converter) of figure 15.3.

The fitness measure for this problem employs two time-domain simulations, each representing a different input signal. The two input signals vary by shape and time scale.

The first input signal starts at 0 Volts at time 0, remains at that level for 50 microseconds, rises to 10 Volts over a period of 100 microseconds, remains at that level for 300 microseconds, falls to 0 Volts over a period of 100 microseconds, and remains at that level for another 450 microseconds. Note that these input signals, specified in terms of voltage, are passed through a voltage-to-current converter with a conversion factor of −66.7 millivolts per Ampere.

The second input signal starts at 0 Volts at time 0, remains at that level for 75 microseconds, rises to 6 2/3 Volts over a period of 300 microseconds, remains at that level for 200 microseconds, falls to 0 Volts over a period of 150 microseconds, and remains at that level for another 275 microseconds.

Each input signal is divided into 30 time steps (over a period of 1 millisecond).

We made two runs of this problem.

For the first run of this problem, fitness is the weighted sum, over each time value for each fitness case, of the absolute value of the difference between the amount of current sunk by the circuit and the desired amount of current given by the time-varying input signal. The difference for each time value for each fitness case is weighted by the product of the reciprocal of the patented circuit's average absolute error for that fitness case and the reciprocal of the number of fitness cases (i.e., two). Thus, a fitness of 1.0 is assigned to the patented circuit.

After the first run was successful, we made a second run. In the first run, there is a weighting associated with each fitness case. For the second run, there is an additional weighting associated with each time value. For the second run, we applied an additional multiplicative weighting of 10.0 for each time value where the absolute error is greater than the patented circuit's average absolute error over the two fitness cases (i.e., 41 millivolts).

15.3.5.4 Fitness Measure for Voltage-Current Conversion Circuit The voltage-current conversion circuit patented by Akira Ikeuchi and Naoshi Tokuda of Mitsumi Electric Co., Ltd. (U.S. patent 6,166,529) takes two time-domain voltage signals as input and is intended to produce a stable current whose magnitude is proportional to the difference between the two input voltage signals. The current is probed at current probe point IOUT0.

Fitness is measured using four time-domain fitness cases involving input signals of various shapes. Each fitness case involves an added and subtracted input part.

For the first fitness case, the subtracted input signal rises linearly from 2 Volts to 10 Volts for a period of 1 second while the added input signal falls linearly from 10 Volts to 2 Volts.

For the second fitness case, the subtracted input signal is a 6.4 Hz sine wave centered at 6 Volts with a minimum value of 2.5 Volts and a maximum value of 9.5 Volts. The added input signal is a 4.5 Hz sine wave centered at 5.5 Volts with a minimum value of 2.0 Volts and a maximum value of 9.0 Volts.

For the third fitness case, both the added input signal and the subtracted signal are 2.0 Hz sine waves centered at 6 Volts with a minimum value of 2.0 Volts and a maximum value of 10.0 Volts.

For the fourth fitness case, the subtracted input signal is a 6.4 Hz sine wave centered at 4 Volts with a minimum value of 3 Volts and a maximum value of 5 Volts while the added input signal is a 4.5 Hz sine wave centered at 4 Volts with a minimum value of 3.0 Volts and a maximum value of 5.0 Volts.

Each input signal is divided into 30 time steps (over a period of 1 second).

The desired output current is the difference between the added input signal and the subtracted input signal multiplied by 1 milliampere per Volt.

For all four fitness cases, the time-varying voltage source VS is a 3.2 Hz sine wave centered at 0 Volts with a minimum value of -1 Volt and a maximum value of $+1$ Volt.

Fitness is the weighted sum, over all time values for the four fitness cases, of the absolute difference between the desired output current and the circuit's actual output current. The weight associated with the time values for each of the four fitness cases is one fourth of the reciprocal of the patented circuit's average absolute error for that fitness case. The values of the average absolute error of the patented circuit for each of the four fitness cases are 41, 36, 26, and 44 microvolts, respectively. Note that when the patented circuit is evaluated using this fitness measure, it is assigned a fitness of 1.0.

15.3.5.5 Fitness Measure for Cubic Function Generator The purpose of the cubic function generator circuit is to produce an output current (probed at VFINAL) that is proportional to the cube of the input current. Stefano Cipriani and Anthony A. Takeshian of Conexant Systems, Inc. received U.S. patent 6,160,427 for their cubic function generator circuit.

The patented circuit is to be "compact" in the sense that it requires a voltage drop across no more than two transistors at any point in the circuit. The compactness constraint is enforced by allowing the evolutionary process access to only a 2-Volt power supply (i.e., just less than three base-to-emitter drops).

Fitness is measured using four time-domain fitness cases involving input signals of various shapes and time scales.

The first fitness case is a ramp that rises from 0 Volts to 1.26 Volts (the cube root of 2) over a period of 1 millisecond.

The second fitness case is a single full cycle of a sine wave ranging between 0 Volts and +1.26 Volts over a period of 1 millisecond.

The third fitness case is a ramp that falls from 1.26 Volts to 0 Volts over a period of 10 milliseconds.

The fourth fitness case is a constant input of 1 Volt lasting 10 milliseconds.

All four of these input signals are passed through a voltage-to-current converter with a conversion factor of 100 microamperes per Volt.

There are 100 time steps associated with each fitness case.

Fitness is the sum, over the four fitness cases and 101 time values for each fitness case, of the absolute value of the weighted difference between the cube of the input voltage signal (which passes through a voltage-to-current converter) and the circuit's actual output (probed at VFINAL, the output of a current-to-voltage converter). The weight for each time value is 1.0 if the absolute value of the difference is less than 7 millivolts (i.e., the average absolute error of the patented circuit of figure 15.32) and the weight is 10.0 otherwise.

The number of hits is defined to be the number of time values for which the weighting is 1.0 (as opposed to 10.0).

15.3.5.6 Fitness Measure for Tunable Integrated Active Filter U.S. patent 6,225,859 covers a tunable integrated active filter that was invented by Robert Irvine and Bernd Kolb of Infineon Technologies AG of Germany. The purpose of the tunable integrated active filter is to perform the function of a lowpass filter whose passband boundary is dynamically specified by a control signal. The circuit has two inputs: a to-be-filtered incoming signal VIN and a control signal ICONTROL.

Because the high-level statement of the behavior for the desired filter circuit is expressed in terms of the circuit's differing response to various frequencies, the output is measured in the frequency domain.

The passband boundary, f, ranges over the following nine values: 441, 588, 784, 1,046, 1,395, 1,861, 2,482, 3,310, and 4,414 Hz. I_c acts as a control signal ICONTROL in the patented circuit. As can be seen from the equations given in section 10.5, the value of 441 Hz corresponds to a value of I_c of 100 microamperes of current and a value of 4,414 Hz corresponds to a value of I_c of 10 milliamperes of current (a reasonable range for the patented circuit).

For each of the nine values of f, the current source ICONTROL takes on a specific value (i.e., the value of the control variable, I_c) as defined by the equation in section 10.5.

For each of the nine values of the free variable f, SPICE is instructed to perform an AC small signal analysis and report the circuit's behavior from $f/100$ to $10f$ (an interval of three decades) with each of these three decades being divided into 20 parts (using a logarithmic scale). Thus, there are a total of 61 sampled frequencies associated with each value of f. Note that the 41st point is f itself.

For the first run of this problem, fitness is composed only of a performance penalty. The second run and subsequent runs also include a parsimony penalty.

The performance penalty is the weighted sum, over the 61 frequencies for each of the nine values of the free variable f, of the absolute weighted deviation between the output of the individual circuit at the probe point VOUT and the output of the target circuit in figure 10.12 at the probe point VOUT.

For all frequencies less than or equal to f, the absolute deviation is calculated in Volts. However, for all frequencies greater than f, the absolute deviation is calculated in decibels.

For all frequencies greater than f, the weighting is 10.0 if the absolute deviation (in decibels) is greater than 3 dB, but 1.0 otherwise.

For the 31st through 40th points of the passband, the weighting is 100,000 if the absolute deviation is greater than 3 dB, but 10,000 otherwise.

For all other points in the passband, the weighting is 10,000 if the absolute deviation is greater than 3 dB, but 1,000 otherwise.

The larger weights for the passband compensate for the fact that deviations measured in Volts are (for the range of voltages used in this problem) less than those measured in decibels.

The larger weighting for the 31st through 40th points of the passband reflects the fact that the challenge in designing many filters lies with the frequency response in the part of the passband just before the start of the transitional region.

The number of hits is defined as the number of points for which the voltage (in decibels) that is produced by the individual circuit at the probe point VOUT is within 3 dB of the voltage (in decibels) produced by the target circuit.

The first run of this problem was successful. However, the best-of-run circuit from the first run has 12 transistors, three capacitors, and one resistor (in addition to the components in the test fixture). In reviewing the patent document, it was apparent that the spirit of the invention involves finding a parsimonious solution to the problem. The inventor's patented circuit (shown in figure 15.35) is parsimonious. Therefore, parsimony was explicitly included as part of the fitness measure on each subsequent run of this problem.

For the second and subsequent runs, fitness is a weighted sum of a parsimony penalty and a performance penalty. The parsimony penalty is equal to the number of components in the circuit. For individuals with 543 or fewer hits (out of 549), fitness is the performance penalty plus 10^{-12} times the parsimony penalty. For individuals with 544 (99%) or more hits (out of 549), fitness is 10^{-18} times the performance penalty plus 10^{-12} times the parsimony penalty. Note that the reason for choosing 99% of the hits as the trigger for parsimony in the fitness measure is that the patented circuit does not attain 100% of the hits for this problem (given our definition of hits).

Table 15.4 Population sizes for various runs of the six problems

Circuit	Population
Low-voltage balun circuit	5,000,000
Mixed analog-digital variable capacitor	2,000,000
High-current load circuit—First run	2,000,000
High-current load circuit—Second run	Same as first run
Voltage-current conversion circuit	5,000,000
Cubic function generator—First run	5,000,000
Cubic function generator—Second run	2,000,000
Tunable integrated active filter—First run	2,000,000
Tunable integrated active filter—Second run	Same as first run
Tunable integrated active filter—Third run	Same as first run
Tunable integrated active filter—Fourth run	Same as first run

15.3.6 Control Parameters

Table 15.4 shows the population size for each of the six problems in this chapter.

15.4 Results for the Six Post-2000 Patented Circuits

15.4.1 Results for Low-Voltage Balun Circuit

The best-of-generation circuit from generation 0 has a fitness of 39.7.

On generation 97 of the balun problem, a circuit emerged that satisfies the problem's high-level requirements. Specifically, the circuit produces (in response to a sinusoidal input signal) sinusoidal signals at both of the output terminals (probed at VOUT0 and VOUT1 in figure 15.1) that have the desired magnitude and phase. Moreover, the difference between these two signals (probed at VDIFF) also has the correct magnitude and phase.

The best-of-run circuit (figure 15.11) from generation 97 has a fitness of 0.429. By way of reference, the patented circuit has a total fitness of 1.72. Thus, the genetically evolved circuit achieves roughly a fourfold improvement over the patented circuit in terms of our fitness measure.

Note that several transistors that are manifestly nonfunctional were deleted from figure 15.11. Similarly, some or all of the nonfunctional transistors have been removed from some of the other genetically evolved circuits in this chapter.

Figure 15.12 shows the time-domain behavior of the best-of-run balun circuit from generation 97 of the first run for both the in-phase and out-of-phase output ports. As can be seen, the genetically evolved circuit divides the input signal into two half-amplitude signals that are 180 degrees out of phase from each other.

Figure 15.13 shows the frequency response (probed at VDIFF) for the best-of-run balun circuit from generation 97 and the circuit described in U.S. patent 6,265,908 (figure 15.15). The ideal response would be a horizontal line at 10 millivolts. As can be seen, the genetically evolved circuit performs better at the higher frequencies.

Looking at the magnitude of the difference (probed at VDIFF) between the out-of-phase and in-phase outputs, the best-of-run circuit is superior to the patented circuit

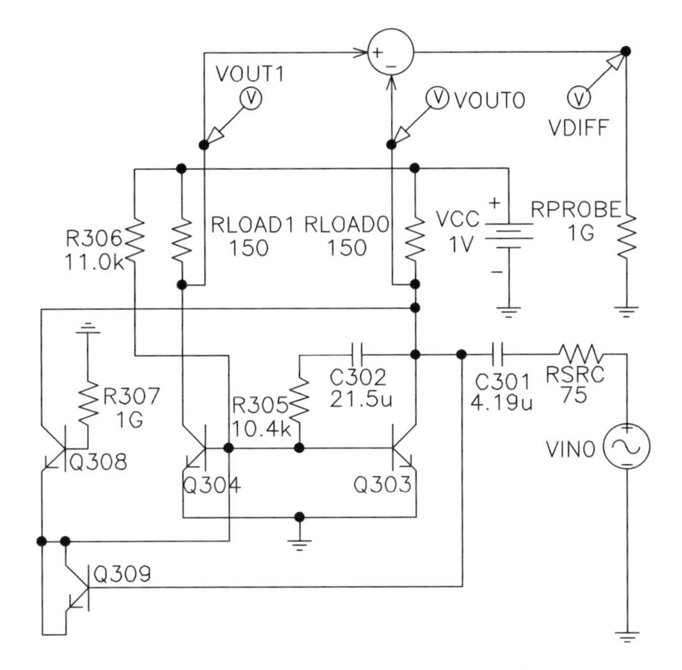

Figure 15.11 Best-of-run balun circuit.

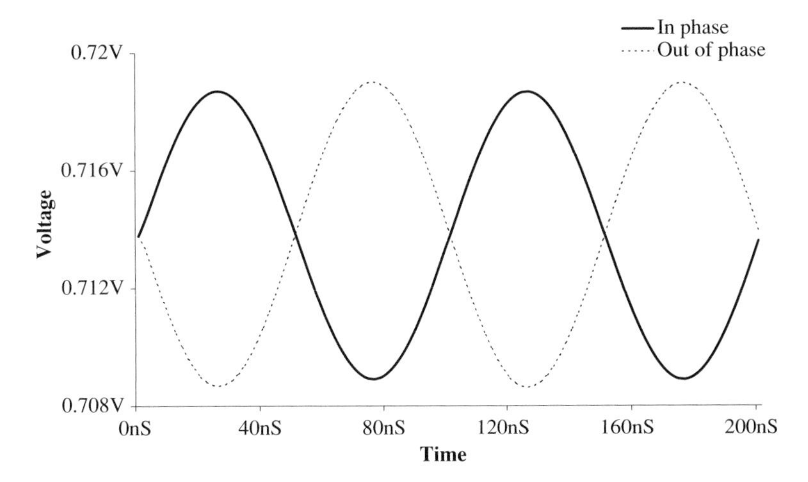

Figure 15.12 Time-domain behavior of the best-of-run balun circuit.

in terms of its frequency response. Specifically, the genetically evolved circuit has an average absolute error of 0.47 dB from 1 MHz to 100 MHz, whereas Lee's patented circuit has an average absolute error of 0.60 dB.

The best-of-run circuit is also superior to the patented circuit in terms of its total harmonic distortion (THD). Specifically, the total harmonic distortion for the

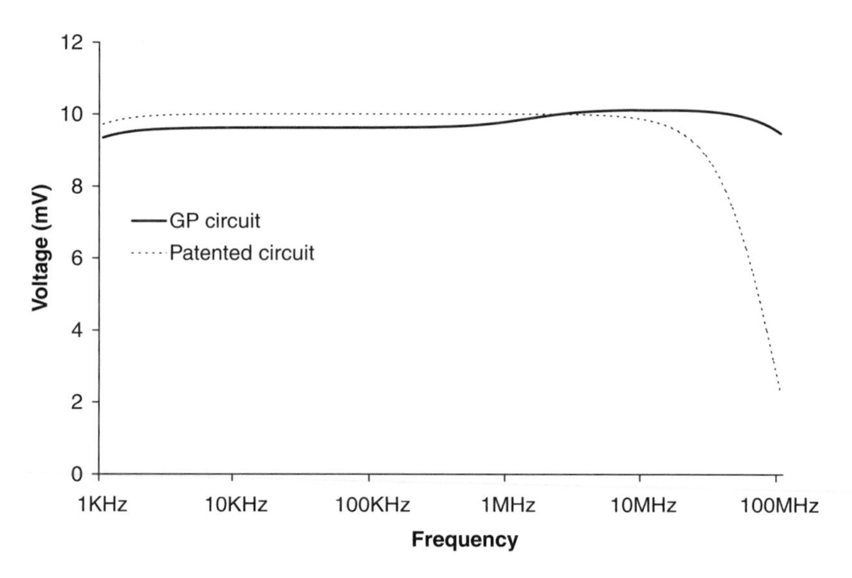

Figure 15.13 Comparison of the frequency response of the best-of-run balun circuit and the patented circuit.

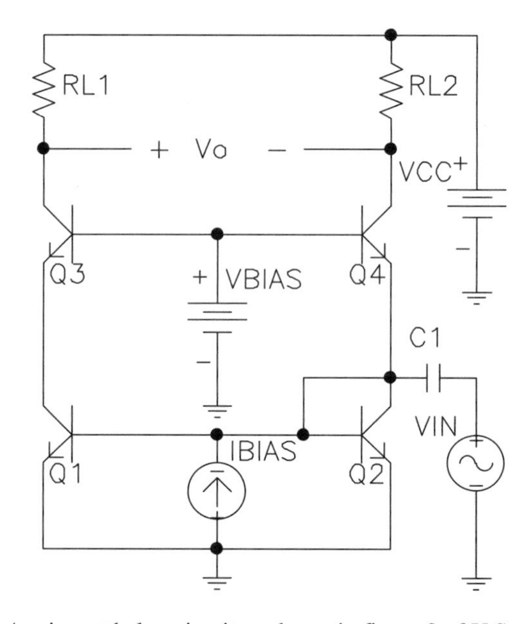

Figure 15.14 A prior art balun circuit as shown in figure 2 of U.S. patent 6,265,908.

genetically evolved circuit is −65.6 dB whereas it is −33.9 dB for the patented circuit.

In his patent, Lee presents a previously known conventional (prior art) balun circuit as "figure 2." Figure 15.14 shows "figure 2" of U.S. patent 6,265,908 (Lee 2001).

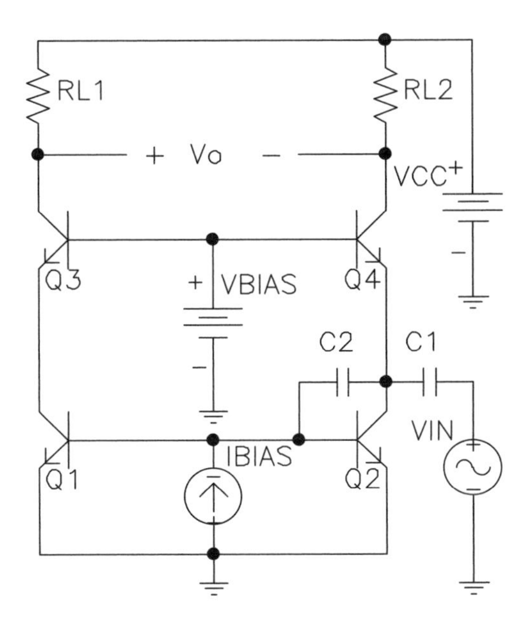

Figure 15.15 Lee's low voltage balun circuit as shown in figure 3 of patent 6,265,908.

Lee presents his invention as "figure 3" in his patent. Figure 15.15 shows "figure 3" of U.S. patent 6,265,908.

Lee (2001) identifies the essence of his invention in the patent documents. The essential difference between Lee's invention and the prior art is a coupling capacitor C2 located between the base and the collector of the transistor Q2. As Lee explains,

"FIG. 2 offers a circuit diagram depicting another conventional balun circuit."

"FIG. 3 represents a circuit diagram of a low voltage balun circuit in accordance with a preferred embodiment of the present invention....

"*The structure of the inventive balun circuit shown in FIG. 3 is identical to that of FIG. 2 except that a capacitor C_2 is further provided thereto. The capacitor C_2 is a coupling capacitor disposed between the base and the collector of the transistor Q_2* and serves to block DC components which may be fed to the base of the transistor Q_2 from the collector of the transistor Q_2." (Emphasis added.)

This essential difference between Lee's invention and the prior art is an integral part of claim 1 of Lee's patent. Claim 1 covers a circuit comprising:

"a set of a first load element, a first device and a second device connected in series in that order between a supply voltage source and the ground;

"a set of second load element, a third device, and a fourth device connected in series in that order between the supply voltage source and the ground,

"wherein each of the first to fourth devices has a control electrode and a first and a second electrodes, first electrodes of the first and the third devices being connected to second electrodes of the second and the fourth devices, respectively;

"a bias voltage source connected to control electrodes of the first and the third devices;

"a bias current source connected to control electrodes of the second and the fourth devices;

"a first capacitor;

"an input voltage source coupled to the second electrode of the second device via the first capacitor; and

"*a second capacitor coupled between the control electrode and the second electrode of the second device.*" (Emphasis added.)

As an aid in mapping claim 1 of Lee's patent to "figure 3" in Lee's patent (and to figure 15.15 in this book), we have paraphrased claim 1 in terms of the labels in Lee's "figure 3" as follows:

a set comprising resistor RL2, transistor Q4 and transistor Q2 connected in series, in that order, between supply voltage source VCC and ground;

a set comprising resistor RL1, transistor Q3 and transistor Q1 connected in series, in that order, between supply voltage source VCC and ground,

wherein the emitters of transistors Q4 and Q3 are connected to the collectors of transistors Q2 and Q1, respectively;

a bias voltage source VBIAS is connected to the bases of transistors Q4 and Q3;

a bias current source I_{bias} is connected to the bases of transistors Q2 and Q1;

a first capacitor C1;

an input voltage source VIN is coupled to the collector of transistor Q2 via the first capacitor C1; and

a second capacitor C2 is coupled between the base and the collector of transistor Q2.

Thus, the best-of-run circuit from generation 97 (figure 15.11) possesses the very feature that Lee identifies as the essence of his invention, namely the second capacitor C2 (called "C302" in figure 15.11).

In addition to having the specific feature that Lee identifies as the essence of his invention, the best-of-run circuit from generation 97 (figure 15.11) has several additional features that are enumerated in claim 1.

First, the genetically evolved circuit has "a first capacitor" with

"an input voltage source coupled to the second electrode of the second device via the first capacitor."

The "first capacitor" is C1 in Lee's circuit (figure 15.15) and is C301 in the genetically evolved circuit (figure 15.11). The "input voltage source" is VIN in Lee's circuit and is VIN0 in the genetically evolved circuit. The "second electrode of the second device" is the collector of transistor Q2 in Lee's circuit and the collector of transistor Q303 in the genetically evolved circuit.

Second, the genetically evolved circuit has

"a set of a first load element,... a second device connected in series in that order between a supply voltage source and the ground."

The "first load element" is RL2 in Lee's circuit (figure 15.15) and is RLOAD0 in the genetically evolved circuit (figure 15.11). The "second device" is transistor Q2 in Lee's circuit and transistor Q303 in the genetically evolved circuit. The "supply voltage source" is VCC in both Lee's circuit and the genetically evolved circuit. The "first load element" and the "second device" of the genetically evolved circuit are indeed "connected in series in that order between a supply voltage source and the ground." However, the genetically evolved circuit does not have "a first device" corresponding to transistor Q4 of Lee's patented circuit (figure 15.15).

Third, in a similar vein, the genetically evolved circuit has

> "a set of second load element, ... a fourth device connected in series in that order between the supply voltage source and the ground."

However, the genetically evolved circuit does not have "a third device" corresponding to transistor Q3 of Lee's patented circuit (figure 15.15).

Thus, even though the best-of-run circuit from generation 97 (figure 15.11) possesses the feature that Lee identifies as the essence of his invention and also possesses the three additional important features of Lee's circuit, it does not infringe Lee's patent because it lacks yet other features enumerated in claim 1.

Lee's circuit satisfied the Patent Office's requirement of being "useful" because balun circuits operating off a 1-Volt power supply are of current commercial interest for practical applications in cellular phones and high-definition television sets. Because the genetically evolved circuit duplicates the functionality of Lee's invention, it too should be considered "useful."

Lee's circuit satisfied the Patent Office's requirement of being "new" because it has a key feature that distinguished it from the prior art. The genetically evolved circuit has this same distinguishing feature.

Moreover, the genetically evolved circuit has several unusual features that might never occur to an experienced electrical engineer. Thus, the genetically evolved circuit is

> "[un]obvious ... to a person having ordinary skill in the art to which said subject matter pertains." (35 *United States Code* 103a)

Moreover, the genetically evolved circuit does not infringe Lee's patent and is therefore "new" in comparison to Lee's patented invention (which is, of course, now part of the prior art). The genetically evolved circuit should therefore satisfy the Patent Office's requirement of being "new" with respect to Lee.

Therefore, the best-of-run evolved circuit from generation 97 may possibly be a patentable new invention. The patentability of this genetically evolved circuit may well depend on whether its key features exist in prior art (other than Lee's).

Our run of the balun problem was made a little less than a year after the patent was issued for Lee's invention.

Both an infringing and a non-infringing circuit qualify as a human-competitive result under criterion A of table 1.2. Thus, we would have been equally pleased, for the purposes of this book, if the run had produced an infringing circuit.

For other purposes, however, a novel solution may be more desirable. One may simply have a scientific interest in producing novel solutions to challenging problems. Suppose, however, that one has an economic interest in designing balun circuits. Alternatively, one might have an interest in patenting a novel circuit in order to gain commercial advantage. Alternatively, one might have an interest in avoiding infringement of an existing patent (either to avoid paying royalties or because the patent holder is unwilling to license a competitor). Although rarely mentioned in textbooks, engineers in many industries spend a considerable amount of time and effort in devising designs that do not infringe the claims of unexpired patents (notably those held by competing companies). Indeed, the avoidance of infringement is the most important design criterion (albeit often not overtly acknowledged) in many real-world engineering design organizations.

In any of these three situations, a fitness measure can be formulated that incorporates the degree to which an individual in the population in a run of genetic programming satisfies the problem's design requirements as well as the degree to which the individual avoids characteristics of the prior art.

In particular, imagine that the clock is turned back to the day when Lee's patent was issued (July 24, 2001). At that moment, the prior art consisted of Lee's just-patented circuit as well as all previously known balun circuits (patented or not). The prior art for balun circuits includes the two circuits presented by Lee himself as figures 1 and 2 of his patent. And, the prior art no doubt includes yet other designs (some of which may have been patented and some of which may simply be in the literature of the field).

If we seek a design that differs from the prior art, a fitness measure can be constructed to include a way to measure the similarity between a circuit employing the prior art and an individual candidate circuit in the population being bred by genetic programming.

There are numerous alternative ways to implement this. One might, for example, use a rule-based system to examine each candidate circuit.

Because circuits can be conveniently represented by labeled graphs, a second approach is to use a graph isomorphism algorithm on a candidate circuit and various template graphs representing the key characteristics of the relevant prior art. For example, the templates for the balun problem would represent the key characteristics of Lee's now-patented balun circuit, older balun circuits (such as those cited by Lee himself as figures 1 and 2 of his patent), and any other previously known balun circuits in the literature. The measure of similarity could be based on the size of the maximal common subgraph between a candidate circuit and a template circuit. The measure of similarity might employ a cost function based on the number of shared nodes and edges and the types of electrical components (and perhaps even ranges of component values) in the graph. An example of such a cost function in connection with novelty-driven evolution for filter circuits may be found in section 27.6 of *Genetic Programming III* (Koza, Bennett, Andre, and Keane 1999a).

In making runs where the goal is to generate numerous different 100%-compliant circuits for post-run examination, the run should not be terminated upon evolution of the first 100%-compliant individual. Instead, the run should harvest all (or a significant subset of) the 100%-compliant circuits. Once genetic programming has successfully created one or more novel solutions to the problem at hand, a design engineer may examine them. Because the evolutionary process is not encumbered by the kind of preconceptions and predispositions that govern human engineers, some of these harvested 100%-compliant circuits may have unexpected features.

15.4.2 Results for Mixed Analog-Digital Variable Capacitor

The best-of-generation circuit from generation 0 has a fitness of 73.5.

In generation 95, a circuit emerged with an average absolute error, over the 16 fitness cases, of 0.808 milliamperes. This is equivalent to 100.6% of the average absolute error (0.803) of the patented circuit. This circuit has a highly non-symmetric topology consisting of a relatively large number of components (10 transistors, 7 capacitors, and 6 resistors). This circuit has a fitness of 7.50.

We also harvested the smallest individual with no more than a certain maximum level of absolute error from each processing node. One of the small individuals from generation 98 (with average absolute error equal to 117.5% of that of the patented circuit) has a topology that closely resembles that of the patented circuit. This designated circuit is shown in figure 15.16.

Figure 15.17 shows the mixed analog-digital variable capacitor circuit (embedded in a test fixture) that we created from the description in U.S. patent 6,013,958.

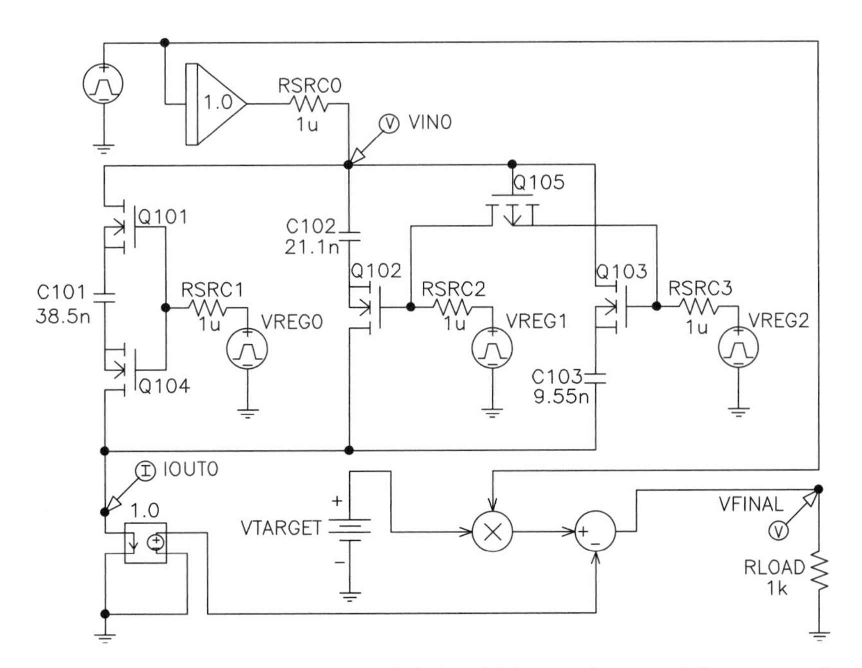

Figure 15.16 Designated mixed analog-digital variable capacitor circuit from generation 98.

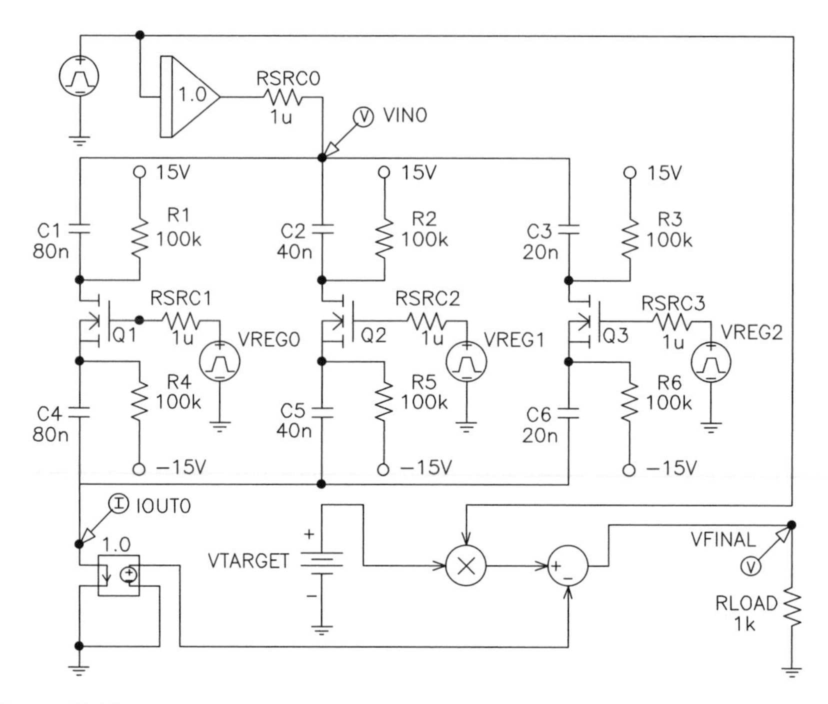

Figure 15.17 Mixed analog-digital variable capacitor circuit of U.S. patent 6,013,958.

Claim 1 of U.S. patent 6,013,958 by Turgut Sefket Aytur of Lucent Technologies Inc. covers:

> "An integrated circuit, for use in a balanced line configuration, said integrated circuit comprising:
>
> "a plurality of combinations of an integrated unit capacitor connected in series with a controlled channel of an integrated transistor switch, the combinations being connected in parallel, wherein at least one of the transistor switches has a control gate coupled to a bit output of a register whereby the capacitance of the interconnect variable capacitor depends on what word is stored in the register, wherein the integrated unit capacitor in at least one of the combinations is coupled to a balanced signal line pair, and wherein at least one of the combinations comprises a transistor switch connected between two integrated unit capacitors connected to a line of said balanced signal line pair."

The patented circuit (figure 15.17) consists of a plurality (three) of "combinations ... connected in parallel" as specified in claim 1.

The first "combination" in the patented circuit (figure 15.17) contains the "integrated unit capacitor" C1 "connected in series with a controlled channel of an integrated transistor switch" Q1. The control is provided by "bit output" VREG0 (0 or 1) from the digital register. Transistor switch Q1 has its "control gate coupled to a bit output of a register" VREG0. Also, "the capacitance of the interconnect variable capacitor depends on what word is stored in the register." Specifically, this capacitance is 0 if the transistor switch Q1 is open (i.e., VREG0 from the digital register is −15 Volts, corresponding to a binary 0) or is 40 nanofarads (the net capacitance of the two 80 nF capacitors, C1 and C4, in series) if the transistor switch Q1 is closed (i.e., VREG0 from the digital register is +15 Volts, corresponding to a binary 1). Moreover, the "combination" "comprises a transistor switch" Q1 "connected between two integrated unit capacitors" (namely capacitors C1 and C4) "connected to a line of said balanced signal line pair."

Similarly, the second "combination" in the patented circuit (figure 15.17) contains capacitor C2, transistor switch Q2, bit output VREG1, and second capacitor C5. The third "combination" in the patented circuit contains capacitor C3, transistor switch Q3, bit output VREG2, and second capacitor C6.

The patented circuit (figure 15.17) also satisfies the requirements of claim 2:

> "The integrated circuit as recited in claim 1, wherein control gates for a different number of combinations are connected to at least one of the bit outputs of the register."

Claim 3 covers,

> "The integrated circuit as recited in claim 1, wherein the numbers are weighted using a weighting scheme selected from the group consisting of binary weighting and unit weighting."

In the patented circuit (figure 15.17), capacitors with "binary weighting" of 20, 40, and 80 nanofarads are associated with the low, middle, and high-order bits, respectively, of the three-bit register probed at "bit outputs" VREG2, VREG1, and VREG0, respectively.

The designated mixed analog-digital variable capacitor circuit from generation 98 (figure 15.16) has many of the key structural features of the patented circuit (figure 15.17).

At a high level, the genetically evolved circuit (figure 15.16) consists of three "combinations ... connected in parallel" as specified in claim 1.

The first "combination" in the genetically evolved circuit (figure 15.16) contains "integrated unit capacitor" C101 "connected in series with a controlled channel of an integrated transistor switch" Q101. The control is provided by "bit output" VREG0 (−15 or +15 Volts) from the digital register. Transistor switch Q101 has its "control gate coupled to a bit output of a register" VREG0. Also, "the capacitance of the interconnect variable capacitor depends on what word is stored in the register." Specifically, this capacitance is 0 if the transistor switch Q101 is open (i.e., VREG0 from the digital register is −15 Volts) and is 38.5 nF (the capacitance of C101) if the transistor switch Q101 is closed (i.e., VREG0 from the digital register is +15 Volts). Note that there is an additional transistor Q104 which is open or closed when Q101 is open or closed, respectively.

Similarly, the second "combination" in the genetically evolved circuit (figure 15.16) contains capacitor C102, transistor switch Q102, and bit output VREG1 from the digital register. The third "combination" in the genetically evolved circuit contains capacitor C103, transistor switch Q103, and bit output VREG2 from the digital register.

Note that deleting Q105 has no effect on the circuit's performance.

Because none of the three "combinations" of the genetically evolved circuit (figure 15.16) contain a transistor switch "connected between two integrated unit capacitors," the genetically evolved circuit does not infringe claim 1.

If a commercial enterprise were seeking a variable capacitor circuit that did not infringe Aytur's patent, the genetically evolved circuit might be a serviceable alternative circuit. That is, genetic programming may have "engineered around" Aytur's patent by creating a functionally equivalent non-infringing alternative.

In the genetically evolved circuit (figure 15.16), capacitors C103, C102, and C101 with values of 9.55, 21.1, and 38.5 nanofarads, respectively, are associated with the low, middle, and high-order bits, respectively, of the three-bit register providing "bit outputs" VREG2, VREG1, and VREG0, respectively. The numerical values of 9.55, 21.1, and 38.5 are close to an exponential sequence. Because the numerical values are so close to an exponential sequence, the genetically evolved circuit would undoubtedly be regarded as equivalent (under the doctrine of equivalences of the patent law) to the "binary weighting" mentioned in claim 3 of the patent. However, claim 3 is a dependent claim and the genetically evolved circuit does not infringe the independent claim with which it is associated (claim 1). Thus, in spite of the fact that the genetically evolved circuit employs yet another key feature of the invention, the genetically evolved circuit does not infringe claim 3.

15.4.3 Results for High-Current Load Circuit

We made two runs of this problem.

15.4.3.1 Results for First Run of High-Current Load Circuit

The best-of-generation circuit from generation 0 has a fitness of 61.7.

Our first run for this problem eventually reached a plateau and produced a circuit that sunk the desired current into one of the negative power supplies (rather than to ground). The fitness of this circuit is 0.713 (i.e., it has only 71.3% of the weighted absolute error of the patented circuit).

It so happens that, for the purpose of testing power supplies, it is preferable that the current be sunk to ground, rather than into the negative power supply. A circuit that sinks the current into the negative power supply is not a valid solution to this problem. We did not include anything in our fitness measure to specify that the current should be sunk to ground.

However, in addition to producing a "cheating" solution, genetic programming also produced a circuit (figure 15.18) in generation 114 that duplicates Daun-Lindberg and Miller's parallel FET transistor structure. This circuit has a fitness (weighted absolute error) of 1.82, or 182% of the weighted absolute error for the patented circuit. The fitness of this best-of-run circuit from generation 114 is 1.82.

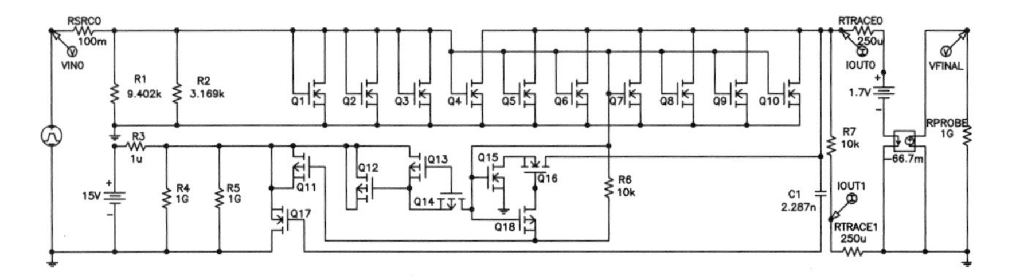

Figure 15.18 Best-of-run high-current load circuit from the first run.

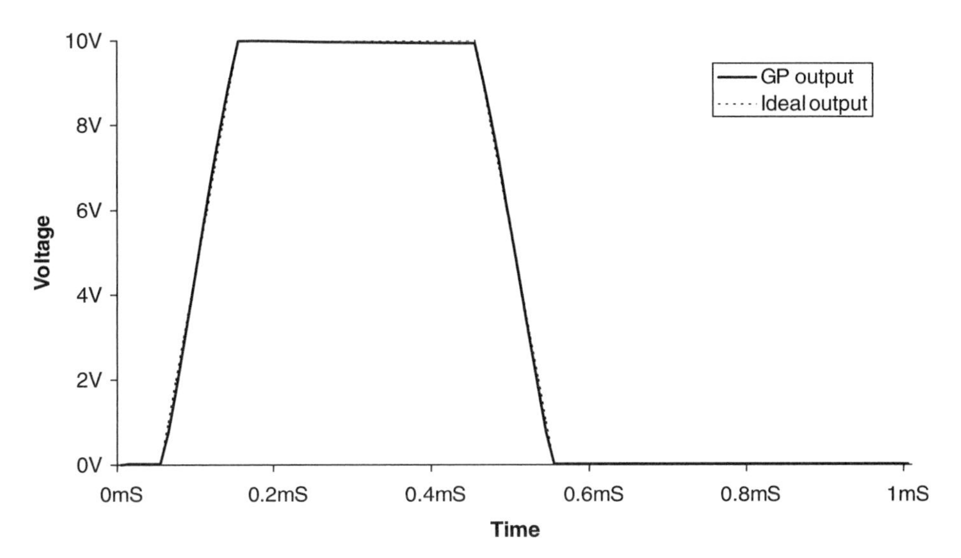

Figure 15.19 Comparison of output of the best-of-run high-current load circuit from the first run for fitness case 1 and the ideal output.

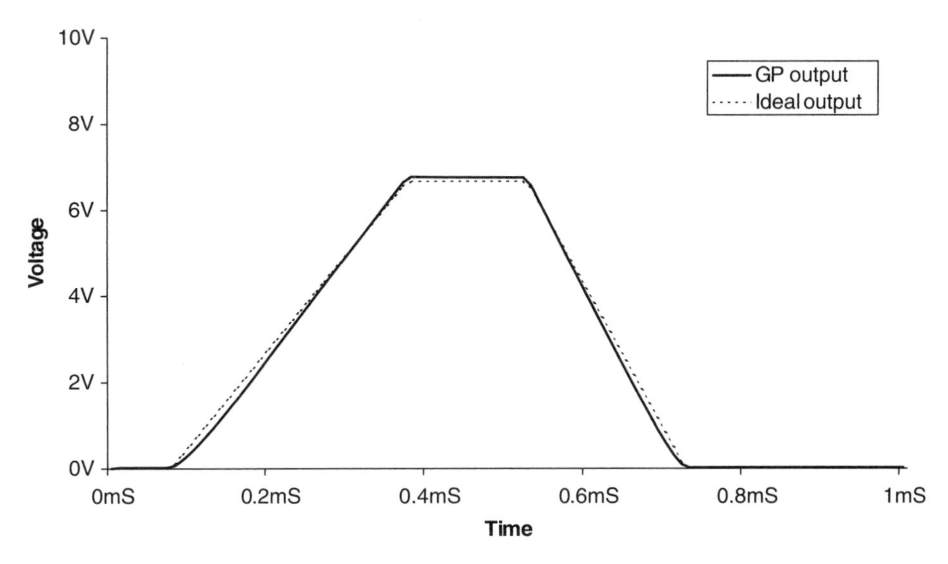

Figure 15.20 Comparison of output of the best-of-run high-current load circuit from the first run for fitness case 2 and the ideal output.

Figure 15.19 compares the output of the best-of-run high-current load circuit from generation 114 for fitness case 1 and the ideal output. As can be seen, the two curves are almost indistinguishable.

Figure 15.20 compares the output of the best-of-run high-current load circuit from generation 114 for fitness case 2 and the ideal output. As can be seen, the two curves are almost indistinguishable.

Averaged over the two fitness cases, the best-of-run circuit from generation 114 of the first run of the high-current load problem has

- 288 millivolts maximum absolute error, and
- 71.85 millivolts average absolute error.

Again, note that the circuit's output is probed at VFINAL (the output of a current-to-voltage converter).

Table 15.5 shows the average absolute error and maximum absolute error (both in millivolts) for the best-of-run circuit from generation 114 of the first run for the two fitness cases.

Figure 15.21 shows the high-current load circuit of U.S. patent 6,211,726.

Table 15.6 gives the average absolute error and maximum absolute error (both in millivolts) for the patented high-current load circuit.

The genetically evolved circuit from the first run shares the following features found in claim 1 of U.S. patent 6,211,726:

> "A variable, high-current, low-voltage, load circuit for testing a voltage source, comprising:
>
> "a plurality of high-current transistors having source-to-drain paths connected in parallel between a pair of terminals and a test load."

Table 15.5 Average absolute error and maximum absolute error for the best-of-run circuit from the first run of the high-current load problem

Fitness case	Average absolute error	Maximum absolute error
10 Volt pulse	50.6	288
6.667 Volt pulse	93.1	276

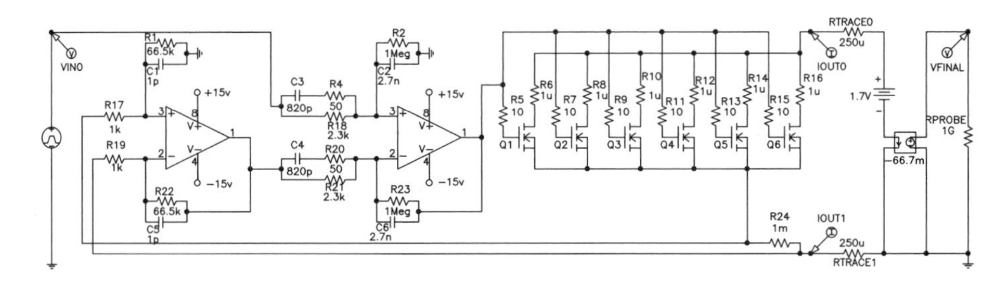

Figure 15.21 High-current load circuit of U.S. patent 6,211,726.

Table 15.6 Average absolute error and maximum absolute error for the patented high-current load circuit

Fitness case	Average absolute error	Maximum absolute error
10 Volt pulse	48.0	1,530
6.667 Volt pulse	34.0	770

However, the remaining elements of claim 1 in U.S. patent 6,211,726 are very specific and the genetically evolved circuit does not read on these remaining elements. In fact, the remaining elements of the genetically evolved circuit bear hardly any resemblance to the patented circuit. In this instance, genetic programming produced a circuit that duplicates the functionality of the patented circuit using a different structure.

15.4.3.2 Results for Second Run of High-Current Load Circuit The first run employed ideal power supplies that could deliver unlimited amounts of power. In order to prevent the creation of "cheating" circuits and to obtain a more accurate circuit, we imposed a limit on the current that can be provided by the positive and negative power supplies in our second run of this problem.

The best-of-generation circuit from generation 0 has a fitness of 500.3.

The best-of-run individual emerged in generation 215 (figure 15.22). This circuit has 92% of the weighted absolute error of the patented circuit. The fitness of this best-of-run circuit from generation 215 is 2.41.

Averaged over the two fitness cases, the best-of-run circuit from generation 215 of the second run of the high-current load problem has

- 120.4 millivolts maximum absolute error, and
- 37.0 millivolts average absolute error.

Table 15.7 shows the average absolute error and maximum absolute error (both in millivolts) for the best-of-run circuit from generation 215 of the second run for the two fitness cases.

As can be seen from tables 15.6 and 15.7, the evolved circuit is superior to the patented circuit in terms of average absolute error on the first fitness case and comparable on the second fitness case. It is significantly better in terms of maximum absolute error on both fitness cases.

Like the evolved circuit from the first run, the evolved circuit from the second run shares the "plurality of high-current transistors having source-to-drain paths connected

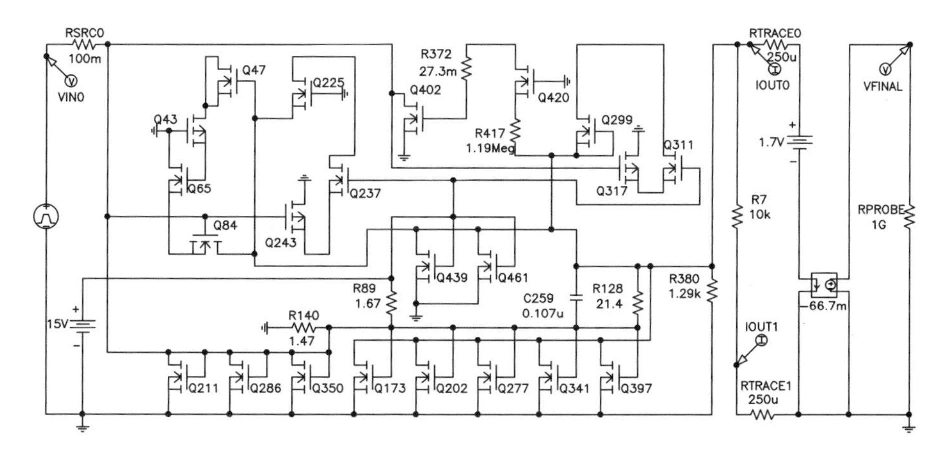

Figure 15.22 Best-of-run high-current load circuit from the second run.

Table 15.7 Average absolute error and maximum absolute error for the best-of-run circuit from the second run of the high-current load problem

Fitness case	Average absolute error	Maximum absolute error
10 Volt pulse	39.9	98.7
6.667 Volt pulse	34.2	120.4

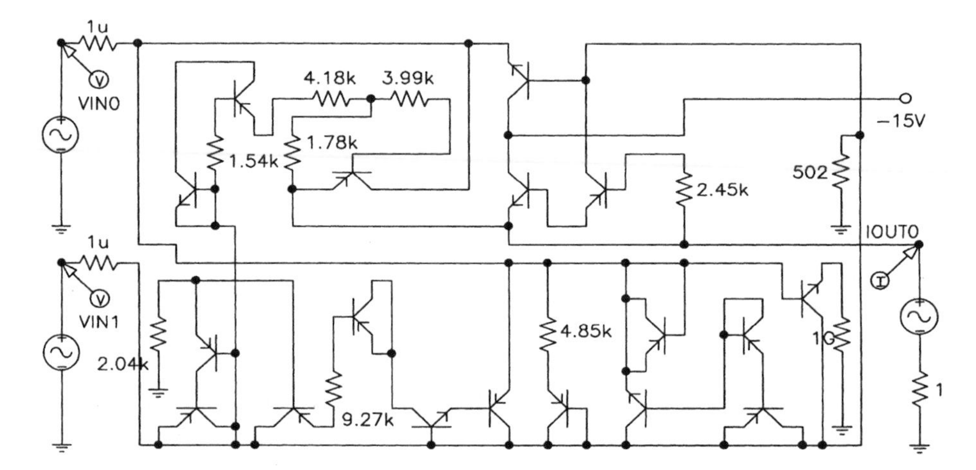

Figure 15.23 Best-of-run circuit for the voltage-current-conversion circuit problem.

in parallel between a pair of terminals and a test load" of U.S. patent 6,211,726. However, the genetically evolved circuit does not read on the remaining portions of the patent claims. Again, genetic programming produced a circuit that duplicates the functionality of the patented circuit using a different structure.

15.4.4 Results for Voltage-Current Conversion Circuit

The best-of-generation circuit from generation 0 has a fitness of 35.1.

A circuit (figure 15.23) emerged in generation 109 that has 62% of the average (weighted) absolute error of the patented circuit. The fitness of this best-of-run circuit from generation 109 is 0.619.

Figure 15.24 compares the voltage produced by the best-of-run voltage-current-conversion circuit from generation 109 with the ideal output for fitness case 1. This voltage is probed at the node between output voltage source VS and resistor RLOAD, and is equal to the current at IOUT0 multiplied by 1 Volt per Ampere. As can be seen, the two curves are almost indistinguishable.

Figure 15.25 compares the voltage produced by the best-of-run voltage-current-conversion circuit from generation 109 with the ideal output for fitness case 2. Again, the two curves are almost indistinguishable.

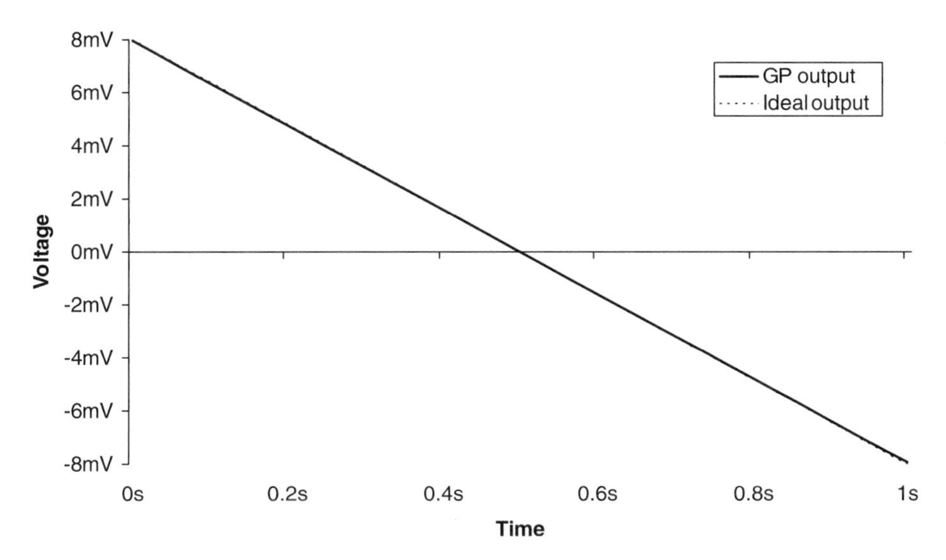

Figure 15.24 Comparison of the voltage produced by the best-of-run voltage-current-conversion circuit with the ideal output for fitness case 1.

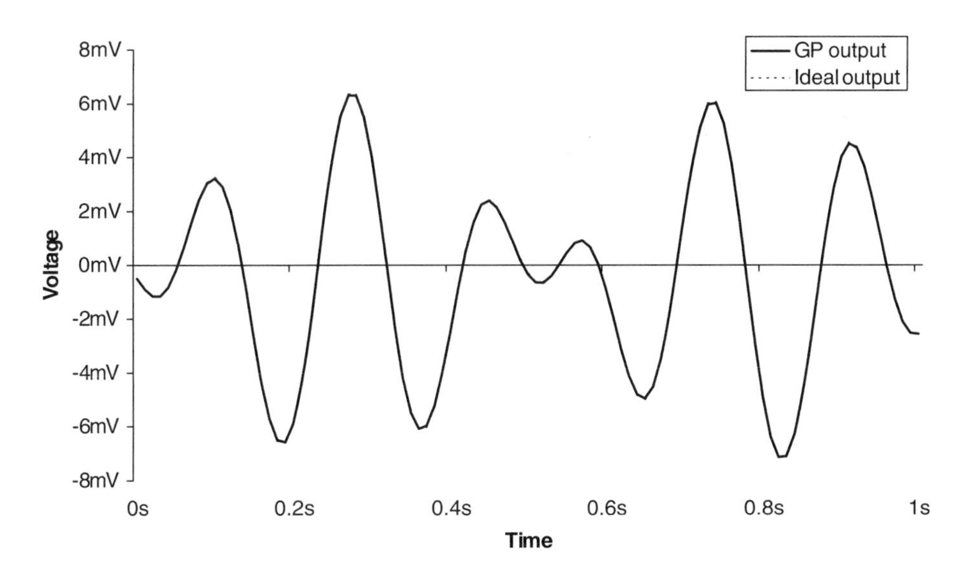

Figure 15.25 Comparison of the voltage produced by the best-of-run voltage-current-conversion circuit with the ideal output for fitness case 2.

Figure 15.26 compares the voltage produced by the best-of-run voltage-current-conversion circuit from generation 109 with the ideal output for fitness case 3. The ideal value here is 0 Volts (as shown by the horizontal line). As can be seen, the voltage produced by the best-of-run circuit has a near-zero average absolute error (13.7 microvolts).

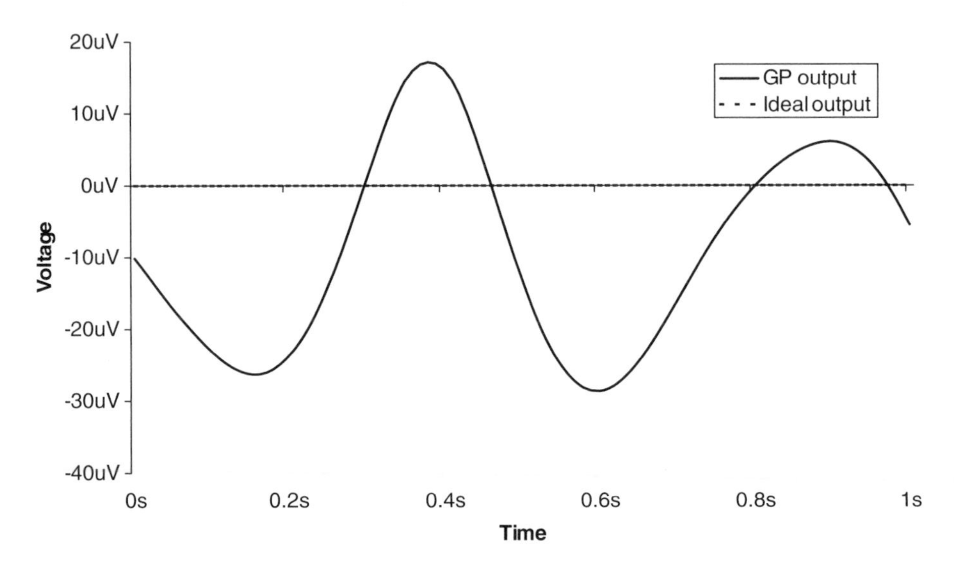

Figure 15.26 Comparison of the voltage produced by the best-of-run voltage-current-conversion circuit with the ideal output for fitness case 3.

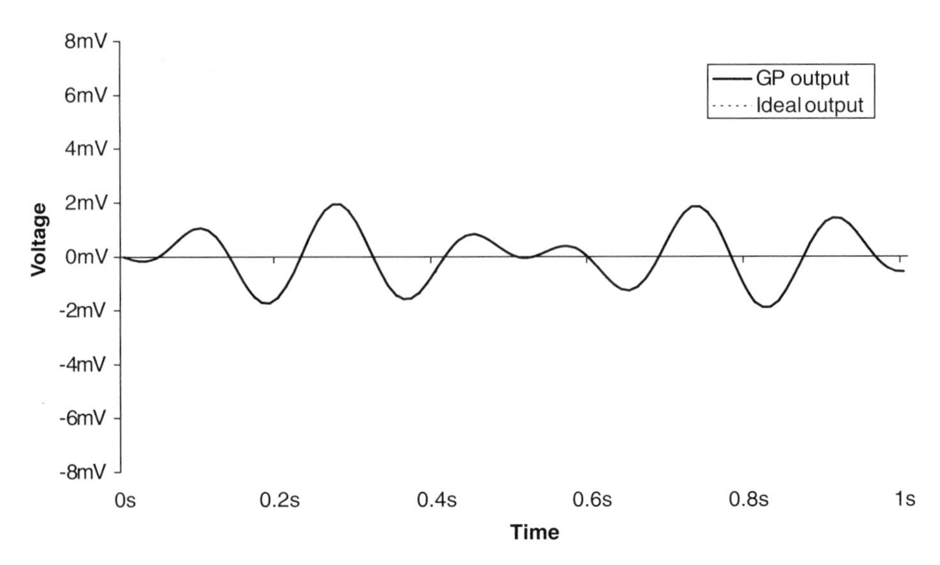

Figure 15.27 Comparison of the voltage produced by the best-of-run voltage-current-conversion circuit with the ideal output for fitness case 4.

Figure 15.27 compares the voltage produced by the best-of-run voltage-current-conversion circuit from generation 109 with the ideal output for fitness case 4. The two curves are almost indistinguishable.

Figure 15.28 shows the voltage-current-conversion circuit of U.S. patent 6,166,529 (Ikeuchi and Tokuda 2000).

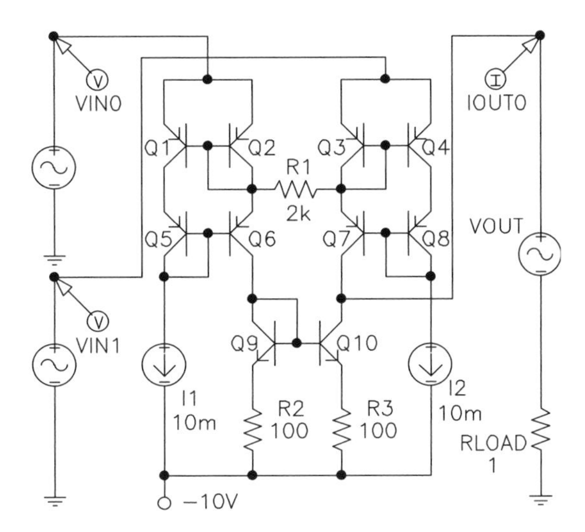

Figure 15.28 The voltage-current-conversion circuit of U.S. patent 6,166,529.

The best-of-run circuit from generation 109 (figure 15.23) for the Voltage-current conversion circuit problem solves the problem in a different manner than the patented circuit (figure 15.28). The best-of-run circuit has a current source proportional to VIN0 and a current sink proportional to VIN1.

Note that the best-of-run circuit from generation 109, like the patented circuit, is "not restricted by the requirement that the input voltages be lower than the power source voltage."

15.4.5 Results for Cubic Function Generator

We made two runs of this problem. After the first successful run (using the commercially popular `2N3904` and `2N3906` transistors), we made a second run using higher frequency transistors.

15.4.5.1 Results for First Run of Cubic Function Generator The best-of-generation circuit from generation 0 has a fitness of 1.87.

The best-of-run evolved circuit (figure 15.29) emerged in generation 182 and has an average absolute error of 4.20 millivolts. The patented circuit has an average absolute error of 6.76 millivolts. That is, the evolved circuit has approximately 59% of the error of the patented circuit over our four fitness cases. The fitness of this best-of-run cubic signal generation circuit from generation 182 of the first run is 0.0125.

Averaged over the four fitness cases, the best-of-run individual from generation 182 from the first run has:

- 4.20 millivolts average absolute error,
- 26.7 millivolts maximum absolute error, and
- 93.6% of the possible number of hits.

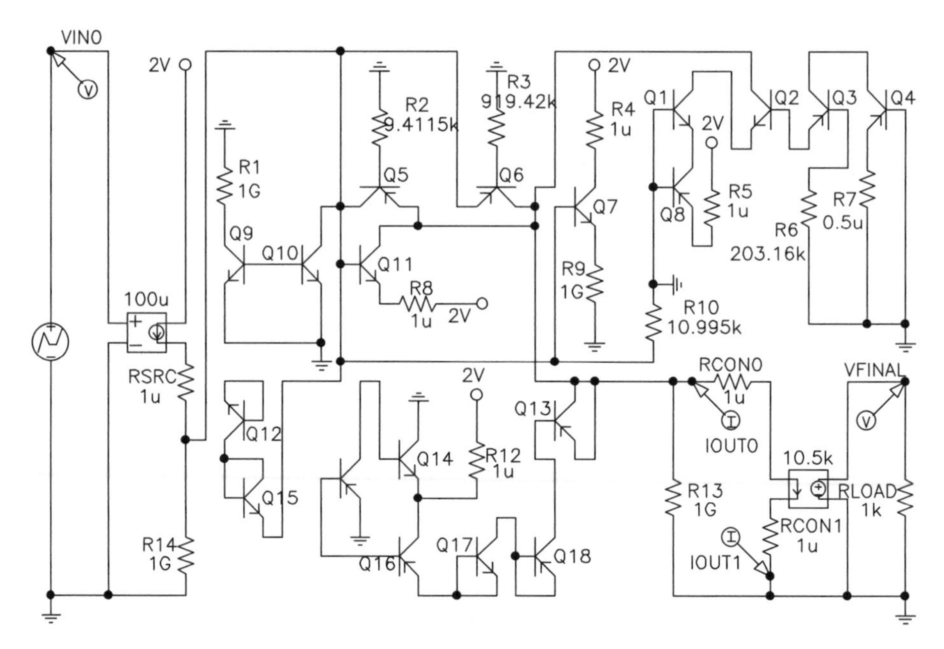

Figure 15.29 Best-of-run cubic signal generation circuit from the first run.

Table 15.8 Average and maximum absolute error for the best-of-run circuit from the first run of the cubic signal generation problem

Fitness case	Average absolute error	Maximum absolute error
Rising ramp	3.26	22.9
Sine wave	4.80	26.7
Falling ramp	3.24	22.8
Constant	5.49	5.49

Table 15.8 shows the average absolute error in millivolts and the maximum absolute error in millivolts for the best-of-run circuit from the first run for all four fitness cases.

Figure 15.30 compares the output produced by the best-of-run cubic signal generation circuit from generation 182 (solid line) to the target cubic curve (dotted line). As can be seen, the two curves are almost indistinguishable.

Figure 15.31 compares the error of the best-of-run cubic signal generation circuit from generation 182 of the first run and the error of the circuit of U.S. patent 6,160,427. As can be seen from the figure, the error produced by the genetically evolved circuit is generally less than that produced by the patented circuit.

Figure 15.32 shows the cubic function generator of U.S. patent 6,160,427. This circuit has nine transistors. The figure includes our test fixture.

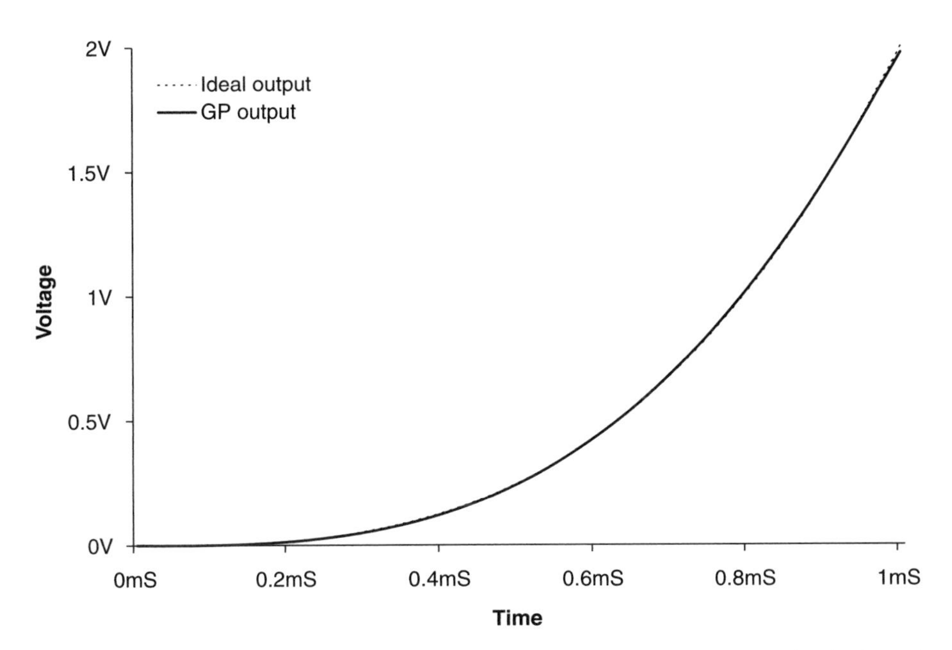

Figure 15.30 Output produced by the best-of-run cubic signal generation circuit from the first run.

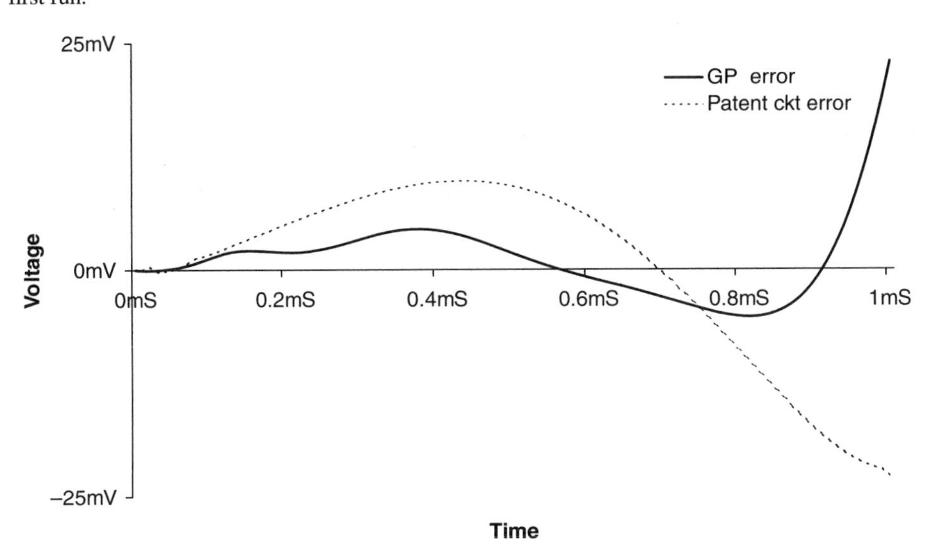

Figure 15.31 Comparison of the error of the best-of-run cubic signal generation circuit from the first run and the error of the circuit of U.S. patent 6,160,427.

Averaged over the four fitness cases, the patented circuit has:

- 6.76 millivolts average absolute error,
- 17.3 millivolts maximum absolute error, and
- 61% of the possible number of hits.

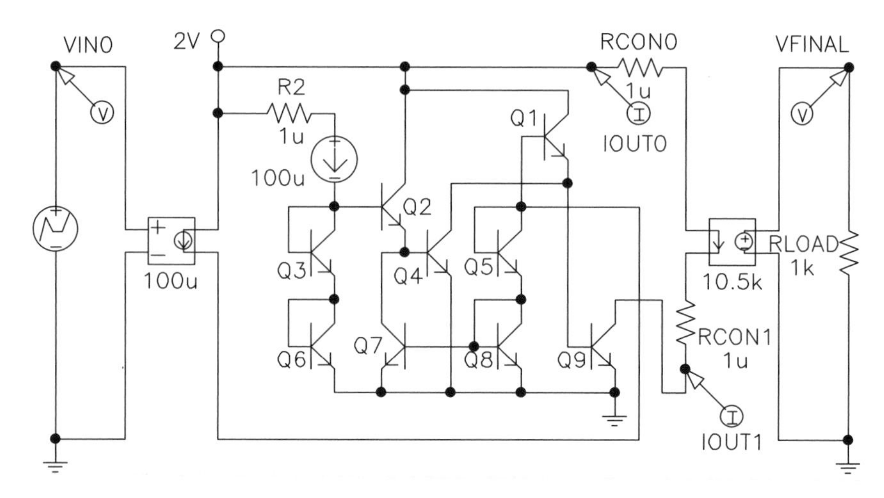

Figure 15.32 Cubic function generator of U.S. patent 6,160,427.

Table 15.9 Average and maximum absolute error for the patented cubic function generator with the 2N3904 and 2N3906 transistors used in the first run of the cubic signal generation problem

Fitness case	Average absolute error	Maximum absolute error
Rising ramp	7.15	17.3
Sine wave	7.45	17.3
Falling ramp	7.15	17.3
Constant	5.33	5.33

Table 15.9 shows the average absolute error and the maximum absolute error (both in millivolts) for the patented cubic function generator using the 2N3904 (*npn*) and 2N3906 (*pnp*) transistors for all four fitness cases.

The claims in U.S. patent 6,160,427 amount to a very specific description of the patented circuit. The genetically evolved circuit does not read on these claims and, in fact, bears hardly any resemblance to the patented circuit. In this instance, genetic programming produced a circuit that duplicates the functionality of the patented circuit using a very different structure.

15.4.5.2 Results for Second Run of Cubic Function Generator After producing a satisfactory cubic function generator in the first run, we reran this problem using higher frequency transistor models—specifically, the HFA3046 (*npn*) in lieu of 2N3904 and HFA3128 (*pnp*) in lieu of 2N3906.

In the second run, the best-of-generation circuit from generation 0 has a fitness of 1.86.

The best-of-run circuit (figure 15.33) emerged in generation 326 of the second run. The fitness of this circuit is 0.0129.

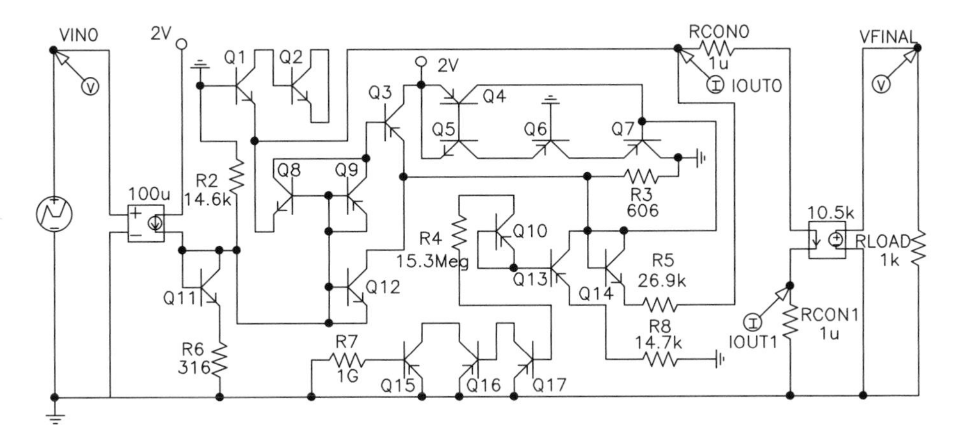

Figure 15.33 Best-of-run circuit from the second run of the cubic signal generation problem.

Table 15.10 Average and maximum absolute error for the best-of-run circuit from the second run of the cubic signal generation problem

Fitness case	Average absolute error	Maximum absolute error
Rising ramp	4.01	7.85
Sine wave	5.33	15.0
Falling ramp	3.92	7.36
Constant	1.18	1.20

Averaged over the four fitness cases, the best-of-run individual from generation 326 of the second run has:

- 3.61 millivolts average absolute error,
- 15.0 millivolts maximum absolute error, and
- 88.6% of the possible number of hits.

Table 15.10 shows the average absolute error in millivolts and the maximum absolute error in millivolts for all four fitness cases for the best-of-run circuit from generation 326 of the second run.

In comparison, when the higher frequency transistor models are used in the patented circuit, averaged over the four fitness cases, the patented circuit has

- 6.63 millivolts average absolute error,
- 67.0 millivolts maximum absolute error, and
- 48% of the possible number of hits.

Note that the definition of the number of hits for this particular problem is based on the average absolute error (7 millivolts) of the patented circuit using the higher frequency

Table 15.11 Average and maximum absolute error for the patented circuit with the high-frequency transistors used in the second run of the cubic signal generation problem

Fitness case	Average absolute error	Maximum absolute error
Rising ramp	10.1	42.1
Sine wave	11.5	67.0
Falling ramp	4.50	30.4
Constant	0.5	0.5

transistors. Hence, it is reasonable that the patented circuit scores about 50% of the possible number of hits.

Table 15.11 gives the average absolute error and maximum absolute error for the patented circuit using the high-frequency transistors for all four fitness cases (i.e., the HFA3046 and HFA3128 transistors).

As can be seen from tables 15.10 and 15.11, the genetically evolved circuit is superior to the patented circuit in terms of both average and maximum absolute error for the first three fitness cases. For the fourth fitness case, the genetically evolved circuit is not superior to the patented circuit, but has a very low value of both maximum and absolute error.

For additional details on the first five problems in this chapter, see Streeter, Keane, and Koza 2002b and 2003.

15.4.6 Tunable Integrated Active Filter

We made four runs of this problem. After the first successful run, we made additional runs in order to obtain a more parsimonious solution, to generate audit trails, and to harvest numerous solutions for comparison to the patented circuit.

15.4.6.1 Results for First Run of Tunable Integrated Active Filter In the first run, the best-of-generation circuit from generation 0 has a fitness of 94.5.

The best-of-run circuit from the first run emerged in generation 70. It has fitness of 1.09 and scores 100% (549 out of 549) hits.

Averaged over the nine in-sample frequencies used during the run of genetic programming, the best-of-run circuit from the first run has

- 54.6 millivolts average absolute error for frequencies in the passband,
- 0.58 dB average absolute error for frequencies to the right of the passband, and
- 100% of the possible number of hits.

Figure 15.34 shows the frequency response for the nine in-sample frequencies for the best-of-run circuit from generation 70 of the first run.

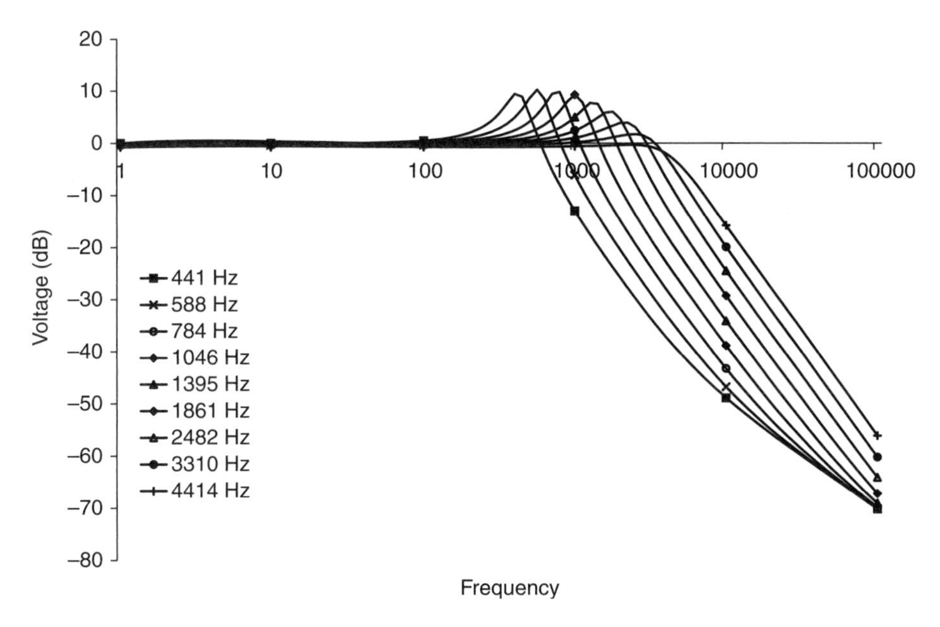

Figure 15.34 In-sample frequency response for the best-of-run circuit from the first run of the tunable integrated active filter problem.

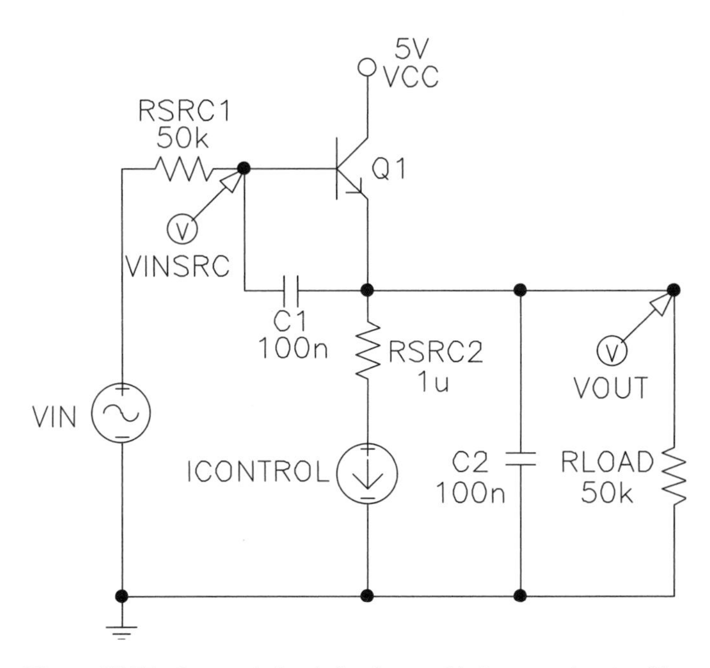

Figure 15.35 Patented circuit for the tunable integrated active filter.

Figure 15.35 shows the patented tunable integrated active filter circuit. Averaged over the nine in-sample frequencies, the patented circuit has

- 146 millivolts average absolute error for frequencies in the passband,
- 0.66 dB average absolute error for frequencies to the right of the passband, and
- 99.1% of the possible number of hits.

Eight out-of-sample frequencies were used to cross-validate the best-of-run circuit from generation 70 of the first run. These frequencies (509, 679, 906, 1,208, 1,611, 2,149, 2,866, and 3,822 Hz) are spaced halfway (on a logarithmic scale) between each consecutive pair of the original nine in-sample frequencies.

Averaged over the eight out-of-sample frequencies, the best-of-run circuit from the first run has

- 55.0 millivolts average absolute error in the passband,
- 0.51 dB average absolute error to the right of the passband, and
- 99.8% of the possible number of hits.

Figure 15.36 shows the frequency response for the eight out-of-sample frequencies for the best-of-run circuit from generation 70 of the first run.

Figure 15.35 shows the patented circuit for the tunable integrated active filter. Averaged over the eight out-of-sample frequencies, the patented circuit has

- 149 millivolts average absolute error for frequencies in the passband,
- 0.62 dB average absolute error for frequencies to the right of the passband, and
- 99.6% of the possible number of hits.

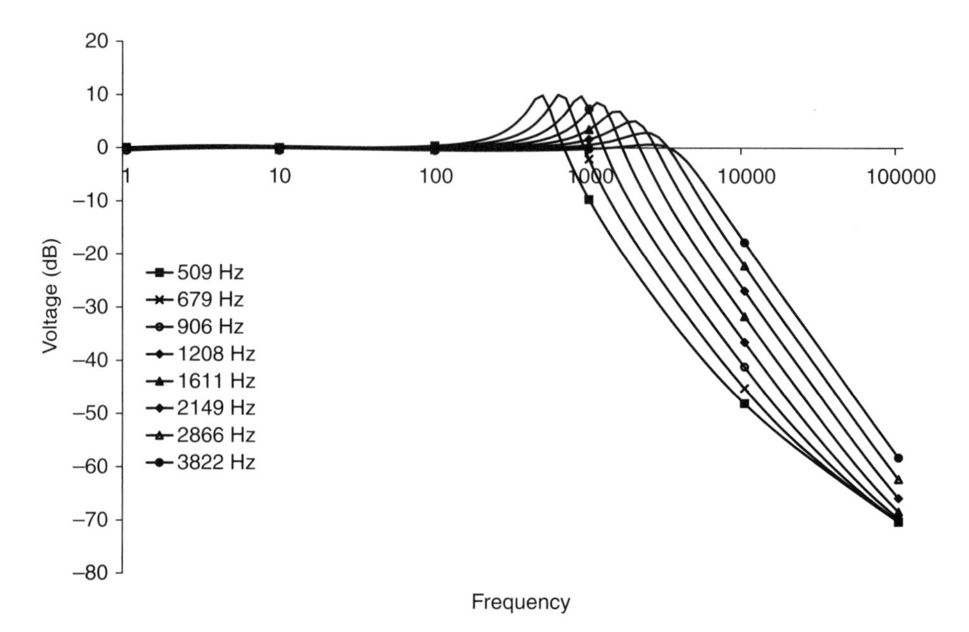

Figure 15.36 Out-of-sample frequency response for the best-of-run circuit from the first run of the tunable integrated active filter problem.

The best-of-run circuit from generation 70 has 12 transistors, three capacitors, and one resistor (in addition to the components in the test fixture).

This genetically evolved circuit is much larger than the circuit (figure 15.35) described in U.S. patent 6,225,859 (Irvine and Kolb 2001) and clearly contains the type of appendix-like features that frequently emerge during the evolutionary process.

Parsimony is an important consideration in circuit design. Unnecessary components occupy space, consume power, and increase manufacturing costs.

One technique for analyzing large circuits created by evolutionary methods is "Muntzing," named after Earl "Madman" Muntz, who used the technique to design inexpensive television receivers in the 1950s. Robert Pease explained "Muntzing" in his "Pease Porridge" column (1992).

> "[H]ow did Muntz get his circuits designed to be so inexpensive?... The story around the industry was that he would wander around to an engineer's workbench and ask, 'How's your new circuit coming?'
>
> "After a short discussion, Earl would say, 'But, you seem to be over-engineering this—I don't think you need this capacitor.' He would reach out with his handy nippers (insulated) that he always carried in his shirt-pocket, and snip out the capacitor in question.
>
> "Well, doggone, the picture was still there! Then he would study the schematic some more, and SNIP, SNIP, SNIP. Muntz had made a good guess of how to simplify and cheapen the circuit.
>
> "Then, usually, he would make one SNIP too many, and the picture or the sound would stop working. He would concede to the designer, 'Well, I guess you have to put that last part back in,' and he would walk away. That was 'Muntzing'—the ability to delete all parts not strictly essential for basic operation."

By "Muntzing" the best-of-run circuit from generation 70, we were able to remove seven transistors, one capacitor, and one resistor. However, the reduced circuit still contained five transistors and was still substantially larger than the patented circuit.

15.4.6.2 Results for Second Run of Tunable Integrated Active Filter After the first run produced a satisfactory (although non-parsimonious) solution, we decided to make an additional run in which parsimony was explicitly incorporated into the fitness measure.

In the second run, the best-of-generation circuit from generation 0 has a fitness of 63.3.

A parsimonious solution emerged in generation 50 of the second run. This best-of-run circuit consists (in addition to the components in the test fixture) of only one transistor and two capacitors (figure 15.37). This circuit has a fitness of 13×10^{-12}.

Averaged over the nine values of frequency, the best-of-run circuit from generation 50 of the second run has

- 72.7 millivolts average absolute error for frequencies in the passband,
- 0.39 dB average absolute error for frequencies to the right of the passband, and
- 99.5% of the possible number of hits.

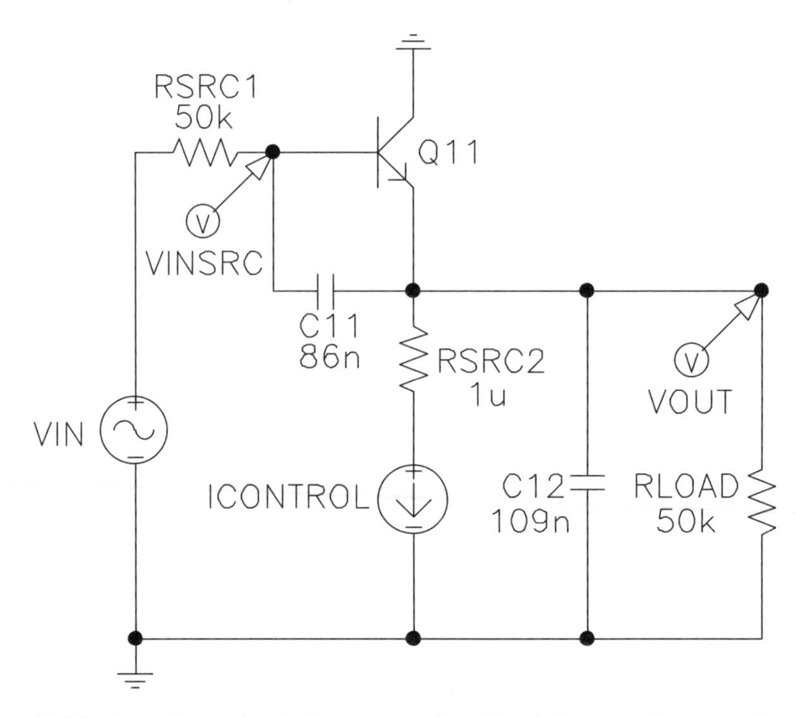

Figure 15.37 Best-of-run circuit from generation 50 of the second run of the tunable integrated active filter problem.

Figure 15.38 shows the frequency response for the nine in-sample frequencies for the best-of-run circuit from generation 50 of the second run.

The best-of-run circuit from generation 50 of the second run scores 99.5% of the possible number of hits. Note that this circuit scores more hits than that scored by the patented circuit. Also, note that the parsimony element of the fitness measure for this problem becomes relevant when 99% of the possible number of hits is achieved.

As can be seen in comparing figures 15.37 and 15.35, the topology of the best-of-run circuit from generation 50 from the second run is identical to the topology of the patented circuit, except that the collector of the transistor in the best-of-run circuit is connected to ground, whereas the collector of the transistor in the patented circuit is connected to V_{cc}. Because the adjustable current source in our test fixture is an ideal current source, it is able to pull current down from ground. Thus, this difference has no effect on the circuit's behavior. Moreover, the sizing of the best-of-run circuit from the second run (86 nanofarads for C11 and 109 nanofarads for capacitor C12) is very close to that of the patented circuit (i.e., 100 nanofarads for both C1 and C2).

The best-of-run circuit from the second run emerged in generation 50 (figure 15.37) and infringes claim 1 of U.S. patent 6,225,859.

Claim 1 of patent 6,225,859 states:

"An integrated low-pass filter, comprising:

"a filter input terminal;

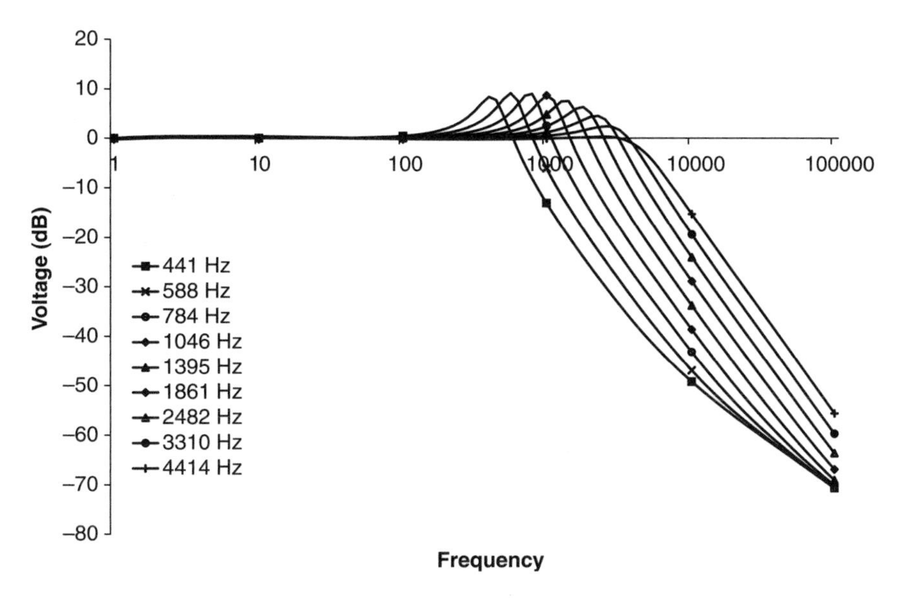

Figure 15.38 In-sample frequency response for the best-of-run circuit from generation 50 of the second run of the tunable integrated active filter problem.

"a transistor having a control terminal, an output terminal, a transistor terminal, and a source impedance;

"a resistor connected between said input terminal and said control terminal of said transistor;

"a first capacitor connected between said resistor and said control terminal of said transistor on one hand and to said output terminal of said transistor on the other hand;

"a filter output terminal connected to said output terminal of said transistor;

"a second capacitor connected between said filter output terminal and a first reference potential; and

"an adjustable current source connected between said filter output terminal and the first reference potential, said transistor terminal of said transistor being connected to a second reference potential, said source impedance defined by the equation:

$$Z_e = \frac{R_1}{\beta} + j\omega R_1 C_1 \frac{V_T}{I_C},$$

"where:

R_1 = resistance of said resistor;
β = differential current amplification;
C_1 = capacitance of said first capacitor;
V_T = temperature voltage; and
I_C = current generated by said adjustable current source."

The genetically evolved circuit reads on each element of claim 1. As an aid in mapping claim 1 of patent 6,225,859 to the best-of-run circuit from generation 50, we have paraphrased claim 1 as follows:

An integrated low-pass filter, comprising:
a filter input terminal VIN;
a transistor having a control terminal (base), an output terminal (emitter), a transistor terminal (collector), and a source impedance;
a resistor RSRC1 connected between said input terminal VIN and said control terminal (base) of said transistor;
a first capacitor C11 connected between said resistor RSRC1 and said control terminal (base) of said transistor on one hand and to said output terminal (emitter) of said transistor on the other hand;
a filter output terminal VOUT connected to said output terminal (emitter) of said transistor;
a second capacitor C12 connected between said filter output terminal VOUT and a first reference potential (ground); and
an adjustable current source ICONTROL connected between said filter output terminal VOUT and the first reference potential (ground), said transistor terminal (collector) of said transistor being connected to a second reference potential (ground in figure 15.37 and VCC in 15.35), said source impedance defined by the equation:

$$Z_e = \frac{R_1}{\beta} + j\omega R_1 C_1 \frac{V_T}{I_C},$$

where:

R_1 = resistance of said resistor RSRC1;
β = differential current amplification;
C_1 = capacitance of said first capacitor C11;
V_T = temperature voltage; and
I_C = current generated by said adjustable current source ICONTROL.

Averaged over the eight out-of-sample frequencies, the best-of-run circuit from generation 50 of the second run has

- 74.9 millivolts average absolute error in the passband,
- 0.35 dB average absolute error for frequencies to the right of the passband, and
- 99.6% of the possible number of hits.

Figure 15.39 shows a pace-setting individual from generation 25 of the second run that is less parsimonious than the eventual best-of-run circuit from generation 50 (figure 15.37). The less parsimonious circuit has two more components (i.e., transistor Q105 and 1-giga-Ohm resistor R10) than the more parsimonious best-of-run circuit

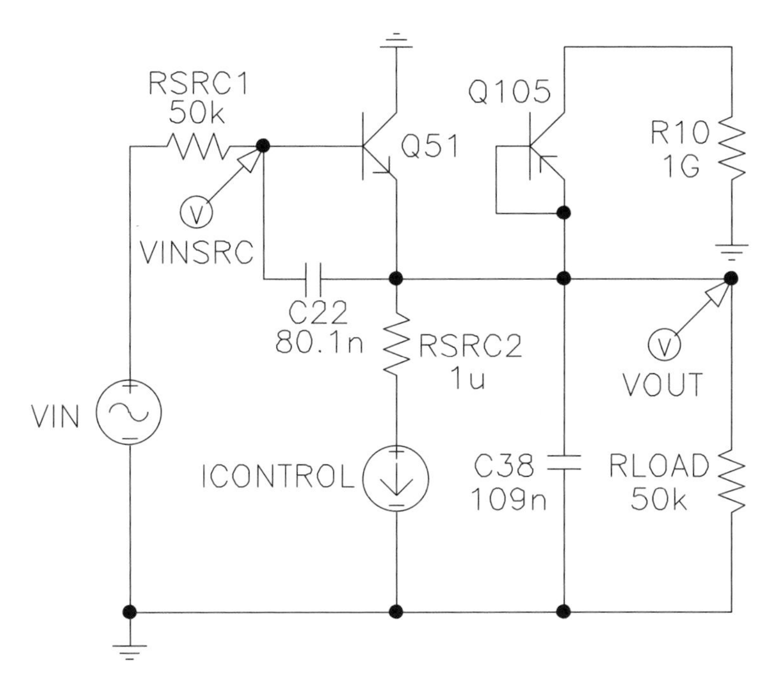

Figure 15.39 Pace-setting circuit from generation 25 of the second run that is less parsimonious than the eventual best-of-run circuit from generation 50 for the tunable integrated active filter problem.

from generation 50. The two additional components have no noticeable effect on the circuit's operation because the resistor's value is so large. If Q105 and R10 are removed from this less parsimonious circuit, the resulting pruned circuit has the same topology as the more parsimonious best-of-run circuit from generation 50 and almost identical sizing. Specifically, the value of capacitor C38 in the less parsimonious circuit (109 nanofarads) is identical to the value of the corresponding capacitor (C12) in the more parsimonious circuit. Moreover, C22 from the less parsimonious circuit has a value of 80.1 nanofarads compared to a value of 86 nanofarads for C11 in the more parsimonious circuit. The less parsimonious circuit of figure 15.39 infringes patent 6,225,859.

15.4.6.3 Results of Third Run of Tunable Integrated Active Filter After getting an infringing solution in the second run, we made a third run in the hope of getting a non-infringing solution. A secondary purpose of this third run was to generate a genealogical audit trail showing how the parents of the satisfactory solution sire the eventual solution. Except for the added auditing code, the code for the third run is the same as the second run.

In the third run, the best-of-generation circuit from generation 0 has a fitness of 89.7.

A satisfactory circuit emerged in generation 38 of the third run. Ignoring the components in the test fixture, the best-of-run circuit from generation 38 has one transistor

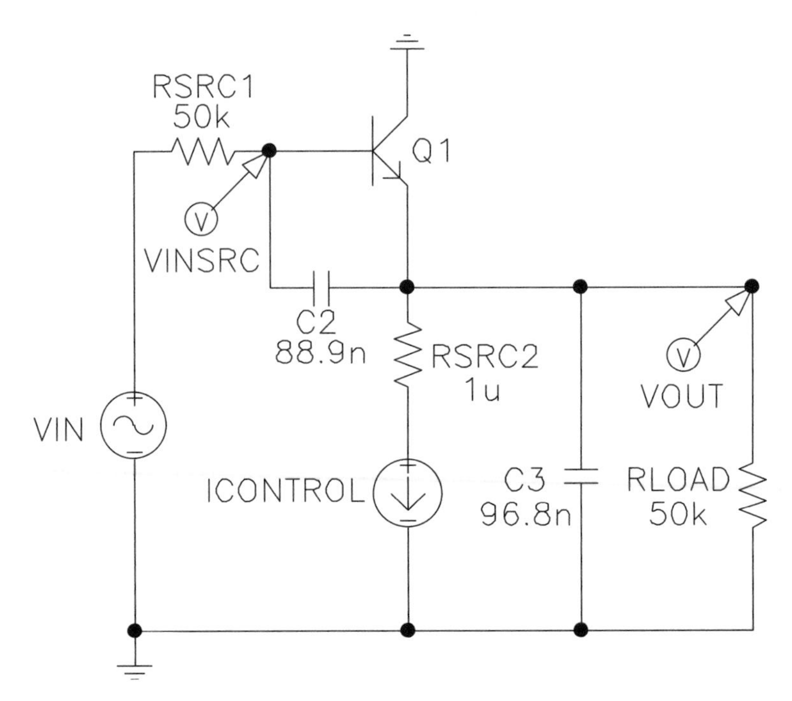

Figure 15.40 Best-of-run circuit from generation 38 of the third run of the tunable integrated active filter problem.

and two capacitors (figure 15.40). This circuit has a fitness of 13×10^{-12}. As can be seen, the best-of-run circuit from generation 38 of the third run has the same topology and approximately the same component values as the best-of-run circuit from the second run (figure 15.37) and hence it infringes U.S. patent 6,225,859. Specifically, C2 from the best-of-run circuit from generation 38 of the third run has a value of 88.9 nanofarads compared to a value of 86 nanofarads for C11 in the best-of-run circuit from the second run. Also, C3 from the best-of-run circuit from generation 38 of the third run has a value of 96.8 nanofarads compared to a value of 109 nanofarads for C12 in the best-of-run circuit from the second run.

The best-of-run circuit from generation 38 of the third run (figure 15.40) is the offspring produced by a crossover of two parents from generation 37.

Figure 15.41 shows the receiving (female) parent from generation 37 from the third run. As can be seen, the receiving parent has the same (i.e., correct) topology as the best-of-run circuit (but different component values).

Figure 15.42 shows the donor (male) parent from generation 37.

The crossover that produced the best-of-run circuit from generation 38 of the third run (figure 15.40) replaced the 48.8-nanofarad capacitor CREMOVED of the receiving

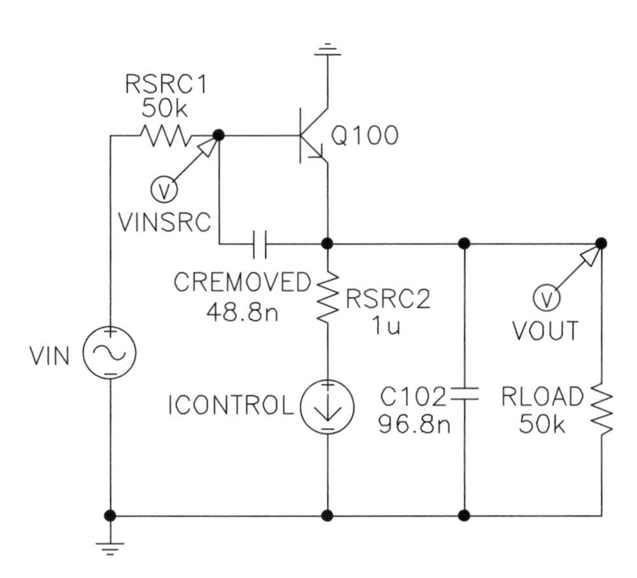

Figure 15.41 Receiving parent from generation 37 of the third run of the tunable integrated active filter problem.

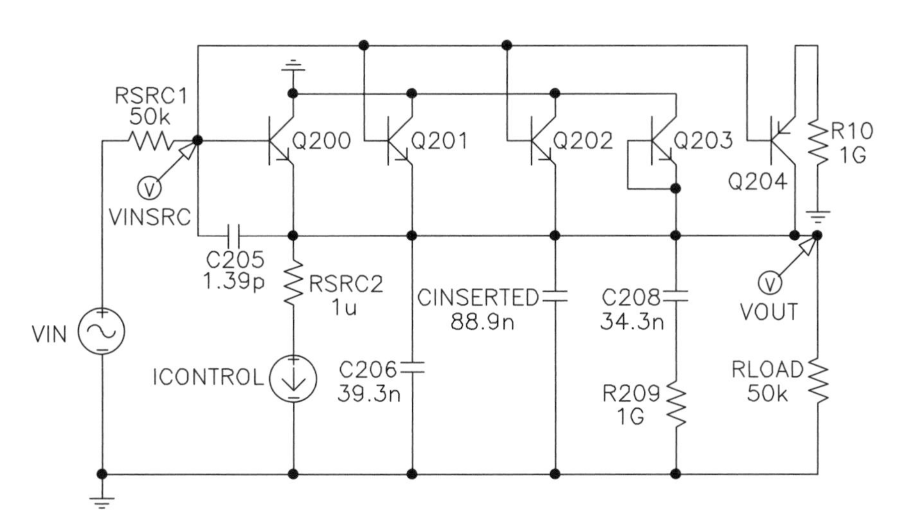

Figure 15.42 Donor parent from generation 37 of the third run of the tunable integrated active filter problem.

parent from generation 37 with the 88.9-nanofarad capacitor CINSERTED of the donor parent from generation 37.

Table 15.12 compares (as an average over the nine values of frequency) the average absolute error in millivolts for frequencies in the passband, the average absolute

Table 15.12 Comparisons of characteristics of the two parents of the best-of-run circuit from generation 38 of the third run of the tunable integrated active filter problem

	Receiving (female) parent from generation 37	Donor (male) parent from generation 37	Offspring in generation 38
Average absolute error in the passband	111 millivolts	336 millivolts	94.9 millivolts
Average absolute error for frequencies to the right of the passband	5.51 dB	12.7 dB	0.85 dB
Percentage of hits	65.8%	64.2%	99.6%

error in decibels for frequencies to the right of the passband, and the percentage of the possible number of hits for

- the receiving (female) parent from generation 37 (figure 15.41),
- the donor (male) parent from generation 37 (figure 15.42), and
- the best-of-run circuit from generation 38 of the third run (figure 15.40).

As can been seen in table 15.12, the receiving and donor parents are each somewhat fit, but have only about two-thirds of the possible number of hits. After the crossover, the offspring (i.e., the best-of-run circuit from generation 38 of the third run) has 99.6% of the possible number of hits and distinctly better average absolute error in the passband (in millivolts) and average absolute error for frequencies to the right of the passband (in decibels).

Figure 15.43 shows the frequency response curves for the best-of-run circuit from generation 38, the RLC model used in the definition of the fitness measure, the receiving parent, and the donor parent when a value of 10 milliamperes is used for the control signal (corresponding to a passband boundary of 4414 Hz). As can be seen, the frequency response curve for the receiving parent (marked by X's) generally follows the curve for the RLC model (marked by filled triangles) but is consistently about 4.5 dB too high. The frequency response curve for the donor parent (marked by hollow circles) follows the target curve very poorly, with an absolute error of 25.4 dB at 100,000 Hz. After being mated, these two parents produce the best-of-run individual (marked filled squares) that closely follows the target curve associated with the RLC model. In fact, for most of the frequencies, the frequency response curves for the best-of-run individual (filled squares) and the RLC model (filled triangles) are indistinguishable.

15.4.6.4 Results of Fourth Run of Tunable Integrated Active Filter The second and third runs of this problem both produced infringing solutions. We then made a fourth run to try again to evolve a non-infringing solution. The fourth run has code to harvest a large number of satisfactory solutions from the various processing nodes of the 1,000-node parallel computer. In order to maximize the chance of producing a non-infringing solution, the run was permitted to continue long after it produced this first satisfactory solution. Except for these changes, the code for the fourth run is the same as the second and third runs.

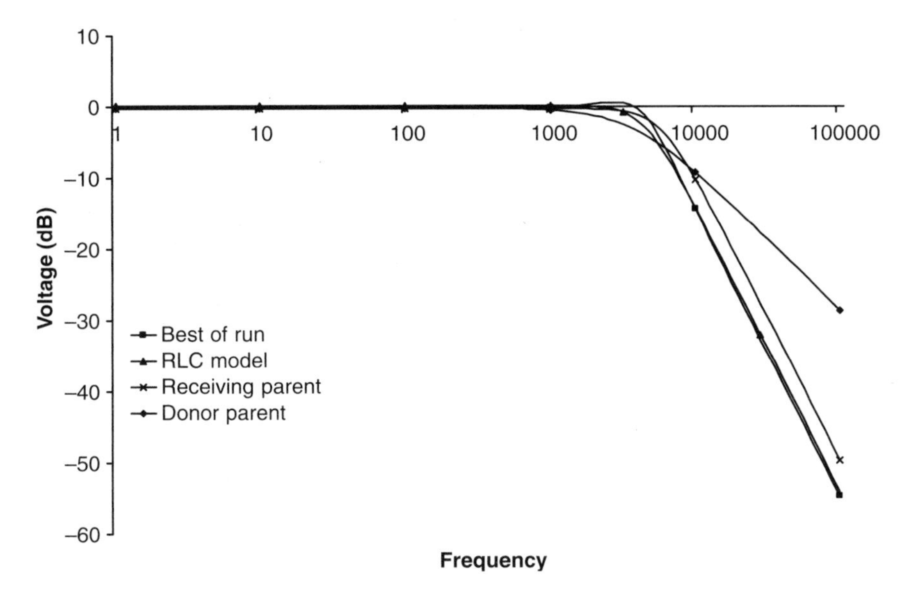

Figure 15.43 Comparison of frequency response for the best-of-run circuit from generation 38 of the third run, the RLC model, the receiving parent from generation 37, and the donor parent from generation 37.

Table 15.13 Comparison of values of two capacitors in patented circuit and the best-of-run circuits from three different runs of the tunable integrated active filter problem

	Generation	Capacitor C1	Capacitor C2
Patented circuit	NA	100	100
Second run	50	86	109
Third run	38	88.9	96.8
Fourth run	27	85	105

In the fourth run, the best-of-generation circuit from generation 0 has a fitness of 104.1.

The first satisfactory circuit emerged in generation 27 of the fourth run. This circuit has the same topology and approximately the same component values as the best-of-run circuits from the second run (figure 15.37) and third run (figure 15.40) and hence it infringes U.S. patent 6,225,859. This circuit has a fitness of 13×10^{-12}.

Table 15.13 compares the values of the two capacitors from the best-of-run circuits from the second, third, and fourth runs that correspond to C1 and C2 of the patented circuit (figure 15.35). All values are in nanofarads.

Averaged over the eight out-of-sample frequencies, the best-of-run circuit from the fourth run has

- 95.6 millivolts average absolute error in the passband,
- 0.89 dB average absolute error to the right of the passband, and
- 99.3% of the possible number of hits.

To our surprise, every one of over a hundred other harvested satisfactory circuits from the fourth run that we examined also infringed U.S. patent 6,225,859. The topology of most of the satisfactory circuits from the fourth run is exactly that of the patented circuit (figure 15.35). The topology of all the other satisfactory circuits consists of the patented circuit plus some additional non-functional circuitry. All these other satisfactory circuits (with their additional non-functional circuitry) read on the claims of patent 6,225,859. That is, all the other satisfactory circuits that we harvested and examined also infringe. This fact suggests that the patented solution to this problem lies in a large basin of attraction (in the search space used by genetic programming).

15.5 Commercial Practicality of Genetic Programming for Automated Circuit Synthesis

The previous sections of this chapter demonstrated that genetic programming can automatically synthesize analog circuits that duplicate the functionality of six circuits that were patented after January 1, 2000.

Table 15.14 tallies the computer time consumed by the 11 runs of the six post-2000 patented circuits in this chapter. For each problem, column 2 of the table shows the total population size, M (for all 1,000 processing nodes of the parallel computer system). Column 3 shows the generation number i that yielded the best-of-run

Table 15.14 Computer time consumed by 11 runs of the six problems involving post-2000 patented inventions

Problem	Population M	Generation i	$M*(i+1)$	Total hours
Low-voltage balun circuit	5,000,000	97	490,000,000	25
Mixed analog-digital variable capacitor circuit	2,000,000	98	198,000,000	88
High-current load circuit—First run	2,000,000	114	230,000,000	134
High-current load circuit—Second run	2,000,000	215	432,000,000	67
Voltage-current conversion circuit	5,000,000	109	550,000,000	83
Cubic function generator—First run	5,000,000	182	915,000,000	206
Cubic function generator—Second run	2,000,000	326	654,000,000	135
Tunable integrated active filter—First run	2,000,000	70	142,000,000	23
Tunable integrated active filter—Second run	2,000,000	50	102,000,000	14
Tunable integrated active filter—Third run	2,000,000	38	78,000,000	12
Tunable integrated active filter—Fourth run	2,000,000	27	56,000,000	6

individual presented in this chapter. Column 4 shows the product of the total population size and the number of generations $(i+1)$ run before the best-of-run individual was encountered. Column 5 shows the length of the run in hours.

As can be seen from table 15.14, the average number of hours for runs involving each of the six post-2000 patented circuits in the table is 25, 88, 99, 83, 170, and 14, respectively. The average of these averages is 80 hours (3.3 days). (We use the average of the averages here because we made four runs of the problem that took the least computer time).

All the problems in this chapter were run on a home-built parallel computer system consisting of 1,000 350-MHz Pentium II processors (described in greater detail in section 17.1.2). This system operates at an overall rate of 3.5×10^{11} Hz. A 3.3-day run represents about 10^{17} cycles (i.e., 100 petacycles).

The relentless iteration of Moore's law (Moore 1996) promises increased availability of computational resources in future years. If available computer capacity continues to double approximately every 18 months over the next decade, a computation requiring 80 hours at the time of this writing (February 2003) will require only about 1% as much computer time (i.e., about 48 minutes) a decade from now.

Aside from the promise of increased availability of computational resources in the future, there are two reasons why we are currently not near the boundary of the current capability of genetic programming to automatically synthesize analog circuits.

First, multiple runs of a probabilistic algorithm are often necessary to solve a problem. However, ignoring partial runs used for debugging purposes, all 11 runs involving the six post-2000 patented circuits produced a satisfactory solution. The success rate of 100% in this chapter is thus unusual with a probabilistic algorithm. This success rate suggests that we are currently not near the boundary of the current capability of genetic programming.

Second, the (largely intentional) inefficiency of the runs in this chapter is a second indication. As previously mentioned, the runs of genetic programming in this book were undertaken with two distinct (and conflicting) orientations. One orientation involves using as little human-supplied knowledge, information, analysis, and intelligence as possible. The other is the engineer's orientation to solve problems as quickly and efficiently as possible.

The dominant orientation of this chapter (that is, the first orientation) is, of course, entirely irrelevant to the practicing engineer who may be interested in the commercial practicality of automatic circuit synthesis by means of genetic programming.

Runs of circuit synthesis problems can be accelerated in the following 10 ways.

First, each of the problems in this chapter could have been solved much more efficiently by using a more customized initial circuit. Our use of the floating embryo on the six post-2000 patented circuits enabled us to minimize the differences in the human-supplied preparatory steps among the six problems in this chapter. This uniformity helps to persuade the reader that the human user need employ only *de minimus* information and domain knowledge in order to launch a run of genetic programming. However, the floating embryo (section 4.7.1.1) used on all six problems in this chapter is manifestly inefficient.

Second, the components that are inserted into a developing circuit need not be as primitive as a single transistor, resistor, or capacitor. Instead, the set of component-creating functions could easily be expanded to include numerous frequently-used combinations of components. Potentially useful combinations of components include voltage gain stages, Darlington emitter-follower sections, current mirrors, cascodes, three-ported voltage divider subcircuits composed of two resistors in series, and three-ported subcircuits consisting of two resistors (or capacitors) with their common point connected to power or ground. For certain problems, the set of primitives could readily be expanded to include higher-level entities, such as filters, op amps, oscillators, voltage-controller current sources, multipliers, and phase-locked loops.

Third, although the runs of the six post-2000 patented circuits in this chapter were intentionally done in a uniform way, a practicing engineer has no reason to enforce such uniformity. For example, we did not use automatically defined functions for any of the six problems in this chapter. However, most practical electrical circuits are replete with reuse. A practicing engineer would recognize that reuse is specifically important in at least two of the six problems in this chapter (namely the mixed analog-digital integrated circuit for variable capacitance and the low-voltage high-current transistor circuit for testing a voltage source). The practicing engineer would have no reason to forgo the manifest benefits of reuse and automatically defined functions.

Fourth, considerable work has been done in recent years to accelerate the convergence characteristics of circuit simulators. As a result, there are numerous commercially available simulators that are considerably more efficient than the one used for the runs in this book. We use a version of the SPICE3 simulator (Quarles, Newton, Pederson, and Sangiovanni-Vincentelli 1994) that we modified in various ways (as described in Koza, Bennett, Andre, and Keane 1999a). Speedups of up to 10-to-1 are reportedly possible today.

Fifth, there are numerous opportunities to incorporate problem-specific knowledge into a run of genetic programming. For example, the developmental process need not start merely with modifiable wires. A substructure of known utility for a particular problem can be hard-wired into the embryo, thereby relieving genetic programming of the need to reinvent it.

Sixth, it is possible to integrate general knowledge of electrical engineering into a run of genetic programming. For example, Sripramong and Toumazou (2002) combine current-flow analysis into their runs of genetic programming for the purpose of automatically synthesizing CMOS amplifiers.

Seventh, the authors believe that the efficiency of runs can be improved by adapting several of the principles set forth in Goldberg's recently published book on genetic algorithms entitled *The Design of Innovation: Lessons from and for Competent Genetic Algorithms* (Goldberg 2002) to the domain of genetic programming.

Eighth, because there usually are multiple competing elements in the fitness measures of problems of circuit synthesis in the real world, the authors believe that the efficiency of runs of genetic programming can be improved by using some of the recently published new techniques of multiobjective optimization (Osyczka 1984; Bagchi 1999; Deb 2001; Coello Coello, Van Veldhuizen, and Lamont 2002; and Zitzler, Deb, Thiele, Coello Coello, and Corne 2001).

Ninth, the authors believe that the efficiency of runs can be improved by adapting some of the innovative ideas in Erick Cantu-Paz's book on genetic algorithms entitled

Efficient and Accurate Parallel Genetic Algorithms (Cantu-Paz 2000) to the domain of genetic programming.

Tenth, there has been an outpouring of theoretical work in the past few years on the theory of genetic algorithms and genetic programming. In particular, the authors believe that many of the insights in *Foundations of Genetic Programming* (Langdon and Poli 2002) can be used to improve the efficiency of runs of genetic programming.

15.6 Human-Competitiveness of the Results for the Six Post-2000 Patented Circuits

All six of the problems in this chapter involve post-2000 patented inventions.

Referring to the eight criteria in table 1.2 for establishing that an automatically created result is competitive with a human-produced result, each of the six genetically evolved circuits in this chapter satisfies the first of the eight criteria:

(A) The result was patented as an invention in the past, is an improvement over a patented invention, or would qualify today as a patentable new invention.

The fact that genetic programming rediscovered both the topology and sizing of six patented circuits, each of which was unobvious "to a person having ordinary skill in the art," establishes that each of the six genetically evolved results satisfy Arthur Samuel's criterion (1983) for artificial intelligence and machine learning:

"[T]he aim [is]... to get machines to exhibit behavior, which if done by humans, would be assumed to involve the use of intelligence."

15.7 Routineness for the Six Post-2000 Patented Circuits

As mentioned in section 1.1.3, a problem-solving method may be measured by its routineness. As also mentioned there, a problem-solving method is routine if it is general and if relatively little human effort is required to get the method to successfully handle new problems within a particular domain and to successfully handle new problems from a different domain.

The main difference from one problem to the next in this chapter is the fitness measure. Each fitness measure is a straightforward translation of the goal of the particular problem. Beyond that, the preparatory steps for each of the six problems are substantially the same except for minor adjustments reflecting each problem's number of input and output ports and the fact that certain problems used a particular power supply voltage or transistor model. These adjustments were designed to force genetic programming to address the particular challenge of the problem and to comply with the spirit of the invention on which the problem was based.

Thus, relatively little effort was required to make the transition from one problem to the next in this chapter. That is, the transition was routine.

15.8 AI Ratio for the Six Post-2000 Patented Circuits

Section 1.1.2 mentions that a method for solving problems may be measured by the ratio of that which is delivered by the operation of the *artificial* system (if any) to the amount of *intelligence* that is supplied by the human employing the method (i.e., its AI ratio).

What is the AI ratio for the solution produced by genetic programming to each of the six problems in this chapter?

The results produced by genetic programming for the six problems are each human-competitive because each problem arose from a previously patented invention. Thus, the solutions produced by genetic programming for each of these six problems have a high amount of "A."

The preparatory steps for the six problems in this chapter are substantially the same as the preparatory steps for previous problems of circuit synthesis in this book (except that a different goal was specified by the fitness measure). Thus, the solutions produced by genetic programming for the six problems in this chapter incorporate a small amount of "I."

The high amount of "A" represented by the results for the six problems in this chapter, in conjunction with the small amount of "I" provided by the human user for each problem, means that the AI ratios are high for the solutions produced by genetic programming.

16

Problems for Which Genetic Programming May Be Well Suited

This chapter discusses the characteristics of problems for which

- the genetic algorithm may be better suited than genetic programming (section 16.1),
- genetic programming may be better suited than the genetic algorithm (section 16.2), and
- genetic methods (such as the genetic algorithm and genetic programming) may be better suited than hill climbing, gradient search, and simulated annealing (section 16.3).

16.1 Characteristics Suggesting the Use of the Genetic Algorithm

Consider the hypothetical problem of finding the optimal tuning for a PID controller (figure 12.1) for controlling a particular plant. This problem entails the discovery of the five numerical values that characterize a PID controller, namely

- the gain, K_p, applied to the signal entering the controller's proportional (P) block,
- the gain, K_i, applied to the signal entering the controller's integrative (I) block,
- the gain, K_d, applied to the signal entering the controller's derivative (D) block,
- the setpoint weighting, b, applied to the reference signal (prior to subtraction of the plant output) used in the difference that is fed into the controller's proportional (P) block, and
- the setpoint weighting, d, applied to the reference signal (prior to subtraction of the plant output) used in the difference that is fed into the controller's derivative (D) block.

As this problem is stated, the controller's topology is prespecified to be the PID topology. Discovery of the controller's size and shape is not part of this problem. Instead, the only issue is the discovery of the five numerical values.

Moreover, there is nothing in the statement of this problem that suggests that there would be any advantage in trying to represent the solution in the form of a computer program.

For these reasons, the genetic algorithm operating on fixed-length strings is a good choice for searching for the required optimal (or near-optimal) combination of five numerical values.

In applying the genetic algorithm to this problem, the user would typically employ a five-part chromosome string in which each part represents one of the five numerical values. At the user's option, each numerical value may be represented in the chromosome as a string of bits or as a floating-point number.

16.2 Characteristics Suggesting the Use of Genetic Programming

Genetic programming may be well suited to a particular problem if one or more of the following characteristics are a major part of a problem:

(1) discovering the size and shape of the solution,
(2) reusing substructures,
(3) discovering the number of substructures,
(4) discovering the nature of the hierarchical references among substructures,
(5) passing parameters to a substructure,
(6) discovering the type of substructures (e.g., subroutines, iterations, loops, recursions, or storage),
(7) discovering the number of arguments possessed by a substructure,
(8) maintaining syntactic validity and locality by means of a developmental process, or
(9) discovering a general solution in the form of a parameterized topology containing free variables.

As can be seen, most these nine characteristics flow from the fact that genetic programming operates in the domain of computer programs.

We now discuss each of these nine characteristics in greater detail.

16.2.1 Discovering the Size and Shape of the Solution

The genetic algorithm operating on fixed-length strings is well suited to the problem in section 16.1 involving the tuning of a PID controller because the size and shape of the solution is not an issue. However, if discovering the size and shape of the solution is a major part of a problem, genetic programming may be better suited to the problem.

To illustrate this point, recall the problem in chapter 6 of designing a wire antenna using genetic programming. This problem was previously solved by Linden and Altshuler using the genetic algorithm operating on fixed-length strings (Linden 1997; Altshuler and Linden 1999). As part of the preparatory steps prior to applying the genetic algorithm to this problem, Linden and Altshuler prespecified that

- all the antenna's wires (i.e., the directors, the reflectors, and the driven elements) would be straight,

- all the antenna's wires would be arranged in parallel (with all their midpoints lying along a single straight line that is parallel to the *X*-axis), and
- the energy source (the transmission line) would be connected to the midpoint of a single straight wire (the driven element).

In other words, Linden and Altshuler prespecified that the solution would have the topology of a Yagi-Uda antenna.

In addition, Linden and Altshuler further prespecified

- the number of directors, and
- the number of reflectors.

After specifying all these things, the only remaining issues are the discovery of the lengths of the wires and the spacing between the wires. As mentioned in section 16.1, the discovery of a set of numerical values is a task for which the genetic algorithm operating on fixed-length strings is well suited.

When genetic programming was applied to this problem in chapter 6, none of the above five aspects of the solution were prespecified. Instead, all five aspects were dynamically discovered during the run of genetic programming. If discovering the size and shape of the solution is a major part of a problem, genetic programming may be a good choice.

As another illustration of this point, recall the problems of circuit synthesis in chapters 4, 5, 10, 11, 14, and 15. In each case, genetic programming automatically discovered both the topology and sizing of the solution, including

- the total number of components in the circuit,
- the type of each component (e.g., resistor, capacitor, transistor) at each location in the circuit,
- the sizing of each component that requires sizing,
- a list of all the connections between the leads of the circuit's components, and
- a list of all the connections between the leads of the circuit's components and all the circuit's external points, including
  - input ports,
  - output ports,
  - power sources, and
  - ground.

Similarly, in the problems of controller synthesis in chapters 3, 9, and 13, the controller's topology was not prespecified. Instead, genetic programming automatically discovered both the topology and tuning during the run. In the one problem of controller synthesis in this book in which the controller's topology was prespecified (chapter 12), the discovery of the size and shape of the four mathematical expressions was a major part of the problem.

Note that genetic programming is not the only way to dynamically determine the size and shape of a problem's solution. Smith (1980) expanded the concept of the genetic algorithm to include variable-length chromosomes. Linden and Altshuler might have employed the genetic algorithm operating on variable-length strings to the antenna problem described in chapter 6. In addition to discovering the lengths and spacing of the parallel straight wires, Smith's form of the genetic algorithm could readily have been used to dynamically determine the number of wires (i.e., the number of directors, the reflectors, and the driven elements). Thus, by availing themselves of Smith's genetic algorithm operating on variable-length strings, Linden and Altshuler could have avoided prespecifying three of the above five aspects of the problem's solution.

16.2.2 Reuse of Substructures

Genetic programming may be well suited to a particular problem if reusing substructures is likely to be important in solving the problem.

Systems in the real world typically contain massive regularity, symmetry, homogeneity, and modularity.

For example, non-trivial analog electrical circuits almost always contain multiple occurrences of certain subcircuits (e.g., voltage gain stages, Darlington emitter-follower sections, current mirrors, cascodes, voltage divider subcircuits). At a higher level, analog circuits often also contain multiple occurrences of various more complex entities, such as filters, op amps, oscillators, voltage-controlled current sources, and phase-locked loops. Similarly, digital circuits almost always contain multiple occurrences of certain standard cells. And, digital circuits typically contain multiple occurrences of higher-level entities (e.g., counters, registers, multiplexers). The design of large circuits would be considerably more difficult (and perhaps even impractical) if the designer had to separately think through the design of each subcircuit from the first principles of electrical engineering on each occasion when the subcircuit is needed. Reuse enables the designer to solve a particular problem once and, thereafter, simply reuse the solution.

Reuse of a subcircuit can be realized in the context of genetic programming in several different ways.

First, the circuit-constructing functions responsible for a useful subcircuit may reside in an automatically defined function (ADF). In this kind of reuse, the automatically defined function resembles a subroutine in an ordinary computer program. Multiple occurrences of the subcircuit result when the automatically defined function is repeatedly invoked. Automatically defined functions are described in detail and illustrated in connection with electrical circuits in *Genetic Programming III: Darwinian Invention and Problem Solving* (Koza, Bennett, Andre, and Keane 1999a, chapters 27 and 28).

Second, the circuit-constructing functions responsible for a useful subcircuit may reside in an automatically defined copy (ADC). An automatically defined copy is an iteration specialized to problems of circuit synthesis. Multiple occurrences of the subcircuit result when the copy body branch (CBB) is repeatedly invoked by the copy control branch (CCB) of the automatically defined copy, as described in *Genetic Programming III* (Koza, Bennett, Andre, and Keane 1999a, chapter 30).

Third, although often overlooked, the ordinary crossover operation may work as a mechanism for reusing a subtree that is responsible for a useful subcircuit. Multiple occurrences of a useful subcircuit may be created when the subtree responsible for the useful subcircuit (being part of a reasonably fit individual) is included in a crossover fragment that the crossover operation inserts into another (reasonably fit) individual that already produces the subcircuit.

The importance of reuse can be illustrated by means of a problem of synthesizing a lowpass filter with a passband boundary of 1,000 Hz.

To make this illustration concrete, we use the same specifications found in section 4.7.

A *T-section* for a lowpass filter (Johnson 1950; Williams and Taylor 1995) is a T-shaped subcircuit containing equally valued inductors on the two arms of the "T" and a capacitor on the vertical segment of the "T."

Figure 16.1 shows a circuit consisting of one T-section. This circuit contains a 105,500-microhenry inductor on each arm of the "T" and a 186-nanofarad capacitor on the vertical segment of the "T."

Figure 16.2 shows the frequency-domain behavior of the circuit (figure 16.1) consisting of one T-section. As can be seen, a single T-section acts as a (very poor) lowpass filter.

One way to construct a lowpass filter satisfying this problem's requirements is by means of a cascade of *identical* T-sections in which each T-section contains one 186-nanofarad capacitor and two equal 105,500-microhenry inductors (Koza, Bennett, Andre, and Keane 1999a, section 30.3.3). Figure 16.3 shows a circuit consisting of a cascade of two identical T-sections.

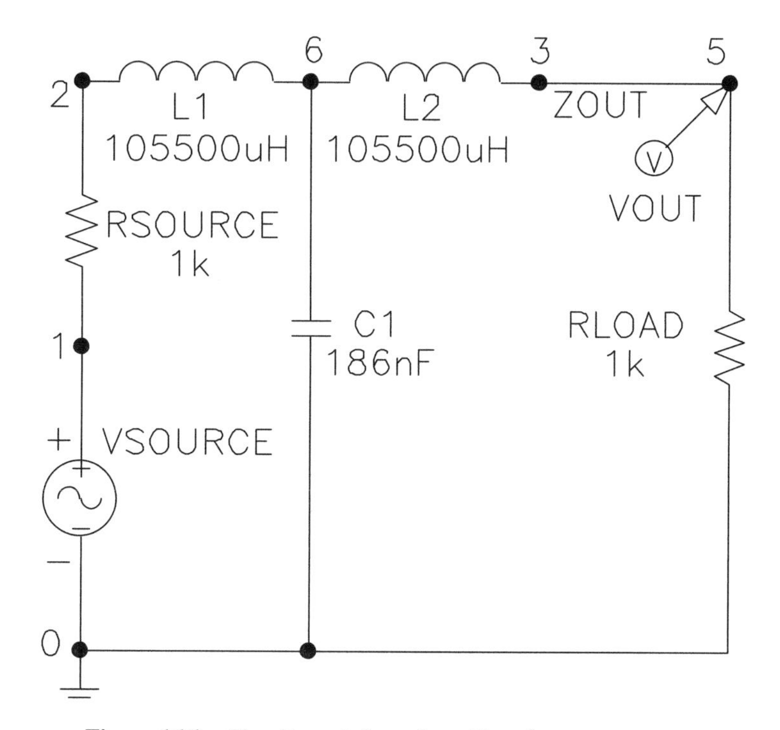

Figure 16.1 Circuit consisting of one T-section.

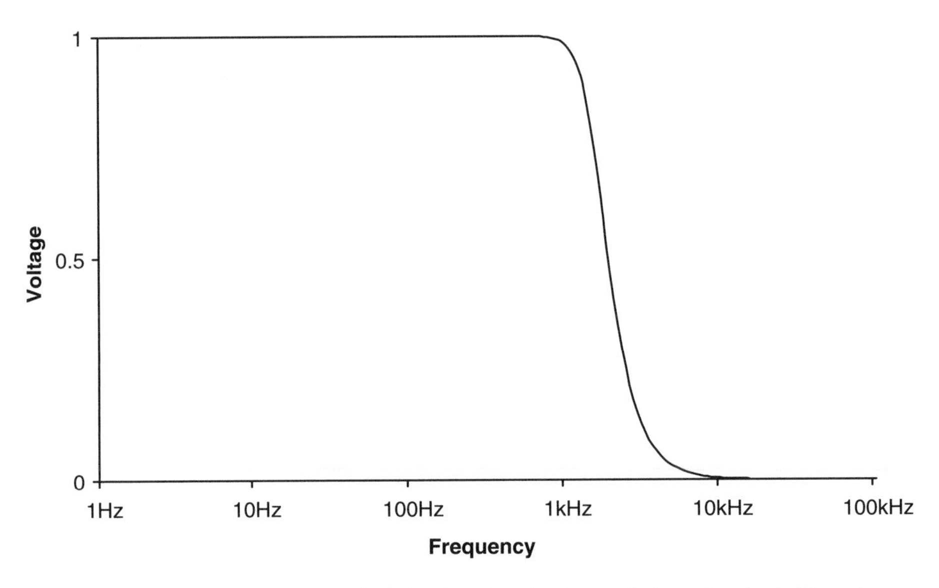

Figure 16.2 Frequency-domain behavior of a circuit consisting of a single T-section.

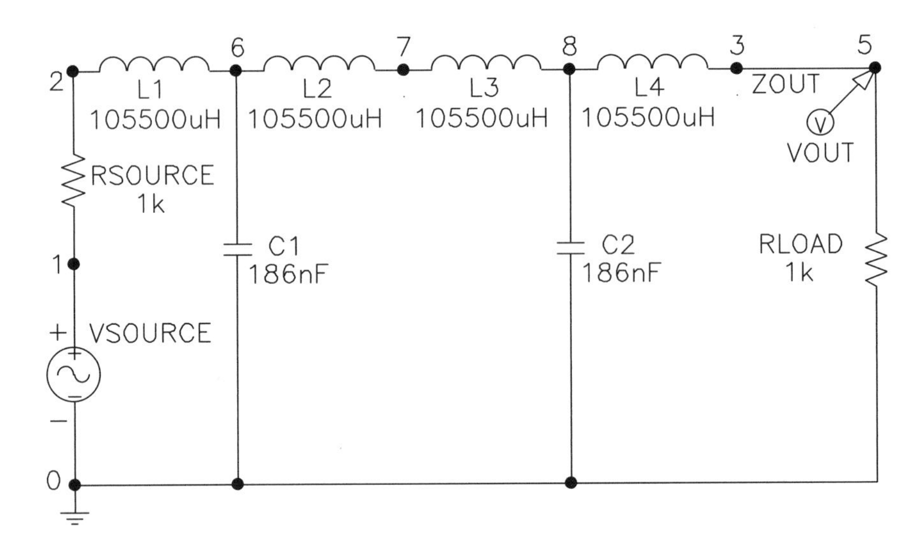

Figure 16.3 Circuit consisting of a cascade of two identical T-sections.

Figure 16.4 shows the frequency-domain behavior of a circuit (figure 16.3) consisting of a cascade of two identical T-sections. Although this circuit does not satisfy the problem's stated requirements, its filtering performance is distinctly better than of the circuit consisting of only a single T-section (figures 16.1 and 16.2).

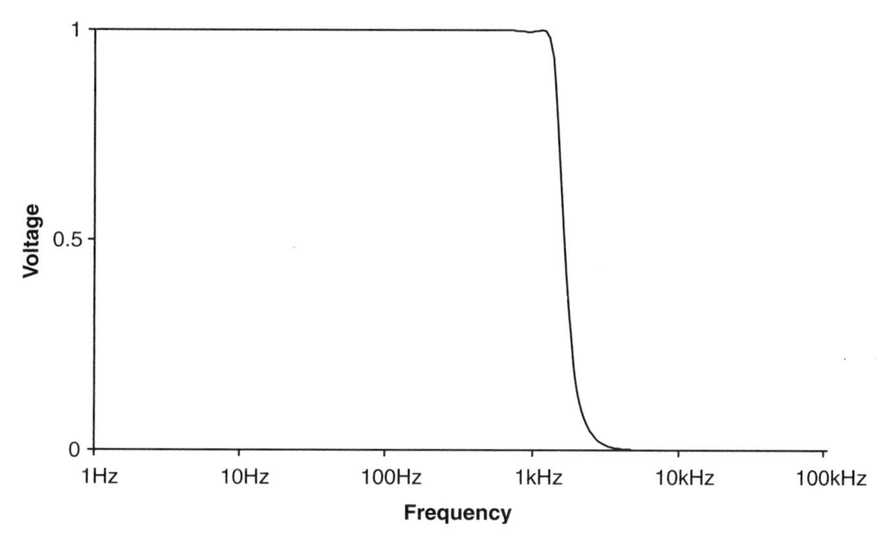

Figure 16.4 Frequency-domain behavior of the circuit consisting of a cascade of two identical T-sections shown in figure 16.3.

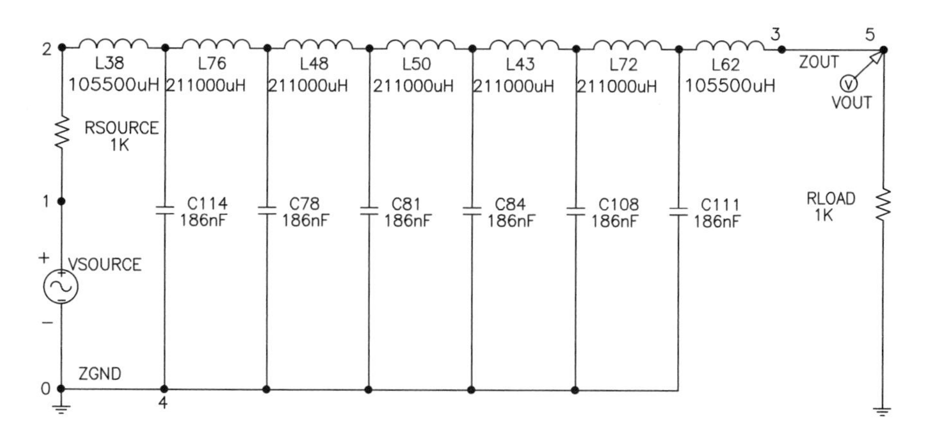

Figure 16.5 Circuit consisting of a cascade of six identical T-sections.

The addition of a third, fourth, and fifth identical T-section further enhances filtering performance. For example, figure 16.5 shows a circuit consisting of a cascade of six identical T-sections. For reasons of space in this figure, each series pair of two 105,500-microhenry inductors is replaced by the equivalent 211,000-microhenry inductor.

Figure 16.6 shows the frequency-domain behavior of a circuit (figure 16.5) consisting of six T-sections. The filtering performance obtained from six T-sections is considerably better than that obtained from fewer T-sections (figures 16.2 and 16.4).

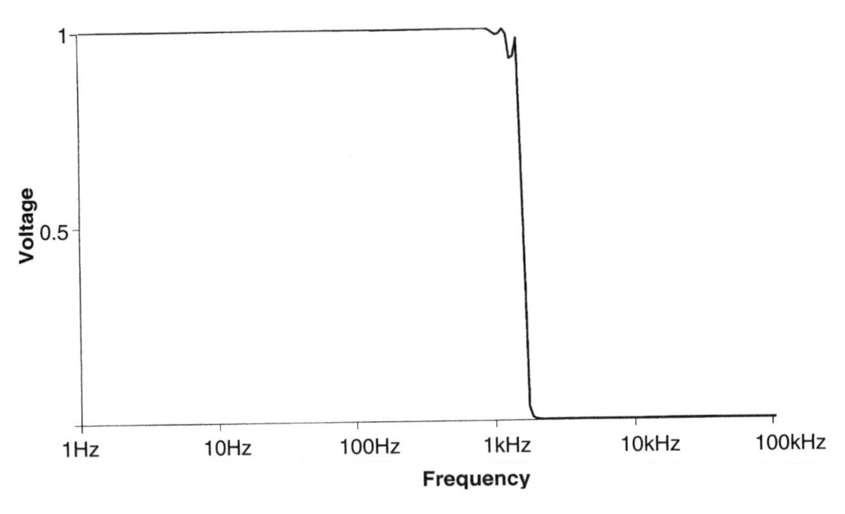

Figure 16.6 Frequency-domain behavior of the circuit consisting of a cascade of six identical T-sections shown in figure 16.5.

In fact, this cascade of six identical T-sections is a 100%-complaint solution to the stated problem (section 4.7).

Now consider how one might approach this problem of circuit synthesis using the genetic algorithm operating on fixed-length strings.

First, the human user would choose the maximum number of components in the circuit. In practice, the user would choose a generous maximum in order to accommodate a circuit of any size that is reasonably likely to solve the stated problem. For reasons of space in the tables below, we use an artificially low maximum (six).

Because the components involved in this problem have only two leads, each component can be conveniently represented by one symbolic variable and three numeric variables. The symbolic variable specifies the type of component (with "L" denoting an inductor, "C" denoting a capacitor, and "W" denoting a nonmodifiable wire). The first of the three numeric variables specifies the component's sizing (with inductors in millihenrys and capacitors in nanofarads). The other two numeric variables specify the nodes to which the component's leads are connected.

All symbolic and numeric variables could, at the user's option, be converted into bits. However, we will proceed herein using a chromosome in which symbolic variables ("L," "C," and "W") appear in chromosome positions of the form $4i+1$ (where i is an integer); floating-point component values appear in positions of the form $4i+2$; and integers representing nodes appear in positions of the forms $4i+3$ and $4i$. That is, we use what can be viewed as a genetic algorithm with a constrained syntactic structure.

Table 16.1 shows one way of representing the circuit consisting of one T-section (shown in figure 16.1) using the genetic algorithm operating on fixed-length strings. The first component is a 105.5-millihenry (105,500-microhenry) inductor whose leads are connected to nodes 2 (the incoming signal) and 6 (an intermediate node). The second component is a 186-nanofarad capacitor whose leads are connected to node 6 and node 0 (ground). The third component is a 105.5 millihenry inductor whose leads are connected to node 6 and node 3 (the circuit's output port). The fourth component is a nonmodifiable wire (with 0 as its component value) whose leads are connected to node 0 and node 7 (another intermediate point). The fifth component is a 100-nanofarad capacitor whose leads are connected to nodes 0 and 7. The sixth component is a 100-nanofarad capacitor whose leads are connected to node 0 and node 8 (yet another intermediate point). Note that the fourth and fifth components form an isolated loop and that the sixth component is a dangling component. Oddities such as this will frequently occur in the initial random generation. However, in practice, isolated loops and dangling wires can be easily edited out of the initial random generation by routine preprocessing that can occur prior to a circuit's evaluation by a simulator, reconfigurable analog chip, or other physical embodiment.

The circuit consisting of a single T-section (represented by the chromosome in table 16.1 and shown in figure 16.1) is a lowpass filter (albeit a very poor one). It is precisely the kind of circuit that one often encounters as a best-of-generation individual in an early generation of a run of either the genetic algorithm or genetic programming. This circuit, consisting of a single T-section (whose three external points are at nodes 2, 0, and 3), does not satisfy the problem's requirements. However, the T-section is a useful building block and represents progress toward the eventual solution.

Table 16.2 shows one possible way of representing the circuit consisting of two T-sections shown in figure 16.3. The first, second, and third components in this chromosome represent the first T-section. Node 6 is the internal point of the first T-section. The three external points of this first T-section are nodes 2, 0, and 7. The fourth, fifth, and sixth components in this chromosome represent the second T-section. The three external points of the second T-section are nodes 7, 0, and 3 (with node 8 being the internal point of the T-section). In particular, the fourth component is a 105.5-millihenry (105,500-microhenry) inductor whose leads are connected to intermediate nodes 7 and 8. The fifth component is a 186-nanofarad capacitor whose leads are connected to nodes 8 and 0. The sixth component is a 105.5-millihenry inductor whose leads are connected to node 8 and 3 (the circuit's output point). Notice that the third component in table 16.2 is connected to node 7 (an intermediate point), as opposed to being connected to node 3 (the circuit's output point) as it was in table 16.1. This change reflects the fact that the output of the first T-section in table 16.2 will be fed into the second T-section. Also note that the final output of the cascade of two identical T-sections is connected to node 3 (the circuit's output point).

Table 16.1 A first illustrative chromosome for the six-component circuit

L	105.5	2	6	**C**	186	6	0	**L**	105.5	6	3	**W**	0	0	7	**C**	100	0	7	**C**	200	8	0

Table 16.2 Chromosome for circuit consisting of two identical T-sections shown in figure 16.3

L	105.5	2	6	**C**	186	6	0	**L**	105.5	6	7	**L**	105.5	7	8	**C**	186	8	0	**L**	105.5	8	3

Although the genetic algorithm operating on fixed-length strings is capable of solving the stated problem, notice that its discovery of the first T-section will not aid in discovering the second T-section. The reasons are that each T-section in the chromosome is defined in terms of particular node numbers and that the three external node numbers associated with each new T-section are necessarily different from those associated with the previous T-section. In particular, after discovering the first T-section (with external points 2, 0, and 7), the genetic algorithm operating on fixed-length strings must separately discover the second T-section (with external points 7, 0, and 3).

In contrast, representations employed by genetic programming have the ability to leverage the once-learned T-section. Consider one way in which this can happen. Suppose that a particular individual in an early generation develops into a circuit with a single T-section. During the early generations of the run, this individual will be relatively fit (when compared to cohorts lacking the T-section or some equally good structure). Thus, this individual will probably be selected (and, indeed, reselected) to participate in crossover. On some of the occasions in which this first individual is selected to participate in crossover, the subtree that is responsible for the T-section will be included in the crossover fragment. Meanwhile, a relatively fit second individual will be selected to mate with the first individual. Possibly, the second individual may be reasonably fit because it too already contains one occurrence of a T-section. Note that because reselection is allowed, the first individual may occasionally mate with itself. Thus, the ordinary crossover operation will sometimes create an offspring with two T-sections. At the moment when this doubling-up occurs, the offspring will probably have distinctly better fitness than its cohorts in the population. The offspring with two T-sections will thus immediately start winning future tournaments in which it participates and will therefore frequently participate in future reproduction, mutation, and crossover operations.

Of course, this scenario does not occur on every crossover. However, about 90% of the genetic operations on a typical run of genetic programming are crossovers. Moreover, a run of genetic programming involves a large population being bred over many generations. Thus, there are numerous opportunities for a doubling-up of T-sections.

Figure 16.7 shows a circuit-constructing program tree that develops into the single T-section shown in figure 16.1. In this figure, the first and third arguments of the three-argument `THREE_GROUND` function (labeled 100) each create a 105,500-microhenry inductor. The second argument of the `THREE_GROUND` function creates a capacitor connected to ground. The first argument of the three-argument `SERIES` function is an `INPUT_0` function that causes a connection to be made between the incoming signal (node 2 in figure 16.1) and the first inductor. The third argument of the `SERIES` function is an `OUPTUT_0` function that causes a connection to be made between the circuit's output point (node 3 in figure 16.1) and the second inductor. The second argument of the `SERIES` function is a `THREE_GROUND` function that causes a connection to be made between the ungrounded lead of the capacitor and the first and second inductor. For simplicity, figure 16.7 shows the eventual component values (105,000 microhenrys and 186 nanofarads) as opposed to the intermediate arguments of the nonlinear mapping (section 3.5.5).

Figure 16.8 shows a circuit-constructing program tree that develops into the two T-sections shown in figure 16.3. The upper left portion of figure 16.8 (i.e., everything

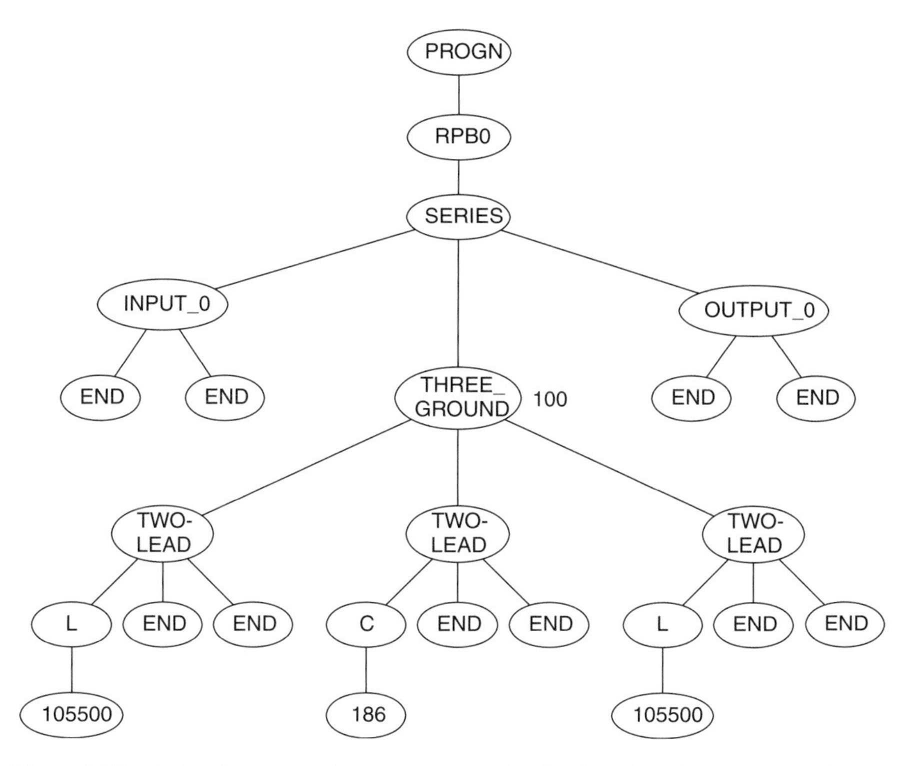

Figure 16.7 A circuit-constructing program tree that develops into the single T-section shown in figure 16.1.

above the point labeled 200) is identical to figure 16.7. As can be seen, the subtree rooted at (that is, below) the point labeled 200 in figure 16.8 is identical to the subtree rooted at the `THREE_GROUND` function labeled 100 in figure 16.7. That is, the subtree that produces the T-section that is responsible (during early generations of the run) for the filtering ability of the individual of figure 16.7 is embedded inside another individual that itself produces a single T-section. Figure 16.8 shows a way by which the ordinary crossover operation can create a circuit consisting of a cascade of two identical T-sections. The circuit consisting of a cascade of two identical T-sections has better fitness than the circuit consisting of just one T-section.

Automatically defined functions provide a second way by which genetic programming can leverage on the already-learned utility of a T-section.

Figure 16.9 shows a circuit-constructing program tree consisting of a result-producing branch (`RPB0`) on the left and an automatically defined function (`ADF0`) on the right. The entire program tree develops into the two T-sections shown in figure 16.3. Automatically defined function `ADF0` develops into one T-section. The result-producing branch (`RPB0`) invokes `ADF0` twice, thereby creating a circuit consisting of two identical T-sections.

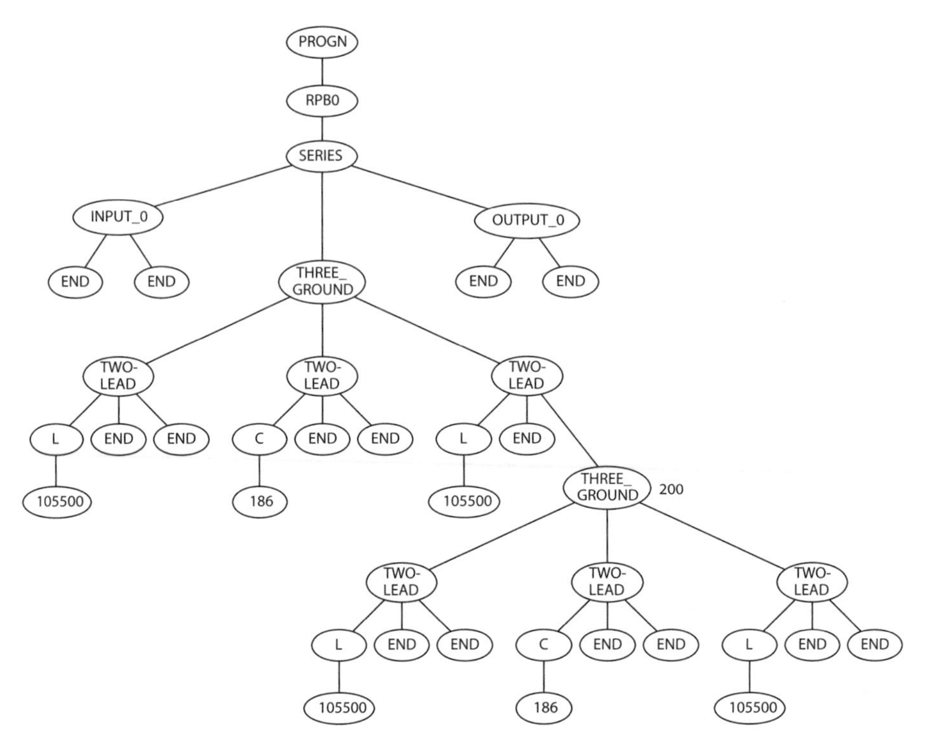

Figure 16.8 A circuit-constructing program tree that develops into the two T-sections shown in figure 16.3.

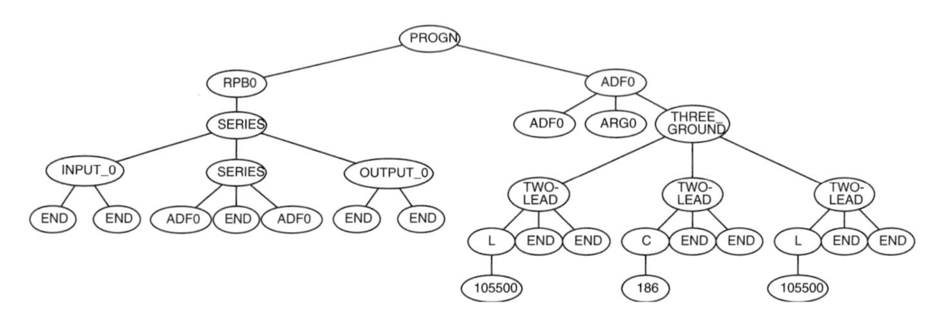

Figure 16.9 Circuit-constructing program tree containing automatically defined function `ADF0` that develops into the two T-sections shown in figure 16.3.

The automatically defined copy (Koza, Bennett, Andre, and Keane 1999a, chapter 30) provides yet another way by which genetic programming can leverage on the already-learned utility of a T-section.

In addition, reuse can also be realized in the context of genetic programming in many additional ways, including by means of automatically defined iterations (ADIs), automatically defined loops (ADLs), automatically defined recursions (ADRs), or automatically defined stores (ADSs) as discussed in *Genetic Programming III* (Koza, Bennett, Andre, and Keane 1999a).

The genetic algorithm operating on fixed-length strings cannot readily leverage a previously discovered useful substructure in the just-described ways. In fact, if the (usual) homologous crossover is employed, there is no way to move a T-section attached to, say, nodes 2, 0, and 3 to another part of the chromosome. Thus, after discovering one T-section, the genetic algorithm operating on fixed-length strings must separately (and laboriously) discover the second one.

Note that the genetic algorithm operating on variable-length strings (Smith 1980) has the same difficulty. It can easily create two identical T-sections attached to the same set of nodes. For example, the sub-string producing a T-section attached to nodes 2, 0, and 3 (table 16.3) could easily be inserted, by crossover, into another chromosome that already contains a T-section attached to nodes 2, 0, and 3.

However, the result would be two T-sections attached to the same group of nodes (that is, 2, 0, and 3). The point is that the genetic algorithm operating on variable-length strings does not readily yield a *series* composition in which the one T-section is attached to nodes 2, 0, and 7 and the second T-section is attached to differently numbered nodes (say, nodes 7, 0, and 3). In contrast, genetic programming has the ability to easily produce, by crossover, either a series or parallel composition of T-sections (with the series composition being the more useful in the immediate situation).

16.2.3 The Number of Substructures

If discovering the number of substructures is an important part of the problem, genetic programming may be well suited to the problem.

Table 16.3 Sub-string producing a T-section

L	105.5	2	6	**C**	186	6	0	**L**	105.5	6	3

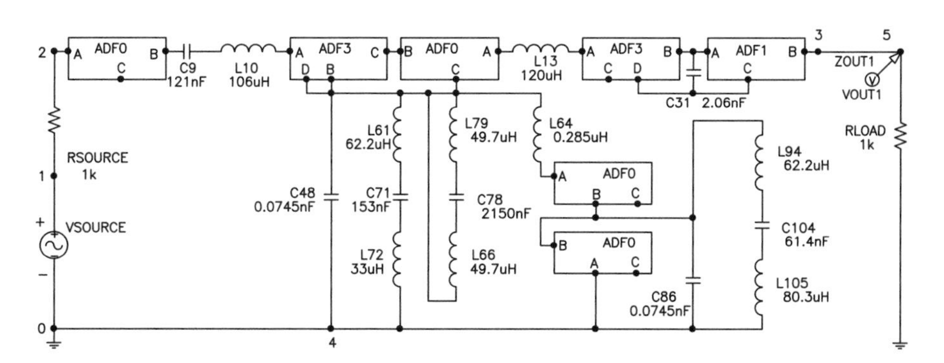

Figure 16.10 Genetically evolved double-bandpass filter circuit.

Consider the problem of automatically synthesizing the double-bandpass filter described in chapter 36 of *Genetic Programming III* (Koza, Bennett, Andre, and Keane 1999a). The architecture-altering operations enable genetic programming to automatically create the architecture of the overall program during the run.

The genetically evolved double-bandpass filter (figure 16.10) contains four occurrences of the three-ported automatically defined function ADF0, two occurrences of the four-ported automatically defined function ADF3, and one occurrence of the three-ported automatically defined function ADF1. The architecture-altering operations dynamically determined, during the run of genetic programming, the number of automatically defined functions that were used in solving this problem.

Automatically defined function ADF0 appears four times in the genetically evolved double-bandpass filter (figure 16.10). Figure 16.11 shows the three-ported subcircuit produced by the execution of ADF0.

Automatically defined function ADF3 appears twice in the genetically evolved double-bandpass filter (figure 16.10). Figure 16.12 shows the four-ported subcircuit produced by the execution of ADF3.

16.2.4 Hierarchical References among the Substructures

If discovering the nature of the hierarchical references among the substructures is a major part of the problem, genetic programming may be appropriate.

The architecture-altering operations enable genetic programming to automatically create the architecture of the overall program during the run, including the hierarchical arrangement of automatically defined functions.

Figure 16.13 shows the call tree for the genetically evolved crossover (woofer-tweeter) filter described in detail in chapter 33 of *Genetic Programming III* (Koza, Bennett, Andre, and Keane 1999a). This call tree shows that the circuit-constructing program tree for this filter consists of three result-producing branches (RPB0, RPB1, and RPB2). It also shows that RPB0 contains a reference to one-argument automatically defined function ADF3 (the number of arguments being shown inside the braces)

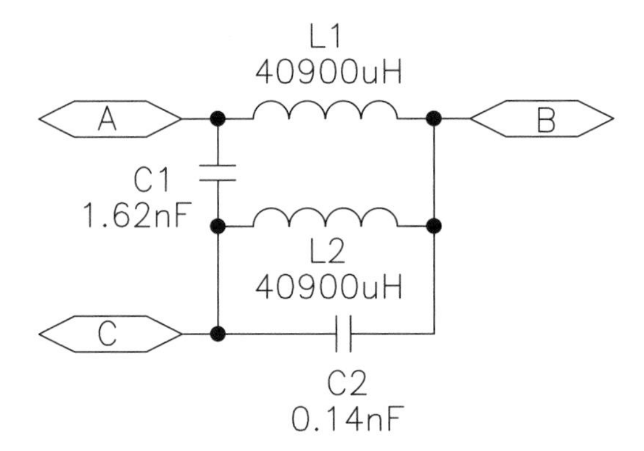

Figure 16.11 This three-ported subcircuit produced by automatically defined function ADF0 appears four times in the genetically evolved double-bandpass filter.

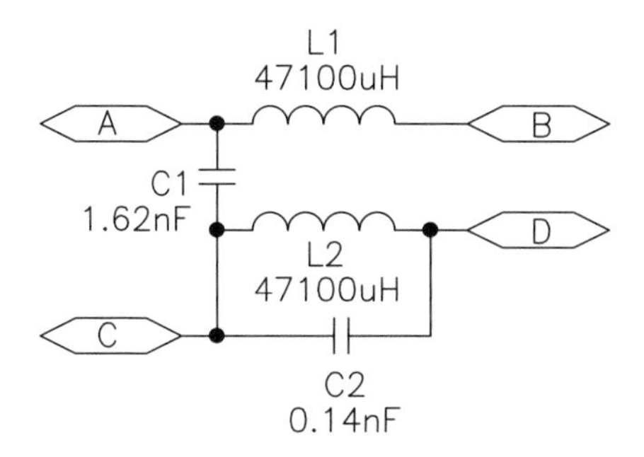

Figure 16.12 This four-ported subcircuit produced by automatically defined function ADF3 appears twice in the genetically evolved double-bandpass filter.

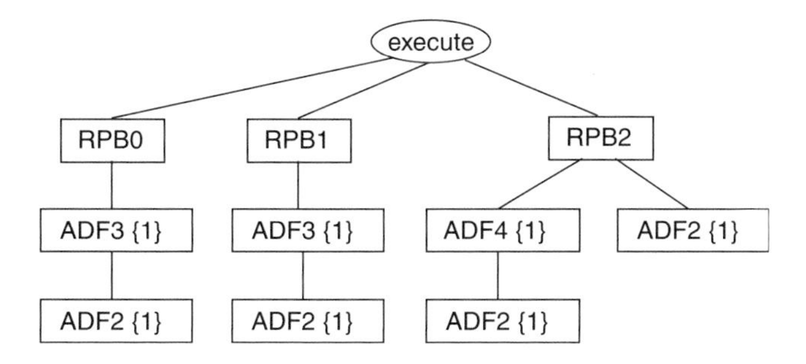

Figure 16.13 Call tree for a genetically evolved crossover (woofer-tweeter) filter.

and that ADF3 invokes one-argument automatically defined function ADF2. Similarly, the figure shows that RPB1 contains a reference to one-argument automatically defined function ADF3 and that ADF3 invokes two-argument automatically defined function ADF2. The figure also shows that RPB2 contains references to one-argument automatically defined functions ADF4 and ADF2 and that ADF4, in turn, invokes one-argument automatically defined function ADF2.

16.2.5 Passing Parameters to Substructures

If passing parameters to a substructure is likely to be useful in solving the problem, genetic programming may be appropriate.

Continuing the discussion from section 16.2.4, consider the subcircuit produced by the three-ported automatically defined function ADF3 (figure 16.14) that appears in the genetically evolved crossover (woofer-tweeter) filter. In addition to the mundane matter of inserting an ordinary 5,130-nanofarad capacitor C112, the execution of automatically defined function ADF3 has the following two noteworthy consequences:

- ADF3 inserts a *parameterized capacitor* C39 whose component value is dependent on the dummy variable ARG0 that is passed into automatically defined function

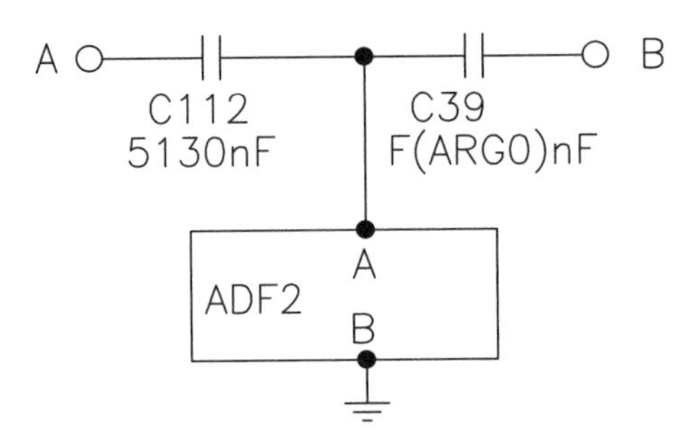

Figure 16.14 Subcircuit produced by the execution of three-ported automatically defined function ADF3 of the genetically evolved crossover (woofer-tweeter) filter.

ADF3 from the program that calls ADF3 (namely the result-producing branch RPB0 shown in figure 16.13).

- ADF3 calls automatically defined function ADF2. A dummy variable ARG0 is passed along from ADF3 to ADF2. In turn, ADF2 creates a *parameterized inductor* whose component value is dependent on this dummy variable.

The passing of a parameter can be seen by examining the genetically evolved LISP S-expressions for the crossover (woofer-tweeter) filter.

The first result-producing branch, RPB0, is shown below. As shown in the underlined portion of this S-expression, automatically defined function ADF3 is called with a numerical argument of 0.619651.

```
(PARALLEL0 (L (+ (- 1.883196E-01 (- -9.095883E-02
5.724576E-01)) (- 9.737455E-01 -9.452780E-01)) (FLIP END))
(SERIES (C (+ (+ -6.668774E-01 -8.770285E-01) 4.587758E-
02) (NOP END)) (SERIES END END (PARALLEL1 END END END END))
(FLIP (SAFE_CUT))) (PAIR_CONNECT_0 END END END) (PAIR_CON-
NECT_0 (L (+ -7.220122E-01 4.896697E-01) END) (L (-
-7.195599E-01 3.651142E-02) (SERIES (C
(+ -5.111248E-01 (- (- -6.137950E-01 -5.111248E-01)
(- 1.883196E-01 (- -9.095883E-02 5.724576E-01)))) END)
(SERIES END END (ADF3 6.196514E-01)) (NOP END))) (NOP END)))
```

Automatically defined function ADF3 is shown below. The dummy variable (ARG0) appears twice in ADF3. Each occurrence is underlined. ARG0 is instantiated with the value of 0.619651 that is obtained from the first result-producing branch, RPB0. Also, note that the value 0.9737455 is passed to automatically defined function ADF2 (shown in bold below). The component value for an inductor created by ADF2 is determined by this value.

```
(C (+ (- (+ (+ (+ 5.630820E-01 (- 9.737455E-01
-9.452780E-01)) (+ ARG0 6.953752E-02)) (- (- 5.627716E-
```

```
02 (+ 2.273517E-01 (+ 1.883196E-01 (+ 9.346950E-02 (+
-7.220122E-01 (+ 2.710414E-02 1.397491E-02)))))) (- (+
(- 2.710414E-02 -2.807583E-01) (+ -6.137950E-01
-8.554120E-01)) (- -8.770285E-01 (- -4.049602E-01
-2.192044E-02))))) (+ (+ 1.883196E-01 (+ (+ (+ (+
9.346950E-02 (+ -7.220122E-01 (+ 2.710414E-02 1.397491E-
02))) (- 4.587758E-02 -2.340137E-01)) 3.226026E-01) (+
-7.220122E-01 (- -9.131658E-01 6.595502E-01))))
3.660116E-01)) 9.496355E-01) (THREE_GROUND_0 (C (+ (- (+
(+ (+ 5.630820E-01 (- 9.737455E-01 -9.452780E-01)) (+ (-
(- -7.195599E-01 3.651142E-02) -9.761651E-01) (- (+ (-
(- -7.195599E-01 3.651142E-02) -9.761651E-01) 6.953752E-
02) 3.651142E-02))) (- (- 5.627716E-02 (- 1.883196E-01
(- -9.095883E-02 5.724576E-01))) (- (+ (- 2.710414E-02
-2.807583E-01) (+ -6.137950E-01 (+ ARG0 6.953752E-02)))
(- -8.770285E-01 (- -4.049602E-01 -2.192044E-02))))) (+
(+ 1.883196E-01 -7.195599E-01) 3.660116E-01)) 9.496355E-
01) (NOP (FLIP (PAIR_CONNECT_0 END END END)))) (FLIP
(SERIES (FLIP (FLIP (FLIP END))) (C (- (+ 6.238477E-01
6.196514E-01) (+ (+ (- (- 4.037348E-01 4.343444E-01)
(+ -7.788187E-01 (+ (+ (- -8.786904E-01 1.397491E-02) (-
-6.137950E-01 (- (+ (- 2.710414E-02 -2.807583E-01)
(+ -6.137950E-01 -8.554120E-01)) (- -8.770285E-01 (-
-4.049602E-01 -2.192044E-02))))) (+ (+ 7.215142E-03
1.883196E-01) (+ 7.733750E-01 4.343444E-01))))) (- (-
-9.389297E-01 5.630820E-01) (+ -5.840433E-02 3.568947E-
01))) -8.554120E-01)) (NOP END)) END)) (FLIP (ADF2
9.737455E-01))))
```

Once a dummy variable (formal parameter) is passed from one part of a genetically evolved hierarchy into another, different instantiations of the dummy variable may be passed.

16.2.6 Type of Substructures

If discovering the type of substructures (e.g., subroutines, iterations, loops, recursions, or storage) is an important aspect of the problem, genetic programming may be appropriate.

The Genetic Programming Problem Solver (GPPS) uses a standardized set of functions and terminals and thereby eliminates the need for the user to prespecify a function set and terminal set for the problem. GPPS uses the architecture-altering operations to dynamically create, duplicate, and delete subroutines and loops during the run of genetic programming. Additionally, in version 2.0 of GPPS, the architecture-altering operations are used to dynamically create, duplicate, and delete recursions and internal storage. Because the architecture of the evolving program is automatically determined during the run, GPPS eliminates the need for the user to specify in advance whether to employ subroutines, loops, recursions, and internal storage in solving a given problem. It similarly eliminates the need for the user to specify the number of arguments possessed by each subroutine.

Consider the problem of evolving a computer program for the even-6-parity problem.

In one run of this problem using GPPS 1.0 (described in section 21.1.2.2 of *Genetic Programming III*), the genetically evolved solution consisted of

- one result-producing branch, and
- two automatically defined loops.

The architecture-altering operations in GPPS determined that there would be two automatically defined loops.

16.2.7 Number of Arguments Possessed by Substructures

If discovering the number of arguments possessed by a substructure is an important part of the problem, genetic programming may be appropriate.

Continuing our discussion of the Genetic Programming Problem Solver (GPPS) from section 16.2.6, in one run on the even-6-parity problem (described in section 21.1.2.1 of *Genetic Programming III*), the genetically evolved solution consisted of

- one result-producing branch,
- one automatically defined loop, and
- two two-argument automatically defined functions.

The architecture-altering operations in GPPS determined that the automatically defined functions would possess two arguments.

16.2.8 The Developmental Process

If a developmental process is a desirable way to represent a solution to the problem, genetic programming may be appropriate.

As previously mentioned, the developmental approach has several advantages in automatically synthesizing complex structures such as circuits.

For one thing, the developmental approach has the specific advantage of preserving locality. Because most of the component-creating, topology-modifying, and development-controlling functions operate on a small local area of the circuit, the subtrees that are transplanted by the crossover operation generally operate locally. Thus, when a crossover replaces a subtree in one individual with a subtree from another individual, it (usually) replaces a local structure in the circuit created by the first individual with a local structure in the circuit created by the second individual. The developmental process works in conjunction with the crossover operation in preserving locality.

The developmental process has the additional advantage of preserving electrical connectivity. There are no unconnected leads in the initial circuit. Each component-creating, topology-modifying, and development-controlling function preserves connectivity at each stage of the developmental process. The result is that there are no unconnected leads in the fully developed circuit.

Also, the developmental approach enables useful parts of a circuit-constructing program tree to be reused (as already illustrated in section 16.2.2 by figures 16.7, 16.8, and 16.9 and tables 16.1 and 16.2). Reuse eliminates the need to "reinvent the wheel" on each occasion when a particular sequence of steps may be useful. Reuse makes it possible to exploit a problem's modularities, symmetries, and regularities (and thereby potentially accelerate the problem-solving process).

The preservation of syntactic validity and executablity of the circuit-constructing program trees during all genetic operations, the preservation of the constrained syntactic structure during all genetic operations, the preservation of electrical connectivity, the preservation of locality during crossover, and the facilitation of reuse together contribute to the efficiency by which developmental genetic programming is able to synthesize circuits.

In contrast, the ordinary crossover operation in the genetic algorithm operating on fixed-length character strings rarely preserves syntactic validity when chromosome strings are used to represent complex structures (e.g., electrical circuits).

When syntactic validity is not preserved, the user is usually faced with three choices concerning the syntactically invalid offspring:

- deletion,
- penalization, or
- repair.

For complex structures such as circuits, deletion is rarely a practical option because the percentage of syntactically valid offspring after a crossover is so low that virtually every individual would be deleted. Penalization is rarely practical for the same reason. Thus, in practice, repair is the most viable option.

The ordinary crossover operation used in the genetic algorithm has the specific advantage that all genetic material in the offspring comes from one or the other parent. This is usually not the case when repair is undertaken. That is, the offspring's genetic material does not come exclusively from its parents.

The parents come from the current generation of the run of genetic programming. As such, the parents are likely to be relatively fit individuals. The benefit of creating offspring from genetic material from the run's current generation is discussed in detail in section 16.3.1.2. The repair process does not confer this benefit on the resultant offspring.

As an example of a crossover operation whose offspring requires repair, consider the circuits defined by the chromosomes in tables 16.4 and 16.5.

The chromosome of the first parent (table 16.4) codes for the single T-section depicted in figure 16.15. This T-section contains two inductors (L1 and L2) and one capacitor (C1). This individual is functionally equivalent to the circuit of figure 16.1 and therefore has the same fitness and the same frequency-domain behavior as shown in figure 16.2. As previously mentioned, a single T-section acts as a (very poor) low-pass filter; however, in an early generation of the run, a single T-section may be among the best individuals in the population at the time.

The chromosome of the second parent (table 16.5) codes for the single T-section depicted in figure 16. This T-section contains two inductors (L3 and L4) and one capacitor (C2). This individual is functionally equivalent to the first parent.

Now consider a crossover operation whose offspring requires repair. In particular, consider a crossover point at the right of the 12th gene. That is, genes 1 through 12 come from the chromosome in table 16.4 and genes 13 through 24 come from the chromosome in table 16.5.

Table 16.6 shows the chromosome of the offspring resulting from this crossover.

Figure 16.17 shows the offspring. The offspring circuit contains three inductors and two capacitors. Components L6 and C3 originate from the first parent (where they were called L1 and C1, respectively, in figure 16.15). Components L5, L7, and C4 originate from the second parent (where they were called L3, L4, and C2 respectively, in figure 16.16). After the crossover, one end (node 9) of the inductor L7 is left dangling. Similarly, one end (node 8) of the capacitor C4 is also dangling. Moreover, the offspring has no connection to the output probe point VOUT. Because the offspring is syntactically invalid, a mechanism must now be invoked to repair it.

There are numerous methods in the literature for repairing a syntactically invalid offspring. Some overtly employ random steps. The ones that do not employ random steps typically employ steps that are so arbitrary that they might as well be random. The point is not that these repair mechanisms do not succeed in creating a syntactically valid offspring, but that they succeed in a manner that resembles mutation.

Table 16.4 Chromosome of first parent

L	105.5	2	6	W	0	6	7	C	186	7	0	W	0	7	8	L	105.5	8	9	W	0	9	3

Table 16.5 Chromosome of second parent

W	0	8	6	W	0	9	3	W	0	7	8	L	105.5	6	9	C	186	8	0	L	105.5	2	7

Table 16.6 Result of a crossover

L	105.5	2	6	W	0	6	7	C	186	7	0	L	105.5	6	9	C	186	8	0	L	105.5	2	7

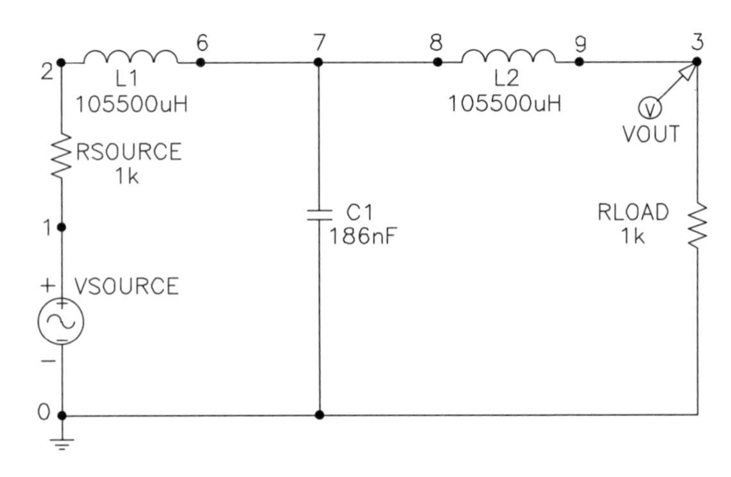

Figure 16.15 First parent.

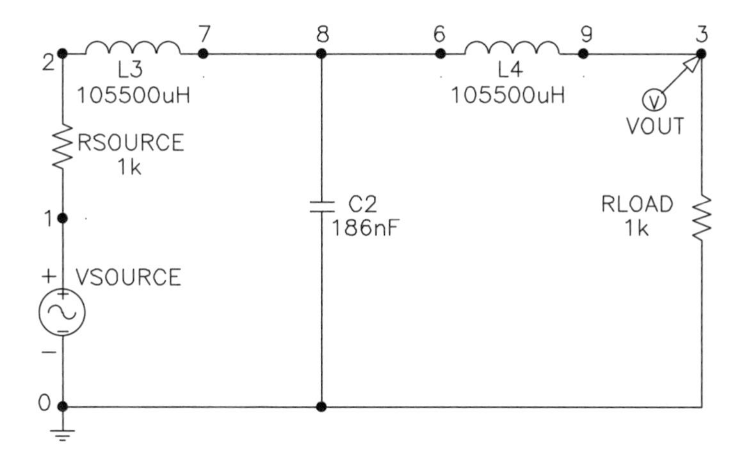

Figure 16.16 Second parent.

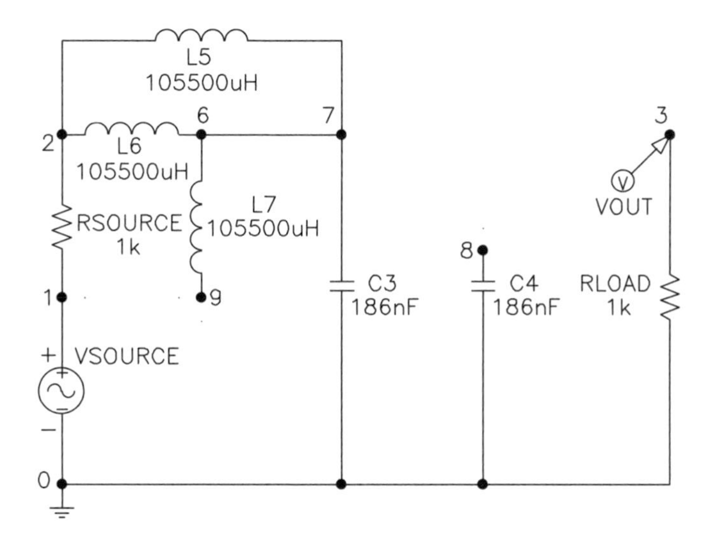

Figure 16.17 Offspring circuit.

When the genetic algorithm operating on fixed-length character strings is being used for problems involving complex structures, the probability is exceedingly small that a syntactically valid chromosome will result from a crossover that is applied directly to the structure. Thus, virtually every crossover requires repair. To the extent that the repair mechanism is either random or arbitrary, virtually every crossover becomes mutational in character.

In summary, the benefits of the developmental process include

- reuse,
- preservation of locality,

- preservation of electrical validity (for circuits),
- preservation of syntactic validity and executability (thereby alleviating the need for repair and the consequent introduction of genetic material from sources other than the two parents from the current generation of the run).

16.2.9 Parameterized Topologies Containing Free Variables

The genetic algorithm operating on fixed-length character strings can easily find the roots of a particular quadratic equation such as $3x^2+4x+5=0$. In applying the genetic algorithm to the problem of finding the real and imaginary parts of this equation's two roots, one could employ a four-part chromosome string in which each part represents one of the four required numerical values. At the user's option, each of the four required values may be represented in the chromosome as a string of bits or as a floating-point number. In either case, the fitness measure for an individual chromosome can be defined as the sum, over the two complex numbers represented by the chromosome, of the distance (in the complex plane) between the result of substituting the complex number into $3x^2+4x+5$ and the origin (0, 0).

Notice that a separate run of the genetic algorithm would be required to solve a different quadratic equation (e.g., $10x^2+2x+7=0$).

In contrast, the well-known mathematical formula

$$x=\frac{-b\pm\sqrt{b^2-4ac}}{2a}$$

is a general solution to the problem of finding the equation's roots. It is applicable to all quadratic equations. This formula contains three free variables (*a, b*, and *c*) representing the three coefficients of the quadratic equation $ax^2+bx+c=0$. Any quadratic equation (including $10x^2+2x+7=0$ and $3x^2+4x+5=0$) can be solved merely by instantiating the free variables *a, b*, and *c* in the formula with the particular numerical values of *a, b*, and *c*. The free variables confer generality on the formula and enable it to provide a solution to the entire category of problems.

In a similar way, free variables (inputs) confer generality on a computer program. Free variables enable a single genetically evolved program to solve an entire category of problems (as opposed to a single instance of a problem).

Thus, if the problem requires the generality conferred by free variables, genetic programming is especially appropriate.

Eleven problems in this book (listed in section 1.3) demonstrate the capability of genetic programming to automatically create, in a single run, a general (parameterized) solution to a problem in the form of a graphical structure whose nodes or edges represent components and where the parameter values of the components are specified by mathematical expressions containing free variables. Chapters 9 and 13 demonstrate the ability of genetic programming to create parameterized topologies for controllers. Chapters 10 and 11 demonstrate this ability for electrical circuits. Four of these 11 problems illustrate the use of conditional developmental operators in conjunction with free variables (chapter 11).

16.3 Characteristics Suggesting the Use of Genetic Methods

This section discusses two differences between the genetic algorithm and genetic programming and certain other search methods (e.g., hill climbing, gradient search, and simulated annealing) that, we believe, make genetic methods particularly advantageous for solving the problems dealt with in this book.

First, the genetic algorithm and genetic programming differ from hill climbing and gradient search in that they are not greedy.

Second, in genetic methods, the recombination (crossover) operation works in conjunction with the population in a way that makes genetic search particularly advantageous.

16.3.1 Non-Greedy Nature of Genetic Methods

Greedy search methods never chose a point that is known to be inferior in preference to a point that is known to be superior. In particular, hill climbing (section 3.4.1) and gradient search (section 3.4.2) pursue a search by greedily moving to a point that is known to be superior to the current point.

Most existing techniques of artificial intelligence and machine learning rely exclusively on greedy methods to move from the current point in the search space to the next point. Greedy methods have the deficiency that they are easily trapped on a local optimum that is not a global optimum of the search space. The temptation to rely on greedy methods is reinforced because many of the toy problems in the literature of the fields of machine learning and artificial intelligence are so simple that they can, in fact, be solved by these methods. However, popularity cannot cure the innate tendency of hill climbing and gradient search to become easily trapped on a local optimum.

Genetic programming and the genetic algorithm differ from hill climbing and gradient search in that they are not greedy. That is, during the course of a genetic search, a point that is known to be inferior will sometimes be preferred to a point that is known to be superior. The fact that genetic methods employ a population (i.e., are parallel searches) gives genetic methods the luxury of retaining both known superior points and known inferior points. The larger the population, the greater the variety of the retained points. The fact that genetic methods employ a population helps to liberate genetic methods from the nearly irresistible temptation to pursue the search only by greedily moving to a point that is known to be superior to the current point. Genetic methods intentionally allocate a certain mathematically principled (small) number of trials to points that are known to be inferior. That is, they do some exploration (while still predominantly pursuing an exploitative strategy). This allocation of trials to known-inferior individuals is not motivated by charity. Instead, the allocation of trials to known-inferior individuals is based on the expectation that it will sometimes unearth a trajectory through the search space that leads to points that ultimately have a higher payoff. The fact that genetic programming operates with a large population enables it to make a certain small number of adventurous moves while predominantly pursuing the most immediately gratifying avenues of advance through the search space.

The non-greedy approach of genetic methods is advantageous because interesting and non-trivial problems almost always have high-payoff points that are inaccessible to hill climbing and gradient search. In fact, the existence of points in the search space that are not accessible to greedy search is a good working definition of non-triviality. Given that problems of machine intelligence are among the most formidable and challenging problems in the field of computer science, it seems especially appropriate that these problems be pursued with non-greedy methods that acknowledge the non-triviality of the space being searched. Indeed, it would be truly extraordinary to be able to solve such formidable and challenging problems by means of greedy search.

It should be noted that simulated annealing (Kirkpatrick, Gelatt, and Vecchi 1983; Aarts and Korst 1989; Salamon, Sibani, and Frost 2002) shares the feature of genetic algorithms and genetic programming of allocating a certain mathematically principled (small) number of trials to points that are known to be inferior. In the case of genetic programming and the genetic algorithm, the exploration of seemingly inferior areas is accomplished by the probabilistic Darwinian selection process. The best individuals in the population are not guaranteed to be selected and the worst individuals are not necessarily excluded. In the case of simulated annealing, the exploration of seemingly inferior areas is accomplished by means of the Metropolis algorithm and the Boltzmann equation. There is, in fact, a similarity in the mathematical underpinnings of simulated annealing and genetic methods (Goldberg 1990; Mahfoud and Goldberg 1995). Simulated annealing resembles a genetic search with a population of one.

Genetic methods also differ from simulated annealing in that they sometimes pass over a point that is known to be superior (whereas simulated annealing never does).

16.3.2 Recombination in Conjunction with the Population in Genetic Methods

The recombination (crossover) operation of the genetic algorithm and genetic programming works in conjunction with the population in a way that appears to be particularly advantageous for solving the problems of interest in this book.

Genetic methods are not, of course, the only search methods that employ parallelism. There are parallel versions of hill climbing (section 3.4.1), gradient search (section 3.4.2), and simulated annealing (section 3.4.3). However, there is no transfer of information among the different threads of such parallel searches. That is, each thread of the parallel search stands alone. Likewise, there are versions of hill climbing, gradient search, and simulated annealing in which the search is iteratively restarted after the search becomes trapped on a local optimum point. However, when the search is restarted, there is usually no forward transfer of information gained from the earlier searches. That is, each restarted run stands alone.

The parallel versions of hill climbing, gradient search, and simulated annealing resemble the simultaneous dropping of a large number of parachutists onto the problem's search space in the hope that one of them, acting alone, will find the global optimum point. The parachutists do their best, given their localized view of the entire search space, to locate a global optimum. Because the parachutists search simultaneously, the parallelism of this type of search assuredly reduces the elapsed time required to locate a global optimum. However, there is no transfer of information among the parachutists in the parallel versions of hill climbing, gradient search, and simulated annealing. Each parachutist generates new candidate points without reference to anything learned by the other parachutists. Hence, no thread of the parallel search is any more efficient at discovering a good point in the search space than if there were no parallelism.

The iterative restart versions of hill climbing, gradient search, and simulated annealing resemble the sequential dropping of one parachutist at a time onto the search space. Each single parachutist makes an effort to discover a global optimum point. The restarting gives the search new life after a parachutist becomes hopelessly trapped on a non-global optimum point. However, after each restart, the single parachutist generates new candidate points without reference to anything learned previously. That is, information is not transferred from earlier searches to later searches. Consequently, except for the fact that the parachutist is restarted at a new starting point, no restarted search is any more efficient than if there were no restarting.

Genetic programming and the genetic algorithm differ from hill climbing, gradient search, and simulated annealing in that important information is exchanged among the threads of the parallel search. This information transfer is accomplished by means of the recombination (crossover) operation. Continuing the analogy with the parachutists, genetic methods simultaneously drop a large number of parachutists onto the search space. However, genetic methods additionally outfit all the parachutists with wireless telephones that enable them to exchange information that is learned by the other parachutists. This transfer of information enables parachutists to be rapidly concentrated in the most promising areas of the search space.

The mechanism for implementing this advantageous transfer of information is explained by making three points:

(1) The statistical distribution of what Turing in 1948 called "a combination of genes" and what Holland in 1975 called a "schema" changes from generation to generation in a way that enables the population to become a storehouse of information and experience about the search space.
(2) The mutation operation in genetic programming and the genetic algorithm can be viewed as a special case of the recombination (crossover) operation.
(3) The recombination operation allocates future trials to individuals that contain promising "combination[s] of genes" or schemata.

16.3.2.1 The Changing Mix of Schemata Starting with the first of these three points, consider a run of the genetic algorithm involving a population (of size M) with chromosome strings of length L over an alphabet of K characters. In generation 0, the probability of occurrence of individuals in the population possessing a specified single gene value at a particular single position along the chromosome is a *uniform* value of $1/K$. The set of individuals having specified gene values at particular positions along the chromosome is called a *schema*. The number, $o(H)$, of specified positions in a schema H is referred to as the *specificity* or *order* of the schema. In generation 0, the probability of occurrence of individuals in the population possessing any combination of specified gene values at $o(H)$ particular positions along the chromosome is a *uniform* value of

$$\frac{1}{K^{o(H)}}.$$

This formula applies to every one of the $(K+1)^L$ possible schemata.

As an example, consider a schema of specificity 3 involving chromosomes of length $L = 5$ over an alphabet size of $K = 4$, in which there is a "3" in position 1 of the chromosome, a "2" in position 3, and a "0" in position 5. This schema may be conveniently represented by 3#2#0, where # is a "don't care" symbol. The probability of occurrence of individuals in the population in generation 0 belonging to this schema (and any other schema of specificity 3) is 1/64.

After generation 0, the probability of occurrence of individuals in the population of a run of the genetic algorithm possessing a specified single gene value at a particular single position of the chromosome will increase or decrease (from its original uniform value) depending on whether the specified gene value is helpful or detrimental (respectively) to the individual's fitness (based on average performance measured with necessarily limited sampling). This change in the statistical makeup of the population after generation 0 occurs because of the probabilistic fitness-based selection of individuals to participate in the genetic operations.

However, the changing probability of occurrence of a single gene is not what is important about genetic search. As Turing said in his 1948 paper entitled "Intelligent Machinery" (Turing 1948, page 12; Ince 1992, page 127; Meltzer and Michie 1969, page 23):

> "There is the genetical or evolutionary search by which a *combination of genes* is looked for, the criterion being the survival value." (Emphasis added.)

In *Adaptation in Natural and Artificial Systems* (1975), Holland formalized the notion of "combination[s] of genes" with the idea of a schema. The important point is that, after generation 0, the probability of occurrence of individuals in the population possessing specified "combination[s] of genes" at particular positions in the chromosome will also increase or decrease depending on whether the combination of genes is helpful or detrimental (respectively) to the individual's fitness (based on average performance measured with necessarily limited sampling).

The result is that the probability of occurrence of individuals in the population possessing any of the $(K + 1)^L$ specified combinations of gene values at particular positions in the chromosome departs from the initial uniform random value prevailing in generation 0. The expected number of individuals belonging to any schema varies, by generation, in accordance with Holland's fundamental theorem of genetic algorithms (1975). As the statistical distribution of schemata changes in later generations, the population becomes a storehouse of memory of previous experience about the search space. Specifically, the population contains an increasing number of occurrences of above-average "combination[s] of genes" (that is, schemata). Further analysis is provided in Stephens and Waelbroeck 1997.

As explained in *Foundations of Genetic Programming* (Langdon and Poli 2002), the population in a run of genetic programming similarly stores previous experience with the search space (using a definition of schema appropriate to the rooted point-labeled trees with ordered branches that are used in genetic programming). The population of individuals in a run of genetic programming also contains an increasing number of occurrences of above-average "combination[s] of genes" (schemata).

That is, for both genetic programming and the genetic algorithm, Darwinian selection and reproduction cause the statistical distribution of various combinations of genes to change from generation to generation.

In this book, there is a cornucopia of examples of building blocks that appear in relatively fit individuals in intermediate generations of runs of genetic programming and that later appear in the run's best-of-run individual.

For example, the subtree

```
(SUB_SIGNAL REFERENCE_SIGNAL PLANT_OUTPUT)
```

appears in relatively fit individuals in intermediate generations of many runs of control problems in this book and later becomes part of the best-of-run individuals (e.g., section 9.2.2). This subtree implements negative feedback. Negative feedback is highly useful in controllers. Harold Black invented negative feedback for electrical circuits in 1927 and also obtained patents covering negative feedback for controllers (chapter 14). Negative feedback is an integral part of the PID controller patented by Callender and Allan Stevenson in 1939.

As another example, a subtree that produces an inductor-capacitor rung is a useful building block in creating a multi-rung ladder filter. This example of a reused substructure is discussed in detail in section 16.2.2 and is further illustrated in sections 5.2.2 and 11.2.2.

As yet another example, a subtree that produces a Darlington emitter-follower section is an extremely useful building block in a wide variety of circuits.

As still another example, there is a subtraction of the output of a PID sub-controller and output of a second sub-controller in all three genetically evolved non-PID controllers in chapter 13. This feature is common to all three runs because it is useful in solving the problem at hand.

Darwinian selection not only causes the frequency of beneficial genes and beneficial "combination[s] of genes" (schemata) to increase from generation to generation, it also causes the frequency of detrimental single genes and detrimental "combination[s] of genes" to decrease.

For example, consider the subtree

```
(TWO-GROUND (OUTPUT_0 END END) END)
```

in connection with problems of circuit synthesis. This subtree is a "poison pill" because it wires the circuit's output probe point to ground.

As another example, inductors are not helpful in the problem of creating the topology, sizing, placement, and routing of the 60 dB amplifier described in section 5.3.1.3. Because the inductor-creating function is in the function set, inductors necessarily appear frequently in the randomly created circuits of generation 0. In the larger circuits in the population at generation 0, it is a virtual certainty that the circuit will contain one or more inductors. Even in the smaller circuits of generation 0, an inductor is as likely as any beneficial component. As the run of this problem progressed, the frequency of occurrence of inductors decreased. At the end of the successful run, there were no inductors in the best-of-run circuit.

As yet another example, consider capacitors in computational circuits (sections 15.3.3, 11.3, 4.6). A computational circuit ideally performs the desired computation in the same way at all times (regardless of what input signals it previously processed). However, a non-degenerate functioning capacitor in a computational circuit typically creates memory (in the form of a stored charge). As runs of genetic programming for computational circuit problems progress, the frequency of occurrence of capacitors decreases. At the end of a successful run, there will generally be no non-degenerate functioning capacitors in the best-of-run circuit.

16.3.2.2 Viewing Mutation and Crossover in a Common Framework Before discussing how the crossover operation employs the changing statistical distribution of schemata in the population to differentially allocate future trials to more promising areas of the search space, it is helpful to first examine the nature of the mutation operation.

The mutation operation in genetic methods operates on a single parental point in the search space. In that respect, the mutation operation in genetic methods corresponds to the modification operation in hill climbing, gradient search, and simulated annealing.

The mutation operation in genetic methods selects one parental individual from the population based on fitness and creates one new offspring program from the single parental individual.

In particular, in the genetic algorithm, a point is randomly chosen in the chromosome; the chromosome value at the chosen mutation point is deleted; a new chromosome value is randomly picked from the same set of possibilities in the same manner as chromosomes were randomly created for the initial population of generation 0; and the new random chromosome value is implanted at the chosen mutation point.

In some implementations of the genetic algorithm, more than one mutation point is chosen along the chromosome and, in other implementations, the new chromosome value is required to be different from the old value. However, neither of these variations changes the essence of the argument below.

In the mutation operation in genetic programming (section 2.2.1), a point is randomly chosen in a single parental program tree; the subtree rooted at the chosen mutation point is deleted from the program tree; a new subtree is randomly grown using the available functions and terminals in the same manner as trees were randomly grown for the initial population of generation 0; and the new random subtree is implanted at the chosen mutation point.

To understand the mutation operation in the genetic algorithm and genetic programming, it is helpful to think of the mutation operation and the crossover operation in a common framework. In particular, it is helpful to view mutation as a special case of crossover.

In genetic programming, the subtree that is grown and implanted by the mutation operation (in lieu of the deleted subtree) can be viewed as having come from a randomly chosen subtree from an individual that was created by the same random process as the randomly created individuals in generation 0. That is, a mutation operation in generation i can be viewed as a crossover in which the remainder comes from a *fitness-selected parent* from *the current generation* i and the crossover fragment

Table 16.7 Comparison of crossover and mutation operations

	Remainder comes from a randomly chosen subtree of a	Fragment comes from a randomly chosen subtree of a
Mutation	fitness-selected parent from current generation i	random parent (as if) from generation 0
Crossover	fitness-selected parent from current generation i	fitness-selected parent from current generation i

consists of a randomly chosen subtree from a *randomly selected parent* that *could have come from generation 0.* In saying this, we do not mean that the particular subtree that is implanted by the mutation operation in generation i was itself necessarily present in generation 0 of the particular run involved. Instead, we mean that the subtree that is implanted by the mutation operation in generation i is produced by the same random process that was originally used to produce individuals in generation 0.

Contrast this situation with the crossover operation in genetic programming. In a crossover at generation i, the remainder comes from a *fitness-selected parent* from *the current generation i* (i.e., the same source as the mutation operation). However, the difference is that the crossover fragment consists of a randomly chosen subtree from a *fitness-selected parent* from *the current generation i.*

Viewing the mutation operation as a special case of the crossover operation puts both operations into a common framework. By doing this, it can be seen that the difference between these two operations is the statistical properties of the *source* of the fragment from which the offspring is constructed. The difference between the two operations concerns whether the fragment is randomly chosen versus fitness-selected and whether the fragment comes from the current generation i or is created in the same way as individuals were created for generation 0. Table 16.7 compares the crossover operation and mutation operation in terms of the source of the crossover fragment and remainder used to construct the offspring. Keep in mind that the mutation operation in genetic methods corresponds to the modification operation in hill climbing, gradient search, and simulated annealing.

The foregoing argument is a variation of the argument made in greater detail in chapter 56 of *Genetic Programming III: Darwinian Invention and Problem Solving* (Koza, Bennett, Andre, and Keane 1999a).

16.3.2.3 Taking Advantage of the Information Stored in the Population in Allocating Future Trials Now consider a run of genetic programming or the genetic algorithm at generation i and focus on an individual that has just been selected to participate in the next genetic operation. The fact that the individual was selected indicates that it is likely to be relatively fit. The fact that the run is not yet over suggests that the ideal solution to the problem has yet to be discovered.

In the case of genetic programming, given that the selected individual is relatively fit, it must contain some sequence of functions operating on certain terminals (that is, some subtree) that contributes to the satisfaction of the problem's requirements. Given that the selected individual is not the ideal solution, there must be some difference between the sequence of functions in the selected individual and the sequence of functions in the ideal solution.

When the mutation operation is performed in generation *i*, the subtree that is inserted into the selected individual has the same statistically random profile as subtrees found in generation 0 of the run. Of course, an inserted subtree from generation 0 may, with a certain small nonzero probability, be exactly what is required to convert the selected individual into a better-performing individual (or even the ideal solution). Ignoring such extremely unlikely situations, how can we characterize subtrees created in the same way that generation 0 was created? The answer is that these subtrees usually come from individuals that have very poor fitness (because all the individuals in generation 0 have very poor fitness). That is, these subtrees can be characterized as coming from individuals whose structural features do not, in general, contribute to the satisfaction of the problem's requirements to any significant degree.

To make this more concrete, when the mutation operation is performed in generation *i*, it inserts genetic material in a way that does not take any advantage of information such as

- a subtree implementing negative feedback is extraordinarily useful in controllers,
- a subtree producing one inductor-capacitor rung is a useful building block in creating a multi-rung ladder filter,
- a subtree producing a Darlington emitter-follower section is an extremely useful building block in a wide variety of circuits,
- an inductor is generally not a helpful building block for synthesizing an amplifier,
- a capacitor is generally not a helpful building block for synthesizing a computational circuit, and
- wiring the circuit's output probe point to ground is a very bad idea.

Now consider what happens when crossover is performed in generation *i*. The subtree that is inserted into the selected individual comes from a cohort from generation *i*. Ignoring the extremely unlikely situations in which an inserted subtree is exactly what is required to convert the selected individual into a better-performing individual or ideal solution, how can we characterize subtrees from cohorts from generation *i*? The answer is that these subtrees usually come from individuals that have relatively high fitness (i.e., fitness in the same general neighborhood as the selected individual). That is, these subtrees can be characterized as coming from individuals that contain certain subtrees that contribute to the satisfaction of the problem's requirements. That is, when crossover is performed, both the crossover fragment and remainder come from individuals that already have relatively high fitness and that partially satisfy the problem's requirements. These subtrees can also be characterized as having a tendency to lack detrimental subtrees. In later generations of a run, these subtrees are unlikely to contain poison pills.

If a given problem is of such a nature that a satisfactory solution can be assembled from subtrees coming from individuals that partially satisfy the problem's requirements, then crossover operates in precisely the way necessary to carry out the required assemblage.

Simulated annealing, hill climbing, and gradient search do not deliver this benefit (in either their point-to-point version or parallel versions). In the point-to-point version of these methods, there is no population at all and hence no opportunity to take advantage of information about other individuals. In the parallel version of these methods, the modification operation simply does not take advantage of any information about the structure of the other individuals.

Note that simulated annealing, hill climbing, and gradient search are similar in this respect to a genetic method employing only mutation. As discussed in chapter 56 of *Genetic Programming III* (Koza, Bennett, Andre, and Keane 1999a), the computational effort required to solve a benchmark problem (involving the synthesis of a lowpass filter) using genetic programming without crossover is considerably greater than the computational effort required to solve it with crossover.

We believe that, for the types of problems that we deal with in this book, the search is facilitated by assembling the solution from subtrees coming from individuals that partially satisfy the problem's requirements. That is, we believe that partial solutions to the problem are helpful in constructing better solutions and that a parallel search employing recombination is better suited to the problems of interest than a search employing only a mutation operation acting on a single individual.

As previously mentioned in section 1.1.4, Turing speculated in 1948 that a "genetical or evolutionary search" for a "combination of genes" might be used to achieve machine intelligence. Turing did not envision the use of either a population or recombination in his early papers on this "genetical or evolutionary search." However, he correctly perceived the inadequacy of conducting the search by means of mutation alone. In discussing the potential for speeding up learning by employing a knowledgeable human to pass judgment on each candidate individual, Turing remarked (Turing 1950, page 456; Ince 1992, page 156):

> "One may hope, however, that this process will be more expeditious than evolution. The survival of the fittest is a slow method for measuring advantages. The experimenter, by exercise of intelligence, should be able to speed it up. *Equally important is the fact that he is not restricted to random mutations. If he can trace a cause for some weakness he can probably think of the kind of mutation which will improve it.*" (Emphasis added.)

In other words, Turing envisioned addressing the inadequacies of mutation by *subtracting* detrimental combinations of genes in a *human-directed, problem-specific*, and *domain-specific* way.

In contrast, Holland's genetic algorithm (1975) addresses the inadequacy of mutation by *adding* promising combinations of genes in an *automated, problem-independent*, and *domain-independent* way.

In Holland's approach, no human is required to "trace a cause" or to "exercise… intelligence" because (as the Schema Theorem establishes) the population at generation i of the run automatically contains the promising "combination[s] of genes" (schemata). The population is the storehouse of this information. Given the existence of the population, the insertion of promising combinations of genes is accomplished by the recombination (crossover) operation. The recombination operation inserts a "combination of genes" into the current individual. The inserted "combination of genes" is obtained from a cohort from the current generation i of the run. Moreover, the recombination operation does this automatically (without resort to a knowledgeable human) and it does it in a problem-independent and domain-independent way.

Of course, not all problems are of the nature that a solution can be assembled from subtrees coming from individuals that partially satisfy the problem's requirements. In their paper entitled "No Free Lunch Theorems for Optimization," Wolpert and Macready (1997) make the point that if one search method is better than another at solving a certain subset of problem instances from a universe of problems with a specified finite search space and a specified finite set of cost values, it is correspondingly worse on that universe's remaining problem instances. The NFL theorem applies to all search methods, including the genetic algorithm, genetic programming, any AI search method, hill climbing, gradient search, simulated annealing, and random search. In other words, over the entire universe of all problem instances, a genetic search method relying primarily on crossover (such as genetic programming or the genetic algorithm) is no better than, say, simulated annealing, hill climbing, gradient search, a genetic method employing only mutation, or, for that matter, random search. Further analysis and discussion is provided in Whitley 2000 and in *The Design of Innovation: Lessons from and for Competent Genetic Algorithms* (Goldberg 2002).

Having said that, the fact is that the entire universe of problem instances is not the same as the subset containing problems of interest. For the problems of interest in this book, we believe that there are useful "building blocks" in the intermediate generations prior to the emergence of the problem's solution and that, because of this, population-based methods employing recombination can advantageously assemble the problem's solution from subtrees coming from individuals that partially satisfy the problem's requirements. That is, we believe that a genetic search method (employing a population in conjunction with recombination) is better suited to the problems of interest in this book than a search method employing only a mutation operation.

17

Parallel Implementation and Computer Time

Techniques of genetic and evolutionary computation generally require significant computational resources to solve non-trivial problems. This statement is generally true *a fortiori* when the results are human-competitive.

Fortunately, because of three factors, the computer time necessary to achieve human-competitive results has become increasingly available in recent years.

First, the speed of commercially available single computers continues to double approximately every 18 months in accordance with Moore's law (Moore 1996). This exponential growth in computational power (equivalent to two orders of magnitude per decade) is currently expected to continue (and possibly accelerate) over the next decade. Petaflop computing (Sterling 1998b; Sterling and Foster 1996a, 1996b; Messina, Sterling, and Smith 1999) is expected to be available in the next few years.

Second, techniques of genetic and evolutionary computation in general (and genetic programming in particular) are especially amenable to efficient (and almost effortless) parallelization (Holland 1975; Robertson 1987; Tanese 1989; Goldberg 1989; Stender 1993; Koza and Andre 1995; Andre and Koza 1995, 1996a, 1996b; Cantu-Paz 2000). Techniques for parallelizing runs of genetic programming are detailed in chapter 62 of *Genetic Programming III: Darwinian Invention and Problem Solving* (Koza, Bennett, Andre, and Keane 1999a).

Third, parallel cluster computer systems can be assembled with relative ease in the Beowulf-style using "Commodity Off-The-Shelf" (COTS) hardware (Sterling 1998a; Sterling, Salmon, Becker, and Savarese 1999; Bennett, Koza, Shipman, and Stiffelman 1999). Two different home-built Beowulf systems (described in the next section) were used for the work described in this book.

The combined effect of these three factors can be seen by comparing the runs of genetic programming in this book with those in the 1999 book *Genetic Programming III*. The 14 human-competitive results in table 61.1 of *Genetic Programming III* were produced (between 1994 and 1999) in runs that averaged about 3.5 days and that consumed an average of 1.5 petacycles (1.5×10^{15} cycles). All but one of the human-competitive results in *Genetic Programming III* were produced by a 64-node Parsytec® parallel computer with an 80-MHz PowerPC 601 microprocessor at each processing node (described in detail in section 62.1 of *Genetic Programming III*). This 1995-vintage parallel system operated, as a whole, at a rate of 5.12 GHz.

We are frequently asked if it is necessary to have a large parallel system to produce human-competitive results. At the time of this writing (February 2003), a single 2.67-GHz laptop computer can be purchased for less than $1,000. This $1,000 2003-vintage laptop runs at approximately half the 5.12-GHz clock rate of the *entire* 1995-vintage 64-node parallel computer system that was responsible for virtually all the results (including the 14 human-competitive results) in the 1999 book *Genetic Programming III*. Ignoring the numerous improvements in microprocessor architecture over the past eight years, a run that took 3.5-days to produce a human-competitive result on the 1995-vintage 64-node parallel computer system can be executed in a mere 7 days on a single $1,000 2003-vintage laptop computer. Moreover, if one had a pressing reason to replicate a run from *Genetic Programming III* in precisely 3.5 days (as opposed to 7 days), one could either combine two $1,000 2003-vintage laptops into a small two-node Beowulf-style cluster or wait 18 months for yet another iteration of Moore's law. In either case, human-competitive results can be readily generated today using genetic programming on a single $1,000 2003-vintage laptop computer. Moreover, as time passes, it will be possible to make runs on a single computer that would today occupy a large parallel computer.

17.1 Computer Systems Used for Work in This Book

A parallel computer system was used on 36 of the 37 problems in this book. A desktop Pentium machine was used for the remaining problem (the genetic network problem in chapter 7). This section describes the two parallel computer systems used for the 36 problems.

17.1.1 Alpha Parallel Computer System

A minority of the problems described in this book were run on a home-built Beowulf-style parallel cluster computer system consisting of 70 processors (each containing a 533-MHz DEC Alpha microprocessor and 64 megabytes of RAM). In most cases, a 66-node version of this system was used (with the four remaining nodes being used for code development and testing). A 56-node version of this system was used on one problem in this book. The processors communicate with one another as if they were arranged in a two-dimensional 7×8 or 6×11 toroidal mesh. The 56-node version of this system operates at an overall rate of 2.9×10^{10} Hz and the 66-node version operates at an overall rate of 3.5×10^{10} Hz. The system has a DEC Alpha computer as host. The processors communicate with one another by a 100 megabit-per-second Ethernet. The processors and the host use the Linux operating system. Details of the construction of this system can be found in *Genetic Programming III*. A picture of the Alpha parallel computer and additional information about it are available at `www.genetic-programming.com`.

For both the Alpha parallel computer system and Pentium parallel computer system (described in section 17.1.2), the distributed "island" version of parallel genetic programming was employed. In the "island" version of parallel genetic programming, semi-isolated subpopulations (demes) of individuals reside at each processing node. The time-consuming evaluation of fitness and the relatively quick genetic operations are performed separately on each processing node. Generations are run asynchronously on each processing node. This approach efficiently uses the computational resources of each processing node because no computational resources are wasted in an attempt to synchronize activity on a global basis. At the end of each generation, a small percentage of the node's subpopulation emigrates to each of the adjacent processing nodes. The emigrants wait in a buffer at their destination node until that processing node reaches the end of its current generation. When the Alpha parallel computer was used, a subpopulation of size $Q = 1{,}000$, 2,000, 5,000, 10,000, or 20,000 resided at each of the $D = 56$ or 66 processing nodes. In each generation, $B = 2\%$ (the migration rate) of the node's subpopulation (selected probabilistically on the basis of fitness) emigrated to each of the four adjacent processors (i.e., a total of 8%). We call this symmetric migration strategy "A" in table B.6 in appendix B. The programming aspects of parallel genetic programming are detailed in *Genetic Programming III*.

17.1.2 Pentium Parallel Computer System

A majority of the problems described in this book were run on a home-built Beowulf-style parallel cluster computer system consisting of 1,000 350-MHz Pentium II processors (each accompanied by 64 megabytes of RAM). That is, the system operates at an overall rate of 3.5×10^{11} Hz. This is 10 times faster than the Alpha system (section 17.1.1). The Pentium system has a 350-MHz Pentium II computer as host. The processing nodes communicate with one another by 100 megabit-per-second Ethernet. The processing nodes and the host use the Linux operating system. This Beowulf system was constructed in a manner that is generally similar to that used for the Alpha system. A picture of the 1,000-Pentium system and additional information about it are available at `www.genetic-programming.com`.

When the 1,000-node Pentium parallel computer is used, distributed genetic programming is employed with a population size of $Q = 100$, 500, 2,000, 5,000, or 10,000 at each of the $D = 1{,}000$ processing nodes. Note that when the 1,000-node Pentium parallel computer was used, the population residing at each processing node was often smaller than when the Alpha parallel computer was used. The reason is that much more difficult and time-consuming problems were generally run on the 1,000-node Pentium parallel computer. For example, we used only a subpopulation size of 100 (i.e., a total population of only 100,000) on the runs seeking improved parameterized controllers (chapters 12 and 13). The reason was that the evaluation of fitness for each individual in the population took about one minute (in chapter 12) or two minutes (in chapter 13).

Two processors are housed in each of the 500 physical boxes of the 1,000-Pentium system. As each processor (asynchronously) completes a generation, emigrants from each subpopulation (selected probabilistically based on fitness) are dispatched to each of the four toroidally adjacent processors. For many runs in this book, the migration rate is 2% (but 10% if the toroidally adjacent node is in the same physical box). We call this symmetric migration strategy "B" in table B.6 in appendix B.

For certain runs in this book, a different migration strategy was used. This migration strategy was generally used on the difficult and time-consuming problems for which the subpopulation on each node was very small (e.g., 100 or 500). In this strategy, groups of 40 processors are treated almost as if they were one processing node. The entire 1,000-node system is then treated as a 5×5 system. That is, the 1,000 processors are hierarchically organized such that there are $5 \times 5 = 25$ high-level groups (each containing 40 processors). If the adjacent node belongs to a different group, the migration rate is 2% and emigrants are probabilistically selected based on fitness. If the adjacent node belongs to the same group, emigrants are selected randomly with a 5% migration rate (10% if the adjacent node is in the same physical box). We call this migration strategy "C" in table B.6 in appendix B.

17.2 Computer Time for Problems in This Book

A total of 41 runs are reported in this book for the 36 problems that were run on one of the parallel computer systems described in sections 17.1.1 and 17.1.2.

Table 17.1 tallies the computer time consumed by the 41 runs employing a parallel computer system. For each run in this table, column 2 shows the total population size, M, for all processing nodes. Column 3 shows the generation number i that yielded the best-of-run individual presented in this book. Column 4 shows the product of the total population size and the number of generations $(i+1)$ that were run before encountering the best-of-run individual. Column 5 shows the number of hours consumed in creating the best-of-run individual. Column 6 shows the number of processing nodes on the parallel computer used. In this column, the numbers 56 and 66 refer to the Alpha system and the number 1,000 denotes the 1,000-Pentium system. The 56-node Alpha system operates at 2.9×10^{10} Hz; the 66-node Alpha system operates at 3.5×10^{10} Hz; and the 1,000-Pentium system operates at 3.5×10^{11} Hz. Column 7 shows the number of cycles per second for all the processing nodes of the particular parallel computer used. Column 8 shows the number of petacycles (10^{15} cycles) consumed by the run (totaled over all the processing nodes).

As can be seen in table 17.1, the 41 runs ran for an average elapsed time of 81.9 hours (3.4 days) and consumed an average of 93.4 petacycles.

The number of petacycles consumed by the runs (column 8 of table 17.1) range over about three orders of magnitude—from the 3.0 petacycles consumed by the relatively simple Zobel network problem (section 10.2) to the 871.9 petacycles (almost 10^{18} cycles) consumed by the four-week run that produced the parsimonious (third) non-PID controller described in section 13.2.3.

If available computer speed continues to double approximately every 18 months in accordance with Moore's law (Moore 1996) for the next decade, a computation running for an elapsed time of 81.9 hours (the average from table 17.1) today will require only about 49 minutes (about 1% as much elapsed time) a decade from now. A computation similar to the most time-consuming run in table 17.1 (the four-week run that produced the third non-PID parameterized controller from section 13.2.3) will require only about 7 hours of elapsed time a decade from now. The runs from chapter 15 involving the six post-2000 patented inventions (table 15.14) ran for an average

Table 17.1 Computer time consumed by 41 runs in this book

Problem	Total population	Generation i	$M^*(i+1)$	Total hours	Nodes	Hz	Petacycles	Section in this book
Two-lag plant	66,000	32	2,178,000	44.5	66	3.5×10^{10}	5.6	3.7
Three-lag plant	66,000	31	2,112,000	32	66	3.5×10^{10}	4.0	3.8
Three-lag plant with five second delay	500,000	126	63,500,000	65	1,000	3.5×10^{11}	81.9	3.9
Non-minimal phase plant	66,000	38	2,574,000	46	66	3.5×10^{10}	5.8	3.10
RC circuit with gain greater than two	660,000	927	612,480,000	24	66	3.5×10^{10}	3.0	4.2
Philbrick circuit	660,000	39	26,400,000	7	66	3.5×10^{10}	0.9	4.3
NAND circuit	132,000	17	2,376,000	10	66	3.5×10^{10}	1.3	4.4
Arithmetic logic unit (ALU) circuit	1,320,000	33	44,880,000	170	66	3.5×10^{10}	21.4	4.5
Square root computational circuit	10,000,000	66	670,000,000	52	1,000	3.5×10^{11}	65.5	4.6
Lowpass filter without an explicit test fixture	1,000,000	211	212,000,000	9	1,000	3.5×10^{11}	11.3	4.7
Lowpass filter with layout	1,120,000	138	155,680,000	28	56	2.9×10^{10}	3.0	5.2
Amplifier with layout	10,000,000	101	1,020,000,000	17	1,000	3.5×10^{11}	21.4	5.3
Yagi-Uda antenna	500,000	90	45,500,000	22	1,000	3.5×10^{11}	27.7	6.5
Metabolic pathway for phospholipid cycle	100,000	225	22,600,000	17	1,000	3.5×10^{11}	21.4	8.6
Metabolic pathway for ketone bodies	100,000	97	9,800,000	20	1,000	3.5×10^{11}	25.2	8.7
Three-lag plant with free variable	500,000	42	21,500,000	23.4	1,000	3.5×10^{11}	29.5	9.1
Controller for two families of plants	100,000	217	21,800,000	40.6	1,000	3.5×10^{11}	51.2	9.2
Zobel network problem (two free variables)	50,000,000	16	8,500,000	2	1,000	3.5×10^{11}	3.0	10.2
Third-order elliptic lowpass filter with a free variable for the modular angle	1,000,000	293	294,000,000	71	1,000	3.5×10^{11}	89.5	10.3
Passive lowpass filter with s free variable for the passband boundary	10,000,000	78	790,000,000	45	1,000	3.5×10^{11}	56.6	10.4
Active lowpass filter with variable passband boundary with free variable	5,000,000	101	501,000,000	151	1,000	3.5×10^{11}	190.3	10.5

(Continued)

Table 17.1 (*Continued*)

Problem	Total population	Generation i	M*$(i+1)$	Total hours	Nodes	Hz	Petacycles	Section in this book
Lowpass/highpass filter with free variables	10,000,000	47	480,000,000	45	1,000	3.5×10^{11}	56.6	11.1
Lowpass/highpass filter with variable passband boundary with a free variable	10,000,000	93	940,000,000	84	1,000	3.5×10^{11}	105.8	11.2
Variable quadratic/cubic computational circuit (one free variable)	1,000,000	241	242,000,000	49	1,000	3.5×10^{11}	61.7	11.3
Variable 40–60 dB amplifier (one free variable)	2,000,000	332	666,000,000	50	1,000	3.5×10^{11}	63.0	11.4
Improved PID tuning rules (four free variables)	100,000	76	7,700,000	107	1,000	3.5×10^{11}	134.6	12.3
First non-PID parameterized controller (two free variables)	100,000	88	8,900,000	320	1,000	3.5×10^{11}	403.2	13.2.1
Second non-PID parameterized controller (two free variables)	100,000	38	3,900,000	397	1,000	3.5×10^{11}	500.2	13.2.2
Third non-PID parameterized controller (two free variables)	100,000	199	20,000,000	692	1,000	3.5×10^{11}	871.9	13.2.3
Reinvention of negative feedback	1,000,000	48	49,000,000	7	1,000	3.5×10^{11}	8.8	14.2
Low-voltage balun circuit	5,000,000	97	490,000,000	25	1,000	3.5×10^{11}	31.5	15.4.1
Mixed analog-digital variable capacitor circuit	2,000,000	98	198,000,000	88	1,000	3.5×10^{11}	110.9	15.4.2
High-current load circuit—First run	2,000,000	114	230,000,000	134	1,000	3.5×10^{11}	168.8	15.4.3.1
High-current load circuit—Second run	2,000,000	215	432,000,000	67	1,000	3.5×10^{11}	84.4	15.4.3.2
Voltage-current conversion circuit	5,000,000	109	550,000,000	83	1,000	3.5×10^{11}	104.6	15.4.4
Cubic function generator—First run	5,000,000	182	915,000,000	206	1,000	3.5×10^{11}	259.6	15.4.5
Cubic function generator—Second run	2,000,000	326	654,000,000	135	1,000	3.5×10^{11}	170.1	15.4.5

(*Continued*)

Table 17.1 (*Continued*)

Problem	Total population	Generation i	M*($i+1$)	Total hours	Nodes	Hz	Petacycles	Section in this book
Tunable integrated active filter—First run	2,000,000	70	142,000,000	23	1,000	3.5×10^{11}	29.0	15.4.6.1
Tunable integrated active filter—Second run	2,000,000	50	102,000,000	14	1,000	3.5×10^{11}	17.6	15.4.6.2
Tunable integrated active filter—Third run	2,000,000	38	78,000,000	12	1,000	3.5×10^{11}	15.1	15.4.6.3
Tunable integrated active filter—Fourth run	2,000,000	27	56,000,000	6	1,000	3.5×10^{11}	7.6	15.4.6.4.
Average	**3,530,000**	**128.7**	**267,000,000**	**81.9**			**93.4**	

elapsed time of 80 hours each (very close to the overall average of 81.9 hours shown in table 17.1). If Moore's law is applied to these runs, a similar run would require only about 48 minutes of elapsed time a decade from now.

Comparing the 41 runs from this book (table 17.1) with the 14 runs involving human-competitive results shown in table 61.1 of *Genetic Programming III*, the elapsed time for the two groups of runs was about the same (3.4 days for this book versus 3.5 days for the earlier book). However, the problems treated in this book were qualitatively more substantial than those in *Genetic Programming III*. The average population size for the 41 runs in this book is 3,530,000 whereas it was only 256,000 in *Genetic Programming III*. The 41 runs in this book entailed about 8 times as many fitness evaluations as those in *Genetic Programming III* (257,000,000 for this book versus 32,800,000 for the earlier book). Moreover, because of the greater complexity of the problems in this book, the 41 runs in this book consumed an average of about 62 times more computational resources than those in *Genetic Programming III* (93.4 petacycles for this book versus 1.5 petacycles for the earlier book). The fitness measures for problems in this book are generally far more time-consuming than those in *Genetic Programming III* because the fitness measures in this book contain a considerably greater number of different elements (up to a high of 193 elements for the third non-PID controller in section 13.2.3) and because the elements of the fitness measures in this book usually involve time-domain simulations (which generally consume far more computer time than the frequency-domain simulations and other relatively fast types of simulation and evaluation found in *Genetic Programming III*).

The data in table 17.1 concerning the duration of each run is approximate for several reasons.

First, no adjustment was made in the durational data for the fact that, in practice, no fitness evaluations are actually performed for the percentage (usually about 9%) of the population that is reproduced in each generation or for the (smaller) percentage of the offspring produced by semantics-preserving architecture-altering operations (e.g., subroutine duplication or branch creation).

Second, table 17.1 shows the generation number during which the best-of-run individual was first created on its processing node. However, because of the asynchronous operation of the processing nodes in the parallel system, other processing nodes are, at that same moment, generally at a somewhat earlier or somewhat later generation number.

Third, it is not uncommon for a small fraction of the 1,000 processing nodes to stop during a particular run because of a deeply-buried problem-specific software flaw that becomes exposed by the execution of the run's hundreds of millions of genetic operations.

Fourth, it is not uncommon for some nodes to be out of commission at any given time because of fan failures (the only hardware problem that we regularly encounter).

Fifth, no adjustment was made to reflect the fact that we frequently commandeer 10 nodes of the 1,000-node system in the middle of a run in order to test a future run's software.

In the last three cases, all individuals on the adversely affected processing nodes as well as all individuals that attempt to emigrate to the affected nodes are simply lost.

None of these five factors are material to the statistics concerning the overall operation of a 1,000-node system.

18

Historical Perspective on Moore's Law and the Progression of Qualitatively More Substantial Results Produced by Genetic Programming

This chapter reviews the qualitative nature of the results produced by genetic programming over the 15-year period between 1987 and 2002. This chapter relates to this book's book fourth main point, namely that genetic programming has delivered a progression of qualitatively more substantial results in synchrony with five approximately order-of-magnitude increases in the expenditure of computer time.

18.1 Five Computer Systems Used in 15-Year Period

Table 1.4 lists the five computer systems used to produce our group's reported work on genetic programming in the 15-year period between 1987 and 2002.

The first entry in table 1.4 is a serial computer. The four subsequent entries are parallel computer systems. The presence of four increasingly powerful parallel computer systems in the table reflects the fact that genetic programming has successfully taken advantage of parallelization as a source of increased computational power.

Column 1 of table 1.4 identifies the computer system. Serial Texas Instruments 25-MHz LISP machines were used for the work in *Genetic Programming* (Koza 1992a) and *Genetic Programming II* (Koza 1994a). A 64-node Transtech® transputer parallel machine (processing bytes at 30 MHz) and a 64-node 80-MHz Parsytec PowerPC® parallel machine were used for the work in *Genetic Programming III* (Koza, Bennett, Andre, and Keane 1999a). These two systems are described in section 62.1 of *Genetic Programming III*. A 70-node Alpha system (described in section 17.1.1 of this book) and a 1,000-node Pentium system (described in section 17.1.2 of this book) were used for the work in this book.

Column 2 of table 1.4 shows the time period in which each computer system was used.

Column 3 of table 1.4 shows the overall number of petacycles (10^{15} cycles) per day for the entire system (i.e., all processing nodes, in the case of the parallel machines). Of course, a computer's clock rate is not a precise indicator of machine performance. For one thing, there are differences in the instruction sets, cache designs, memory bandwidth, and a myriad of other hardware details between the various machines. Moreover, differences in operating systems, compilers, programming languages (e.g., LISP, C, Java) and our own genetic programming software also play a role. Nonetheless, for purposes of the qualitative discussion in this chapter, the number of petacycles per day for the entire system is a reasonable proxy for a system's overall performance.

Column 4 shows the speed-up between each successive computer system in table 1.4 (with the base case being the LISP machine used between 1987 and 1994).

Column 5 shows, for each system listed in table 1.4, the speed-up over the LISP machine (the table's first entry).

Column 6 of table 1.4 identifies the book in which the human-competitive results (if any) are reported.

Column 7 of table 1.4 shows the number of human-competitive results (as defined in table 1.2 and itemized in table 1.3 of this book) produced by each computer system. Two of the 28 human-competitive results in the table could arguably be assigned to a different system because they were each run successfully on different computer systems. Specifically, the first run of the transmembrane identification problem was first run on the LISP machine and later run on the Parsytec parallel system. The lowpass filter problem was run at various times on the Parsytec, Alpha, and Pentium parallel systems.

Table 1.4 shows the following:

- There is approximately an order-of-magnitude speed-up (column 4) between each successive computer system in the table. Note that, according to Moore's law (Moore 1996), exponential increases in computer power correspond approximately to *constant* periods of time.
- There is a 13,900-to-1 speed-up (column 5) between the most recent machine (the 1,000-node parallel computer system used for most of the work in this book) and the earliest computer system in the table (the serial LISP machine).
- The slower early machines generated few or no human-competitive results, whereas the faster more recent machines generated numerous human-competitive results.

18.2 Qualitative Nature of Results Produced by the Five Computer Systems

The four successive order-of-magnitude increases in computer power shown in table 1.4 resulted in a succession of qualitatively more substantial results, as described below.

The results reported in the first book, *Genetic Programming: On the Programming of Computers by Means of Natural Selection* (Koza 1992a) were produced entirely on a serial LISP machine. The problems in the 1992 book were generally "toy" problems from the literature of the fields of artificial intelligence and machine learning in the 1980s and early 1990s. This work established the breadth of application of genetic programming by demonstrating that genetic programming could solve a wide variety of problems from a wide variety of fields.

The work in the second book *Genetic Programming II: Automatic Discovery of Reusable Programs* (Koza 1994a) was also produced entirely on a serial LISP machine. The first human-competitive result in table 1.3 (involving the transmembrane segment identification problem) appeared in that book.

The work in the third book, *Genetic Programming III: Darwinian Invention and Problem Solving* (Koza, Bennett, Andre, and Keane 1999a) was produced on the 64-node Transtech transputer parallel machine and the 64-node 80-MHz Parsytec parallel machine. The slower 64-node Transtech transputer machine produced one human-competitive result (involving motifs for families of proteins). Twelve human-competitive results were produced on the faster 64-node 80-MHz Parsytec machine.

The first two human-competitive results (produced by the first two computer systems listed in table 1.4) can be characterized as classification problems operating on one-dimensional discrete data (protein sequences). Both of these results outperformed previous human-created solutions. Neither involved patented inventions.

The first of the 12 human-competitive results produced by the third computer system (the 64-node 80-MHz Parsytec machine) involved cellular automata. In this problem, the evaluation of fitness of each candidate involved processing discrete data in two dimensions (i.e., 149 cells of the one-dimensional cellular automata considered over 600 time steps). This result outperformed previous human-created solutions. Thus, insofar as human-competitive results are concerned, the increased computational resources of the third computer system facilitated an increase, from one to two, in the dimensionality of the discrete data being processed.

Nine of the 12 other human-competitive results produced by the third computer system involved continuous-time electrical signals (analyzed primarily in the frequency domain). Thus, insofar as human-competitive results are concerned, the increased computational resources of the third computer system facilitated the transition in the character of the data from discrete data to continuous signals.

Ten of the 12 human-competitive results (table 1.3) produced by the third computer system infringe, improve upon, or duplicate the functionality of patents that were issued in the first two-thirds of the 20th century. Thus, insofar as human-competitive results are concerned, the increased computational resources of the third computer system facilitated the transition to problems involving previously issued (but non-recent) patents.

Concerning the present book, the first of the two human-competitive results produced by the fourth computer system (the 70-node Alpha parallel machine) involved time-domain simulations (i.e., the rediscovery of the patented PID-D2 controller in section 3.7). Time-domain simulations of continuous signals are, in general, considerably more time-consuming than frequency-domain simulations. Thus, insofar as human-competitive results are concerned, the increased computational resources of the fourth computer system extended the nature of the analysis being performed from the frequency domain to the computationally more intensive time domain.

The second of the two human-competitive results produced by the fourth computer system added the automated creation of placement and routing (layout) to the synthesis of topology and sizing of analog electrical circuits (chapter 5). The increased computational resources of the fourth computer system facilitated the transition between the synthesis of discrete-component circuits to the synthesis of laid-out circuits.

Ten of the 12 human-competitive results in the present book were produced by the fifth computer system listed in table 1.4 (the 1,000-node Pentium II parallel machine).

The fifth computer system made it possible to dramatically increase the number and complexity of the simulations. For example, the fitness evaluation for the run that yielded the parsimonious third non-PID controller (described in section 13.2.3) included eight different time-domain simulations for each of 24 different plants (i.e., 192 separate runs of the simulator). The evaluation of fitness took an average of over two minutes per individual. This evaluation was performed on each of 20,000,000 candidate individuals encountered during the run.

Eleven of the results (in chapters 9, 10, 11, and 13) involve parameterized topologies—that is, the automatic creation, in a single run, of a general (parameterized) solution to a problem in the form of a graphical structure whose nodes or edges represent components and where the parameter values of the components are specified by mathematical expressions containing free variables. The computer time required for the evaluation of fitness of each individual in the population by problems involving parameterized topologies is an integral multiple of the time required for the corresponding problem without free variables. The increased computational resources of the fifth computer system facilitated the transition to problems involving free variables.

All of the 14 human-competitive results produced by the fifth computer system listed in table 1.4 involve patents. Six of these patents were issued after January 1, 2000. Thus, insofar as human-competitive results are concerned, the increased computational resources of the fifth computer system advanced the period represented by the patents from first two-thirds of the 20th century to the 21st century.

Four successive order-of-magnitude increases in computer power are explicitly shown in table 1.4. An additional order-of-magnitude increase was achieved by the expedient of making extraordinarily long runs on the largest machine in the table (the 1,000-node Pentium® II parallel machine). The length of the run that produced the genetically evolved controller described in section 13.2.3 was 28.8 days—almost an order-of-magnitude increase (9.3 times) over the 3.4-day average for other problems described in this book (table 17.1). A patent application was filed for the controller produced by this four-week run (Keane, Koza, and Streeter 2002a). This genetically evolved controller outperforms controllers employing the widely used Ziegler-Nichols tuning rules and the recently developed Åström-Hägglund tuning rules. If the final 9.3-to-1 increase in table 1.5 is counted as an additional speed-up, the overall speed-up between the first and last entries in the table is 130,660-to-1. In other words, the transition from merely duplicating the functionality of 21st-century patented inventions to generating patentable new inventions corresponds to another order-of-magnitude increase in the expenditure of computational resources.

18.3 Effect of Order-of-Magnitude Increases in Computer Power on the Qualitative Nature of the Results Produced by Genetic Programming

Table 1.5 is organized around the five order-of-magnitude increases in the expenditure of computing power resulting from the transitions described in the previous section (section 18.2). Column 4 of table 1.5 characterizes the qualitative nature of the results produced by genetic programming. The table as a whole shows the progression of qualitatively more substantial results produced by genetic programming in terms of five order-of-magnitude increases in the expenditure of computational resources.

Taking a broad view of this succession of order-of-magnitude increases in computer power, we can say the following:

- The Texas Instruments LISP machine (the base case for all subsequent comparative measurements of computer power) produced solutions to several dozen toy problems of the 1980s and early 1990s from the fields of artificial intelligence and machine learning.
- The 9-to-1 increase in computer power associated with the 64-node Transtech transputer parallel machine yielded two human-competitive results that were not patent-related.
- The 22-to-1 increase in computer power associated with the 64-node 80-MHz Parsytec parallel machine yielded numerous human-competitive results involving 20th-century patented inventions.
- The combined 69-to-1 increase in computer power associated with the next two computer systems (the 70-node 533-MHz Alpha parallel machine and 1,000-node 350-MHz Pentium II parallel machine) yielded numerous human-competitive results involving 21st-century patented inventions.
- The 9-to-1 increase in computer power resulting from running the 1,000-node 350-MHz Pentium II parallel machine for 28.8 days (as compared to the average of 3.4 days shown in table 17.1) yielded one of the controllers claimed as a new invention in the patent application filed on July 12, 2002 (Keane, Koza, and Streeter 2002a).

The order-of-magnitude increases in computer power shown in table 1.5 correspond closely (albeit not perfectly) with the following progression of qualitatively more substantial results produced by genetic programming:

- toy problems,
- human-competitive results not related to patented inventions,
- 20th-century patented inventions,
- 21st-century patented inventions, and
- patentable new inventions.

In other words, genetic programming is able to take advantage of the exponentially increasing computational power made available by iterations of Moore's law—that is, it is Mooreware.

These results establish this book's fourth main point: Genetic programming has delivered a progression of qualitatively more substantial results in synchrony with five approximately order-of-magnitude increases in the expenditure of computer time.

19

Conclusion

For the following reasons, we believe that we have established this book's four main points.

19.1 Genetic Programming Now Routinely Delivers High-Return Human-Competitive Machine Intelligence

This book's first main point is: Genetic programming now routinely delivers high-return human-competitive machine intelligence.

We defined the term "human-competitive" by means of the eight criteria in table 1.2. We listed 36 human-competitive results (of which we are aware) that have been produced by genetic programming in table 1.3. We justified the rating of "human-competitive" for new results in this book in sections 3.7.3, 4.3.3, 4.4.3, 5.2.3, 9.2.3, 12.4, 13.3, 14.3, and 15.6.

In section 1.1.2, we mentioned that a method purporting to produce machine intelligence can be measured by its AI ratio—that is, the ratio of that which is delivered by the automated operation of the *artificial* method to the amount of *intelligence* that is supplied by the human applying the method to a particular problem. In numerous places throughout this book, we evaluated (in a qualitative way) the numerator and denominator of the AI ratio for particular genetically evolved results (sections 3.7.4, 3.8.4, 3.9.4, 3.10.4, 4.2.4, 4.3.5, 4.4.5, 4.5.4, 4.6.4, 4.7.4, 5.2.5, 5.3.4, 6.7, 7.4.2, 8.6.2, 8.7.2, 9.1.4, 9.2.5, 10.2.4, 10.3.4, 10.4.4, 10.5.4, 11.1.4, 11.2.4, 12.6, 13.5, 14.5, and 15.8). The numerator of the AI ratio for all 41 results presented in this book is generally high because the result is either human-competitive or, at the least, very substantial. The denominator of the AI ratio for all 41 results is generally low because the human user pre-supplies only *de minimus* knowledge, information, analysis, and intelligence prior to launching a run of genetic programming. Thus, it can be said that genetic programming has produced numerous "high-return" results.

We mentioned in section 1.1.3 that an automated problem-solving method can be measured in terms of its routineness. Routineness means that the method is general and, additionally, that relatively little human effort is required to get the method to successfully handle new problems within a particular domain and to successfully handle new problems from a different domain. In numerous places throughout this book, we demonstrated

- the routineness of the transition from one problem to another problem in the same domain (sections 3.8.3, 3.9.3, 3.10.3, 4.3.4. 4.4.4, 4.5.3, 4.6.3, 4.7.3, 5.3.3, 8.7.1, 9.2.4, 10.3.3. 10.4.3. 10.5.3, 11.2.3, 13.4. 14.4. and 15.7),
- the routineness of the transition from one domain to the next (sections 4.2.3, 5.2.4, 6.6, 7.4.1, 8.6.1, and 12.5),
- the routineness of the transition from a non-parameterized version of a problem to a parameterized version (sections 9.1.3 and 10.2.3), and
- the routineness of the transition from a problem involving parameterized topologies without conditional developmental operators to a problem involving parameterized topologies with them (section 11.1.3).

19.2 Genetic Programming Is an Automated Invention Machine

This book's second main point is: Genetic programming is an automated invention machine.

There are now 23 instances (shown in tables C.1 and C.2 in appendix C) where genetic programming has duplicated the functionality of a previously patented invention, infringed a previously issued patent, or created a patentable new invention. Specifically, there are 15 instances where genetic programming has created an entity that either infringes or duplicates the functionality of a previously patented 20th-century invention, six instances where genetic programming has done the same with respect to an invention patented after January 1, 2000 (chapter 15), and two instances where genetic programming has created a patentable new invention (chapters 12 and 13). The two new inventions are general-purpose controllers that outperform controllers employing the widely used Ziegler-Nichols tuning rules and the recently developed Åström-Hägglund tuning rules. As discussed in section 1.2, the inventions generated by genetic programming exhibit the kind of creativity and illogical discontinuity from previous human work that is required to obtain a patent. There are numerous examples in this book where genetic programming has unearthed a novel and creative solution to a problem.

As available computer power continues to increase in accordance with Moore's Law, we anticipate that the use of genetic programming as an automated invention machine will become more and more common and that increasing complex entities will be invented by means of genetic programming.

19.3 Genetic Programming Can Automatically Create Parameterized Topologies

This book's third main point is: Genetic programming can automatically create a general solution to a problem in the form of a parameterized topology.

Eleven problems demonstrate that genetic programming can automatically create, in a single run, a general (parameterized) solution to a problem in the form of a graphical structure whose nodes or edges represent components and where the parameter values of the components are specified by mathematical expressions containing free variables. Chapters 9 and 13 demonstrate the ability of genetic programming to create parameterized topologies for controllers. Chapters 10 and 11 demonstrate this ability for electrical circuits.

19.4 Genetic Programming Has Delivered Qualitatively More Substantial Results in Synchrony with Increasing Computer Power

This book's fourth main point is: Genetic programming has delivered a progression of qualitatively more substantial results in synchrony with five approximately order-of-magnitude increases in the expenditure of computer time.

In section 1.4, we raised a series of questions concerning any proposed approach to machine intelligence.

- Is the method formulated with sufficient precision to enable it to be implemented (or is it vagueware)?
- Has the method been successfully demonstrated on a specific single problem (or is it promiseware)?
  - Was the method applied to a difficult demonstrative problem (or is it toyware)?
  - Did the method top out after succeeding on a single demonstrative problem?
- Has the method solved multiple problems (or is it soloware)?
  - Are the multiple problems difficult?
  - Did the method top out at this stage?
- Has the method solved problems from multiple domains (or is it nicheware)?
  - Are the domains difficult?
  - Did the method top out at this stage?
- Were the results human-competitive?
- Can the method profitably take advantage of the increased computational power available by means of parallel processing (or is it serialware)?
- Or, is the method Mooreware—able to take advantage of the exponentially increasing computational power made available by the relentless iteration of Moore's law?

Genetic Programming: On the Programming of Computers by Means of Natural Selection (Koza 1992a) demonstrated that genetic programming is not vagueware, promiseware, soloware, or nicheware.

The numerous human-competitive results discussed in this book establish that it is not toyware.

The fact that genetic programming has been successfully run on multiple parallel computer systems shows that it is not serialware.

In chapter 18, we noted that five successive order-of-magnitude increases in computer power (shown in table 1.5) correspond closely (albeit not perfectly) with the following progression of qualitatively more substantial results produced by genetic programming:

- toy problems,
- human-competitive results not related to patented inventions,
- 20th-century patented inventions,
- 21st-century patented inventions, and
- patentable new inventions.

In other words, genetic programming is able to take advantage of the exponentially increasing computational power made available by iterations of Moore's law—that is, it is Mooreware.

As far as we know, genetic programming is, at the present time, unique among methods of artificial intelligence and machine learning in terms of its duplication of numerous previously patented results, unique in its generation of patentable new results, unique in the breadth and depth of problems solved, unique in its demonstrated ability to produce parameterized topologies, and unique in its delivery of routine high-return, human-competitive machine intelligence.

Appendix A: Functions and Terminals

Table A.1 shows the name and a reference for functions and terminals (i.e., functions of arity 0) used in this book.

Table A.1 Functions and terminals

Name	Arity	Full name	Section in this book
+	2	Numeric addition	2.2.1
−	2	Numeric subtraction	2.2.1
*	2	Numeric multiplication	2.2.1
%	2	Protected numeric division	2.2.1
>	2	Greater-than comparative function	7.2.1
<	2	Less-than comparative function	7.2.1
2N3904	0	2N3904 transistor model	10.5.1.4
2N3906	0	2N3906 transistor model	10.5.1.4
ABS_SIGNAL	1	Insert an absolute value block into a controller	3.5.1
ADD_3_SIGNAL	3	Insert a 3-input addition block into a controller	3.5.1
ADD_SIGNAL	2	Insert an addition block into a controller	3.5.1
ADF	Variable	Automatically defined function	3.5.4
AH	0	Connect to Åström-Hägglund controller	13.1.2
B_C_E, …, E_C_B	0	Transistor lead permutation terminals	10.5.1.4
BIFURCATE_POSITIVE	0	Bifurcate positive end of wire in inserting transistor	10.5.1.4

(*Continued*)

Table A.1 (*Continued*)

Name	Arity	Full name	Section in this book
BIFURCATE_NEGATIVE	0	Bifurcate negative end of wire in inserting transistor	10.5.1.4
C	2	Capacitor-inserting function	4.2.1.3
C-LAYOUT	2	Geographically-aware capacitor-inserting function	5.2.1.3
C_NEW	1	Create a capacitor	4.7.1.3
C00002	0	Concentration of cofactor ATP	8.5.3
C00116	0	Concentration of glycerol	8.5.3
C00162	0	Concentration of fatty acid	8.5.3
C00165	0	Concentration of diacyl-glycerol	8.5.3
CAP_LEVEL	0	Represents gene expression level in a genetic network	7.3.3
CONSTANT_0	0	Connect to time-domain signal with constant 0 value	3.5.2
CONTROLLER_OUTPUT	0	Connect to a controller's output	3.5.2
CR_1_1	4	1-subtrate, 1-product reactor block	8.4.1.1
CR_1_2	5	1-subtrate, 2-product reactor block	8.4.1.1
CR_2_1	5	2-subtrate, 1-product reactor block	8.4.1.1
CR_2_2	6	2-subtrate, 2-product reactor block	8.4.1.1
DELAY	1	Insert a delay block into a controller	3.5.1
DIFFERENTIAL_INPUT_INTEGRATOR	2	Insert a differential input integrator block into a controller	3.5.1
DIFFERENTIATOR	1	Insert a differentiator block into a controller	3.5.1
DIV_NUMERIC	2	Protected numeric division for controllers	9.1.1.3
DIV_SIGNAL	2	Insert a division block into a controller	3.5.1
DOWN_OR_RIGHT	0	Parallel-divide down or right	10.1.2
DRAW	2	Drawing turtle function	6.3.1
END	0	Development-ending function	4.2.1.4
F	0	Free variable representing passband boundary	10.3.1.3
F1	0	Free variable representing passband boundary	11.1.1.3

(*Continued*)

Table A.1 (*Continued*)

Name	Arity	Full name	Section in this book
F2	0	Free variable representing stopband boundary	11.1.1.3
FIRST_PRODUCT	1	Selects the first product produced by a chemical reaction	8.4.1.1
FLIP	1	Flip (polarity-reversing) function	4.2.1.3
GAIN	2	Insert a gain block into a controller	3.5.1
GLUCOSE_LEVEL	0	Represents glucose level in a genetic network	7.3.3
HFA3046	0	HFA3046 transistor model	15.3.3
HFA3128	0	HFA3128 transistor model	15.3.3
IF	3	If operator	7.2.1
IF_POSITIVE	3	Insert an if-positive block into a controller	3.5.1
IFGTZ_DEVELOPMENTAL	3	If greater than zero developmental operator	11.1
INPUT_0	2	Connect to a circuit's input	4.7.1.1
INTEGRATOR	1	Insert an integrator block into a controller	3.5.1
INT1, INT2, INT3	0	Intermediate substance terminals	8.5.3
INVERTER	1	Insert an inverter block into a controller	3.5.1
IRF511	0	IRF511 transistor model	15.3.3
IRF9230	0	IRF9230 transistor model	15.3.3
IRFZ44	0	IRFZ44 transistor model	15.3.3
KU	0	Free variable representing plant's ultimate gain	9.2.1.2
L	0	Free variable representing plant's dead time	9.2.1.2
L	2	Inductor-inserting function	10.3.1.4
L1	0	Free variable representing inductance of a specific inductor	10.1.2.4
L_NEW	1	Create an inductor	4.7.1.3
L-LAYOUT	2	Geographically aware inductor-inserting function	5.2.1.3
LACTOSE_LEVEL	0	Represents lactose level in a genetic circuit	7.3.3
LAG	2	Insert a lag block into a controller	3.5.1
LAG2	3	Insert a second order lag block into a controller	3.5.1
LANDMARK	1	Landmark turtle function	6.3.1
LEAD	2	Insert a lead block into a controller	3.5.1
LEFT_1, ..., LEFT_4	0	Takeoff point reference terminals	13.1.2

(*Continued*)

Table A.1 (*Continued*)

Name	Arity	Full name	Section in this book
LIMITER	3	Insert a limiter block into a controller	3.5.1
LOG_F	0	Free variable representing logarithm of passband boundary for filter	10.5.1.4
MTP50P03HDL	0	MTP50P03HDL transistor model	15.3.3
MULT_SIGNAL	2	Insert a multiplication block into a controller	3.5.1
NODE	2	Connect distant points in a circuit	10.1.1
NOP	1	No-operation function	4.2.1.3
OUTPUT_0	2	Connect to a circuit's output	4.7.1.1
PAIR_CONNECT_0	3	Connects a pair of distant points in a circuit	4.2.1.3
PAIR_CONNECT_1	3	Connects a pair of distant points in a circuit	4.2.1.3
PARALLEL-LAYOUT-LEFT	4	Geographically aware parallel division function	5.2.1.3
PARALLEL-LAYOUT-RIGHT	4	Geographically aware parallel division function	5.2.1.3
PARALLEL0	4	Parallel-division function, version 0	4.2.1.3
PARALLEL1	4	Parallel-division function, version 1	4.2.1.3
PARALLEL_NEW	4	New version of parallel division function	10.1.2
PLANT_OUTPUT	0	Connect to a controller's plant output	3.5.2
POW	2	Power function	12.2.3
PROGN	Variable	Connective function	6.3.1
Q	6	Transistor-inserting function	10.1.5
Q_DIODE_NPN	1	Insert an NPN diode into a circuit	4.4.1.4
Q_DIODE_PNP	1	Insert a PNP diode into a circuit	4.4.1.4
Q_GND_EMIT_NPN	1	Insert an NPN transistor whose emitter is connected to ground	4.4.1.4
Q_GND_EMIT_PNP	1	Insert a PNP transistor whose emitter is connected to ground	4.4.1.4
Q_POS5V_COLL_NPN	1	Insert an NPN transistor whose collector is connected to a 5 V power supply	4.4.1.4
Q_POS5V_EMIT_PNP	1	Insert a PNP transistor whose emitter is connected to a 5 V power supply	4.4.1.4
Q_THREE_NPN0, ..., Q_THREE_NPN11	3	Insert an NPN transistor into a circuit	4.4.1.4

(*Continued*)

Table A.1 (*Continued*)

Name	Arity	Full name	Section in this book
Q_THREE_PNP0, ..., Q_THREE_PNP11	3	Insert a PNP transistor into a circuit	4.4.1.4
R	2	Resistor-inserting function	4.2.1.3
R1	0	Free variable representing inductance of a specific resistor	10.1.2.4
$\Re$	0	Random floating-point constants in specified range	3.5.5.1
$\Re_{\text{integer}}$	0	Random integer constants between 0 and 99	6.3.2
$\Re_{\text{p}}$	0	Random perturbable floating-point value in specified range	3.5.5.2
$\Re_{\text{real}}$	0	Random floating-point constants between 0.0 and 1.0	6.3.2
R_NEW	1	Create a resistor	4.7.1.3
REFERENCE_SIGNAL	0	Connect to a controller's reference signal	3.5.2
REPEAT	2	Repeat turtle function	6.3.1
REPRESSOR_LEVEL	0	Represents repressor level in a genetic network	7.3.3
RETAINING_ THREE_GROUND0	3	Connect to ground	4.3.1.4
RETAINING_ THREE_GROUND1	3	Connect to ground	4.3.1.4
RETAINING_ THREE_POS5V_0	3	Connect to 5 V power supply	4.4.1.4
RETAINING_ THREE_POS5V_0	3	Connect to 5 V power supply	4.4.1.4
REXP	1	Protected exponential function	9.2.1.3
RIGHT_1, ..., RIGHT_4	0	Takeoff point reference terminals	13.1.2
RLOG	1	Protected natural logarithm function	9.2.1.3
SAFE_CUT	0	Safe-cut developmental function	4.2.1.4
SECOND_PRODUCT	1	Selects the second product produced by a chemical reaction	8.4.1.1
SERIES	3	Series division function	4.2.1.3
SERIES-LAYOUT	3	Geographically aware series division function	5.2.1.3
SIN_THETA	0	Free variable representing sine of a filter's modular angle	10.2.1.4
SUB_SIGNAL	2	Insert a subtraction block into a controller	3.5.1
TAKEOFF	1	New takeoff point function	13.1.1
TARGET	0	Free variable representing identity of target circuit (for squaring/ cubing problem and 40/60 dB amplifier problem)	11.3.1.4
TAU	0	Free variable representing plant's time constant	9.1.1.2

(*Continued*)

Table A.1 (*Continued*)

Name	Arity	Full name	Section in this book
THREE_GROUND	3	Three-argument connection to ground	10.2.1.3
TU	0	Free variable representing plant's ultimate period	9.2.1.2
TURN-RIGHT	1	Turn right turtle function	6.3.1
TR	0	Free variable representing plant's rise time	9.2.1.2
TWO_GROUND	2	Two-argument connection to ground	10.2.1.3
TWO_LEAD	3	Two-leaded-component-inserting function	10.1.4
TWO_NEG15V	2	−15 Volt reference voltage source function	11.3.1.3
TWO_POS1V	2	+1 Volt reference voltage source function	11.3.1.3
TWO_POS2V	2	+2 Volt reference voltage source function	11.3.1.3
TWO_POS5V	2	+5 Volt reference voltage source function	11.3.1.3
TWO_POS15V	2	+15 Volt reference voltage source function	11.3.1.3
UP_OR_LEFT	0	Parallel-divide up or left	10.1.2
VIA-TO-GROUND-NEG-LEFT-LAYOUT	3	Geographically aware via to ground function	5.2.1.3
VIA-TO-GROUND-NEG-RIGHT-LAYOUT	3	Geographically aware via to ground function	5.2.1.3
VIA-TO-GROUND-POS-T-LAYOUT	3	Geographically aware via to ground function	5.2.1.3
VIA-TO-GROUND-POS-RIGHT-LAYOUT	3	Geographically aware via to ground function	5.2.1.3
X	0	Represents value on the *X*-axis for symbolic regression problems	2.2.1

Appendix B: Control Parameters

Broadly speaking, we have used substantially the same (almost certainly non-optimal) choices of control parameters from problem to problem over a period of years. Although particular problems in this book could possibly be solved more efficiently by means of a different choice of control parameters, we believe that our policy of substantial consistency in the choice of control parameters helps the reader eliminate superficial concerns that the demonstrated success of genetic programming depends on shrewd or fortuitous choices of the control parameters. That is, the results produced by genetic programming are not the fruit of intricate and astute tailoring of control parameters to a particular problem.

This book continues the policy of *Genetic Programming* (Koza 1992a), *Genetic Programming II* (Koza 1994a), and *Genetic Programming III* (Koza, Bennett, Andre, and Keane 1999a) of using a fixed set of default values for most of the minor control parameters throughout the book. Thus, unless otherwise indicated for a specific problem, the values of all control parameters for all problems in this book are the values specified in *Genetic Programming III* (Koza, Bennett, Andre, and Keane 1999a).

In addition to the values of the control parameters inherited from *Genetic Programming III*, tables B.1 and B.2 present the percentages of genetic operations that are used on or before and after generation 5 of runs of the 10 problems in this book employing the architecture-altering operations. Tables B.3 and B.4 present this information for seven additional problems employing the architecture-altering operations. B.5 presents this information for all the remaining problems. Because the architecture-altering operations were not used on any of the problems in table B.5, one table of percentages is applicable to all generations.

Table B.6 shows the migration strategy (A, B, or C) used for the 41 runs in this book employing a parallel computer system. Strategy A is described in section 17.1.1 and strategies B and C are described in section 17.1.2.

The information in these tables can be summarized as follows: Not noteworthy. Many of the differences simply reflect adjustments necessary to accommodate the presence or absence of the architecture-altering operations. Other differences (such as the migration strategy) arise from some another choice (e.g., the computer system or the population size). Most of the other differences reflect the chronology of our work and the evolution of our thinking on how to maximize the efficiency runs of genetic programming in general (as opposed to any special exigency of a particular problem). At least one of the differences (where the column does not add up to 100%) reflects a typographical error made during programming.

Table B.1 Percentages of operations before generation 5 for 10 problems

	Two-lag plant (section 3.7)	Three-lag plant (section 3.8)	Three-lag plant with five-second delay (section 3.9), non-minimal phase plant (section 3.10), and three-lag plant with a free variable (section 9.1)	Controller for two families of plants (section 9.2)	RC circuit with gain greater than two (section 4.2), Philbrick circuit (section 4.3)	ALU circuit (section 4.5)	Amplifier with layout (section 5.3)
Crossover on internal points	68%	0%	0%	0%	0%	0%	0%
Crossover on terminals	10%	0%	0%	0%	0%	0%	0%
Crossover on non-numeric internal points	0%	45%	45%	46%	50%	50%	50%
Crossover on non-numeric terminals	0%	9%	9%	9%	10%	10%	10%
Crossover on numeric terminals	0%	5%	5%	9%	10%	10%	10%
Reproduction	10%	9%	9%	9%	9%	9%	9%
Subtree mutation	1%	1%	1%	1%	1%	1%	1%
Numeric constant mutation	0%	20%	20%	20%	20%	20%	20%
Subroutine duplication	5%	5%	5%	2%	0%	0%	0%
Argument duplication	0%	0%	0%	0%	0%	0%	0%

Subroutine deletion	1%	1%	1%	2%	0%	0%	0%
Argument deletion	0%	0%	0%	0%	0%	0%	0%
Subroutine creation	5%	5%	5%	2%	0%	0%	0%
Argument creation	0%	0%	0%	0%	0%	0%	0%
Iteration creation	0%	0%	0%	0%	0%	0%	0%
Loop creation	0%	0%	0%	0%	0%	0%	0%
Recursion creation	0%	0%	0%	0%	0%	0%	0%
Storage creation	0%	0%	0%	0%	0%	0%	0%
Max. points for RPB	150	150	150	150	800	300	300
Max. points for ADF	100	100	100	100	NA	NA	NA

Table B.2 Percentages of operations after generation 5 for 10 problems

	Two-lag plant (section 3.7)	Three-lag plant (section 3.8)	Three-lag plant with five-second delay (section 3.9), non-minimal phase plant (section 3.10), and three-lag plant with a free variable (section 9.1)	Controller for two families of plants (section 9.2)	RC circuit with gain greater than two (section 4.2), Philbrick circuit (section 4.3)	ALU circuit (section 4.5)	Amplifier with layout (section 5.3)
Crossover on internal points	76%	0%	0%	0%	0%	0%	0%
Crossover on terminals	10%	0%	0%	0%	0%	0%	0%
Crossover on non-numeric internal points	0%	47%	49%	46%	50%	50%	50%
Crossover on non-numeric terminals	0%	9%	9%	9%	10%	10%	10%
Crossover on numeric terminals	0%	9%	9%	9%	10%	10%	10%
Reproduction	10%	9%	9%	9%	9%	9%	9%
Subtree mutation	1%	1%	1%	1%	1%	1%	1%
Numeric constant mutation	0%	20%	20%	20%	20%	20%	20%
Subroutine duplication	1%	1%	1%	2%	0%	0%	0%
Argument duplication	0%	0%	0%	0%	0%	0%	0%

Subroutine deletion	1%	1%	1%	2%	0%	0%	0%
Argument deletion	0%	0%	0%	0%	0%	0%	0%
Subroutine creation	1%	1%	1%	2%	0%	0%	0%
Argument creation	0%	0%	0%	0%	0%	0%	0%
Iteration creation	0%	0%	0%	0%	0%	0%	0%
Loop creation	0%	0%	0%	0%	0%	0%	0%
Recursion creation	0%	0%	0%	0%	0%	0%	0%
Storage creation	0%	0%	0%	0%	0%	0%	0%
Max. points for RPB	150	150	150	150	800	300	300
Max. points for ADF	100	100	100	100	NA	NA	NA

Table B.3 Percentages of operations before generation 5 for seven problems

	Analog NAND circuit (section 4.4)	Lowpass filter with layout (section 5.2)	Square root computational circuit (section 4.6)	Passive lowpass filter with variable passband boundary (section 10.4)	Lowpass/highpass filter with free variables (section 11.1), Lowpass/highpass filter with variable passband boundary with a free variable (section 11.2)	Yagi-Uda antennas problem (section 6.5)
Crossover on internal points	79%	79%	0%	0%	0%	60%
Crossover on terminals	10%	10%	0%	0%	0%	10%
Crossover on non-numeric internal points	0%	0%	60%	49%	49%	0%
Crossover on non-numeric terminals	0%	0%	10%	10%	10%	0%
Crossover on numeric terminals	0%	0%	0%	0%	0%	0%
Reproduction	10%	10%	9%	9%	9%	9%
Subtree mutation	1%	1%	1%	1%	1%	1%
Numeric constant mutation	0%	0%	20%	20%	20%	20%
Subroutine duplication	0%	0%	0%	5%	5%	0%
Argument duplication	0%	0%	0%	0%	0%	0%

Subroutine deletion	0%	0%	0%	1%	1%	0%
Argument deletion	0%	0%	0%	0%	0%	0%
Subroutine creation	0%	0%	0%	5%	5%	0%
Argument creation	0%	0%	0%	0%	0%	0%
Iteration creation	0%	0%	0%	0%	0%	0%
Loop creation	0%	0%	0%	0%	0%	0%
Recursion creation	0%	0%	0%	0%	0%	0%
Storage creation	0%	0%	0%	0%	0%	0%
Max. points for RPB	300	600	500	300	??	500
Max. points for ADF	NA	NA	NA	100	??	NA

Table B.4 Percentages of operations after generation 5 for seven problems

	Analog NAND circuit (section 4.4)	Lowpass filter with layout (section 5.2)	Square root computational circuit (section 4.6)	Passive lowpass filter with variable passband boundary (section 10.4)	Lowpass/highpass filter with free variables (section 11.1), Lowpass/highpass filter with variable passband boundary with a free variable (section 11.2)	Yagi-Uda antennas problem (section 6.5)
Crossover on internal points	79%	79%	0%	0%	0%	60%
Crossover on terminals	10%	10%	0%	0%	0%	10%
Crossover on non-numeric internal points	0%	0%	60%	57%	57.50%	0%
Crossover on non-numeric terminals	0%	0%	10%	10%	10%	0%
Crossover on numeric terminals	0%	0%	0%	0%	0%	0%
Reproduction	10%	10%	9%	9%	9%	9%
Subtree mutation	1%	1%	1%	1%	1%	1%
Numeric constant mutation	0%	0%	20%	20%	20%	20%
Subroutine duplication	0%	0%	0%	1%	1%	0%
Argument duplication	0%	0%	0%	0%	0%	0%

Subroutine deletion	0%	0%	0%	1%	0.50%	0%
Argument deletion	0%	0%	0%	0%	0%	0%
Subroutine creation	0%	0%	0%	1%	1%	0%
Argument creation	0%	0%	0%	0%	0%	0%
Iteration creation	0%	0%	0%	0%	0%	0%
Loop creation	0%	0%	0%	0%	0%	0%
Recursion creation	0%	0%	0%	0%	0%	0%
Storage creation	0%	0%	0%	0%	0%	0%
Max. points for RPB	300	600	500	300	??	500
Max. points for ADF	NA	NA	NA	100	??	NA

Table B.5 Percentages of operations for the problems not using the architecture-altering operations

	Genetic network for *lac* operon (chapter 7)	Metabolic pathway for phospholipid cycle (section 8.6), Metabolic pathway for ketone bodies (section 8.7)	Lowpass filter without an explicit test fixture (section 4.7), Zobel network problem (two free variables (section 10.2), Third-order elliptic lowpass filter with variable modular angle (section 10.3), Active lowpass filter with variable passband boundary with free variable (section 10.5), Variable Quadratic/ cubic computational circuit (section 11.3), Variable 40–60 dB amplifier (section 11.4), Improved PID tuning rules (four freevariables) (section 12.3), all three runs of non-PID parameterized controller (with four free variables), Problem of reducing amplifier distortion by means of negative feedback (section 14.2), all 11 runs of the six problems involving post-2000 patented inventions in chapter 15
Crossover on internal points	63%	58.50%	63%
Crossover on terminals	7%	6.50%	7%
Crossover on non-numeric internal points	0%	0%	0%
Crossover on non-numeric terminals	0%	0%	0%
Crossover on numeric terminals	0%	0%	0%
Reproduction	9%	9%	9%
Subtree mutation	1%	1%	1%
Numeric constant mutation	20%	20%	20%
Subroutine duplication	0%	0%	0%
Argument duplication	0%	0%	0%
Subroutine deletion	0%	0%	0%
Argument deletion	0%	0%	0%
Subroutine creation	0%	0%	0%
Argument creation	0%	0%	0%
Iteration creation	0%	0%	0%
Loop creation	0%	0%	0%
Recursion creation	0%	0%	0%
Storage creation	0%	0%	0%
Max. points for RPB	1,000	500	500
Max. points for ADF	NA	NA	NA

Table B.6 Migration strategy used on the 41 runs

Migration strategy	Problems
A	Two-lag plant (section 3.7), Three-lag plant (section 3.8), Non-minimal phase plant (section 3.10), RC circuit with gain greater than two (section 4.2), Philbrick circuit (section 4.3), NAND circuit (section 4.4), Arithmetic logic unit (ALU) circuit (section 4.5), Layout of lowpass filter (section 5.2)
B	Three-lag plant with five second delay (section 3.9), Lowpass/highpass filter with free variable (section 11.1), Lowpass/highpass filter with variable passband boundary with a free variable (section 11.2)
C	Three-lag plant with free variable (section 9.1), Controller for two families of plants (section 9.2), Square root computational circuit (section 4.6), Lowpass filter without an explicit test fixture (section 4.7), Layout of amplifier (section 5.3), Yagi–Uda antenna (section 6.5), Metabolic pathway for phospholipid cycle (section 8.6), Metabolic pathway for ketone bodies (section 8.7), Zobel network problem (two free variables) (section 10.2), Third-order elliptic lowpass filter with variable modular angle (section 10.3), Passive lowpass filter with variable passband boundary (section 10.4), Active lowpass filter with variable passband boundary with free variable (section 10.5), Variable Quadratic/cubic computational circuit (section 11.3), Variable 40–60 dB amplifier (section 11.4), Improved PID tuning rules (four free variables) (section 12.3), First non-PID parameterized controller (two free variables) (section 13.2.1), Second non-PID parameterized controller (two free variables) (section13.2.2), Third non-PID parameterized controller (two free variables) (section 13.2.3), Negative feedback (section 14.2), Low-voltage balun circuit (section 15.4.1), Mixed analog-digital variable capacitor circuit (section 15.4.2), High-current load circuit (sections 15.4.3.1 and 15.4.3.2), Voltage–current conversion circuit (section 15.4.4), Cubic function generator (sections 15.4.5 and 15.4.5), Tunable integrated active filter (sections 15.4.6.1, 15.4.6.2. 15.4.6.3, and 15.4.6.4)

The current published literature in the fields of genetic algorithms and genetic programming provides a considerable amount of conflicting advice on the topic of choosing control parameters. Moreover, it is rarely clear how one should go about translating a problem's high-level statement in English into particular choices of control parameters. Thus, it is not clear how a user of genetic programming would go about shrewdly or optimally tailoring the control parameters for a particular problem even if that were the intent. Of course, if one has sufficient understanding of the dynamics of the evolutionary process and sufficient insight concerning the fitness landscape of an unseen new problem, there is nothing wrong with doing such tailoring if the primary objective is to solve the particular problem at hand.

Appendix C: Patented or Patentable Inventions Generated by Genetic Programming

Tables C.1 and C.2 provide additional information on the 23 patent-related results (of the 36 human-competitive results produced by genetic programming in table 1.3).

Table C.1 provides additional information on the 21 results that relate to previously patented inventions. Eleven of the 21 results in table C.1 infringe previously issued patents and 10 duplicate the functionality of previously patented inventions in a non-infringing way. The first 10 entries in table C.1 refer to problems that were solved in *Genetic Programming III: Darwinian Invention and Problem Solving* (Koza, Bennett, Andre, and Keane 1999a). The last 11 entries in table C.1 are described in this book. The last six entries in table C.1 relate to patents for analog circuits that were issued after January 1, 2000.

Four of the 21 entries in the body of table C.1 are marked "See text." These entries relate to groups of previously patented inventions (as opposed to single patents) that are described in detail in *Genetic Programming III: Darwinian Invention and Problem Solving* (Koza, Bennett, Andre, and Keane 1999a). Concerning computational circuits (the 7th entry in table C.1), dozens of different computational circuits have been patented, including, for example, square root circuits (Newbold 1962) and logarithmic circuits (Green 1958). Concerning electronic thermometers (the 8th entry in table C.1), at least two dozen temperature-sensing circuits have been patented, including, for example, ones by Haeusler (1976) and Massey (1970). Concerning voltage reference circuits (the 9th entry in table C.1), Robert C. Dobkin and Robert J. Widlar of National Semiconductor Corporation received U.S. patent 3,617,859 for the voltage reference circuit (Dobkin and Widlar 1971). Subsequent to the renowned Dobkin-Widlar circuit, other patents have been issued for voltage reference circuits, including U.S. patent 3,743,923 to Goetz Wolfgang Steudel of RCA Corporation (Steudel 1973). Hundreds of patents have been issued for amplifiers (the 10th entry in table C.1).

Table C.2 shows the two inventions generated by genetic programming for which a patent application has been filed.

Table C.1 Twenty-one previously patented inventions reinvented by genetic programming

	Invention	Date	Inventor	Place	Patent	Reference
1	Darlington emitter-follower section	1953	Sidney Darlington	Bell Telephone Laboratories	2,663,806	Section 42.3 of *Genetic Programming III*
2	Ladder filter	1917	George Campbell	American Telephone and Telegraph	1,227,113	Section 25.15.1 of *Genetic Programming III* and section 5.2 of this book
3	Crossover filter	1925	Otto Julius Zobel	American Telephone and Telegraph	1,538,964	Section 32.3 of *Genetic Programming III*
4	"*M*-derived half section" filter	1925	Otto Julius Zobel	American Telephone and Telegraph	1,538,964	Section 25.15.2 of *Genetic Programming III*
5	Cauer (elliptic) topology for filters	1934 – 1936	Wilhelm Cauer	University of Gottingen	1,958,742, 1,989,545	Section 27.3.7 of *Genetic Programming III*
6	Sorting network	1962	Daniel G. O'Connor and Raymond J. Nelson	General Precision, Inc.	3,029,413	Sections 21.4.4, 23.6, and 57.8.1 of *Genetic Programming III*
7	Computational circuits	See text	See text	See text	See text	Section 47.5.3 of *Genetic Programming III*
8	Electronic thermometer	See text	See text	See text	See text	Section 49.3 of *Genetic Programming III*
9	Voltage reference circuit	See text	See text	See text	See text	Section 50.3 of *Genetic Programming III*
10	60 and 96 dB amplifiers	See text	See text	See text	See text	Section 45.3 of *Genetic Programming III*
11	Second-derivative controller	1942	Harry Jones	Brown Instrument Company	2,282,726	Section 3.7 of this book
12	Philbrick circuit	1956	George Philbrick	George A. Philbrick Researches	2,730,679	Section 4.3 of this book
13	NAND circuit	1971	David H. Chung and Bill H. Terrell	Texas Instruments Incorporated	3,560,760	Section 4.4 of this book
14	PID (proportional, integrative, and derivative) controller	1939	Albert Callender and Allan Stevenson	Imperial Chemical Limited	2,175,985	Section 9.2 of this book

(*Continued*)

Table C.1 (*Continued*)

	Invention	Date	Inventor	Place	Patent	Reference
15	Negative feedback	1937	Harold S. Black	American Telephone and Telegraph	2,102,670, 2,102,671	Chapter 14 of this book
16	Low-voltage balun circuit	2001	Sang Gug Lee	Information and Communications University	6,265,908	Section 15.4.1 of this book
17	Mixed analog-digital variable capacitor circuit	2000	Turgut Sefket Aytur	Lucent Technologies Inc.	6,013,958	Section 15.4.2 of this book
18	High-current load circuit	2001	Timothy Daun-Lindberg and Michael Miller	International Business Machines Corporation	6,211,726	Section 15.4.3 of this book
19	Voltage-current conversion circuit	2000	Akira Ikeuchi and Naoshi Tokuda	Mitsumi Electric Co., Ltd.	6,166,529	Section 15.4.4 of this book
20	Cubic function generator	2000	Stefano Cipriani and Anthony A. Takeshian	Conexant Systems, Inc.	6,160,427	Section 15.4.5 of this book
21	Tunable integrated active filter	2001	Robert Irvine and Bernd Kolb	Infineon Technologies AG	6,225,859	Section 15.4.6 of this book

Table C.2 Two patentable inventions created by genetic programming

	Claimed invention	Date of patent application	Inventors	Reference
1	Improved general-purpose tuning rules for a PID controller	July 12, 2002	Martin A. Keane, John R. Koza, and Matthew J. Streeter	Section 12.3 of this book
2	Improved general-purpose non-PID controllers	July 12, 2002	Martin A. Keane, John R. Koza, and Matthew J. Streeter	Section 13.2 of this book

Bibliography

Aarts, Emile and Korst, Jan. 1989. *Simulated Annealing and Boltzmann Machines*. Chichester: John Wiley and Sons.

Aaserud, O. and Nielsen, I. Ring. 1995. Trends in current analog design: A panel debate. *Analog Integrated Circuits and Signal-Processing*. 7(1)5–9.

Abelson, Harold and diSessa, Andrea. 1980. *Turtle Geometry*. Cambridge, MA: The MIT Press.

Altshuler, Edward E. and Linden, Derek S. 1998. *Process for the Design of Antennas Using Genetic Algorithm*. U.S. patent 5,719,794. Applied for on July 19, 1995. Issued on February 17, 1998.

Altshuler, Edward E. and Linden, Derek S. 1999. Design of wire antennas using genetic algorithms. In Rahmat-Samii, Yahya and Michielssen, Eric (editors). *Electromagnetic Optimization by Genetic Algorithms*. New York, NY: John Wiley and Sons. Chapter 8. Pages 211–248.

Andersson, Bjorn, Svensson, Per, Nordin, Peter, and Nordahl, Mats. 1999. Reactive and memory-based genetic programming for robot control. In Poli, Riccardo, Nordin, Peter, Langdon, William B., and Fogarty, Terence C. 1999. *Genetic Programming: Second European Workshop. EuroGP'99. Proceedings*. Lecture Notes in Computer Science. Volume 1598. Berlin, Germany: Springer-Verlag. Pages 161–172.

Andre, David, Bennett III, Forrest H, and Koza, John R. 1996. Discovery by genetic programming of a cellular automata rule that is better than any known rule for the majority classification problem. In Koza, John R., Goldberg, David E., Fogel, David B., and Riolo, Rick L. (editors). 1996. *Genetic Programming 1996: Proceedings of the First Annual Conference, July 28–31, 1996, Stanford University*. Cambridge, MA: MIT Press. Pages 3–11.

Andre, David and Koza, John R. 1995. Parallel genetic programming on a network of transputers. In Rosca, Justinian (editor). *Proceedings of the Workshop on Genetic Programming: From Theory to Real World Applications*. University of Rochester. National Resource Laboratory for the Study of Brain and Behavior. Technical Report 95-2. June 1995. Pages 111–120.

Andre, David and Koza, John R. 1996a. Parallel genetic programming: A scalable implementation using the transputer architecture. In Angeline, P. J. and Kinnear, K. E. Jr. (editors). 1996. *Advances in Genetic Programming 2*. Cambridge, MA: The MIT Press.

Andre, David and Koza, John R. 1996b. A parallel implementation of genetic programming that achieves super-linear performance. In Arabnia, Hamid R. (editor). *Proceedings of the International Conference on Parallel and Distributed Processing Techniques and Applications*. Athens, GA: CSREA. Volume III. Pages 1163–1174.

Andre, David and Teller, Astro. 1999. Evolving team Darwin United. In Asada, Minoru and Kitano, Hiroaki (editors). *RoboCup-98: Robot Soccer World Cup II*. Lecture Notes in Computer Science. Volume 1604. Berlin: Springer-Verlag. Pages 346–352.

Angeline, Peter J. and Kinnear, Kenneth E. Jr. (editors). 1996. *Advances in Genetic Programming 2*. Cambridge, MA: The MIT Press.

Angeline, Peter J. 1997. An alternative to indexed memory for evolving programs with explicit state representations. In Koza, John R., Deb, Kalyanmoy, Dorigo, Marco, Fogel, David B., Garzon, Max, Iba, Hitoshi, and Riolo, Rick L. (editors). *Genetic Programming 1997: Proceedings of the Second Annual Conference, July 13–16, 1997, Stanford University*. San Francisco, CA: Morgan Kaufmann. Pages 423–430.

Angeline, Peter J. 1998a. Multiple interacting programs: A representation for evolving complex behaviors. *Cybernetics and Systems*. 29(8)779–806.

Angeline, Peter J. 1998b. Evolving predictors for chaotic time series. In Rogers, S., Fogel, D., Bezdek, J., and Bosacchi, B. (editors). *Proceedings of SPIE (Volume 3390): Application and Science of Computational Intelligence*, Bellingham, WA: The International Society for Optical Engineering. Pages 170–180.

Angeline, Peter J. and Fogel, David B. 1997. An evolutionary program for the identification of dynamical systems. In Rogers, S. (editor). *Proceedings of SPIE (Volume 3077): Application and Science of Artificial Neural Networks III*. Bellingham, WA: The International Society for Optical Engineering. Pages 409–417.

Arkin, Adam, Shen, Peidong, and Ross, John. 1997. A test case of correlation metric construction of a reaction pathway from measurements. *Science*. 277. Pages 1275–1279. August 29, 1997.

Armstrong, Edwin Howard. 1914. *Wireless Receiving System*. U.S. patent 1,113,149. Filed October 29, 1913. Issued October 6, 1914.

Åström, Karl J. and Hägglund, Tore. 1995. *PID Controllers: Theory, Design, and Tuning*. Second Edition. Research Triangle Park, NC: Instrument Society of America.

Aytur, Turgut Sefket. 2000. *Integrated Circuit with Variable Capacitor*. U.S. patent 6,013,958. Filed July 23, 1998. Issued January 11, 2000.

Babanezhad, J. N. and Temes, G. C. 1986. Analog MOS computational circuits. *Proceedings of the IEEE Circuits and System International Symposium*. Piscataway, NJ: IEEE Press. Pages 1156–1160.

Babovic, Vladan. 1996. *Emergence, Evolution, Intelligence: Hydroinformatics*. Rotterdam, The Netherlands: Balkema Publishers.

Bagchi, Tapan P. 1999. *Multiobjective Scheduling by Genetic Algorithms*. Boston: Kluwer Academic Publishers.

Balanis, Constantine A. 1982. *Antenna Theory: Analysis and Design*. New York, NY: John Wiley and Sons.

Banzhaf, Wolfgang, Daida, Jason, Eiben, A. E., Garzon, Max H., Honavar, Vasant, Jakiela, Mark, and Smith, Robert E. (editors). 1999. *GECCO-99: Proceedings of the Genetic and Evolutionary Computation Conference, July 13–17, 1999, Orlando, Florida USA*. San Francisco, CA: Morgan Kaufmann.

Banzhaf, Wolfgang, Nordin, Peter, Keller, Robert E., and Francone, Frank D. 1998. *Genetic Programming: An Introduction*. San Francisco, CA: Morgan Kaufmann and Heidelberg: dpunkt.

Banzhaf, Wolfgang, Nordin, Peter, Keller, Richard, and Olmer, Markus. 1997. Generating adaptive behavior for a real robot using function regression with genetic programming. In Koza, John R., Deb, Kalyanmoy, Dorigo, Marco, Fogel, David B., Garzon, Max, Iba, Hitoshi, and Riolo, Rick L. (editors). *Genetic Programming 1997: Proceedings of the Second Annual Conference, July 13–16, 1997, Stanford University*. San Francisco, CA: Morgan Kaufmann. Pages 35–43.

Banzhaf, Wolfgang, Poli, Riccardo, Schoenauer, Marc, and Fogarty, Terence C. 1998. *Genetic Programming: First European Workshop. EuroGP'98. Paris, France, April 1998 Proceedings*. Lecture Notes in Computer Science. Volume 1391. Berlin, Germany: Springer-Verlag.

Barnum, H., Bernstein, H.J. and Spector, Lee. 2000. Quantum circuits for OR and AND of ORs. *Journal of Physics A: Mathematical and General*. 33(45)8047–8057. November 17, 2000.

Bennett III, Forrest H. and Koza, John R. 2002. *Method and Apparatus for Automatic Synthesis, Placement and Routing of Complex Structures*. U.S. patent 6,424,959. Filed June 17, 1999. Issued July 23, 2002.

Bennett III, Forrest H, Koza, John R., Andre, David, and Keane, Martin A. 1996. Evolution of a 60 decibel op amp using genetic programming. In Higuchi, Tetsuya, Iwata, Masaya, and Lui, Weixin (editors). *Proceedings of International Conference on Evolvable Systems: From Biology to Hardware (ICES-96)*. Lecture Notes in Computer Science, Volume 1259. Berlin: Springer-Verlag. Pages 455–469.

Bennett III, Forrest H, Koza, John R., Yu, Jessen, and Mydlowec, William. 2000. Automatic synthesis, placement, and routing of an amplifier circuit by means of genetic programming. In Miller, Julian, Thompson, Adrian, Thomson, Peter, and Fogarty, Terence C. (editors). 2000. *Evolvable Systems: From Biology to Hardware. Third International Conference, ICES 2000, Edinburgh, Scotland, UK, April 2000 Proceedings*. Lecture Notes in Computer Science. Volume 1801. Berlin, Germany: Springer-Verlag. Pages 1–10.

Bennett III, Forrest H, Koza, John R., Keane, Martin A., Yu, Jessen, Mydlowec, William, and Stiffelman, Oscar. 1999. Evolution by means of genetic programming of analog circuits that perform digital functions. In Banzhaf, Wolfgang, Daida, Jason, Eiben, A. E., Garzon, Max H., Honavar, Vasant, Jakiela, Mark, and Smith, Robert E. (editors). 1999. *GECCO-99: Proceedings of the Genetic and Evolutionary Computation Conference, July 13–17, 1999, Orlando, Florida USA*. San Francisco, CA: Morgan Kaufmann. Pages 1477–1483.

Bennett III, Forrest H, Koza, John R., Shipman, James, and Stiffelman, Oscar. 1999. Building a parallel computer system for $18,000 that performs a half peta-flop per day. In Banzhaf, Wolfgang, Daida, Jason, Eiben, A. E., Garzon, Max H., Honavar, Vasant, Jakiela, Mark, and Smith, Robert E. (editors). 1999. *GECCO-99: Proceedings of the Genetic and Evolutionary Computation Conference, July 13–17, 1999, Orlando, Florida USA*. San Francisco, CA: Morgan Kaufmann. Pages 1484–1490.

Black, Harold S. 1928. *Translating System*. U.S. patent 1,686,792. Filed February 3, 1925. Issued October 9, 1928.

Black, Harold S. 1935. *Wave Translation System*. U.S. patent 2,003,282. Filed August 8, 1928. Issued June 4, 1935.

Black, Harold S. 1937a. *Wave Translation System*. U.S. patent 2,102,670. Filed August 8, 1928. Issued December 21, 1937.

Black, Harold S. 1937b. *Wave Translation System*. U.S. patent 2,102,671. Filed April 22, 1932. Issued December 21, 1937.

Black, Harold S. 1977. Inventing the negative feedback amplifier. *IEEE Spectrum*. December 1977. Pages 55–60.

Blickle, Tobias. 1997. *Theory of Evolutionary Algorithms and Application to System Synthesis*. TIK-Schriftenreihe Nr. 17. Zurich, Switzerland: vdf Hochschul Verlag AG an der ETH Zuerich.

Boutin, Noel. 2002. Use time-domain analysis of Zobel network. *EDN*. July 27, 2002. Page 86.

Bower, James M. and Bolouri, Hamid. 2000. *Computational Modeling of Genetic and Biochemical Networks*. Cambridge, MA: MIT Press.

Boyd, S. P. and Barratt, C. H. 1991. *Linear Controller Design: Limits of Performance*. Englewood Cliffs, NJ: Prentice Hall.

Bryson, Arthur E., and Ho, Yu-Chi. 1975. *Applied Optimal Control*. New York: Hemisphere Publishing.

Burke, Gerald J. 1992. *Numerical Electromagnetics Code—NEC-4: Method of Moments—User's Manual*. Lawrence Livermore National Laboratory report UCRL-MA-109338. Livermore, CA: Lawrence Livermore National Laboratory.

Callender, Albert and Stevenson, Allan Brown. 1939. *Automatic Control of Variable Physical Characteristics*. U.S. patent 2,175,985. Filed February 17, 1936 in the United States. Filed February 13, 1935 in Great Britain. Issued October 10, 1939 in the United States.

Campbell, George A. 1917. *Electric Wave Filter*. Filed July 15, 1915. U.S. patent 1,227,113. Issued May 22, 1917.

Cantu-Paz, Erick. 2000. *Efficient and Accurate Parallel Genetic Algorithms*. Boston: Kluwer Academic Publishers.

Cauer, Wilhelm. 1934. *Artificial Network*. U.S. patent 1,958,742. Filed June 8, 1928 in Germany. Filed December 1, 1930 in the United States. Issued May 15, 1934.

Cauer, Wilhelm. 1935. *Electric Wave Filter*. U.S. patent 1,989,545. Filed June 8, 1928. Filed December 6, 1930 in the United States. Issued January 29, 1935.

Cauer, Wilhelm. 1936. *Unsymmetrical Electric Wave Filter*. U.S. patent 2,048,426. Filed November 10, 1932 in Germany. Filed November 23, 1933 in the United States. Issued July 21, 1936.

Chung, David H. and Terrell, Bill H. 1971. *Logic NAND Gate Circuits*. U.S. patent 3,560,760. Filed February 2, 1970. Issued February 2, 1971.

Cipriani, Stefano and Takeshian, Anthony A. 2000. *Compact Cubic Function Generator*. U.S. patent 6,160,427. Filed September 4, 1998. Issued December 12, 2000.

Coello Coello, Carlos A., Van Veldhuizen, David A., and Lamont, Gary B. 2002. *Evolutionary Algorithms for Solving Multi-Objective Problems*. Boston: Kluwer Academic Publishers.

Cohn, John M., Garrod, David J., Rutenbar, Rob A., and Carley, L. Richard. 1994. *Analog Device-Level Layout Automation*. Boston: Kluwer.

Collado-Vides, Julio and Hofestadt, Ralf. 2002. *Gene Regulation and Metabolism*. Cambridge, MA: The MIT Press.

Comisky, William, Yu, Jessen, and Koza, John. 2000. Automatic synthesis of a wire antenna using genetic programming. *Late Breaking Papers at the 2000 Genetic and Evolutionary Computation Conference, Las Vegas, Nevada*. Pages 179–186.

Crawford, L. S., Cheng, V. H. L., and Menon, P. K. 1999. Synthesis of flight vehicle guidance and control laws using genetic search methods. *Proceedings of 1999 Conference on Guidance, Navigation, and Control*. Reston, VA: American Institute of Aeronautics and Astronautics. Paper AIAA-99-4153.

Darlington, Sidney. 1953. *Semiconductor Signal Translating Device*. U.S. patent 2,663,806. Filed May 9, 1952. Issued December 22, 1953.

Daun-Lindberg, Timothy Charles and Miller, Michael Lee. 2001. *Low Voltage High-Current Electronic Load*. U.S. patent 6,211,726. Filed June 28, 1999. Issued April 3, 2001.

Deb, Kalyanmoy. 2001. *Multi-Objective Optimization using Evolutionary Algorithms*. Boston: Kluwer Academic Publishers.

Dewell, Larry D. and Menon, P. K. 1999. Low-thrust orbit transfer optimization using genetic search. *Proceedings of 1999 Conference on Guidance, Navigation, and Control*. Reston, VA: American Institute of Aeronautics and Astronautics. Paper AIAA-99-4151.

D'haeseleer, Patrik, Wen, Xiling, Fuhrman, Stefanie, and Somogyi, Roland. 1999. Linear modeling of mRNA expression levels during CNS development and injury. In Altman, Russ B. Dunker, A. Keith, Hunter, Lawrence, Klein, Teri E., and Lauderdale, Kevin (editors). *Pacific Symposium on Biocomputing '99*. Singapore: World Scientific. Pages 41–52.

Dobkin, Robert C. and Widlar, Robert J. 1971. *Electrical Regulator Apparatus Including a Zero-Temperature Coefficient Voltage Reference Circuit*. U.S. patent 3,617,859. Filed May 23, 1970. Issued November 2, 1971.

Dorf, Richard C. and Bishop, Robert H. 1998. *Modern Control Systems*. Eighth edition. Menlo Park, CA: Addison-Wesley.

Drechsler, Rolf. 1998. *Evolutionary Algorithms for VLSI CAD*. Boston: Kluwer Academic Publishers.

Foster, James A., Lutton, Evelyne, Miller, Julian, Ryan, Conor, and Tettamanzi, Andrea G. B. (editors). 2002. *Genetic Programming: 5th European Conference, EuroGP 2002, Kinsale, Ireland, April 2002 Proceedings*. Berlin: Springer-Verlag.

Garey, Michael R. and Johnson, David S. 1979. *Computers and Intractability: A Guide to the Theory of NP-Completeness*. New York, NY: W. H. Freeman.

Getreu, Ian. 2002. Productivity tools for analog/mixed-signal designs: Ready for prime time? *Electronic Design*. June 10, 2002. Page 40.

Gilbert, Barrie. 1968. A precise four-quadrant multiplier with subnanosecond response. *IEEE Journal of Solid-State Circuits*. Volume SC-3. Number 4. December 1968. Pages 365–373.

Gilbert, Barrie. 1979. *Multiplier Circuit*. U.S. patent 4,156,283. Filed October 3, 1977. Issued May 22, 1979.

Goddard, Robert. 1915. *Method of and Apparatus for Producing Electrical Impulses or Oscillations*. U.S. patent 1,159,209. Filed August 1, 1912. Issued November 2, 1915.

Goldberg, David E. 1989. *Genetic Algorithms in Search, Optimization, and Machine Learning*. Reading, MA: Addison-Wesley.

Goldberg, David E. 1990. A note on Boltzmann tournament selection for genetic algorithms and population-oriented simulated annealing. *Complex Systems*. 4(4)445–460.

Goldberg, David E. 2002. *The Design of Innovation: Lessons from and for Competent Genetic Algorithms*. Boston: Kluwer Academic Publishers.

Green, Milton. 1958. *Logarithmic Converter Circuit*. U.S. patent 2,861,182. Filed June 16, 1953. Issued November 18, 1958.

Grimbleby, J. B. 1995. Automatic analogue network synthesis using genetic algorithms. *Proceedings of the First International Conference on Genetic Algorithms in Engineering Systems: Innovations and Applications (GALESIA)*. London: Institution of Electrical Engineers. Pages 53–58.

Gruau, Frederic. 1992a. *Cellular Encoding of Genetic Neural Networks*. Technical report 92–21. Laboratoire de l'Informatique du Parallélisme. Ecole Normale Supérieure de Lyon. May 1992.

Gruau, Frederic. 1992b. Genetic synthesis of Boolean neural networks with a cell rewriting developmental process. In Schaffer, J. D. and Whitley, Darrell (editors). *Proceedings of the Workshop on Combinations of Genetic Algorithms and Neural Networks 1992*. Los Alamitos, CA: The IEEE Computer Society Press.

Haeusler, Jochen, 1976. *Arrangement for Measuring Temperatures*. U.S. patent 3,943,434. Filed February 6, 1974. Issued March 9, 1976.

Haupt, Randy L. 1994. Thinned arrays using genetic algorithms. *IEEE Transactions on Antennas and Propagation*. Volume 42: Pages 993–999.

Higuchi, Tetsuya, Iwata, Masaya, and Lui, Weixin (editors). 1997. *Evolvable Systems: From Biology to Hardware: First International Conference, ICES-96, Tsukuba, Japan, October 1996 Proceedings*. Lecture Notes in Computer Science, Volume 1259. Berlin: Springer-Verlag.

Higuchi, Tetsuya, Niwa, Tatsuya, Tanaka, Toshio, Iba, Hitoshi, de Garis, Hugo, and Furuya, Tatsumi. 1993a. Evolving hardware with genetic learning: A first step towards building a Darwin machine. In Meyer, Jean-Arcady, Roitblat, Herbert L. and Wilson, Stewart W. (editors). *From Animals to Animats 2: Proceedings of the Second International Conference on Simulation of Adaptive Behavior*. Cambridge, MA: The MIT Press. 1993. Pages 417–424.

Higuchi, Tetsuya, Niwa, Tatsuya, Tanaka, Toshio, Iba, Hitoshi, de Garis, Hugo, and Furuya, Tatsumi. 1993b. *Evolvable Hardware–Genetic-Based Generation of Electric Circuitry at Gate and Hardware Description Language (HDL) Levels*. Electrotechnical Laboratory technical report 93-4. Tsukuba, Japan: Electrotechnical Laboratory.

Holland, John H. 1975. *Adaptation in Natural and Artificial Systems: An Introductory Analysis with Applications to Biology, Control, and Artificial Intelligence*. Ann Arbor, MI: University of Michigan Press. Second edition. Cambridge, MA: The MIT Press 1992.

Hsu, Feng-Hsiung. 2002. *Behind Deep Blue: Building the Computer That Defeated the World Chess Champion*. Princeton, NJ: Princeton University Press.

Iba, Hitoshi. 1996. *Genetic Programming*. Tokyo: Tokyo Denki University Press. In Japanese.

Ikeuchi, Akira and Tokuda, Naoshi. 2000. *Voltage-Current Conversion Circuit*. U.S. patent 6,166,529. Filed February 24, 2000 in the United States. Issued December 26, 2000 in the United States. Filed March 10, 1999 in Japan.

Ince, D. C. (editor). 1992. *Mechanical Intelligence: Collected Works of A. M. Turing*. Amsterdam: North Holland.

Irvine, Robert and Kolb, Bernd. 2001. *Integrated Low-Pass Filter*. U.S. patent 6,225,859. Filed September 14, 1998. Issued May 1, 2001.

Jacob, Christian. 1997. *Principia Evolvica: Simulierte Evolution mit Mathematica*. Heidelberg, Germany: dpunkt.verlag.

Jacob, Christian. 2001. *Illustrating Evolutionary Computation with Mathematica*. San Francisco: Morgan Kaufmann.

Jamshidi, Mo, Coelho, Leandro dos Santos, Krohling, Renato A., and Fleming, Peter J. 2003. *Robust Control Systems with Genetic Algorithms*. Boca Raton, FL: CRC Press.

Johnson, Kenneth S. 1926. *Electric-Wave Transmission*. U.S. patent 1,611,916. Filed March 9, 1923. Issued December 28, 1926.

Johnson, Walter C. 1950. *Transmission Lines and Networks*. New York: NY: McGraw-Hill.

Johnson, J. Michael and Rahmat-Samii, Yahya. 1999. Genetic algorithms and method of moments (GA/MOM) for the design of integrated antennas. *IEEE Transactions on Antennas and Propagation*. 47(10)1606–1614. October 1999.

Jones, Eric A. 1999. *Genetic Design of Antennas and Electronic Circuits*. Ph.D. Thesis. Department of Electrical and Computer Engineering. Duke University.

Jones, Harry S. 1942. *Control Apparatus*. U.S. patent 2,282,726. Filed October 25, 1939. Issued May 12, 1942.

Keane, Martin A., Koza, John R., and Rice, James P. 1993. Finding an impulse response function using genetic programming. *Proceedings of the 1993 American Control Conference*. Evanston, IL: American Automatic Control Council. Volume III. Pages 2345–2350.

Keane, Martin A., Koza, John R., and Streeter, Matthew J. 2002a. *Improved General-Purpose Controllers*. U.S. patent application filed July 12, 2002.

Keane, Martin A., Koza, John R., and Streeter, Matthew J. 2002b. Automatic synthesis using genetic programming of an improved general-purpose controller for industrially representative plants. In Stoica, Adrian, Lohn, Jason, Katz, Rich, Keymeulen, Didier and Zebulum, Ricardo (editors) 2002. *Proceedings of 2002 NASA/DoD Conference on Evolvable Hardware*. Los Alamitos, CA: IEEE Computer Society. Pages 113–122.

Keane, Martin A., Yu, Jessen, and Koza, John R. 2000. Automatic synthesis of both the topology and tuning of a common parameterized controller for two families of plants using genetic programming. In Whitley, Darrell, Goldberg, David, Cantu-Paz, Erick, Spector, Lee, Parmee, Ian, and Beyer, Hans-Georg (editors). *GECCO-2000: Proceedings of the Genetic and Evolutionary Computation Conference, July 10–12, 2000, Las Vegas, Nevada*. San Francisco: Morgan Kaufmann. Pages 496–504.

Keymeulen, Didier, Stoica, Adrian, Lohn, Jason, and Zebulum, Ricardo Salem (editors). 2001. *Proceedings of the Third NASA/DOD Workshop on Evolvable Hardware, Pasadena, California, July 12–14, 2001*. Los Alamitos, CA. IEEE Computer Society.

Kinnear, Kenneth E. Jr. (editor). 1994. *Advances in Genetic Programming*. Cambridge, MA: MIT Press.

Kirkpatrick, S., Gelatt, C. D., and Vecchi, M. P. 1983. Optimization by simulated annealing. *Science* 220. Pages 671–680.

Kitano, Hiroaki. 1990. Designing neural networks using genetic algorithms with graph generation system. *Complex Systems*. 4 (1990) 461–476.

Kitano, Hiroaki. 2001. *Foundations of Systems Biology*. Cambridge, MA: The MIT Press.

Koza, John R. 1988. *Nonlinear Genetic Algorithms for Solving Problems*. U.S. patent application filed May 20, 1988.

Koza, John R. 1989. Hierarchical genetic algorithms operating on populations of computer programs. In *Proceedings of the 11th International Joint Conference on Artificial Intelligence*. San Mateo, CA: Morgan Kaufmann. Volume I. Pages 768–774.

Koza, John R. 1990a. *Genetic Programming: A Paradigm for Genetically Breeding Populations of Computer Programs to Solve Problems*. Stanford University Computer Science Department technical report STAN-CS-TR-90-1314. June 1990.

Koza, John R. 1990b. *Non-Linear Genetic Algorithms for Solving Problems*. U.S. patent 4,935,877. Filed May 20, 1988. Issued June 19, 1990.

Koza, John R. 1992a. *Genetic Programming: On the Programming of Computers by Means of Natural Selection*. Cambridge, MA: MIT Press.

Koza, John R. 1992b. *Non-Linear Genetic Algorithms for Solving Problems by Finding a Fit Composition of Functions*. U. S. patent 5,136,686. Filed March 28, 1990. Issued August 4, 1992.

Koza, John R. 1992c. Hierarchical automatic function definition in genetic programming. In Whitley, Darrell (editor). 1993. *Foundations of Genetic Algorithms 2*. San Mateo, CA: Morgan Kaufmann Publishers. Pages 297–318.

Koza, John R. 1992d. A genetic approach to finding a controller to back up a tractor-trailer truck. In *Proceedings of the 1992 American Control Conference*. Evanston, IL: American Automatic Control Council. Volume III. Pages 2307–2311.

Koza, John R. 1993. Discovery of rewrite rules in Lindenmayer systems and state transition rules in cellular automata via genetic programming. *Symposium on Pattern Formation (SPF-93), Claremont, California. February 13, 1993*. A copy of this presented, but otherwise unpublished, paper is available at `http://www.smi.stanford.edu/people/koza`.

Koza, John R. 1994a. *Genetic Programming II: Automatic Discovery of Reusable Programs*. Cambridge, MA: MIT Press.

Koza, John R. 1994b. *Genetic Programming II Videotape: The Next Generation*. Cambridge, MA: MIT Press.

Koza, John R. 1994c. *Architecture-Altering Operations for Evolving the Architecture of a Multi-Part Program in Genetic Programming*. Stanford University Computer Science Department technical report STAN-CS-TR-94-1528. October 21, 1994.

Koza, John R. 1995a. Evolving the architecture of a multi-part program in genetic programming using architecture-altering operations. In McDonnell, John R., Reynolds, Robert G., and Fogel, David B. (editors). 1995. *Evolutionary Programming IV: Proceedings of the Fourth Annual Conference on Evolutionary Programming*. Cambridge, MA: The MIT Press. Pages 695–717.

Koza, John R. 1995b. Gene duplication to enable genetic programming to concurrently evolve both the architecture and work-performing steps of a computer program. *Proceedings of the 14th International Joint Conference on Artificial Intelligence*. San Francisco: Morgan Kaufmann. Pages 734–740.

Koza, John R. 1995c. Two ways of discovering the size and shape of a computer program to solve a problem. In Eshelman, Larry J. (editor). *Proceedings of the Sixth International Conference on Genetic Algorithms*. San Francisco: Morgan Kaufmann. Pages 287–294.

Koza, John R. and Andre, David. 1995. *Parallel Genetic Programming on a Network of Transputers*. Stanford University Computer Science Department technical report STAN-CS-TR-95-1542. January 30, 1995.

Koza, John R., Andre, David, and Tackett, Walter Alden. 1994. *Simultaneous Evolution of the Architecture of a Multi-Part Program to Solve a Problem Using Architecture Altering Operations*. U.S. patent application filed August 4, 1994.

Koza, John R., Andre, David, and Tackett, Walter Alden. 1998. *Simultaneous Evolution of the Architecture of a Multi-Part Program to Solve a Problem Using Architecture Altering Operations*. U. S. patent 5,742,738. Filed August 4, 1994. Issued April 21, 1998.

Koza, John R., Andre, David, and Tackett, Walter Alden. 2000. *Simultaneous Evolution of the Architecture of a Multi-Part Program to Solve a Problem Using Architecture Altering Operations*. U. S. patent 6,058,385. Filed March 7, 1997. Issued May 2, 2000.

Koza, John R., Banzhaf, Wolfgang, Chellapilla, Kumar, Deb, Kalyanmoy, Dorigo, Marco, Fogel, David B., Garzon, Max H., Goldberg, David E., Iba, Hitoshi, and Riolo, Rick. (editors). 1998. *Genetic Programming 1998: Proceedings of the Third Annual Conference*. San Francisco, CA: Morgan Kaufmann.

Koza, John R., and Bennett III, Forrest H. 1999. Automatic synthesis, placement, and routing of electrical circuits by means of genetic programming. In Spector, Lee, Langdon, William B., O'Reilly, Una-May, and Angeline, Peter (editors). *Advances in Genetic Programming 3*. Cambridge, MA: MIT Press. Chapter 6. Pages 105–134.

Koza, John R., Bennett III, Forrest H, Andre, David, and Keane, Martin A. 1996a. Automated design of both the topology and sizing of analog electrical circuits using genetic programming. In Gero, John S. and Sudweeks, Fay (editors). *Artificial Intelligence in Design '96*. Dordrecht: Kluwer Academic Publishers. Pages 151–170.

Koza, John R., Bennett III, Forrest H, Andre, David, and Keane, Martin A. 1996b. Automated WYWIWYG design of both the topology and component values of analog electrical circuits using genetic programming. In Koza, John R., Goldberg, David E., Fogel, David B., and Riolo, Rick L. (editors). *Genetic Programming 1996: Proceedings of the First Annual Conference, July 28–31, 1996, Stanford University*. Cambridge, MA: The MIT Press. Pages 123–131.

Koza, John R., Bennett III, Forrest H, Andre, David, and Keane, Martin A. 1996c. Reuse, parameterized reuse, and hierarchical reuse of substructures in evolving electrical circuits using genetic programming. In Higuchi, Tetsuya, Iwata, Masaya, and Liu, Weixin (editors). 1997. *Proceedings of International Conference on Evolvable Systems: From Biology to Hardware (ICES-96)*. Lecture Notes in Computer Science, Volume 1259. Berlin: Springer-Verlag. Berlin: Springer-Verlag. Pages 312–326.

Koza, John R., Bennett III, Forrest H, Andre, David, and Keane, Martin A. 1996d. Toward evolution of electronic animals using genetic programming. In Langton, Christopher G. and Shimohara, Katsunori (editors). 1997. *Artificial Life V: Proceedings of the Fifth International Workshop on the Synthesis and Simulation of Living Systems*. Cambridge, MA: The MIT Press. Pages 327–334.

Koza, John R., Bennett III, Forrest H, Andre, David, and Keane, Martin A. 1996e. Four problems for which a computer program evolved by genetic programming is competitive with human performance. *Proceedings of the 1996 IEEE International Conference on Evolutionary Computation*. IEEE Press. Pages 1–10.

Koza, John R., Bennett III, Forrest H, Andre, David, and Keane, Martin A. 1996f. *Method and Apparatus for Automated Design of Electrical Circuits Using Genetic Programming*. U.S. patent application filed February 20, 1996.

Koza, John R., Bennett III, Forrest H, Andre, David, and Keane, Martin A. 1999a. *Genetic Programming III: Darwinian Invention and Problem Solving*. San Francisco, CA: Morgan Kaufmann.

Koza, John R., Bennett III, Forrest H, Andre, David, and Keane, Martin A. 1999b. *Method and Apparatus for Automated Designs of Complex Structures using Genetic Programming*. Filed February 20. 1996. U. S. patent 5,867,397. Issued February 2, 1999.

Koza, John R., Bennett III, Forrest H, Andre, David, and Keane, Martin A. 1999c. *Genetic Programming Problem Solver with Automatically Defined Stores, Loops, and Recursions*. U.S. patent application filed April 12, 1999.

Koza, John R., Bennett III, Forrest H, Andre, David, and Keane, Martin A. 2002. *Method and Apparatus for Automated Design of Complex Structures using Genetic Programming*. U.S. patent 6,360,191. Filed February 20, 1996 and January 5, 1999. Issued March 19, 2002.

Koza, John R., Bennett III, Forrest H, Andre, David, and Keane, Martin A. 2003. *Genetic Programming Problem Solver with Automatically Defined Stores, Loops, and Recursions*. U.S. patent 6,532,453. Filed April 12, 1999. Issued March 11, 2003.

Koza, John R., Bennett III, Forrest H, Andre, David, Keane, Martin A., and Brave, Scott. 1999. *Genetic Programming III Videotape: Human-Competitive Machine Intelligence*. San Francisco, CA: Morgan Kaufmann.

Koza, John R., Bennett III, Forrest H, Andre, David, Keane, Martin A., and Dunlap, Frank. 1997. Automated synthesis of analog electrical circuits by means of genetic programming. *IEEE Transactions on Evolutionary Computation*. 1(2). Pages 109–128.

Koza, John R., Bennett, Forrest H, III, Hutchings, Jeffrey L., Bade, Stephen L., Keane, Martin A., and Andre, David. 1997. Evolving sorting networks using genetic programming and the rapidly reconfigurable Xilinx 6216 field-programmable gate array. *Proceedings of the 31st Asilomar Conference on Signals, Systems, and Computers*. Piscataway, NJ: IEEE Press. Pages 404–410.

Koza, John R., Bennett, Forrest H, III, Hutchings, Jeffrey L., Bade, Stephen L., Keane, Martin A., and Andre, David. 1998. Evolving computer programs using rapidly reconfigurable field-programmable gate arrays and genetic programming. *Proceedings of the ACM Sixth International Symposium on Field Programmable Gate Arrays*. New York: ACM Press. Pages 209–219.

Koza, John R., Bennett III, Forrest H, Keane, Martin A., and Andre, David. 1997. Automatic programming of a time-optimal robot controller and an analog electrical circuit to implement the robot controller by means of genetic programming. In *Proceedings of 1997 IEEE International Symposium on Computational Intelligence in Robotics and Automation*. Los Alamitos, CA: Computer Society Press. Pages 340–346.

Koza, John R., Bennett III, Forrest H, Keane, Martin A., Yu, Jessen, Mydlowec, William, and Stiffelman, Oscar. 1999. Searching for the impossible using genetic programming. In Banzhaf, Wolfgang, Daida, Jason, Eiben, A. E., Garzon, Max H., Honavar, Vasant, Jakiela, Mark, and Smith, Robert E. (editors). 1999. *GECCO-99: Proceedings of the Genetic and Evolutionary Computation Conference, July 13–17, 1999, Orlando, Florida USA*. San Francisco, CA: Morgan Kaufmann. Pages 1083–1091.

Koza, John R., Bennett III, Forrest H, and Stiffelman, Oscar. 1999a. Genetic programming as a Darwinian invention machine. In Poli, Riccardo, Nordin, Peter, Langdon, William B., and Fogarty, Terence C. 1999. *Genetic Programming: Second European Workshop. EuroGP'99. Proceedings*. Lecture Notes in Computer Science. Volume 1598. Berlin, Germany: Springer-Verlag. Pages 93–108.

Koza, John R., Bennett III, Forrest H, and Stiffelman, Oscar. 1999b. *An Invention Machine that Automatically Creates Novel Designs*. U.S. patent application filed April 13, 2000.

Koza, John R., Deb, Kalyanmoy, Dorigo, Marco, Fogel, David B., Garzon, Max, Iba, Hitoshi, and Riolo, Rick L. (editors). 1997. *Genetic Programming 1997: Proceedings of the Second Annual Conference, July 13–16, 1997, Stanford University*. San Francisco, CA: Morgan Kaufmann.

Koza, John R., Goldberg, David E., Fogel, David B., and Riolo, Rick L. (editors). 1996. *Genetic Programming 1996: Proceedings of the First Annual Conference, July 28–31, 1996, Stanford University*. Cambridge, MA: MIT Press.

Koza, John R., and Keane, Martin A. 1990a. Cart centering and broom balancing by genetically breeding populations of control strategy programs. In *Proceedings of International Joint Conference on Neural Networks, Washington, January 15–19, 1990*. Hillsdale, NJ: Lawrence Erlbaum. Volume I, Pages 198–201.

Koza, John R., and Keane, Martin A. 1990b. Genetic breeding of nonlinear optimal control strategies for broom balancing. In *Proceedings of the Ninth International Conference on Analysis and Optimization of Systems. Antibes, France, June, 1990*. Berlin: Springer-Verlag. Pages 47–56.

Koza, John R., Keane, Martin A., Bennett III, Forrest H, Yu, Jessen, Mydlowec, William, and Stiffelman, Oscar. 1999. Automatic creation of both the topology and parameters for a robust controller by means of genetic programming. *Proceedings of the 1999 IEEE International Symposium on Intelligent Control, Intelligent Systems, and Semiotics*. Piscataway, NJ: IEEE. Pages 344–352.

Koza, John R., Keane, Martin A., and Streeter, Matthew J. 2003. Evolving inventions. *Scientific American*. February 2003. 288(2) 52–59.

Koza, John R., Keane, Martin A., Yu, Jessen, Bennett III, Forrest H, and Mydlowec, William. 2000. Automatic creation of human-competitive programs and controllers by means of genetic programming. *Genetic Programming and Evolvable Machines*. Volume 1. Number 1/2. Pages 121–164.

Koza, John R., Keane, Martin A., Yu, Jessen, Bennett III, Forrest H, Mydlowec, William, and Stiffelman, Oscar. 1999. Automatic synthesis of both the topology and parameters for a robust controller for a non-minimal phase plant and a three-lag plant by means of genetic programming. *Proceedings of 1999 IEEE Conference on Decision and Control*. Pages 5292–5300.

Koza, John R., Keane, Martin A., Yu, Jessen, Bennett III, Forrest H, and Mydlowec, William. 2003. *Method and Apparatus for Automatic Synthesis of Controllers*. U.S. patent application filed September 10, 1999. Application number 09/393,863. Allowed October 30, 2002.

Koza, John R., Keane, Martin A., Yu, Jessen, and Mydlowec, William. 2000. Automatic synthesis of electrical circuits containing a free variable using genetic programming. In Whitley, Darrell, Goldberg, David, Cantu-Paz, Erick, Spector, Lee, Parmee, Ian, and Beyer, Hans-Georg (editors). *GECCO-2000: Proceedings of the Genetic and Evolutionary Computation Conference, July 10–12, 2000, Las Vegas, Nevada*. San Francisco: Morgan Kaufmann. Pages 551–557.

Koza, John R., Keane, Martin A., Yu, Jessen, Mydlowec, William, and Bennett, Forrest H III. 2000a. Automatic synthesis of both the topology and parameters for a controller for a three-lag plant with a five-second delay using genetic programming. In Cagnoni, Stafano et al. (editors). *Real World Applications of Evolutionary Computing. EvoWorkshops 2000. EvoIASP, Evo SCONDI, EvoTel, EvoSTIM, EvoRob, and EvoFlight, Edinburgh, Scotland, UK, April 2000, Proceedings*. Lecture Notes in Computer Science. Volume 1803. Berlin, Germany: Springer-Verlag. Pages 168–177.

Koza, John R., Keane, Martin A., Yu, Jessen, Mydlowec, William, and Bennett, Forrest H III. 2000b. Automatic synthesis of both the control law and parameters for a controller for a three-lag plant with five-second delay using genetic programming and simulation techniques. In *Proceedings of the 2000 American Control Conference, Chicago, Illinois, June 28–30, 2000*. Evanston, IL: American Automatic Control Council. Pages 453–459.

Koza, John R., Mydlowec, William, Lanza, Guido, Yu, Jessen, and Keane, Martin A. 2000a. *Reverse Engineering and Automatic Synthesis of Metabolic Pathways from Observed Data Using Genetic Programming*. Stanford Medical Informatics Technical Report SMI-2000-0851. November 7, 2000. `http://smi-web.stanford.edu/pubs/SMI_Abstracts/SMI-2000-0851.html`

Koza, John R., Mydlowec, William, Lanza, Guido, Yu, Jessen, and Keane, Martin A. 2000b. Reverse engineering of metabolic pathways from observed data using genetic programming. In Altman, Russ B. Dunker, A. Keith, Hunter, Lawrence, Lauderdale, Kevin, and Klein, Teri (editors). *Pacific Symposium on Biocomputing'99*. Singapore: World Scientific. Pages 434–445.

Koza, John R., Mydlowec, William, Lanza, Guido, Yu, Jessen, and Keane, Martin A. 2001a. Automated reverse engineering of metabolic pathways by means of genetic programming. In

Kitano, Hiroaki. 2001. *Foundations of Systems Biology*. Cambridge, MA: The MIT Press. Pages 95–121.

Koza, John R., Mydlowec, William, Lanza, Guido, Yu, Jessen, and Keane, Martin A. 2001b. Automatic synthesis of both the topology and sizing of metabolic pathways using genetic programming. In Spector, Lee, Goodman, E., Wu, A., Langdon, William B., Voigt, H.-M., Gen, M., Sen, S., Dorigo, Marco, Pezeshk, S., Garzon, Max, and Burke, E. (editors). 2001. *Proceedings of the Genetic and Evolutionary Computation Conference, GECCO-2001*. San Francisco, CA: Morgan Kaufmann. Pages 57–65.

Koza, John R., and Rice, James P. 1991. Genetic generation of both the weights and architecture for a neural network. In *Proceedings of International Joint Conference on Neural Networks, Seattle, July 1991*. Los Alamitos, CA: IEEE Press. Volume II. Pages 397–404.

Koza, John R., and Rice, James P. 1992a. *Genetic Programming: The Movie*. Cambridge, MA: MIT Press.

Koza, John R., and Rice, James P. 1992b. *A Non-Linear Genetic Process for Data Encoding and for Solving Problems Using Automatically Defined Functions*. U.S. patent application filed May 11, 1992.

Koza, John R., and Rice, James P. 1992c. *A Non-Linear Genetic Process for Use with Plural Co-Evolving Populations*. U. S. patent No. 5,148,513. Filed September 18, 1990. Issued September 15, 1992.

Koza, John R., and Rice, James P. 1994a. *A Non-Linear Genetic Process for Data Encoding and for Solving Problems Using Automatically Defined Functions*. U. S. patent application filed May 11, 1992. U. S. patent No. 5,343,554. Issued August 30, 1994.

Koza, John R., and Rice, James P. 1994b. *A Non-Linear Genetic Process for Data Encoding and for Solving Problems Using Automatically Defined Functions*. U.S. patent 5,343,554. Filed May 11, 1992. Issued August 30, 1994.

Koza, John R., and Rice, James P. 1995. *Process for Problem Solving Using Spontaneously Emergent Self-Replicating and Self-Improving Entities*. U. S. patent application filed June 16, 1992. U. S. patent No. 5,390,282. Issued February 14, 1995.

Koza, John R., Rice, James P., and Roughgarden, Jonathan. 1992. Evolution of food foraging strategies for the Caribbean *Anolis* lizard using genetic programming. *Adaptive Behavior*. Volume 1, number 2, pages 47–74.

Koza, John R., Yu, Jessen, Keane, Martin A., and Mydlowec, William. 2000a. Evolution of a controller with a free variable using genetic programming. In Poli, Riccardo, Banzhaf, Wolfgang, Langdon, William B., Miller, Julian, Nordin, Peter, and Fogarty, Terence C. 2000. *Genetic Programming: European Conference, EuroGP 2000, Edinburgh, Scotland, UK, April 2000, Proceedings*. Lecture Notes in Computer Science. Volume 1802. Berlin, Germany: Springer-Verlag. Pages 91–105.

Koza, John R., Yu, Jessen, Keane, Martin A., and Mydlowec, William. 2000b. Use of conditional developmental operators and free variables in automatically synthesizing generalized circuits using genetic programming. *Proceedings of the Second NASA/DoD Workshop on Evolvable Hardware, July 13–15 2000, Palo Alto, California*. Los Alamitos, CA: IEEE Computer Society Press. Pages 5–15.

Kruiskamp, Marinum Wilhelmus. 1996. *Analog Design Automation using Genetic Algorithms and Polytopes*. Eindhoven, The Netherlands: Data Library Technische Universiteit Eindhoven.

Kruiskamp, Marinum Wilhelmus and Leenaerts, Domine. 1995. DARWIN: CMOS opamp synthesis by means of a genetic algorithm. *Proceedings of the 32nd Design Automation Conference*. New York, NY: Association for Computing Machinery. Pages 433–438.

Laing, Shoudan, Fuhrman, Stefanie, and Somogyi, Roland. 1998. REVEAL: A general reverse engineering algorithm for inference of genetic network architecture. In Altman, Russ B.

Dunker, A. Keith, Hunter, Lawrence, and Klein, Teri E. (editors). *Pacific Symposium on Biocomputing '98*. Singapore: World Scientific. Pages 18–29.

Langdon, William B. 1998. *Genetic Programming and Data Structures: Genetic Programming + Data Structures = Automatic Programming!* Amsterdam: Kluwer.

Langdon, W. B., Cantu-Paz, E., Mathias, K., Roy, R., Davis, D., Poli, R., Balakrishnan, K., Honavar, V., Rudolph, G., Wegener, J., Bull, L., Potter, M. A., Schultz, A. C., Miller, J. F., Burke, E., and Jonoska, N. (editors). 2002. *Proceedings of the 2002 Genetic and Evolutionary Computation Conference*. San Francisco, CA: Morgan Kaufmann.

Langdon, William B. and Poli, Riccardo. 2002. *Foundations of Genetic Programming*. Berlin: Springer-Verlag.

Lanza, Guido, Mydlowec, William, and Koza, John R. 2000. Automatic creation of a genetic network for the *lac* operon from observed data by means of genetic programming. Poster paper accepted at First International Conference on Systems Biology in Tokyo on November 14–16, 2000.

Lee, Sang Gug. 2001. *Low Voltage Balun Circuit*. U.S. patent 6,265,908. Filed December 15, 1999. Issued July 24, 2001.

Lee, Thomas H. 1998. *The Design of CMOS Radio-Frequency Integrated Circuits*. Cambridge: Cambridge University Press.

Linden, Derek S. 1997. *Automated Design and Optimization of Wire Antennas Using Genetic Algorithms*. Ph.D. Thesis. Department of Electrical Engineering and Computer Science. Massachusetts Institute of Technology.

Lindenmayer, Aristid. 1968. Mathematical models for cellular interactions in development, I & II. *Journal of Theoretical Biology*. Volume 18. Pages 280–315.

Liu, Yong, Tanaka, Kiyoshi, Iwata, Masaya, Higuchi, Tetsuya, and Yasunaga, Moritoshi (editors). 2001. *Evolvable Systems: From Biology to Hardware, 4th International Conference, ICES 2001, Tokyo, Japan, October 2001 Proceedings*. Lecture Notes in Computer Science, Volume 2210. Berlin: Springer-Verlag.

Lohn, Jason, Stoica, Adrian, Keymeulen, Didier, and Colombano, Silvano (editors). 2000. *Proceedings of the Second NASA/DoD Workshop on Evolvable Hardware, July 13–15 2000, Palo Alto, California*. Los Alamitos, CA: IEEE Computer Society Press.

Loomis, William F. and Sternberg, Paul W. 1995. Genetic networks. *Science*. Pages 269–649. August 4, 1995.

Luke, Sean. 1998. Genetic programming produced competitive soccer softbot teams for RoboCup97. In Koza, John R., Banzhaf, Wolfgang, Chellapilla, Kumar, Deb, Kalyanmoy, Dorigo, Marco, Fogel, David B., Garzon, Max H., Goldberg, David E., Iba, Hitoshi, and Riolo, Rick. (editors). *Genetic Programming 1998: Proceedings of the Third Annual Conference, July 22–25, 1998, University of Wisconsin, Madison, Wisconsin*. San Francisco, CA: Morgan Kaufmann. Pages 214–222.

Macbeth, Ian. 2002. FPAAs: Synthesis by construction. *EDN*. November 28, 2002. Pages 44.

Mahfoud, S. W. and Goldberg, David E. 1995. Parallel recombinative simulated annealing: A genetic algorithm. *Parallel Computing*. Amsterdam: Elsevier Science. Volume 21. Pages 1–28.

Man, K. F., Tang, K. S., Kwong, S., and Halang, W. A. 1997. *Genetic Algorithms for Control and Signal-Processing*. London: Springer-Verlag.

Man, K. F., Tang, K. S., Kwong, S., and Halang, W. A. 1999. *Genetic Algorithms: Concepts and Designs*. London: Springer-Verlag.

Marcano, Diogenes and Duran, Filinto. 1999. Synthesis of linear and planar arrays using genetic algorithms. In Rahmat-Samii, Yahya and Michielssen, Eric (editors). *Electromagnetic Optimization by Genetic Algorithms*. New York, NY: John Wiley and Sons. Chapter 6. Pages 157–179.

Marenbach, Peter, Bettenhausen, Kurt D., and Freyer, Stephan. 1996. Signal path oriented approach for generation of dynamic process models. In Koza, John R., Goldberg, David E., Fogel, David B., and Riolo, Rick L. (editors). *Genetic Programming 1996: Proceedings of the First Annual Conference, July 28–31, 1996, Stanford University*. Cambridge, MA: MIT Press. Pages 327–332.

Massey, John. 1970. *Compensated Resistance Bridge-Type Electrical Thermometer*. U.S. patent 3,541,857. Filed November 27, 1968. Issued November 24, 1970.

Maziasz, Robert L. and Hayes, John P. 1992. *Layout Minimization of CMOS Cells*. Boston: Kluwer.

Mazumder, Pinaki and Rudnick, Elizabeth M. (editors). 1999. *Genetic Algorithms for VLSI Design, Layout and Test Automation*. Upper Saddle River, NJ: Prentice Hall.

McAdams, Harley H. and Shapiro, Lucy. 1995. Circuit simulation of genetic networks. *Science*. Volume 269. Pages 650–656. August 4, 1995.

Mendes, Pedro and Kell, Douglas B. 1998. Nonlinear optimization of biochemical pathways: Applications to metabolic engineering and parameter estimation. *Bioinformatics*. 14(10)869–883.

Menon, P. K., Yousefpor, M., Lam, T., and Steinberg, M. L. 1995. Nonlinear flight control system synthesis using genetic programming. *Proceedings of 1995 Conference on Guidance, Navigation, and Control*. Reston, VA: American Institute of Aeronautics and Astronautics. Pages 461–470.

Messina, Paul, Sterling, Thomas, and Smith, Paul H. (editors). 1999. *Petaflops II: Second Conference on Enabling Technologies for Peta(fl)ops Computing, February 15–19, 1999, Santa Barbara*.

Miller, Julian, Thompson, Adrian, Thomson, Peter, and Fogarty, Terence C. (editors). 2000. *Evolvable Systems: From Biology to Hardware. Third International Conference, ICES 2000, Edinburgh, Scotland, UK, April 2000 Proceedings*. Lecture Notes in Computer Science. Volume 1801. Berlin, Germany: Springer-Verlag.

Miller, Julian, Tomassini, Marco, Lanzi, Pier Luca, Ryan, Conor, Tettamanzi, Andrea G. B., and Langdon, William B. (editors). 2001. *Genetic Programming: 4th European Conference, EuroGP 2001, Lake Como, Italy, April 2001 Proceedings*. Berlin: Springer.

Mittenthal, Jay E., Ao Yuan, Bertrand Clarke, and Scheeline, Alexander. 1998. Designing metabolism: Alternative connectivities for the pentose phosphate pathway. *Bulletin of Mathematical Biology*. Volume 60. Pages 815–856.

Moore, Gordon E. 1996. Can Moore's law continue indefinitely? *Computerworld Leadership Series*. 2(6)2–7. July 15, 1996.

Moretti, Gabe. 2002. The next wave: Synthesis tools help with mixed-signal designs. *EDN*. November 28, 2002. Pages 43–50.

Mydlowec, William and Koza, John. 2000. Use of time-domain simulations in automatic synthesis of computational circuits using genetic programming. *Late Breaking Papers at the 2000 Genetic and Evolutionary Computation Conference, Las Vegas, Nevada*. Pages 187–197.

Newbold, William F. 1962. *Square Root Extracting Integrator*. U.S. patent 3,016,197. Filed September 15, 1958. Issued January 9, 1962.

Newborn, Monty. 2002. *Deep Blue: An Artificial Intelligence Milestone*. New York: Springer.

Nordin, Peter. 1997. *Evolutionary Program Induction of Binary Machine Code and Its Application*. Munster, Germany: Krehl Verlag.

O'Connor, Daniel G. and Nelson, Raymond J. 1962. *Sorting System with N-Line Sorting Switch*. U.S. patent 3,029,413. Issued April 10, 1962.

Ogata, Katsuhiko. 1997. *Modern Control Engineering*. Third edition. Upper Saddle River, NJ: Prentice Hall.

Ohr, Stephan. 2002. Anadigm fields reconfigurable analog device. *EE Times*. September 2, 2002.

Olsson, Jan Roland. 1994a. Inductive functional programming using incremental program transformation. *Artificial Intelligence*. Volume 74. Pages 55–81.

Olsson, Jan Roland. 1994b. *Inductive Functional Programming Using Incremental Program Transformation*. Dr. Scient. thesis. University of Oslo.

O'Neill, Michael and Ryan, Conor. 2003. *Grammatical Evolution: Evolutionary Automatic Programming in an Arbitrary Language*. Boston: Kluwer Academic Publishers.

Osyczka, A. 1984. *Multicriterion Optimization in Engineering with FORTRAN programs*. Ellis Horwood Limited.

Pease, Robert. 1992. What's all this Muntzing stuff, anyhow? *Electronic Design*. July 23, 1992.

Pease, Robert. 1996. What's all this R-C filter stuff, anyhow? *Electronic Design*. March 18, 1996.

Pease, Robert. 1999. What's all this logarithmic stuff, anyhow? *Electronic Design*. August 19, 1999.

Philbrick, George A. 1956. *Delayed Recovery Electric Filter Network*. Filed May 18, 1951. U.S. patent 2,730,679. Issued January 10, 1956.

Poli, Riccardo, Nordin, Peter, Langdon, William B., and Fogarty, Terence C. 1999. *Genetic Programming: Second European Workshop, EuroGP'99. Proceedings*. Lecture Notes in Computer Science. Volume 1598. Berlin, Germany: Springer-Verlag.

Poli, Riccardo, Banzhaf, Wolfgang, Langdon, William B., Miller, Julian, Nordin, Peter, and Fogarty, Terence C. 2000. *Genetic Programming: European Conference, EuroGP 2000, Edinburgh, Scotland, UK, April 2000, Proceedings*. Lecture Notes in Computer Science. Volume 1802. Berlin, Germany: Springer-Verlag.

Prusinkiewicz, Przemyslaw and Hanan, James. 1980. *Lindenmayer Systems, Fractals, and Plants*. New York: Springer-Verlag.

Prusinkiewicz, Przemyslaw, and Lindenmayer, Aristid. 1990. *The Algorithmic Beauty of Plants*. New York: Springer-Verlag.

Ptashne, Mark. 1992. *A Genetic Switch: Phage λ and Higher Organisms*. Second Edition. Cambridge, MA: Cell Press and Blackwell Scientific Publications.

Quarles, Thomas, Newton, A. R., Pederson, D. O., and Sangiovanni-Vincentelli, A. 1994. *SPICE 3 Version 3F5 User's Manual*. Department of Electrical Engineering and Computer Science, University of California. Berkeley, CA. March 1994.

Rahmat-Samii, Yahya and Michielssen, Eric (editors). 1999. *Electromagnetic Optimization by Genetic Algorithms*. New York, NY: John Wiley and Sons.

Rechenberg, Ingo. 1965. *Cybernetic solution path of an experimental problem*. Royal Aircraft Establishment, Ministry of Aviation, Library Translation 1112. Farnborough.

Rechenberg, Ingo. 1973. *Evolutionsstrategie: Optimierung Technischer Systeme nach Prinzipien der Biolgischen Evolution*. Stuttgart-Bad Cannstatt: Verlag Frommann-Holzboog.

Riolo, Rich and Worzel, William. 2003. *Genetic Programming: Theory and Practice*. Boston: Kluwer Academic Publishers.

Robertson, George. 1987. Parallel implementation of genetic algorithms in a classifier system. In Davis, Lawrence (editor). *Genetic Algorithms and Simulated Annealing* London: Pittman.

Ryan, Conor. 1999. *Automatic Re-engineering of Software Using Genetic Programming*. Amsterdam: Kluwer Academic Publishers.

Salamon, Peter, Sibani, Paolo, and Frost, Richard. 2002. *Facts, Conjectures, and Improvements for Simulated Annealing*. Philadelphia: Society for Industrial and Applied Mathematics.

Samuel, Arthur L. 1983. AI: Where it has been and where it is going. *Proceedings of the Eighth International Joint Conference on Artificial Intelligence*. Los Altos, CA: Morgan Kaufmann. Pages 1152–1157.

Sanchez, Eduardo and Tomassini, Marco (editors). 1996. *Towards Evolvable Hardware*. Lecture Notes in Computer Science, Volume 1062. Berlin: Springer-Verlag.

Schaeffer, Jonathan. 1997. *One Jump Ahead: Challenging Human Supremacy in Checkers*. New York: Springer.

Sechen, Carl. 1988. *VLSI Placement and Global Routing using Simulated Annealing*. Boston, MA: Kluwer.

Sheingold, Daniel H. (editor). 1976. *Nonlinear Circuits Handbook*. Norwood, MA: Analog Devices, Inc.

Sipper, Moshe, Mange, Daniel, and Perez-Uribe, Andres (editors). 1998. *Evolvable Systems: From Biology to Hardware. Second International Conference, ICES 98, Lausanne, Switzerland, September 1998 Proceedings*. Lecture Notes in Computer Science 1478. Berlin: Springer-Verlag.

Smith, Steven F. 1980. *A Learning System Based on Genetic Adaptive Algorithms*. Ph.D. dissertation. Pittsburgh, PA: University of Pittsburgh.

Song, Bang-Sup and Harjani, Ramesh. 1995. In Chen, Wai-Kai (editor). *The Circuits and Filters Handbook*. Boca Raton, FL: CRC Press. Pages 2072–2127.

Spector, Lee, Barnum, Howard, and Bernstein, Herbert J. 1998. Genetic programming for quantum computers. In Koza, John R., Banzhaf, Wolfgang, Chellapilla, Kumar, Deb, Kalyanmoy, Dorigo, Marco, Fogel, David B., Garzon, Max H., Goldberg, David E., Iba, Hitoshi, and Riolo, Rick. (editors). *Genetic Programming 1998: Proceedings of the Third Annual Conference*. San Francisco, CA: Morgan Kaufmann. Pages 365–373.

Spector, Lee, Barnum, Howard, and Bernstein, Herbert J. 1999. Quantum computing applications of genetic programming. In Spector, Lee, Langdon, William B., O'Reilly, Una-May, and Angeline, Peter (editors). *Advances in Genetic Programming 3*. Cambridge, MA: The MIT Press. Pages 135–160.

Spector, Lee, Barnum, Howard, Bernstein, Herbert J., and Swamy, N. 1999. Finding a better-than-classical quantum AND/OR algorithm using genetic programming. In IEEE. *Proceedings of 1999 Congress on Evolutionary Computation*. Piscataway, NJ: IEEE Press. Pages 2239–2246.

Spector, Lee, and Bernstein, Herbert J. 2003. Communication capacities of some quantum gates, discovered in part through genetic programming. In Shapiro, Jeffrey H. and Hirota, Osamu (editors). *Proceedings of the Sixth International Conference on Quantum Communication, Measurement, and Computing*. Paramus, NJ: Rinton Press.

Spector, Lee, Goodman, E., Wu, A., Langdon, William B., Voigt, H. -M., Gen, M., Sen, S., Dorigo, Marco, Pezeshk, S., Garzon, Max, and Burke, E. (editors). 2001. *Proceedings of the Genetic and Evolutionary Computation Conference, GECCO-2001*. San Francisco, CA: Morgan Kaufmann.

Spector, Lee, Langdon, William B., O'Reilly, Una-May, and Angeline, Peter (editors). 1999. *Advances in Genetic Programming 3*. Cambridge, MA: The MIT Press.

Spector, Lee and Stoffel, Kilian. 1996a. Ontogenetic programming. In Koza, John R., Goldberg, David E., Fogel, David B., and Riolo, Rick L. (editors). 1996. *Genetic Programming 1996: Proceedings of the First Annual Conference, July 28–31, 1996, Stanford University*. Cambridge, MA: MIT Press. Pages 394–399.

Spector, Lee and Stoffel, Kilian. 1996b. Automatic generation of adaptive programs. In Maes, Pattie, Mataric, Maja J., Meyer, Jean-Arcady, Pollack, Jordan, and Wilson, Stewart W. (editors). 1996. *From Animals to Animats 4: Proceedings of the Fourth International Conference on Simulation of Adaptive Behavior*. Cambridge, MA: The MIT Press. Pages 476–483.

Sripramong, Thanwa and Toumazou, Christofer. 2002. The invention of CMOS amplifiers using genetic programming and current-flow analysis. *IEEE Transactions on Computer-Aided Design of Integrated Circuits and Systems*. 21(11). November 2002. Pages 1237–1252.

Stender, Joachim (editor). 1993. *Parallel Genetic Algorithms*. Amsterdam: IOS Publishing.

Stephens, C. R. and Waelbroeck, H. 1997. Effective degrees of freedom in genetic algorithms and the block hypothesis. In Back, Thomas (editor). 1997. *Genetic Algorithms: Proceedings of the Seventh International Conference*. San Francisco, CA: Morgan Kaufmann. Pages 34–40.

Sterling, Thomas L. 1998a. Beowulf-class clustered computing: Harnessing the power of parallelism in a pile of PCs. In Koza, John R., Banzhaf, Wolfgang, Chellapilla, Kumar, Deb, Kalyanmoy, Dorigo, Marco, Fogel, David B., Garzon, Max H., Goldberg, David E., Iba, Hitoshi, and Riolo, Rick L. (editors). *Genetic Programming 1998: Proceedings of the Third Annual Conference, July 22–25, 1998, University of Wisconsin, Madison, Wisconsin*. San Francisco, CA: Morgan Kaufmann. Pages 883–887.

Sterling, Thomas L. 1998b. *Proceedings of Petaflops-Systems Operations Working Review, Bodega Bay, California, June 1–5, 1998*.

Sterling, Thomas L. and Foster, Ian. 1996a. *Proceedings of Petaflops Architecture Workshop (PAWS '96), April 22–25, 1996*.

Sterling, Thomas L. and Foster, Ian. 1996b. *Proceedings of Petaflops System Software Summer Study (Peta Soft '96), June 17–21, 1996*.

Sterling, Thomas L., Salmon, John, and Becker, Donald J., and Savarese, Daniel F. 1999. *How to Build a Beowulf: A Guide to Implementation and Application of PC Clusters*. Cambridge, MA: MIT Press.

Steudel, Goetz Wolfgang. 1973. *Reference Voltage Generator and Regulator*. U.S. patent 3,743,923. Filed December 2, 1971. Issued July 3, 1973.

Stoica, Adrian, Keymeulen, Didier, and Lohn, Jason (editors). 1999. *Proceedings of the First NASA/DOD Workshop on Evolvable Hardware, Pasadena, California, July 19–21, 1999*. Los Alamitos, CA. IEEE Computer Society.

Stoica, Adrian, Lohn, Jason, Katz, Rich, Keymeulen, Didier and Zebulum, Ricardo Salem (editors). 2002. *Proceedings of 2002 NASA/DoD Conference on Evolvable Hardware*. Los Alamitos, CA: IEEE Computer Society.

Stoica, Adrian, Zebulum, Ricardo, and Keymeulen, Didier. 2001. Polymorphic electronics. In Liu, Yong, Tanaka, Kiyoshi, Iwata, Masaya, Higuchi, Tetsuya, and Yasunaga, Moritoshi (editors). *Evolvable Systems: From Biology to Hardware, 4th International Conference, ICES 2001, Tokyo, Japan, October 2001 Proceedings*. Lecture Notes in Computer Science, Volume 2210. Berlin: Springer-Verlag. Pages 291–302.

Streeter, Matthew J., Keane, Martin A., and Koza, John R. 2002a. Iterative refinement of computational circuits using genetic programming. In Langdon, W. B., Cantu-Paz, E., Mathias, K., Roy, R., Davis, D., Poli, R., Balakrishnan, K., Honavar, V., Rudolph, G., Wegener, J., Bull, L., Potter, M. A., Schultz, A. C., Miller, J. F., Burke, E., and Jonoska, N. (editors). 2002. *Proceedings of the 2002 Genetic and Evolutionary Computation Conference*. San Francisco, CA: Morgan Kaufmann. Pages 877–884.

Streeter, Matthew J., Keane, Martin A., and Koza, John R. 2002b. Routine duplication of post-2000 patented inventions by means of genetic programming. In Foster, James A., Lutton, Evelyne, Miller, Julian, Ryan, Conor, and Tettamanzi, Andrea G. B. (editors). 2002. *Genetic Programming: 5th European Conference, EuroGP 2002, Kinsale, Ireland, April 2002 Proceedings*. Berlin: Springer. Pages 26–36.

Streeter, Matthew J., Keane, Martin A., and Koza, John R. 2003. Automatic synthesis using genetic programming of both the topology and sizing for five post-2000 patented analog and mixed analog-digital circuits. *Proceedings of 2003 Southwest Symposium on Mixed-Signal Design, February 23–25, 2003, Las Vegas, Nevada, U.S.A.* Pages 5–10.

Stutzman, Warren. L. and Thiele, Gary A. 1998. *Antenna Theory and Design*. Second edition. New York, NY: John Wiley and Sons.

Sweriduk, G. D., Menon, P. K., and Steinberg, M. L. 1998. Robust command augmentation system design using genetic search methods. *Proceedings of 1998 Conference on Guidance, Navigation, and Control*. Reston, VA: American Institute of Aeronautics and Astronautics. Pages 286–294.

Sweriduk, G. D., Menon, P. K., and Steinberg, M. L. 1999. Design of a pilot-activated recovery system using genetic search methods. *Proceedings of 1998 Conference on Guidance, Navigation, and Control*. Reston, VA: American Institute of Aeronautics and Astronautics.

Tanese, Reiko. 1989. *Distributed Genetic Algorithm for Function Optimization*. Ph.D. dissertation. Department of Electrical Engineering and Computer Science. University of Michigan.

Teller, Astro. 1996a. *Evolving Programmers: SMART Mutation*. Technical Report CMU-CS-96. Computer Science Department, Carnegie Mellon University.

Teller, Astro. 1996b. Evolving programmers: The co-evolution of intelligent recombination operators. In Angeline, Peter J. and Kinnear, Kenneth E. Jr. (editors). *Advances in Genetic Programming 2*. Cambridge, MA: The MIT Press.

Teller, Astro. 1998. *Algorithm Evolution with Internal Reinforcement for Signal Understanding*. Ph.D. Thesis. Computer Science Department. Carnegie Mellon University. Pittsburgh, Pennsylvania.

Teller, Astro. 1999. The internal reinforcement of evolving algorithms. In Spector, Lee, Langdon, William B., O'Reilly, Una-May, and Angeline, Peter (editors). 1999. *Advances in Genetic Programming 3*. Cambridge, MA: The MIT Press.

Teller, Astro, and Veloso, Manuela. 1995a. *Learning Tree Structured Algorithms for Orchestration into an Object Recognition System*. Technical Report CMU-CS-95-101. Computer Science Department, Carnegie Mellon University.

Teller, Astro, and Veloso, Manuela. 1995b. Program evolution for data mining. In Louis, Sushil (editor). Special Issue on Genetic Algorithms and Knowledge Bases. *The International Journal of Expert Systems*. JAI Press. (3)216–236.

Teller, Astro, and Veloso, Manuela. 1995c. A controlled experiment: evolution for learning difficult problems. *Proceedings of Seventh Portuguese Conference on Artificial Intelligence*. Springer-Verlag. Pages 165–76.

Teller, Astro, and Veloso, Manuela. 1995d. Algorithm Evolution for Face Recognition: What Makes a Picture Difficult? *Proceedings of the IEEE International Conference on Evolutionary Computation*. IEEE Press.

Teller, Astro, and Veloso, Manuela. 1995e. Language Representation Progression in PADO. *Proceedings of AAAI Fall Symposium on Artificial Intelligence*. Menlo Park, CA: AAAI Press.

Teller, Astro and Veloso, Manuela. 1996. PADO: A new learning architecture for object recognition. In Ikeuchi, Katsushi and Veloso, Manuela (editors). *Symbolic Visual Learning*. Oxford University Press. Pages 81–116.

Teller, Astro, and Veloso, Manuela. 1997. Neural programming and an internal reinforcement policy. In Yao, Xin, Kim, Jong-Hwan, and Furuhashi, T. (editors). *Simulated Evolution and Learning*. Lecture Notes in Artificial Intelligence. Volume 1285. Heidelberg, Germany: Springer-Verlag. Pages 279–286.

Thompson, Adrian. 1996. Silicon evolution. In Koza, John R., Goldberg, David E., Fogel, David B., and Riolo, Rick L. (editors). 1996. *Genetic Programming 1996: Proceedings of the First Annual Conference, July 28–31, 1996, Stanford University*. Cambridge, MA: MIT Press. Pages 444–452.

Thompson, Adrian. 1998. *Hardware Evolution: Automatic Design of Electronic Circuits in Reconfigurable Hardware by Artificial Evolution*. Conference of Professors and Heads of Computing / British Computer Society Distinguished Dissertation series. Berlin: Springer-Verlag.

Tomita, Masaru, Hashimoto, Kenta, Takahashi, Kouichi, Shimizu, Thomas Simon, Matsuzaki, Yuri, Miyoshi, Fumihiko, Saito, Kanako, Tanida, Sakura, Yugi, Katsuyuki, Venter, J. Craig, Hutchison, Clyde A. III. 1999. E-CELL: Software environment for whole cell simulation. *Bioinformatics*. Volume 15 (1)72–84.

Turing, Alan M. 1948. Intelligent machinery. Reprinted in Ince, D. C. (editor). 1992. *Mechanical Intelligence: Collected Works of A. M. Turing*. Amsterdam: North Holland. Pages 107–127. Also reprinted in Meltzer, B. and Michie, D. (editors). 1969. *Machine Intelligence 5*. Edinburgh: Edinburgh University Press.

Turing, Alan M. 1950. Computing machinery and intelligence. *Mind*. 59(236)433–460. Reprinted in Ince, D. C. (editor). 1992. *Mechanical Intelligence: Collected Works of A. M. Turing*. Amsterdam: North Holland. Pages 133–160.

Uda, S. 1926. Wireless beam of short electric waves. *Journal of the IEE (Japan)*. March 1926. Pages 273–282.

Uda, S. 1927. Wireless beam of short electric waves. *Journal of the IEE (Japan)*. March 1927. Pages 1209–1219.

Ullman, Jeffrey D. 1984. *Computational Aspects of VLSI*. Rockville, MD: Computer Science Press.

Van Valkenburg, M. E. 1982. *Analog Filter Design*. Fort Worth, TX: Harcourt Brace Jovanovich.

Villagran, Victor and Sbarbaro, Daniel. 1998. A new approach for turning PID controller based on iterative learning. *Proceedings of the 1998 IEEE International Conference on Control Applications*. Volume I. Pages 139–143.

Vladimirescu, Andrei. 1994. *The SPICE Book*. New York, NY: John Wiley and Sons.

Voit, Eberhard O. 2000. *Computational Analysis of Biochemical Systems*. Cambridge: Cambridge University Press.

Wakerly, John F. 1990. *Digital Design Principles and Practices*. Englewood Cliffs, NJ: Prentice Hall.

Webb, Edwin C. 1992. *Enzyme Nomenclature 1992: Recommendations of the Nomenclature Committee of the International Union of Biochemistry and Molecular Biology*. San Diego, CA: Academic Press.

Whitley, Darrell, Goldberg, David, Cantu-Paz, Erick, Spector, Lee, Parmee, Ian, and Beyer, Hans-Georg (editors). 2000. *GECCO-2000: Proceedings of the Genetic and Evolutionary Computation Conference, July 10–12, 2000, Las Vegas, Nevada*. San Francisco: Morgan Kaufmann.

Whitley, Darrell, Gruau, Frederic, and Preatt, Larry. 1995. Cellular encoding applied to neurocontrol. In Eshelman, Larry J. (editor). *Proceedings of the Sixth International Conference on Genetic Algorithms*. San Francisco, CA: Morgan Kaufmann. Pages 460–467.

Whitley, Darrell. 2000. Functions as permutations: Regarding no free lunch, Walsh analysis and summary statistics. In Schoenauer, Marc, Deb, Kalyanmoy, Gunter, Rudolph, Yao, Xin, Lutton, Evelyne, Merelo, Juan Julian, and Schwefel, Hans-Paul (editors). 2000. *Parallel Problem Solving from Nature: 6th International Conference, Paris, France, September 2000 Proceedings*. Berlin: Springer. Page 169–178.

Williams, Arthur B. and Taylor, Fred J. 1995. *Electronic Filter Design Handbook*. Third Edition. New York, NY: McGraw-Hill.

Witczak, Marcin. 2003. *Identification and Fault Detection of Non-Linear Dynamic Systems*. University of Zielona Gora Press.

Wolpert, D. H. and Macready, W. G. 1997. No free lunch theorems for optimization. *IEEE Transactions on Evolutionary Computation*. 1(1) 67– 82. April 1997.

Wong, D. F., Leong, H. W., and Liu. C. L. 1988. *Simulated Annealing for VLSI Design*. Boston, MA: Kluwer.

Wong, Man Leung and Leung, Kwong Sak. 2000. *Data Mining Using Grammar Based Genetic Programming and Applications*. Amsterdam: Kluwer Academic Publishers.

Yagi, H. 1928. Beam transmission of ultra short waves. *Proceedings of the IRE*. Volume 26: Pages 714–741. June 1928.

Yu, Jessen, Keane, Martin A., and Koza, John R. 2000. Automatic design of both topology and tuning of a common parameterized controller for two families of plants using genetic programming. In *Proceedings of Eleventh IEEE International Symposium on Computer-Aided Control System Design (CACSD) Conference and Ninth IEEE International Conference on Control Applications (CCA) Conference, Anchorage, Alaska, September 25–27, 2000*.

Yuh, Chiou-Hwa, Bolouri, Hamid, and Davidson, Eric H. 1998. Genomic cis-regulatory logic: Experimental and computational analysis of a sea urchin gene. *Science*. 279. Pages 1896–1902.

Zebulum, Ricardo Salem, Pacheco, Marco Aurelio C., and Vellasco, Marley Maria B. R. 2002. *Evolutionary Electronics: Automatic Design of Electronic Circuits and Systems by Genetic Algorithms*. Boca Raton, FL: CRC Press.

Ziegler, J. G. and Nichols, N. B. 1942. Optimum settings for automatic controllers. *Transactions of ASME*. (64) 759–768.

Zitzler, Eckart, Deb, Kalyanmoy, Thiele, Lothar, Coello Coello, Carlos A., and Corne, David (editors). 2001. *Evolutionary Multi-Criterion Optimization, First International Conference, EMO 2001, Zurich, Switzerland, March 2001, Proceedings*. Lecture Notes in Computer Science. Volume 1993. Berlin, Germany: Springer-Verlag.

Zobel, Otto Julius. 1926. *Electrical Network and Method of Transmitting Electric Currents*. Filed August 9, 1922. U.S. patent 1,603,305. Issued October 19, 1926.

Zobel, Otto Julius. 1925. *Wave Filter*. Filed January 15, 1921. U.S. patent 1,538,964. Issued May 26, 1925.

Zobel, Otto Julius. 1928. Distortion correction in electrical circuits with constant resistance networks. *Bell Systems Technical Journal*. July 1928. Page 438.

Acknowledgment Concerning Figures

Figures 2.1, 4.8, 4.12, 10.4, 16.5, 16.10, 16.11, 16.12, 16.13, and 16.14 of this book were reprinted from *Genetic Programming III: Darwinian Invention and Problem Solving* by John R. Koza, Forrest H Bennett III, David Andre and Martin A. Keane, pages 38, 582, 835, 403, 589, 666, 667, 667, 637, and 641, respectively, Copyright 1999, with permission from Elsevier.

Index